Advanced Engineering Mathematics with Mathematica®

Advanced Engineering Mathematics with Mathematica®

Edward B. Magrab

CRC Press
Taylor & Francis Group
Boca Raton London New York

CRC Press is an imprint of the
Taylor & Francis Group, an **informa** business

CRC Press
Taylor & Francis Group
6000 Broken Sound Parkway NW, Suite 300
Boca Raton, FL 33487-2742

Printed on acid-free paper

International Standard Book Number-13: 978-0-3678-9325-5 (Hardback)

Visit the Taylor & Francis Web site at
http://www.taylorandfrancis.com

and the CRC Press Web site at
http://www.crcpress.com

eResource material is available for this title at https://www.crcpress.com/9780367893255

For

June Coleman Magrab

Contents

Preface

Aims of the Book

Advanced engineering mathematics refers to the use of analytical and numerical solution methods to solve boundary value problems encountered in modeling and analyzing engineering applications.

The material presented in this book introduces a set of topics that are most often used to solve engineering boundary value problems in heat transfer, vibrations, fluid and solid mechanics, and data acquisition and signal processing. In this regard, the following methods are emphasized: the separation of variables and the generation and application of orthogonal functions and the Laplace transform. Along the way, we introduce and apply the relevant aspects of complex variables, matrices and determinants, Fourier series and Fourier transforms, solution techniques for ordinary differential equations, and, in particular, the Sturm–Liouville system, and the procedures used to make ordinary and partial differential equations found in engineering non-dimensional. To show the importance of the material and its diverse application, several multi-physics applications from the literature are presented.

For a large amount of the material presented, it is shown how the derivations and examples can be verified with Mathematica, thereby integrating symbol programming with the analytical work. All numerical calculations are done with Mathematica. From this integration of the analytical techniques and symbolic computer methods, one can infer at which point in the solution process it is most advantageous to switch to obtaining the results numerically. A benefit of integrating Mathematica with the analytical work is that the reader should be able to solve correctly the vast majority of the exercises, since Mathematica can be used to verify the results. The book can be used without using Mathematica, but any benefits derived from its use may be lost.

Integrating Mathematica Procedures

In order to introduce and employ analytical solution techniques while attempting to integrate the capabilities of Mathematica to solve or verify many of the results one has to strike a balance among analytical detail, numerical methods, analytical short cuts and "tricks," and the use of Mathematica itself. In an attempt to reach some kind of balance, the following observations were taken into account.

- One must attain a certain level of mathematical knowledge and its associated terminology in order to correctly formulate the problem; that is, one must be able to read, interpret, and implement mathematical techniques, methods, and manipulations. These skills must be acquired before one can use Mathematica. In addition, Mathematica cannot solve every problem directly, so fluency with analytical mathematical methods is necessary.
- Where appropriate, the analytical methods and Mathematica should be fully integrated with the material, not just when numerical results are desired.

- A limited but focused range of topics is covered, which permits a consistent and straightforward integration of Mathematica with the material, and because of the structure of Mathematica the Mathematica procedures can be made both concise and readable. In a few of the more complex examples, Mathematica is used to obtain some intermediate results before proceeding with the analytical solution.
- Discussions of numerical methods and procedures *per se* have been eliminated, and by inference Mathematica is the numerical procedure.
- Recognize that the needs of practicing engineers require both the necessary foundations to turn physical models into mathematical ones and the ability to obtain solutions to these models.
- Ensure that the Mathematica procedures minimally interrupt the flow of the material. In this book, the procedures are cited in the appropriate places in the text, but the procedures themselves appear at the end of the chapter.

Features

- From the beginning, emphasis is placed on orthogonal vectors and orthogonal functions, the conditions necessary for their generation, and their very important role in solving classes of linear ordinary and partial differential equations.
- Fourier series and Fourier transforms are introduced as a precursor to generating and using orthogonal functions to solve ordinary and partial differential equations and to emphasize their utility in understanding certain fundamental aspects of data acquisition and signal analysis.
- Presents a concise review of obtaining solutions to ordinary differential equations and a more extensive discussion of the Sturm–Liouville equation and the generation of orthogonal functions. This material is a prerequisite to obtaining solutions to partial differential equations using the separation of variables technique and the Laplace transform method.
- Numerous and widely varied examples are given to show how to make ordinary and partial differential equations found in engineering non-dimensional. Indirectly, it is shown how the nondimensionalization greatly simplifies the appearance of the equation, making, in some cases, the choice of a solution method easier to determine.
- A unique last chapter provides detailed solutions to six multi-physics examples taken from the literature. The solutions to these examples use the techniques presented in the previous chapters. This selection of applications illustrates how the material in the preceding eight chapters can be used to solve a variety of problems from different engineering disciplines.
- Applies the Laplace transform to the spatial variable and shows that this application has great utility and several advantages.
- There is an extensive use of tables to summarize results to make them easily accessible for use in the chapter in which they appear and in subsequent chapters. In several chapters, the tables contain a summary of the differential equations and boundary/initial conditions that are going to be used in the examples that follow.
- Over 185 Mathematica procedures have been created and integrated throughout the text to verify most of the analytic results and to generate the numerical results. It is shown how the Mathematica language permits a straightforward extension to the

use of their built-in finite element analysis capability to solve boundary value problems for which analytical solutions cannot be obtained. The use of this capability, however, still requires knowledge of boundary value terminology and formulation. By continually verifying the vast majority of the results with Mathematica, the reader should eventually feel comfortable with letting Mathematica do some of the symbolic manipulation thereby relieving one of having to perform complex algebraic operations and the necessary validation.

Contents and Organization

The book is organized into nine chapters. In the first chapter, we introduce fundamental matrix operations and the rules governing their manipulation, the role of determinants in matrix operations, how to obtain eigenvalues and eigenvectors of square matrices, the necessary conditions for eigenvectors to be orthogonal, and the linear independence of orthogonal eigenvectors. The determination of eigenvalues and the generation of orthogonal eigenvectors is the central idea behind a general solution method used on classes of ordinary and partial differential equations that are presented in Chapters 6 to 9.

In Chapter 2, we introduce basic complex number operations, Euler's formula, obtaining zeros of functions and residues, analytic functions, and Cauchy's integral formula. Complex numbers are used to represent Fourier series as shown in Chapter 3 and are useful in determining solutions to certain classes of ordinary and partial differential equations as shown in Chapters 4, 5, and 9. The Cauchy integral formula is used to obtain results when the Laplace transform technique is employed in Chapter 8.

In Chapter 3, we introduce the Fourier series and the Fourier transform and show, when the independent variable is time, the duality between the time and the frequency domain. We also present the Fourier transform and the short-time Fourier transform in the context of signal analysis and sampling and show several important practical applications. The central ideas behind the Fourier series and its orthogonal properties are extended and applied to obtaining the solutions to linear ordinary and partial differential equations found in engineering in Chapters 6 to 9.

In Chapter 4, we review several methods that can be used to obtain solutions to first-, second-, and higher-order linear and certain nonlinear ordinary differential equations. For first order differential equations, solutions methods for separable and exact differential equations are presented, and solutions are obtained for the Bernoulli equation. For second-order differential equations, such techniques as reduction of order, conversion to a system of first-order differential equations, method of undetermined coefficients, variation of parameters, and the generation of orthogonal functions to solve a system of second-order equations are presented. In the process of applying these methods, tables of solutions to several different ordinary differential equations that will be useful in subsequent chapters are created. Several examples are given to illustrate how to place ordinary differential equations found in engineering into non-dimensional form. The phase plane plot and the superposition of direction fields are introduced as a way to visualize the solutions of autonomous second-order linear ordinary differential equations.

In Chapter 5, the power series and the method of Frobenius are introduced as a means to solve a class of second-order linear ordinary differential equations. General solutions using power series are obtained for both ordinary points and regular singular points. Then these general results are used to obtain the solutions to Bessel's equation, the modified Bessel equation, the spherical Bessel equation, Legendre's equation, the associated Legendre equation, and the hypergeometric equation. For Bessel functions of the first and second kind, relations regarding their derivatives, integrals, and limiting cases are determined. Numerous general classes of differential equations for which Bessel functions are the solution are presented and then summarized in a table.

In Chapter 6, the solutions to a Sturm–Liouville equation, which is a special form of a homogeneous second-order ordinary differential equation and a certain fourth-order ordinary differential equation, is shown to be orthogonal when the solutions to these equations satisfy a certain set of homogeneous boundary conditions. Special cases of the Sturm–Liouville equation are shown to be Bessel equations, spherical Bessel equations, Legendre equations, hypergeometric equations, certain second-order equations with constant coefficients, and a special form of a fourth-order differential equation. The case of the non-homogeneous Sturm–Liouville equation is also examined. It will be seen in subsequent chapters that these Sturm–Liouville equations appear when the separation of variables technique is used to solve classes of linear partial differential equations and the ability to generate orthogonal functions is the reason for the success of this technique.

In Chapter 7, the separation of variables method is introduced and employed to obtain the general solutions to the Laplace equation, the Helmholtz equation, the wave equation, the diffusion equation, and the Poisson equation in three coordinate systems: Cartesian, polar cylindrical, and spherical. The general solutions for each of these equations in each of these coordinates systems are then applied to numerous boundary value problems encountered in engineering. In addition, several forms of the bi-harmonic equation in Cartesian and polar coordinate systems are examined and, when possible, their solutions obtained. Numerous tables are utilized in this chapter to organize and summarize results and to summarize the differential equations and boundary/initial conditions that are going to be used in the examples for each type of equation that follows. Several examples are given to illustrate how to place partial differential equations into non-dimensional form.

In Chapter 8, the Laplace transform is introduced as another means to solve linear ordinary and partial differential equations. A number of Laplace transform pairs are derived by using the definition of the Laplace transform, and the convolution integral. The Laplace transform of periodic functions is obtained. The inversion integral is shown to be an important tool in obtaining the inverse Laplace transform for many practical engineering problems. The advantages of the Laplace transform are demonstrated by examining several engineering applications.

In Chapter 9, many of the results of the preceding chapters are applied to six multi-physics applications taken from fluid mechanics, solid mechanics, electromagnetics, and heat transfer. Each multi-physics application is complete, and both linear and nonlinear examples have been selected.

Classroom Tested

The material has been used as class notes in many offerings of a first-year graduate elective offered in the Department of Mechanical Engineering at the University of Maryland, College Park, Maryland. The topics selected and the level of detail presented are based, in part, on students' feedback.

In the End

After working through the material in this book, it is the hoped that the reader will have attained a good understanding of the various standard analytical solution methods to solve boundary value problems so that one:

Knows when it is appropriate to apply them and when it is necessary to use approximate or numerical solution methods, that is, when such complexities as nonlinearity and irregular geometry limit the applicability of the analytical methods.

Understands how to integrate the various solution methods with a symbolic computer program and the important role that a symbolic program can play in obtaining solutions, particularly when they can be used to handle complex algebraic manipulation and to obtain partial analytical results.

Recognizes when these solution methods are being used in published articles, perhaps making the articles a little easier to follow.

Edward B. Magrab
College Park, Maryland

Author

Edward B. Magrab is Emeritus Professor in the Department of Mechanical Engineering at the University of Maryland at College Park. He has extensive experience in analytical and experimental analysis of vibrations and acoustics, served as an engineering consultant to numerous companies, and authored or coauthored a number of books on vibrations, noise control, instrumentation, integrated product design, and Mathematica®. He is a Life Fellow of the American Society of Mechanical Engineers.

1 Matrices, Determinants, and Systems of Equations

1.1 DEFINITIONS

A *matrix* is a rectangular array of numbers (or symbols or expressions) consisting of m rows and n columns. Such an array is called a matrix of order $(m \times n)$ [or simply an $(m \times n)$ matrix], and is denoted as

$$A = \begin{pmatrix} a_{11} & a_{12} & \cdots & a_{1n} \\ a_{21} & a_{22} & \cdots & a_{2n} \\ \vdots & & & \\ a_{m1} & a_{m2} & \cdots & a_{mn} \end{pmatrix} \tag{1.1}$$

where the a_{ij} are the elements of the array. The first subscript denotes the row, and the second subscript denotes the column in which the element appears. When the number of rows equals the number of columns, that is, when $m=n$, the matrix is called a *square matrix* of order n.

Operations with matrices are subject to certain rules, which will be indicated as we progress. It should be realized that a matrix is not a single quantity; it is a representation of an array of entities.

For a (2×3) matrix, Eq. (1.1) is created with Mathematica procedure M1.1.

The *transpose* of a matrix is obtained by interchanging the rows and columns and the transpose operation is denoted with a superscript T. Thus, the transpose of the $(m \times n)$ matrix A is

$$\begin{aligned} B = A^T &= \begin{pmatrix} a_{11} & a_{12} & \cdots & a_{1n} \\ a_{21} & a_{22} & \cdots & a_{2n} \\ \vdots & & & \\ a_{m1} & a_{m2} & \cdots & a_{mn} \end{pmatrix}^T \\ &= \begin{pmatrix} b_{11} = a_{11} & b_{12} = a_{21} & \cdots & b_{1m} = a_{m1} \\ b_{21} = a_{12} & b_{22} = a_{22} & \cdots & b_{2m} = a_{m2} \\ \vdots & & & \\ b_{n1} = a_{1n} & b_{n2} = a_{2n} & \cdots & b_{nm} = a_{mn} \end{pmatrix} \end{aligned} \tag{1.2}$$

where B is now an $(n \times m)$ matrix.

For a (2×3) matrix, Eq. (1.2) is created with Mathematica procedure M1.2.

A *symmetric matrix* is a *square* matrix in which $a_{ij}=a_{ji}$. Thus, for a symmetric matrix

$$A = A^T \tag{1.3}$$

A *skew-symmetric matrix* is a square matrix in which $a_{ij}=-a_{ji}$, which implies that $a_{ii}=0$. Thus, $\boldsymbol{A}=-\boldsymbol{A}^T$

A *column matrix* is a matrix with only one column, that is, an $(m\times 1)$ matrix. This matrix is denoted as

$$\boldsymbol{a}=\begin{Bmatrix} a_{11} \\ a_{21} \\ \vdots \\ a_{m1} \end{Bmatrix}=\begin{Bmatrix} a_1 \\ a_2 \\ \vdots \\ a_m \end{Bmatrix} \tag{1.4}$$

A column matrix is frequently referred to as a *column vector.*

A *row matrix* is a matrix with only one row, that is, a $(1\times m)$ matrix. This matrix is denoted as

$$\begin{aligned}\boldsymbol{b}=\boldsymbol{a}^T=\begin{Bmatrix} a_{11} \\ a_{21} \\ \vdots \\ a_{m1} \end{Bmatrix}^T &=\{b_{11}=a_{11} \quad b_{12}=a_{21} \quad \cdots \quad b_{1m}=a_{m1}\} \\ &=\{b_1 \quad b_2 \quad \cdots \quad b_m\}\end{aligned} \tag{1.5}$$

A (3×1) column matrix given by Eq. (1.4) and the corresponding (1×3) row matrix given by Eq. (1.5) are created with Mathematica procedure M1.3.

A *diagonal matrix* is a square matrix in which all the elements are zero except those on the principal diagonal, that is, $a_{ij}=a_{ii}\delta_{ij}$, where δ_{ij} is the Kronecker delta.* Thus, Eq. (1.1) becomes

$$\boldsymbol{A}_\mathrm{D}=\begin{pmatrix} a_{11} & 0 & \cdots & 0 \\ 0 & a_{22} & \cdots & 0 \\ \vdots & & & \\ 0 & 0 & \cdots & a_{nn} \end{pmatrix} \tag{1.6}$$

A (3×3) diagonal matrix given by Eq. (1.6) is created with Mathematica procedure M1.4.

A special case of a diagonal matrix is the *identity matrix*, which is a diagonal matrix with all the elements along the principal diagonal equal to one, that is, $a_{ij}=\delta_{ij}$. This matrix is denoted as $\boldsymbol{I}$ and is given by

$$\boldsymbol{I}=\begin{pmatrix} 1 & 0 & \cdots & 0 \\ 0 & 1 & \cdots & 0 \\ \vdots & & & \\ 0 & 0 & \cdots & 1 \end{pmatrix} \tag{1.7}$$

A (3×3) identity matrix given by Eq. (1.7) is created with Mathematica procedure M1.5.

---

* $\delta_{ij}=0,\ i\neq j$ and $\delta_{ij}=1,\ i=j$.

A *null matrix* is a matrix whose every element is zero. It will be used here with the following shorthand notation in matrix equations

$$0 = \begin{pmatrix} 0 & 0 & \cdots & 0 \\ 0 & 0 & \cdots & 0 \\ \vdots & & & \\ 0 & 0 & \cdots & 0 \end{pmatrix} \tag{1.8}$$

Two matrices $\boldsymbol{A}$ and $\boldsymbol{B}$ of the same order are *equal* if and only if $a_{ij} = b_{ij}$ for every i and j.

1.2 MATRIX OPERATIONS

The sum or the difference of two matrices $\boldsymbol{A}$ and $\boldsymbol{B}$ that have the same order is denoted by

$$\boldsymbol{C} = \boldsymbol{A} \pm \boldsymbol{B} \tag{1.9}$$

and the elements of $\boldsymbol{C}$ are given by the sum or difference of each corresponding element, that is, $c_{ij} = a_{ij} \pm b_{ij}$ for every i and j.

Since the elements of $\boldsymbol{C}^T$ are given by c_{ji}, it is seen that

$$\boldsymbol{C}^T = \left(\boldsymbol{A} \pm \boldsymbol{B}\right)^T = \boldsymbol{A}^T \pm \boldsymbol{B}^T \tag{1.10}$$

For (2×2) matrices $\boldsymbol{A}$ and $\boldsymbol{B}$, Eqs. (1.9) and (1.10) are verified with Mathematica procedure M1.6.

Example 1.1

Consider the two matrices

$$\boldsymbol{A} = \begin{pmatrix} 1 & -2 & 0 \\ 4 & 5 & -7 \end{pmatrix}, \quad \boldsymbol{B} = \begin{pmatrix} 3 & -8 & 4 \\ -4 & 1 & 11 \end{pmatrix}$$

Then, their sum is

$$\boldsymbol{C} = \boldsymbol{A} + \boldsymbol{B} = \begin{pmatrix} 4 & -10 & 4 \\ 0 & 6 & 4 \end{pmatrix} \tag{a}$$

Equation (a) is verified with Mathematica procedure M1.7.

The *multiplication* of two matrices $\boldsymbol{A}$ and $\boldsymbol{B}$ to form the product $\boldsymbol{AB}$ is possible only if the number of columns of $\boldsymbol{A}$ is equal to the number of rows of $\boldsymbol{B}$, that is, if $\boldsymbol{A}$ is an $(m \times k)$ matrix and $\boldsymbol{B}$ is a $(k \times n)$ matrix. Such matrices are said to be *conformable*. For conformable matrices, the inner product dimension, k in this case, is the same. Then the product

$$\boldsymbol{C} = \boldsymbol{AB} \tag{1.11}$$

is an $(m \times n)$ $[=(m\times k)(k\times n)]$ matrix, where

$$c_{ij} = \sum_{p=1}^{k} a_{ip}b_{pj} = a_{i1}b_{1j} + a_{i2}b_{2j} + \cdots + a_{ik}b_{kj} \tag{1.12}$$

Notice that in this case, the multiplication $\boldsymbol{BA}$ is not possible because the number of columns in $\boldsymbol{B}$ does not equal the number of rows of $\boldsymbol{A}$, that is, since $n \neq m$, the inner product dimensions $(k\times n)(m\times k)$ are not equal and therefore the matrices are not conformable. In view of Eq. (1.12), Eq. (1.11) can be written as

$$\boldsymbol{C} = \boldsymbol{AB} = \begin{pmatrix} \sum_{p=1}^{k} a_{1p}b_{p1} & \sum_{p=1}^{k} a_{1p}b_{p2} & \cdots & \sum_{p=1}^{k} a_{1p}b_{pn} \\ \sum_{p=1}^{k} a_{2p}b_{p1} & \sum_{p=1}^{k} a_{2p}b_{p2} & \cdots & \sum_{p=1}^{k} a_{2p}b_{pn} \\ \vdots & \vdots & & \vdots \\ \sum_{p=1}^{k} a_{mp}b_{p1} & \sum_{p=1}^{k} a_{mp}b_{p2} & \cdots & \sum_{p=1}^{k} a_{mp}b_{pn} \end{pmatrix} \tag{1.13}$$

Example 1.2

Consider the two matrices

$$\boldsymbol{A} = \begin{pmatrix} 2 & 5 \\ -1 & 3 \\ 4 & -2 \end{pmatrix}, \quad \boldsymbol{B} = \begin{pmatrix} 6 & 8 \\ 1 & -7 \end{pmatrix}$$

Then, their product is

$$\boldsymbol{C} = \boldsymbol{AB} = \begin{pmatrix} 2\times 6 + 5\times 1 & 2\times 8 - 5\times 7 \\ -1\times 6 + 3\times 1 & -1\times 8 - 3\times 7 \\ 4\times 6 - 2\times 1 & 4\times 8 + 2\times 7 \end{pmatrix} = \begin{pmatrix} 17 & -19 \\ -3 & -29 \\ 22 & 46 \end{pmatrix} \tag{a}$$

Equation (a) is verified with Mathematica procedure M1.8. Note that the product $\boldsymbol{BA}$ is not possible because the matrices are not conformable.

For the case of two (2×2) matrices

$$\boldsymbol{A} = \begin{pmatrix} a_{11} & a_{12} \\ a_{21} & a_{22} \end{pmatrix} \quad \boldsymbol{B} = \begin{pmatrix} b_{11} & b_{12} \\ b_{21} & b_{22} \end{pmatrix} \tag{1.14}$$

Eq. (1.13) gives

$$AB = \begin{pmatrix} a_{11} & a_{12} \\ a_{21} & a_{22} \end{pmatrix} \begin{pmatrix} b_{11} & b_{12} \\ b_{21} & b_{22} \end{pmatrix} = \begin{pmatrix} a_{11}b_{11} + a_{12}b_{21} & a_{11}b_{12} + a_{12}b_{22} \\ a_{21}b_{11} + a_{22}b_{21} & a_{21}b_{12} + a_{22}b_{22} \end{pmatrix} \tag{1.15}$$

Note, however, that

$$BA = \begin{pmatrix} b_{11} & b_{12} \\ b_{21} & b_{22} \end{pmatrix} \begin{pmatrix} a_{11} & a_{12} \\ a_{21} & a_{22} \end{pmatrix} = \begin{pmatrix} a_{11}b_{11} + a_{21}b_{12} & a_{12}b_{11} + a_{22}b_{12} \\ a_{11}b_{21} + a_{21}b_{22} & a_{12}b_{21} + a_{22}b_{22} \end{pmatrix} \tag{1.16}$$

Therefore, in general, $\boldsymbol{AB} \neq \boldsymbol{BA}$.

For (2×2) matrices $\boldsymbol{A}$ and $\boldsymbol{B}$, Eqs. (1.15) and (1.16) are verified with Mathematica procedure M1.9.

If $\boldsymbol{A}$ and $\boldsymbol{B}$ are diagonal matrices, that is, $a_{ij}=a_{ii}\delta_{ij}$, and $b_{ij}=b_{ii}\delta_{ij}$, then from Eq. (1.12) $c_{ij}=c_{ii}\delta_{ij}$. In this case, from Eqs. (1.15) and (1.16), we find

$$AB = \begin{pmatrix} a_{11} & 0 \\ 0 & a_{22} \end{pmatrix} \begin{pmatrix} b_{11} & 0 \\ 0 & b_{22} \end{pmatrix} = \begin{pmatrix} a_{11}b_{11} & 0 \\ 0 & a_{22}b_{22} \end{pmatrix} = BA \tag{1.17}$$

For (2×2) diagonal matrices $\boldsymbol{A}$ and $\boldsymbol{B}$, Eq. (1.17) is verified with Mathematica procedure M1.10.

For the special case when $\boldsymbol{A}=\boldsymbol{a}$ is a column vector and $\boldsymbol{B}=\boldsymbol{b}$ is a column vector, then

$$\boldsymbol{a}^T \boldsymbol{b} = \{a_1 \quad a_2 \quad \cdots \quad a_m\} \begin{Bmatrix} b_1 \\ b_2 \\ \vdots \\ b_m \end{Bmatrix} = \sum_{k=1}^{m} a_k b_k \tag{1.18}$$

Equation (1.18) is called the *inner product*. When $\boldsymbol{a}=\boldsymbol{b}$, Eq. (1.18) becomes

$$\|\boldsymbol{a}\|^2 = \boldsymbol{a}^T \boldsymbol{a} = \sum_{k=1}^{m} a_k^2 \tag{1.19}$$

The quantity $\|\boldsymbol{a}\|$ is called the *norm* of a vector.

For the pair of (1×3) matrices, the results given by Eqs. (1.18) and (1.19) are verified with Mathematica procedure M1.11.

On the other hand, when $\boldsymbol{b}$ is a $(m \times 1)$ column vector and $\boldsymbol{a}$ is a $(1 \times n)$ row vector, we have from Eq. (1.12) that

$$\boldsymbol{c} = \boldsymbol{ba} = \begin{Bmatrix} b_{11} \\ b_{21} \\ \vdots \\ b_{m1} \end{Bmatrix} \{a_{11} \quad a_{12} \quad \cdots \quad a_{1n}\} \rightarrow c_{ij} = \sum_{k=1}^{1} b_{ik} a_{kj} = b_{i1} a_{1j} \tag{1.20}$$

or

$$\boldsymbol{c} = \begin{pmatrix} b_{11}a_{11} & b_{11}a_{12} & \cdots & b_{11}a_{1n} \\ b_{21}a_{11} & b_{21}a_{12} & \cdots & b_{21}a_{1n} \\ \vdots & & & \vdots \\ b_{m1}a_{11} & b_{m1}a_{12} & \cdots & b_{m1}a_{1n} \end{pmatrix} \tag{1.21}$$

When $\boldsymbol{b}$ is a (3×1) column vector and $\boldsymbol{a}$ is a (1×2) row vector, Eqs. (1.20) and (1.21) are verified with Mathematica procedure M1.12.

Multiplying matrix $\boldsymbol{A}$ by a scalar g, which is denoted as $g\boldsymbol{A}$, indicates that each element in the matrix is multiplied by g, that is,

$$g\boldsymbol{A} = \boldsymbol{A}g = g\begin{pmatrix} a_{11} & a_{12} & \cdots & a_{1n} \\ a_{21} & a_{22} & \cdots & a_{2n} \\ \vdots & & & \\ a_{m1} & a_{m2} & \cdots & a_{mn} \end{pmatrix} = \begin{pmatrix} ga_{11} & ga_{12} & \cdots & ga_{1n} \\ ga_{21} & ga_{22} & \cdots & ga_{2n} \\ \vdots & & & \\ ga_{m1} & ga_{m2} & \cdots & ga_{mn} \end{pmatrix} \tag{1.22}$$

It follows from Eq. (1.9) that $g(\boldsymbol{A} + \boldsymbol{B}) = g\boldsymbol{A} + g\boldsymbol{B}$.

For a (2×2) matrix, the result given by Eq. (1.22) is verified with Mathematica procedure M1.13.

The properties of matrix products are further examined in Section 1.5.

1.3 DETERMINANTS

Determinants are defined only for square matrices. For the (2×2) matrix,

$$\boldsymbol{A} = \begin{pmatrix} a_{11} & a_{12} \\ a_{21} & a_{22} \end{pmatrix}$$

the determinant is defined as

$$\det \boldsymbol{A} = \begin{vmatrix} a_{11} & a_{12} \\ a_{21} & a_{22} \end{vmatrix} = a_{11}a_{22} - a_{12}a_{21} \tag{1.23}$$

If $\boldsymbol{A}$ is a diagonal matrix, then $a_{ij} = a_{ii}\delta_{ij}$ and, from Eq. (1.23), $\det\boldsymbol{A} = a_{11}a_{22}$.

For a (3×3) matrix,

$$\boldsymbol{A} = \begin{pmatrix} a_{11} & a_{12} & a_{13} \\ a_{21} & a_{22} & a_{23} \\ a_{31} & a_{32} & a_{33} \end{pmatrix}$$

the determinant is defined as

$$\begin{aligned} \det \boldsymbol{A} &= a_{11}\begin{vmatrix} a_{22} & a_{23} \\ a_{32} & a_{33} \end{vmatrix} - a_{12}\begin{vmatrix} a_{21} & a_{23} \\ a_{31} & a_{33} \end{vmatrix} + a_{13}\begin{vmatrix} a_{21} & a_{22} \\ a_{31} & a_{32} \end{vmatrix} \\ &= a_{11}\left(a_{22}a_{33} - a_{23}a_{32}\right) - a_{12}\left(a_{21}a_{33} - a_{23}a_{31}\right) + a_{13}\left(a_{21}a_{32} - a_{22}a_{31}\right) \end{aligned} \tag{1.24}$$

If $\boldsymbol{A}$ is a diagonal matrix, then $a_{ij}=a_{ii}\delta_{ij}$ and, from Eq. (1.24), $\det\boldsymbol{A}=a_{11}a_{22}a_{33}$.

For a (2×2) matrix, the results given by Eqs. (1.22) and (1.23), respectively, are verified with Mathematica procedure M1.14.

As a general rule, the sign of a_{1j}, $j=2, 3, \ldots, n$ alternates, that is, $(-1)^{j+1}a_{1j}$. For example, for $j=4$

$$\det \boldsymbol{A} = a_{11}\begin{vmatrix} a_{22} & a_{23} & a_{24} \\ a_{32} & a_{33} & a_{34} \\ a_{42} & a_{43} & a_{44} \end{vmatrix} - a_{12}\begin{vmatrix} a_{21} & a_{23} & a_{24} \\ a_{31} & a_{33} & a_{34} \\ a_{41} & a_{43} & a_{44} \end{vmatrix} + a_{13}\begin{vmatrix} a_{21} & a_{22} & a_{24} \\ a_{31} & a_{32} & a_{34} \\ a_{41} & a_{42} & a_{44} \end{vmatrix} - a_{14}\begin{vmatrix} a_{21} & a_{22} & a_{23} \\ a_{31} & a_{32} & a_{33} \\ a_{41} & a_{42} & a_{43} \end{vmatrix} \quad (1.25)$$

Each (3×3) determinant in Eq. (1.25) is further reduced by using Eqs. (1.23) and (1.24). It is seen from Eq. (1.25) that if $\boldsymbol{A}$ is a diagonal matrix, then the values of the determinant are the product of its diagonal elements: $\det\boldsymbol{A}=a_{11}a_{22}a_{33}a_{44}$.

We note the following general properties of determinants. If $\boldsymbol{A}$ and $\boldsymbol{B}$ are square matrices of order n, then

$$\det \boldsymbol{A} = \det \boldsymbol{A}^T \quad (1.26a)$$

$$\det \alpha\boldsymbol{A} = \alpha^n \det \boldsymbol{A} \quad (1.26b)$$

$$\det \boldsymbol{AB} = \det \boldsymbol{A} \det \boldsymbol{B} \quad (1.26c)$$

$$\det (\boldsymbol{A}+\boldsymbol{B}) \neq \det \boldsymbol{A} + \det \boldsymbol{B} \quad (1.26d)$$

Notice from the second equation of Eq. (1.26) that if $\alpha = -1$, then $\det(-\boldsymbol{A}) = (-1)^n\det\boldsymbol{A}$.

Proofs of Eq. (1.26) for (2×2) matrices are as follows.

(1) Eq. (1.26a)

$$\det \boldsymbol{A} = \det \boldsymbol{A}^T$$

$$a_{11}a_{22} - a_{12}a_{21} = a_{11}a_{22} - a_{21}a_{12}$$

(2) Eq. (1.26b)

$$\det \alpha\boldsymbol{A} = \det\begin{pmatrix} \alpha a_{11} & \alpha a_{12} \\ \alpha a_{21} & \alpha a_{22} \end{pmatrix} = \alpha^2 a_{11}a_{22} - \alpha^2 a_{12}a_{21} = \alpha^2 \det \boldsymbol{A}$$

Therefore, for an $(n\times n)$ matrix, $\det\alpha\boldsymbol{A}=\alpha^n\det\boldsymbol{A}$.

(3) Eq. (1.26c)

$$\begin{aligned}\det \boldsymbol{AB} &= \det\begin{pmatrix} a_{11}b_{11}+a_{12}b_{21} & a_{11}b_{12}+a_{12}b_{22} \\ a_{21}b_{11}+a_{22}b_{21} & a_{21}b_{12}+a_{22}b_{22} \end{pmatrix} \\ &= \left(a_{11}b_{11}+a_{12}b_{21}\right)\left(a_{21}b_{12}+a_{22}b_{22}\right) - \left(a_{11}b_{12}+a_{12}b_{22}\right)\left(a_{21}b_{11}+a_{22}b_{21}\right) \\ &= a_{12}a_{21}b_{12}b_{21} - a_{11}a_{22}b_{12}b_{21} - a_{12}a_{21}b_{11}b_{22} + a_{11}a_{22}b_{11}b_{22} \\ &= \left(a_{12}a_{21}-a_{11}a_{22}\right)\left(b_{12}b_{21}-b_{11}b_{22}\right) = \det \boldsymbol{A} \det \boldsymbol{B}\end{aligned}$$

(4) Eq. (1.26d)

$$\det(\boldsymbol{A}+\boldsymbol{B}) = a_{22}b_{11} - a_{21}b_{12} - b_{12}b_{21} - a_{12}\left(a_{21}+b_{21}\right) + b_{11}b_{22} + a_{11}\left(a_{22}+b_{22}\right)$$

$$\det \boldsymbol{A} + \det \boldsymbol{B} = -a_{12}a_{21} + a_{11}a_{22} - b_{12}b_{21} + b_{11}b_{22}$$

$$\det(\boldsymbol{A}+\boldsymbol{B}) - \det \boldsymbol{A} - \det \boldsymbol{B} = a_{22}b_{11} - a_{21}b_{12} - a_{12}b_{21} + a_{11}b_{22} \neq 0$$

For a (2×2) matrix, the results given by Eq. (1.26) are verified with Mathematica procedure M1.15.

We now determine the various conditions for which a determinant can equal zero. Returning to Eq. (1.23), it is seen that the determinant equals zero when $a_{11}a_{22}=a_{12}a_{21}$, which can be written as

$$\frac{a_{11}}{a_{21}} = \frac{a_{12}}{a_{22}} \quad \text{or} \quad \frac{a_{11}}{a_{12}} = \frac{a_{21}}{a_{22}} \tag{1.27}$$

Therefore, we see that $\det\boldsymbol{A}=0$ when $a_{11}=\alpha a_{21}$ and $a_{12}=\alpha a_{22}$ or when $a_{11}=\beta a_{12}$ and $a_{21}=\beta a_{22}$. In the former case, this is equivalent to multiplying the elements of the second row by α and replacing the elements of the first row with these elements. The latter case is equivalent to multiplying the elements of the second column by β and replacing the elements in the first column with these elements. In these respective cases, the determinants become

$$\begin{vmatrix} a_{11} & a_{12} \\ a_{21} & a_{22} \end{vmatrix} \rightarrow \begin{vmatrix} \alpha a_{21} & \alpha a_{22} \\ a_{21} & a_{22} \end{vmatrix} = \alpha a_{21}a_{22} - \alpha a_{21}a_{22} = 0$$

and

$$\begin{vmatrix} a_{11} & a_{12} \\ a_{21} & a_{22} \end{vmatrix} \rightarrow \begin{vmatrix} \beta a_{12} & a_{12} \\ \beta a_{22} & a_{22} \end{vmatrix} = \beta a_{12}a_{22} - \beta a_{12}a_{22} = 0$$

When the elements of a matrix row (column) are proportional to the corresponding elements of another row (column), these rows (columns) are *linearly dependent*. As shown for the (2×2) matrix, the determinant of such a matrix is zero. The same procedure can be used to show that extension to higher-order determinants produces the same result, that the determinant of matrices with linearly dependent rows (columns) is zero. These results are verified for a (3×3) matrix with Mathematica procedure M1.16.

There are several other ways for a determinant to equal zero. Three such possibilities are as follows

$$\boldsymbol{A} = \begin{pmatrix} a_{11} & a_{12} & a_{13} \\ a_{11}+c & a_{12}+c & a_{13}+c \\ a_{11}+d & a_{12}+d & a_{13}+d \end{pmatrix} \tag{1.28}$$

or

$$\boldsymbol{A} = \begin{pmatrix} a_{11} & a_{11}+c & a_{11}+d \\ a_{21} & a_{21}+c & a_{21}+d \\ a_{31} & a_{31}+c & a_{31}+d \end{pmatrix} \tag{1.29}$$

or

$$A = \begin{pmatrix} a_{11} & a_{12} & a_{13} \\ a_{11} + a & a_{12} + b & a_{13} + c \\ a_{11} + ka & a_{12} + kb & a_{13} + kc \end{pmatrix} \tag{1.30}$$

where a, b, c, d, and k are constants. These constants can be positive or negative. In these cases, $\det A = 0$. These results are verified for a (3×3) matrix and a (4×4) matrix with Mathematica procedure M1.17.

The number of rows of a matrix $\boldsymbol{A}$ that are linearly dependent is indicated by the *rank of a matrix*, which is the maximum number of *linearly independent* row vectors of $\boldsymbol{A}$. If $\boldsymbol{A}$ is an $(n \times n)$ matrix and the rank of the matrix is n, then there are n linearly independent row vectors in the matrix. Thus, when the rank of an $(n \times n)$ matrix $\boldsymbol{A}$ is n the determinant of $\boldsymbol{A} \neq 0$ and, as stated in the next section, the inverse of the matrix exists.

1.4 MATRIX INVERSE

The *inverse* of an $(n \times n)$ matrix $\boldsymbol{A}$, which is denoted as $\boldsymbol{A}^{-1}$, is defined as that matrix for which

$$\boldsymbol{A}^{-1}\boldsymbol{A} = \boldsymbol{A}\boldsymbol{A}^{-1} = \boldsymbol{I} \tag{1.31}$$

The inverse of a matrix exists if

$$\det \boldsymbol{A} \neq 0$$

This is always the case if the rank of $\boldsymbol{A}$ is n. If $\det \boldsymbol{A} = 0$, the matrix is called a *singular* matrix. It is noted that since $\det(\boldsymbol{AB}) = \det \boldsymbol{A} \det \boldsymbol{B}$ [recall Eq. (1.26)], we find that

$$\det(\boldsymbol{A}^{-1}\boldsymbol{A}) = 1 = \det \boldsymbol{A}^{-1} \det \boldsymbol{A}$$

and therefore,

$$\det \boldsymbol{A}^{-1} = \frac{1}{\det \boldsymbol{A}} \tag{1.32}$$

For a (3×3) matrix, the result given by Eq. (1.32) is verified with Mathematica procedure M1.18.

The inverse of a 2×2 matrix is determined by

$$\boldsymbol{A}^{-1} = \begin{pmatrix} a_{11} & a_{12} \\ a_{21} & a_{22} \end{pmatrix}^{-1} = \frac{1}{a_{11}a_{22} - a_{12}a_{21}} \begin{pmatrix} a_{22} & -a_{12} \\ -a_{21} & a_{11} \end{pmatrix} \tag{1.33}$$

Notice that $\det \boldsymbol{A} = a_{11}a_{22} - a_{12}a_{21}$, and, therefore, as previously stated, for the inverse to exist, $\det \boldsymbol{A} \neq 0$.

Equation (1.33) is verified with Mathematica procedure M1.19.

To verify that Eq. (1.33) is correct, we use Eq. (1.33) and perform the multiplication

$$\begin{pmatrix} a_{11} & a_{12} \\ a_{21} & a_{22} \end{pmatrix}^{-1} \begin{pmatrix} a_{11} & a_{12} \\ a_{21} & a_{22} \end{pmatrix} = \frac{1}{a_{11}a_{22} - a_{12}a_{21}} \begin{pmatrix} a_{22} & -a_{12} \\ -a_{21} & a_{11} \end{pmatrix} \begin{pmatrix} a_{11} & a_{12} \\ a_{21} & a_{22} \end{pmatrix}$$

$$= \frac{1}{a_{11}a_{22} - a_{12}a_{21}} \begin{pmatrix} c_{11} & c_{12} \\ c_{21} & c_{22} \end{pmatrix}$$

where, from Eq. (1.12)

$$c_{ij} = \sum_{p=1}^{2} a_{ip}a_{pj} = a_{i1}a_{1j} + a_{i2}a_{2j}$$

Therefore,

$$c_{11} = a_{22}a_{11} - a_{12}a_{21}$$

$$c_{12} = a_{22}a_{12} - a_{12}a_{22} = 0$$

$$c_{21} = -a_{21}a_{11} + a_{11}a_{21} = 0$$

$$c_{22} = -a_{21}a_{12} + a_{11}a_{22}$$

and

$$\begin{pmatrix} a_{11} & a_{12} \\ a_{21} & a_{22} \end{pmatrix}^{-1} \begin{pmatrix} a_{11} & a_{12} \\ a_{21} & a_{22} \end{pmatrix} = \begin{pmatrix} \dfrac{a_{11}a_{22} - a_{12}a_{21}}{a_{11}a_{22} - a_{12}a_{21}} & 0 \\ 0 & \dfrac{a_{11}a_{22} - a_{12}a_{21}}{a_{11}a_{22} - a_{12}a_{21}} \end{pmatrix}$$

$$= \begin{pmatrix} 1 & 0 \\ 0 & 1 \end{pmatrix} = \boldsymbol{I}$$

In the special case where the matrix is diagonal, that is, $a_{ij} = a_{ii}\delta_{ij}$, Eq. (1.33) simplifies to

$$\begin{pmatrix} a_{11} & 0 \\ 0 & a_{22} \end{pmatrix}^{-1} = \frac{1}{a_{11}a_{22}} \begin{pmatrix} a_{22} & 0 \\ 0 & a_{11} \end{pmatrix} = \begin{pmatrix} 1/a_{11} & 0 \\ 0 & 1/a_{22} \end{pmatrix} \tag{1.34}$$

Equation (1.34) is verified with Mathematica procedure M1.20.

1.5 PROPERTIES OF MATRIX PRODUCTS

From Eqs. (1.15) and (1.16), it is seen that, in general, the product $\boldsymbol{AB} \neq \boldsymbol{BA}$. However, for *conformable* matrices (matrices where the inner product dimension is the same), multiplication is *associative*, that is,

$$(\boldsymbol{CA})\boldsymbol{B} = \boldsymbol{C}(\boldsymbol{AB}) \tag{1.35}$$

since, based on the inner products,

$$((m\times k)(k\times n))(n\times p) = (m\times k)((k\times n)(n\times p))$$
$$(m\times n)(n\times p) = (m\times k)(k\times p)$$
$$(m\times p) = (m\times p)$$

These conformable matrices are also *distributive*, that is,

$$(\boldsymbol{A}+\boldsymbol{B})\boldsymbol{C} = \boldsymbol{AC}+\boldsymbol{BC} \tag{1.36}$$

From the definition of the transpose of a matrix, it follows that

$$(\boldsymbol{AB})^T = \boldsymbol{B}^T\boldsymbol{A}^T \tag{1.37}$$

The proof of Eq. (1.37) is as follows

$$\boldsymbol{AB} = \boldsymbol{C} \rightarrow c_{ij} = \sum_p a_{ip}b_{pj}$$
$$(\boldsymbol{AB})^T = \boldsymbol{C}^T \rightarrow c_{ji} = \sum_p a_{jp}b_{pi} = \sum_p b_{pi}a_{jp} \rightarrow \boldsymbol{B}^T\boldsymbol{A}^T$$

Equation (1.37) for $\boldsymbol{A}$ of order (2×3) and $\boldsymbol{B}$ of order (3×2) is verified with Mathematica procedure M1.21.

The following product plays an important role in the matrix manipulation process. Let $\boldsymbol{x}$ be an $(n\times 1)$ column vector and $\boldsymbol{A}$ be a square matrix of order n. Then the product

$$c = \boldsymbol{x}^T\boldsymbol{Ax} = \sum_{k=1}^{n}\sum_{j=1}^{n} a_{kj}x_kx_j \tag{1.38}$$

is a scalar, since manipulation of its inner products yields

$$(n\times 1)^T(n\times n)(n\times 1) = (1\times n)(n\times n)(n\times 1) = (1\times n)(n\times 1) = 1\times 1 = 1$$

Equation (1.38) is called the *quadratic form*. When $a_{kj}=a_{jk}$, Eq. (1.38) is called a *symmetric quadratic form*.

For matrices $\boldsymbol{A}$ of order (3×3) and $\boldsymbol{x}$ of order (3×1), the fact that Eq. (1.38) results in a scalar quantity is verified with Mathematica procedure M1.22.

Caution: For two $(n\times n)$ matrices, the equation $\boldsymbol{AB}=0$ does not generally imply that $\boldsymbol{A}=0$ or $\boldsymbol{B}=0$ unless the rank of each matrix is n. Also, the equation $\boldsymbol{AB}=\boldsymbol{AC}$, where $\boldsymbol{B}$ and $\boldsymbol{C}$ are of the same order, does not imply that $\boldsymbol{B}=\boldsymbol{C}$ because $\boldsymbol{A}$, $\boldsymbol{B}$, and $\boldsymbol{C}$ do not have to be square matrices, only conformable; therefore, one may not be able to multiply the equation by the inverse of $\boldsymbol{A}$ unless $\boldsymbol{A}$ is an $(n\times n)$ matrix of rank n.

1.6 EIGENVALUES OF A SQUARE MATRIX

Consider the simple algebraic equation $ax=b$ where a and b are constants. One can solve for x only if $a\neq 0$. Suppose that $b=0$. The condition under which $x=0$ is still $a\neq 0$. However, if $a=0$, then x is arbitrary. Now consider the equation $(a-\lambda)x=0$ where a and λ are constants. When $a\neq\lambda$, $x=0$. However, when $a=\lambda$, x is arbitrary. In what follows, λ will be called the eigenvalue and the corresponding expression for x will be called the eigenvector.

The eigenvalues of an $(n\times n)$ matrix $\boldsymbol{A}$ are given by the solutions of

$$\det\left(\boldsymbol{A}-\lambda\boldsymbol{I}\right)=0 \tag{1.39}$$

A matrix of order $(n\times n)$ has n eigenvalues λ_k, $k=1, 2, \ldots, n$, which can be zero, real, complex, and repeating. Note that Eq. (1.39) results in an nth-order polynomial in λ. This polynomial is called the *characteristic polynomial* or *characteristic equation*. To show this, consider a (2×2) matrix. Then

$$\begin{aligned}\det\left(\boldsymbol{A}-\lambda\boldsymbol{I}\right)&=\det\left(\begin{pmatrix}a_{11} & a_{12}\\ a_{21} & a_{22}\end{pmatrix}-\begin{pmatrix}\lambda & 0\\ 0 & \lambda\end{pmatrix}\right)\\ &=\det\begin{pmatrix}a_{11}-\lambda & a_{12}\\ a_{21} & a_{22}-\lambda\end{pmatrix}\\ &=\left(a_{11}-\lambda\right)\left(a_{22}-\lambda\right)-a_{12}a_{21} \qquad (1.40)\\ &=\lambda^2-\lambda\left(a_{11}+a_{22}\right)+a_{11}a_{22}-a_{12}a_{21}\\ &=\lambda^2-\mathrm{tr}\,\boldsymbol{A}\lambda+\det\,\boldsymbol{A}=0\end{aligned}$$

where tr$\boldsymbol{A}$ indicates the trace of $\boldsymbol{A}$, which is the sum of its diagonal elements, that is,

$$\mathrm{tr}\,\boldsymbol{A}=a_{11}+a_{22} \tag{1.41}$$

The solution of Eq. (1.40) is

$$\lambda_{1,2}=\frac{\mathrm{tr}\,\boldsymbol{A}}{2}\pm\sqrt{\left(\frac{\mathrm{tr}\,\boldsymbol{A}}{2}\right)^2-\det\boldsymbol{A}} \tag{1.42}$$

which are the eigenvalues. If det$\boldsymbol{A}<0$, then there are two real eigenvalues, one positive and one negative. If det$\boldsymbol{A}>0$ and $(\mathrm{tr}\boldsymbol{A})^2>4\det\boldsymbol{A}$, then there are still two real eigenvalues, but both are of the same sign as tr$\boldsymbol{A}$. If det$\boldsymbol{A}>0$ and $(\mathrm{tr}\boldsymbol{A})^2<4\det\boldsymbol{A}$, then there are two complex eigenvalues which are complex conjugates and where the real part equals tr$\boldsymbol{A}/2$. [See Eqs. (2.1) and (2.2) for the definitions of a complex number and its complex conjugate.]. If det$\boldsymbol{A}>0$ and tr$\boldsymbol{A}=0$, there are two imaginary eigenvalues. These observations are important in determining when the solutions to second-order differential equations are stable (bounded) or unstable (unbounded) and are discussed in Section 4.2.11.

Some general relations involving eigenvalues are

$$\sum_{k=1}^{n} \lambda_k = \sum_{k=1}^{n} a_{kk} = \text{tr}\, \boldsymbol{A}$$
$$\prod_{k=1}^{n} \lambda_k = \det \boldsymbol{A} \tag{1.43}$$

From the second equation of Eq. (1.43), we see that if $\det\boldsymbol{A}=0$, then at least one of the eigenvalues must be zero.

Equation (1.43) for $\boldsymbol{A}$ of order (2×2) is verified with Mathematica procedure M1.23.

1.7 SOLUTIONS TO A SYSTEM OF EQUATIONS: EIGENVALUES, EIGENVECTORS, AND ORTHOGONALITY

Consider the following linear system of two equations with two unknowns x_1 and x_2

$$a_{11}x_1 + a_{12}x_2 = b_1$$
$$a_{21}x_1 + a_{22}x_2 = b_2 \tag{1.44}$$

Using the first equation in Eq. (1.44) to solve for x_1 and the second equation to solve for x_2, it is found that

$$x_1 = \frac{1}{a_{11}}\left[b_1 - a_{12}x_2\right]$$
$$x_2 = \frac{1}{a_{22}}\left[b_2 - a_{21}x_1\right]$$

Substituting the second equation into the first equation gives

$$x_1 = \frac{1}{a_{11}}\left[b_1 - a_{12}\left\{\frac{1}{a_{22}}\left[b_2 - a_{21}x_1\right]\right\}\right]$$
$$= \frac{1}{a_{11}a_{22}}\left[a_{22}b_1 - a_{12}b_2 + a_{12}a_{21}x_1\right]$$
$$\left(a_{11}a_{22} - a_{12}a_{21}\right)x_1 = a_{22}b_1 - a_{12}b_2$$

Reversing the substitution process, we arrive at

$$\left(a_{11}a_{22} - a_{12}a_{21}\right)x_2 = a_{11}b_2 - a_{21}b_1$$

The necessary condition to determine x_1 and x_2 requires that $a_{11}a_{22} - a_{12}a_{21} \neq 0$, that is, the determinant $\neq 0$.

We now change notation and place Eq. (1.44) in matrix form. Then

$$\begin{pmatrix} a_{11} & a_{12} \\ a_{21} & a_{22} \end{pmatrix} \begin{Bmatrix} x_1 \\ x_2 \end{Bmatrix} = \begin{Bmatrix} b_1 \\ b_2 \end{Bmatrix} \rightarrow \boldsymbol{Ax} = \boldsymbol{b} \tag{1.45}$$

Solving for x_1 and x_2 using Cramer's rule, it is found that

$$\begin{aligned} x_1 &= \frac{1}{\det \boldsymbol{A}} \begin{vmatrix} b_1 & a_{12} \\ b_2 & a_{22} \end{vmatrix} = \frac{a_{22}b_1 - a_{12}b_2}{a_{11}a_{22} - a_{12}a_{21}} \\ x_2 &= \frac{1}{\det \boldsymbol{A}} \begin{vmatrix} a_{11} & b_1 \\ a_{21} & b_2 \end{vmatrix} = \frac{a_{11}b_2 - a_{21}b_1}{a_{11}a_{22} - a_{12}a_{21}} \end{aligned} \tag{1.46}$$

Equation (1.46) is verified with Mathematica procedure M1.24.

Therefore, extending these results to a system of N equations with at least one $b_j \neq 0$ with N unknowns x_j this system will have a solution if and only if the determinant of the coefficients does not equal zero, that is,

$$\det \boldsymbol{A} \neq 0$$

In practice, a system of N equations of the form given by Eq. (1.45) is most readily solved by available computer programs, in our case Mathematica. The algebraic details given here are necessary to support the introduction of eigenvalues, eigenvectors, and orthogonality.

Consider a system of N homogeneous equations with N unknowns x_j, that is, one in which $b_j = 0$, $j = 1, 2, \ldots, N$. If the determinant of the coefficients does not equal zero, then $x_j = 0$, $j = 1, 2, \ldots, N$. For the case where $N = 2$, it is seen from the right-hand side of Eq. (1.46) that

$$\begin{aligned} x_1 &= \frac{a_{22}b_1 - a_{12}b_2}{\det \boldsymbol{A}} \rightarrow \frac{0}{\det \boldsymbol{A}} = \frac{0}{a_{11}a_{22} - a_{12}a_{21}} \\ x_2 &= \frac{a_{11}b_2 - a_{21}b_1}{\det \boldsymbol{A}} \rightarrow \frac{0}{\det \boldsymbol{A}} = \frac{0}{a_{11}a_{22} - a_{12}a_{21}} \end{aligned}$$

Now consider the following system of equations

$$\begin{pmatrix} a_{11} - \lambda & a_{12} \\ a_{21} & a_{22} - \lambda \end{pmatrix} \begin{Bmatrix} x_1 \\ x_2 \end{Bmatrix} = 0$$

which contains three unknowns: x_1, x_2, and λ. Based on the previous discussion, this system of equations has a non-trivial solution (at least one $x_j \neq 0$) only when

$$\det \begin{pmatrix} a_{11} - \lambda & a_{12} \\ a_{21} & a_{22} - \lambda \end{pmatrix} = 0$$

One always solves λ, which is called the eigenvalue. In this case, the eigenvalues are determined from the characteristic equation given by Eq. (1.42). Then the system of equations can be written in matrix form as

$$\begin{pmatrix} a_{11}-\lambda_n & a_{12} \\ a_{21} & a_{22}-\lambda_n \end{pmatrix} \begin{Bmatrix} x_1^{(n)} \\ x_2^{(n)} \end{Bmatrix} = \begin{Bmatrix} 0 \\ 0 \end{Bmatrix} \rightarrow \boldsymbol{A}\boldsymbol{x}^{(n)} = \lambda_n \boldsymbol{x}^{(n)} \tag{1.47}$$

From these equations, there are four ways to relate $x_1^{(n)}$ and $x_2^{(n)}$. They are

$$\begin{aligned} x_1^{(n)} &= -\frac{a_{12}}{a_{11}-\lambda_n} x_2^{(n)} \quad \text{or} \quad x_2^{(n)} = -\frac{a_{11}-\lambda_n}{a_{12}} x_1^{(n)} \\ x_1^{(n)} &= -\frac{a_{22}-\lambda_n}{a_{21}} x_2^{(n)} \quad \text{or} \quad x_2^{(n)} = -\frac{a_{21}}{a_{22}-\lambda_n} x_1^{(n)} \end{aligned} \tag{1.48}$$

where the first pair is obtained from the first row of Eq. (1.47) and the second pair from the second row. Thus, it is seen that the best that one can do is determine the eigenvalues λ_n and $x_1^{(n)}$ in terms of $x_2^{(n)}$ or vice versa. In general, for an $(N \times N)$ matrix, one solves for λ_n and $(N-1)$ values of $x_j^{(n)}, j=1, 2, \ldots, N$ and $j \neq k$, in terms of $x_k^{(n)}$ at each value of λ_n. The values of $x_j^{(n)}, j=1, 2, \ldots, N$ for each value of n, $n=1, 2, \ldots, N$, comprise the elements of a vector called the *eigenvector*. Frequently, $k=1$ and for notational convenience one sets $x_1^{(n)}=1$ or $k=N$ and one sets $x_n^{(N)}=1$. Lastly, it is seen that the eigenvector represents a relative quantity (a shape), that is, one whose components are scaled, in this case to $x_1^{(n)}$.

To illustrate this, we shall use the second equation of the first row of Eq. (1.48). Then

$$x_1^{(n)} = 1$$

$$x_2^{(n)} = \frac{\lambda_n - a_{11}}{a_{12}}$$

These expressions constitute the components of the eigenvector corresponding to λ_n. These components are frequently placed in a matrix of eigenvectors as follows

$$\boldsymbol{\Phi} = \left\{ \boldsymbol{x}^{(1)} \quad \boldsymbol{x}^{(2)} \right\} = \begin{pmatrix} 1 & 1 \\ \dfrac{\lambda_1 - a_{11}}{a_{12}} & \dfrac{\lambda_2 - a_{11}}{a_{12}} \end{pmatrix} \tag{1.49}$$

If we had used the first equation of the second row and set $x_2^{(n)} = 1$, then

$$\boldsymbol{\Phi} = \left\{ \boldsymbol{x}^{(1)} \quad \boldsymbol{x}^{(2)} \right\} = \begin{pmatrix} \dfrac{\lambda_1 - a_{22}}{a_{21}} & \dfrac{\lambda_2 - a_{22}}{a_{21}} \\ 1 & 1 \end{pmatrix} \tag{1.50}$$

Equation (1.50) is verified with Mathematica procedure M1.25.

We now consider the following (3×3) matrix.

$$\begin{pmatrix} a_{11}-\lambda & a_{12} & a_{13} \\ a_{21} & a_{22}-\lambda & a_{23} \\ a_{31} & a_{32} & a_{33}-\lambda \end{pmatrix} \begin{Bmatrix} x_1 \\ x_2 \\ x_3 \end{Bmatrix} = 0 \tag{1.51}$$

By setting the determinant of the coefficients of Eq. (1.51) to zero, we can find the three eigenvalues λ_n, $n = 1, 2, 3$, which are assumed to be distinct. Then substituting the eigenvalues into Eq. (1.51), we obtain

$$\begin{pmatrix} a_{11}-\lambda_n & a_{12} & a_{13} \\ a_{21} & a_{22}-\lambda_n & a_{23} \\ a_{31} & a_{32} & a_{33}-\lambda_n \end{pmatrix} \begin{Bmatrix} x_1^{(n)} \\ x_2^{(n)} \\ x_3^{(n)} \end{Bmatrix} = 0 \tag{1.52}$$

To obtain the eigenvectors, we somewhat arbitrarily use the last two rows of Eq. (1.52) to determine $x_2^{(n)}$ and $x_3^{(n)}$ in terms of $x_1^{(n)}$ as follows

$$\begin{pmatrix} a_{22}-\lambda_n & a_{23} \\ a_{32} & a_{33}-\lambda_n \end{pmatrix} \begin{Bmatrix} x_2^{(n)} \\ x_3^{(n)} \end{Bmatrix} = \begin{Bmatrix} -a_{21}x_1^{(n)} \\ -a_{31}x_1^{(n)} \end{Bmatrix} \tag{1.53}$$

Solving Eq. (1.53) for $x_2^{(n)}$ and $x_3^{(n)}$, we obtain

$$\begin{aligned} x_2^{(n)} &= \frac{1}{D_n}\begin{vmatrix} -a_{21}x_1^{(n)} & a_{23} \\ -a_{31}x_1^{(n)} & a_{33}-\lambda_n \end{vmatrix} = c_{2n}x_1^{(n)} \\ x_3^{(n)} &= \frac{1}{D_n}\begin{vmatrix} a_{22}-\lambda_n & -a_{21}x_1^{(n)} \\ a_{32} & -a_{31}x_1^{(n)} \end{vmatrix} = c_{3n}x_1^{(n)} \end{aligned} \tag{1.54}$$

where

$$\begin{aligned} c_{2n} &= \frac{1}{D_n}\left[a_{23}a_{31} - a_{21}\left(a_{33}-\lambda_n\right)\right] \\ c_{3n} &= \frac{1}{D_n}\left[a_{21}a_{32} - a_{31}\left(a_{22}-\lambda_n\right)\right] \\ D_n &= \begin{vmatrix} a_{22}-\lambda_n & a_{23} \\ a_{32} & a_{33}-\lambda_n \end{vmatrix} = \left(a_{22}-\lambda_n\right)\left(a_{33}-\lambda_n\right) - a_{23}a_{32} \end{aligned} \tag{1.55}$$

Then the matrix of eigenvectors is

$$\begin{aligned} \hat{\Phi} &= \left\{\boldsymbol{x}^{(1)} \quad \boldsymbol{x}^{(2)} \quad \boldsymbol{x}^{(3)}\right\} \\ &= \begin{pmatrix} x_1^{(1)} & x_1^{(2)} & x_1^{(3)} \\ x_2^{(1)} & x_2^{(2)} & x_2^{(3)} \\ x_3^{(1)} & x_3^{(2)} & x_3^{(3)} \end{pmatrix} \\ &= \begin{pmatrix} x_1^{(1)} & x_1^{(2)} & x_1^{(3)} \\ c_{21}x_1^{(1)} & c_{22}x_1^{(2)} & c_{23}x_1^{(3)} \\ c_{31}x_1^{(1)} & c_{32}x_1^{(2)} & c_{33}x_1^{(3)} \end{pmatrix} \end{aligned} \tag{1.56}$$

Since each eigenvector is proportional to the same quantity and this quantity is an undetermined one, without loss of information we can write Eq. (1.56) as

$$\boldsymbol{\Phi} = \begin{pmatrix} 1 & 1 & 1 \\ c_{21} & c_{22} & c_{23} \\ c_{31} & c_{32} & c_{33} \end{pmatrix} \tag{1.57}$$

This is equivalent to having performed the following matrix operation

$$\boldsymbol{\Phi} = \begin{pmatrix} x_1^{(1)} & x_1^{(2)} & x_1^{(3)} \\ c_{21}x_1^{(1)} & c_{22}x_1^{(2)} & c_{23}x_1^{(3)} \\ c_{31}x_1^{(1)} & c_{32}x_1^{(2)} & c_{33}x_1^{(3)} \end{pmatrix} \begin{pmatrix} 1/x_1^{(1)} & 0 & 0 \\ 0 & 1/x_1^{(2)} & 0 \\ 0 & 0 & 1/x_1^{(3)} \end{pmatrix} \tag{1.58}$$

Generalizing Eq. (1.56), we see that for an $(N \times N)$ matrix

$$\hat{\boldsymbol{\Phi}} = \left\{ \boldsymbol{x}^{(1)} \quad \boldsymbol{x}^{(2)} \quad \cdots \quad \boldsymbol{x}^{(N)} \right\} = \begin{pmatrix} x_1^{(1)} & x_1^{(2)} & \cdots & x_1^{(N)} \\ x_2^{(1)} & x_2^{(2)} & \cdots & x_2^{(N)} \\ \vdots & \vdots & \ddots & \vdots \\ x_N^{(1)} & x_N^{(2)} & \cdots & x_N^{(N)} \end{pmatrix} \tag{1.59}$$

Then, to divide each element in the kth column by the corresponding element in the pth row and kth column, we introduce the diagonal matrix

$$\boldsymbol{\Phi}_D = \begin{pmatrix} 1/x_p^{(1)} & 0 & \cdots & 0 \\ 0 & 1/x_p^{(2)} & \cdots & 0 \\ \vdots & \vdots & \ddots & \vdots \\ 0 & 0 & \cdots & 1/x_p^{(N)} \end{pmatrix} \tag{1.60}$$

and form the matrix product

$$\boldsymbol{\Phi} = \hat{\boldsymbol{\Phi}}\boldsymbol{\Phi}_D = \hat{\boldsymbol{\Phi}} \begin{pmatrix} 1/x_p^{(1)} & 0 & \cdots & 0 \\ 0 & 1/x_p^{(2)} & \cdots & 0 \\ \vdots & \vdots & \ddots & \vdots \\ 0 & 0 & \cdots & 1/x_p^{(N)} \end{pmatrix}$$

For $p = 1$, we have

$$\boldsymbol{\Phi} = \begin{pmatrix} 1 & 1 & \cdots & 1 \\ x_2^{(1)}/x_1^{(1)} & x_2^{(2)}/x_1^{(2)} & \cdots & x_2^{(N)}/x_1^{(N)} \\ \vdots & \vdots & \ddots & \vdots \\ x_N^{(1)}/x_1^{(1)} & x_N^{(2)}/x_1^{(2)} & \cdots & x_N^{(N)}/x_1^{(N)} \end{pmatrix} \tag{1.61}$$

This equation is verified for a (3×3) matrix with Mathematica procedure M1.26.

Another way to arrange the elements of $\hat{\boldsymbol{\Phi}}$ is to divide the elements of a column vector by that column's norm obtained from Eq. (1.19), which in the present notation is

$$\left\|\boldsymbol{x}^{(n)}\right\| = \sqrt{\left(\boldsymbol{x}^{(n)}\right)^T \boldsymbol{x}^{(n)}} = \sqrt{\sum_{k=1}^{N}\left(x_k^{(n)}\right)^2} \quad n = 1,2,\ldots,N \tag{1.62}$$

Thus, if we re-write Eq. (1.60) as

$$\Phi_D = \begin{pmatrix} 1/\left\|x^{(1)}\right\| & 0 & \cdots & 0 \\ 0 & 1/\left\|x^{(2)}\right\| & \cdots & 0 \\ \vdots & \vdots & \ddots & \vdots \\ 0 & 0 & \cdots & 1/\left\|x^{(N)}\right\| \end{pmatrix} \tag{1.63}$$

then, from Eqs. (1.59) and (1.63),

$$\Phi = \hat{\Phi}\Phi_D = \begin{pmatrix} x_1^{(1)}/\left\|x^{(1)}\right\| & x_1^{(2)}/\left\|x^{(2)}\right\| & \cdots & x_1^{(N)}/\left\|x^{(N)}\right\| \\ x_2^{(1)}/\left\|x^{(1)}\right\| & x_2^{(2)}/\left\|x^{(2)}\right\| & \cdots & x_2^{(N)}/\left\|x^{(N)}\right\| \\ \vdots & \vdots & \ddots & \vdots \\ x_N^{(1)}/\left\|x^{(1)}\right\| & x_N^{(2)}/\left\|x^{(2)}\right\| & \cdots & x_N^{(N)}/\left\|x^{(N)}\right\| \end{pmatrix} \tag{1.64}$$

The column vectors of Eq. (1.64) are normalized so that $\left\|\boldsymbol{x}^{(n)}\right\| = 1,\ n = 1,2,\ldots,N$.

Example 1.3

We shall determine the eigenvalues and eigenvectors of

$$\boldsymbol{A} = \begin{pmatrix} 5 & 1 \\ -2 & 2 \end{pmatrix}$$

Then

$$\det\left(\begin{pmatrix} 5 & 1 \\ -2 & 2 \end{pmatrix} - \lambda\begin{pmatrix} 1 & 0 \\ 0 & 1 \end{pmatrix}\right) = \det\begin{pmatrix} 5-\lambda & 1 \\ -2 & 2-\lambda \end{pmatrix} = 0$$

$$\lambda^2 - 7\lambda + 12 = 0$$

Therefore, $\lambda_1 = 4$ and $\lambda_2 = 3$, and from Eq. (1.49) the eigenvectors are

$$\Phi = \left\{\boldsymbol{x}^{(1)} \quad \boldsymbol{x}^{(2)}\right\} = \begin{pmatrix} 1 & 1 \\ \dfrac{\lambda_1 - 5}{1} & \dfrac{\lambda_2 - 5}{1} \end{pmatrix} = \begin{pmatrix} 1 & 1 \\ -1 & -2 \end{pmatrix}$$

This result is verified with Mathematica procedure M1.27.

Example 1.4

We shall determine the eigenvalues and eigenvectors of

$$\boldsymbol{A} = \begin{pmatrix} 5 & 7 & -5 \\ 0 & 4 & -1 \\ 2 & 8 & -3 \end{pmatrix}$$

The eigenvalues are determined from

$$\det\begin{pmatrix} 5-\lambda & 7 & -5 \\ 0 & 4-\lambda & -1 \\ 2 & 8 & -3-\lambda \end{pmatrix} = -\lambda^3 + 6\lambda^2 - 11\lambda + 6 = 0$$

Using Mathematica procedure M1.28, the solution is $\lambda_1 = 1$, $\lambda_2 = 2$, and $\lambda_3 = 3$. Then, from Eqs. (1.54) and (1.55), it is found that

$$\frac{x_2^{(n)}}{x_1^{(n)}} = c_{2n} = \frac{-2}{(4-\lambda_n)(-3-\lambda_n)+8}$$

$$\frac{x_3^{(n)}}{x_1^{(n)}} = c_{3n} = \frac{-2(4-\lambda_n)}{(4-\lambda_n)(-3-\lambda_n)+8}$$

Therefore,

$$\frac{x_2^{(1)}}{x_1^{(1)}} = \frac{1}{2}, \quad \frac{x_2^{(2)}}{x_1^{(2)}} = 1, \quad \frac{x_2^{(3)}}{x_1^{(3)}} = -1$$

$$\frac{x_3^{(1)}}{x_1^{(1)}} = \frac{3}{2}, \quad \frac{x_3^{(2)}}{x_1^{(2)}} = 2, \quad \frac{x_3^{(3)}}{x_1^{(3)}} = -1$$

Using Eq. (1.57), the matrix of eigenvectors becomes

$$\Phi = \begin{pmatrix} 1 & 1 & 1 \\ 1/2 & 1 & -1 \\ 3/2 & 2 & -1 \end{pmatrix} \rightarrow \begin{pmatrix} 1 & 1 & 1 \\ 1/2 & 1 & -1 \\ 3/2 & 2 & -1 \end{pmatrix}\begin{pmatrix} 2 & 0 & 0 \\ 0 & 1 & 0 \\ 0 & 0 & -1 \end{pmatrix}$$

$$= \begin{pmatrix} 2 & 1 & -1 \\ 1 & 1 & 1 \\ 3 & 2 & 1 \end{pmatrix}$$

In other words, in Eq. (1.60) we have set $p = 2$ and, therefore, $x_2^{(1)} = 1/2$, $x_2^{(2)} = 1$, and $x_2^{(3)} = -1$. These results are verified with Mathematica procedure M1.29.

Two eigenvectors are *orthogonal* if

$$\left(\boldsymbol{x}^{(n)}\right)^T \boldsymbol{x}^{(k)} = 0 \quad n \neq k \tag{1.65}$$

As indicated in Eq. (1.47), the eigenvectors are the non-trivial solutions to

$$\boldsymbol{A}\boldsymbol{x}^{(n)} = \lambda_n \boldsymbol{x}^{(n)} \tag{1.66}$$

To determine under what conditions orthogonal eigenvectors can be obtained, we multiply Eq. (1.66) by $\left(\boldsymbol{x}^{(n)}\right)^T$ and obtain

$$\left(\boldsymbol{x}^{(k)}\right)^T \boldsymbol{A}\boldsymbol{x}^{(n)} = \lambda_n \left(\boldsymbol{x}^{(k)}\right)^T \boldsymbol{x}^{(n)} \tag{1.67}$$

Reversing the superscripts in Eq. (1.67) gives

$$\left(\boldsymbol{x}^{(n)}\right)^T \boldsymbol{A}\boldsymbol{x}^{(k)} = \lambda_k \left(\boldsymbol{x}^{(n)}\right)^T \boldsymbol{x}^{(k)} \tag{1.68}$$

where $\lambda_n \neq \lambda_k$. Taking the transpose of Eq. (1.68) yields

$$\begin{aligned}
\left(\left(\boldsymbol{x}^{(n)}\right)^T \boldsymbol{A}\boldsymbol{x}^{(k)}\right)^T &= \left(\lambda_k \left(\boldsymbol{x}^{(n)}\right)^T \boldsymbol{x}^{(k)}\right)^T \\
\left(\boldsymbol{x}^{(k)}\right)^T \left(\left(\boldsymbol{x}^{(n)}\right)^T \boldsymbol{A}\right)^T &= \lambda_k \left(\boldsymbol{x}^{(k)}\right)^T \boldsymbol{x}^{(n)} \\
\left(\boldsymbol{x}^{(k)}\right)^T \boldsymbol{A}^T \boldsymbol{x}^{(n)} &= \lambda_k \left(\boldsymbol{x}^{(k)}\right)^T \boldsymbol{x}^{(n)}
\end{aligned} \tag{1.69}$$

Subtracting Eq. (1.69) from Eq. (1.67) gives

$$(\lambda_n - \lambda_k)\left(\boldsymbol{x}^{(k)}\right)^T \boldsymbol{x}^{(n)} = \left(\boldsymbol{x}^{(k)}\right)^T \left(\boldsymbol{A} - \boldsymbol{A}^T\right)\boldsymbol{x}^{(n)} \quad \lambda_n \neq \lambda_k \tag{1.70}$$

Therefore, in order for the eigenvectors $\boldsymbol{x}^{(n)}$ and $\boldsymbol{x}^{(k)}$ to be orthogonal, $\boldsymbol{A}$ must be symmetric, that is, $\boldsymbol{A} = \boldsymbol{A}^T$. If $\boldsymbol{A}$ is symmetric, then from Eqs. (1.68) and (1.70),

$$\left(\boldsymbol{x}^{(n)}\right)^T \boldsymbol{x}^{(k)} = 0 \rightarrow \left(\boldsymbol{x}^{(n)}\right)^T \boldsymbol{A}\boldsymbol{x}^{(k)} = 0 \quad n \neq k \tag{1.71}$$

and

$$\left(\boldsymbol{x}^{(k)}\right)^T \boldsymbol{x}^{(i)} = \delta_{ki} \left\|\boldsymbol{x}^{(i)}\right\|^2 \tag{1.72}$$

where

$$\left\|\boldsymbol{x}^{(i)}\right\|^2 = \left(x_1^{(i)}\right)^2 + \left(x_2^{(i)}\right)^2 + \cdots + \left(x_n^{(i)}\right)^2 \tag{1.73}$$

and $\left\|\boldsymbol{x}^{(i)}\right\|$ is the norm of the vector [recall Eq. (1.19)].

We now examine the following matrix operation for a (2×2) matrix $\boldsymbol{\Phi}$ given by Eq. (1.59) for $N = 2$

$$\begin{aligned}\Phi^T A\Phi &= \{x^{(1)} \quad x^{(2)}\}^T A\{x^{(1)} \quad x^{(2)}\} \\ &= \{x^{(1)} \quad x^{(2)}\}^T \{Ax^{(1)} \quad Ax^{(2)}\} \\ &= \begin{pmatrix} \left(x^{(1)}\right)^T Ax^{(1)} & \left(x^{(1)}\right)^T Ax^{(2)} \\ \left(x^{(2)}\right)^T Ax^{(1)} & \left(x^{(2)}\right)^T Ax^{(2)} \end{pmatrix}\end{aligned} \tag{1.74}$$

If $A = A^T$, that is, A is a symmetric matrix, then $x^{(n)}$ are orthogonal and using Eqs. (1.69), (1.71), and (1.72), Eq. (1.74) becomes

$$\Phi^T A\Phi = \begin{pmatrix} \left(x^{(1)}\right)^T Ax^{(1)} & 0 \\ 0 & \left(x^{(2)}\right)^T Ax^{(2)} \end{pmatrix} = \begin{pmatrix} \lambda_1 \left\|x^{(1)}\right\|^2 & 0 \\ 0 & \lambda_2 \left\|x^{(2)}\right\|^2 \end{pmatrix} \tag{1.75}$$

Thus, it is seen that this matrix operation with orthogonal vectors uncoupled the system of equations as indicated by the diagonal matrix. This technique is used to solve a system of coupled ordinary differential equations as shown in Section 4.2.8. From Eq. (1.64), it is seen that one can normalize the eigenvectors by

$$\hat{x}^{(i)} = x^{(i)} \big/ \left\|x^{(i)}\right\|$$

so that

$$\left(\hat{x}^{(k)}\right)^T \hat{x}^{(i)} = \delta_{ki} \tag{1.76}$$

These normalized orthogonal eigenvectors are called *orthonormal* eigenvectors.

Two vectors are said to be *linearly independent* provided that neither is a constant multiple of the other. Recall the discussion following Eq. (1.27). In the present case, we more formally state that n eigenvectors of a symmetric ($n \times n$) matrix are linearly independent if the only way to satisfy

$$\sum_{i=1}^{n} c_i x^{(i)} = c_1 x^{(1)} + c_2 x^{(2)} + \cdots + c_n x^{(n)} = 0 \tag{1.77}$$

is if $c_i = 0$ for all i. If we multiply Eq. (1.77) by $\left(x^{(k)}\right)^T$, we obtain

$$\sum_{i=1}^{n} \left(x^{(k)}\right)^T c_i x^{(i)} = \sum_{i=1}^{n} c_i \left(x^{(k)}\right)^T x^{(i)} = \sum_{i=1}^{n} c_i \delta_{ki} \left\|x^{(i)}\right\|^2 = c_i \left\|x^{(i)}\right\|^2 = 0 \tag{1.78}$$

where we have used Eq. (1.19). Since $\left\|x^{(i)}\right\|^2 \neq 0$, $c_i = 0$. Therefore, the eigenvectors of a symmetric matrix are linearly independent.

An important consequence of linear independence is that one can express the vector x as the sum of the orthogonal eigenvectors, that is,

$$x = \sum_{i=1}^{n} c_i x^{(i)} = c_1 x^{(1)} + c_2 x^{(2)} + \cdots + c_n x^{(n)} \tag{1.79}$$

We shall be using the concept behind this result, which is the generation of orthogonal functions, in Section 6.1.4 and Chapter 7.

Example 1.5

With the help of Mathematica procedure M1.30, we shall determine the eigenvalues and eigenvectors for a symmetric matrix and non-symmetric matrix, show that the eigenvectors are orthogonal for the symmetric matrix and not orthogonal for the non-symmetric matrix, and verify Eqs. (1.74) and (1.75). The two matrices are

$$A_s = \begin{pmatrix} 1 & 2.3 & 3 \\ 2.3 & 1.5 & 6 \\ 3 & 6 & 9 \end{pmatrix} \quad A_{as} = \begin{pmatrix} 1 & 2.3 & 3 \\ 3 & 1.5 & 6 \\ 7 & 8 & 9 \end{pmatrix}$$

where A_s is the symmetric matrix and A_{as} the non-symmetric matrix.

The eigenvalues for the symmetric matrix are $\lambda_{s,1} = 13.45$, $\lambda_{s,2} = -1.981$, $\lambda_{s,3} = 0.03041$; those for the non-symmetric matrix are $\lambda_{as,1} = 15.11$, $\lambda_{as,2} = -2.568$, $\lambda_{as,3} = -1.044$. The respective *normalized* matrices of eigenvectors are

$$\Phi_s = \begin{pmatrix} -0.288 & 0.2674 & 0.9195 \\ -0.4734 & -0.8745 & 0.106 \\ -0.8324 & 0.4048 & -0.3784 \end{pmatrix}$$

and

$$\Phi_{as} = \begin{pmatrix} -0.2546 & 0.1089 & -0.8589 \\ -0.4365 & -0.8472 & 0.454 \\ -0.8629 & 0.52 & 0.237 \end{pmatrix}$$

To verify that the columns Φ_s Φ_{as} have been normalized, we consider the first column Φ_s and find

$$\|x^{(1)}\|^2 = (-0.288)^2 + (-0.4734)^2 + (-0.8324)^2$$

$$= 0.0829 + 0.224 + 0.693 = 1$$

Upon using Eq. (1.74) for the symmetric matrix, we obtain

$$\Phi_s^T A_s \Phi_s = \begin{pmatrix} 13.45 & 0 & 0 \\ 0 & -1.981 & 0 \\ 0 & 0 & 0.03041 \end{pmatrix}$$

and for the non-symmetric matrix

$$\Phi_{as}^T A_{as} \Phi_{as} = \begin{pmatrix} 15.11 & 0.2739 & 0.192 \\ -1.612 & -2.568 & 0.3704 \\ -2.781 & 0.9117 & -1.044 \end{pmatrix}$$

Lastly, we verify the orthogonality of the eigenvectors of the symmetric matrix as

$$\left(\boldsymbol{x}^{(1)}\right)^T \boldsymbol{x}^{(2)} = 0, \quad \left(\boldsymbol{x}^{(1)}\right)^T \boldsymbol{x}^{(3)} = 0, \quad \left(\boldsymbol{x}^{(2)}\right)^T \boldsymbol{x}^{(3)} = 0$$
$$\left(\boldsymbol{x}^{(1)}\right)^T \boldsymbol{x}^{(1)} = 1, \quad \left(\boldsymbol{x}^{(2)}\right)^T \boldsymbol{x}^{(2)} = 1, \quad \left(\boldsymbol{x}^{(3)}\right)^T \boldsymbol{x}^{(3)} = 1$$

and the non-orthogonality of the non-symmetric matrix as

$$\left(\boldsymbol{x}^{(1)}\right)^T \boldsymbol{x}^{(2)} = -0.1066, \quad \left(\boldsymbol{x}^{(1)}\right)^T \boldsymbol{x}^{(3)} = -0.184, \quad \left(\boldsymbol{x}^{(2)}\right)^T \boldsymbol{x}^{(3)} = -0.355$$
$$\left(\boldsymbol{x}^{(1)}\right)^T \boldsymbol{x}^{(1)} = 1, \quad \left(\boldsymbol{x}^{(2)}\right)^T \boldsymbol{x}^{(2)} = 1, \quad \left(\boldsymbol{x}^{(3)}\right)^T \boldsymbol{x}^{(3)} = 1$$

MATHEMATICA PROCEDURES

```
Note: In order to make the Mathematica procedures that follow a
little more readable, the Clear statements have been omitted.

(*M1.1*)
A=Array[a_ToString[#1]<>ToString[#2] &,{2,3}]//MatrixForm

(*M1.2*)
A=Transpose[Array[a_ToString[#1]<>ToString[#2] &,{2,3}]];
A//MatrixForm
Transpose[A]//MatrixForm

(*M1.3*)
B=Array[b_ToString[#1]<>ToString[#2] &,{3,1}];
B//MatrixForm
Transpose[B]//MatrixForm

(*M1.4*)
DiagonalMatrix[{a_11,a_22,a_33}]//MatrixForm

(*M1.5*)
IdentityMatrix[3]//MatrixForm

(*M1.6*)
A=Array[a_ToString[#1]<>ToString[#2] &,{2,2}];
B=Array[b_ToString[#1]<>ToString[#2] &,{2,2}];
A+B//MatrixForm
Transpose[(A+B)]//MatrixForm
Transpose[A]+Transpose[B]//MatrixForm
```

```
(*M1.7*)
A={{1,-2,0},{4,5,-7}};
B={{3,-8,4},{-4,1,11}};
A+B//MatrixForm

(*M1.8*)
A={{2,5},{-1,3},{4,-2}};
B={{6,8},{1,-7}};
A.B//MatrixForm

(*M1.9*)
A=Array[a_{ToString[#1]<>ToString[#2]} &,{2,2}];
B=Array[b_{ToString[#1]<>ToString[#2]} &,{2,2}];
Print["AB=",A.B//MatrixForm]
Print["BA=",B.A//MatrixForm]

(*M1.10*)
A=DiagonalMatrix[{a_{11},a_{22}}];
B=DiagonalMatrix[{b_{11},b_{22}}];
A.B//MatrixForm

(*M1.11*)
Ar=Array[a_{ToString[#1]} &,{3,1}];
Bc=Array[b_{ToString[#1]} &,{3,1}];
Print["a^Tb=",Transpose[Ar].Bc//MatrixForm]
Print["a^Ta=",Transpose[Ar].Ar//MatrixForm]

(*M1.12*)
Ar=Array[a_{ToString[#1]<>ToString[#2]} &,{1,2}];
Bc=Array[b_{ToString[#1]<>ToString[#2]} &,{3,1}];
Print["ba=",Bc.Ar//MatrixForm]

(*M1.13*)
A=Array[a_{ToString[#1]<>ToString[#2]} &,{2,2}]
MatrixForm[A]
B=Array[b_{ToString[#1]<>ToString[#2]} &,{2,2}]
MatrixForm[B]
g A//MatrixForm
Simplify[g (A+B)]==Simplify[g A+g B]

(*M1.14*)
Det[Array[a_{ToString[#1]<>ToString[#2]} &,{2,2}]]
Simplify[Det[Array[a_{ToString[#1]<>ToString[#2]} &,{3,3}]]]

(*M1.15*)
A=Array[a_{ToString[#1]<>ToString[#2]} &,{2,2}];
B=Array[b_{ToString[#1]<>ToString[#2]} &,{2,2}];
Det[A]==Det[Transpose[A]]
Simplify[Det[A.B]==Det[A] Det[B]]
Simplify[Det[g A]==g^2 Det[A]]
Simplify[Det[A+B]== Det[A]+Det[B]]
```

```
(*M1.16*)
Acol={{a11,b a11,a13},{a21,b a21,a23},{a31,b a31,a33}};
Arow={{a11,a12,a13},{b a11,b a12,b a13},{a31,a32,a33}};
Print["A=",MatrixForm[Acol]//MatrixForm]
Print["B=",MatrixForm[Arow]//MatrixForm]
Print["Det[A]=",Simplify[Det[Acol]]]
Print["Det[B]=",Simplify[Det[Arow]]]

(*M1.17*)
A={{a11,a12,a13},{a11+c,a12+c,a13+c},{a11+d,a12+d,,a13+d}};
Print["A=",A//MatrixForm]
Print["Det[A]=",Det[A]]
B={{a11,a11+c,a11+d},{a21,a21+c,a21+d},{a31+d,a31"+d,,a31+d}};
Print["B=",B//MatrixForm]
Print["Det[B]=",Det[B]]
Cc={{a11,a12,a13},{a11+a,a12+b,a13+c},{a11+k a,a12+k b,a13+k c}};
Print["C=",Cc//MatrixForm]
Print["Det[C]=",Det[Cc]]
A4={{a11,a12,a13,a14},{a11+c,a12+c,a13+c,a14+c},
    {a11+d,a12+d,a13+d,a14+d},{a11+e,a12+e,a13+e,a14+e}};
Print["A4=",A4//MatrixForm]
Print["Det[A4]=",Det[A4]]
B4={{a11,a11+c,a11+d,a11+e},{a21,a21+c,a21+d,a21+e},
    {a31,a31+c,a31+d,a31+e},{a41,a41+c,a41+e,a41+e}};
Print["B4=",B4//MatrixForm]
Print["Det[B4]=",Det[B4]]
C4={{a11,a12,a13,a14},{a11+a,a12+b,a13+c,a14+d},
    {a11+k a,a12+k b,a13+k c,a14+k d},{a11+g a,a12+g b,
    a13+g c,a14+g d}};
Print["C4=",C4//MatrixForm]
Print["Det[C4]=",Det[C4]]

(*M1.18*)
A=Array[aToString[#1]<>ToString[#2] &,{3,3}];
Simplify[Det[Inverse[A]]]==Simplify[1/Det[A]]

(*M1.19*)
A=Array[aToString[#1]<>ToString[#2] &,{2,2}]
Inverse[A]//MatrixForm

(*M1.20*)
Inverse[DiagonalMatrix[{a11,a22}]]//MatrixForm

(*M1.21*)
A=Array[aToString[#1]<>ToString[#2] &,{2,3}];
B=Array[bToString[#1]<>ToString[#2] &,{3,2}];
Transpose[A.B]==Transpose[B].Transpose[A]

(*M1.22*)
A=Array[aToString[#1]<>ToString[#2] &,{3,3}];
Xx=Array[xToString[#1]<>ToString[#2] &,{3,1}];
Dimensions[Transpose[Xx].A.Xx]
```

```
(*M1.23*)
A=Array[a_{ToString[#1]<>ToString[#2]} &,{2,2}];
lam=Eigenvalues[A];
Simplify[lam [[1]]+ lam [[2]]]==Tr[A]
Simplify[lam [[1]] lam [[2]]]==Det[A]

(*M1.24*)
A=Array[a_{ToString[#1]<>ToString[#2]} &,{2,2}];
B=Array[b_{ToString[#1]} &,{2,1}];
LinearSolve[A,B]

(*M1.25*)
A=Array[a_{ToString[#1]<>ToString[#2]} &,{2,2}];
Transpose[Eigenvectors[A]]//MatrixForm

(*M1.26*)
A={{a11,a12,a13},{a21,a22,a23},{a31,a32,a33}};
A.DiagonalMatrix[{1/a11,1/a12,1/a13}]//MatrixForm

(*M1.27*)
Transpose[Eigenvectors[{{5,1},{-2,2}}]]//MatrixForm

(*M1.28*)
g={{5,7,-5},{0,4,-1},{2,8,-3}};
Eigenvalues[g]

(*M1.29*)
g={{5,7,-5},{0,4,-1},{2,8,-3}};
Transpose[Eigenvectors[g]]//MatrixForm

(*M1.30*)
gg[n_]:=Style[x,Bold]^"("<>ToString[n]<>")"
orth[n_,m_,c_]:=(q=Chop[Transpose[Transpose[
    {c[[All,n]]}]].c[[All,m]]][[1]];
    Row[{"(",gg[n],")T",gg[m],"=",NumberForm[q,4]}])
lam[n_,lams_]:=Row[{λ_{"s,"<>ToString[n]},"=",NumberForm[lams[[n]],4]}]
lama[n_,lamas_]:=Row[{λ_{"as,"<>ToString[n]},"=",
    NumberForm[lamas[[n]],4]}]
Asym={{1,2.3,3},{2.3,1.5,6},{3,6,9}};
Aasym={{1,2.3,3},{3,1.5,6},{7,8,9}};
Print["A_s=",Asym//MatrixForm]
Print["A_as=",Aasym//MatrixForm]
{lams,xns}=Eigensystem[Asym];
{lamas,xnas}=Eigensystem[Aasym];
xns=Transpose[xns];
xnas=Transpose[xnas];
Print["Ps=",NumberForm[xns//MatrixForm,4]]
Print["P_as=",NumberForm[xnas//MatrixForm,4]]
Print[lam[1,lams]," ",lam[2,lams],"     ",lam[3,lams]]
Print[lama[1,lamas],"     ",lama[2,lamas],"     ",lama[3,lamas]]
Print["P_s^T A_s P_s",NumberForm[Chop[Transpose[xns].Asym.xns]
    //MatrixForm,4]]
```

```
Print["Pas^T Aas Pas",NumberForm[Chop[Transpose[xnas].Aasym.xnas]
   //MatrixForm,4]]
Print["For As]"]
Print[orth[1,2,xns],"       ",orth[1,3,xns],"      ",orth[2,3,xns]]
Print[orth[1,1,xns],"       ",orth[2,2,xns],"      ",orth[3,3,xns]]
Print["For Aas"]
Print[orth[1,2,xnas],"      ",orth[1,3,xnas],"     ",orth[2,3,xnas]]
Print[orth[1,1,xnas],"      ",orth[2,2,xnas],"     ",orth[3,3,xnas]]
```

EXERCISES

Section 1.2

1.1 A matrix is an orthogonal matrix if

$$\boldsymbol{X}^T\boldsymbol{X}=\boldsymbol{I}$$

Is the matrix

$$\boldsymbol{X}=\frac{1}{2}\begin{pmatrix}-1 & -1\\ 1 & -1\\ -1 & 1\\ 1 & 1\end{pmatrix}$$

an orthogonal matrix?

1.2 If

$$\boldsymbol{A}=\begin{pmatrix}1 & -1\\ 2 & -1\end{pmatrix}\quad \boldsymbol{B}=\begin{pmatrix}1 & 1\\ 4 & -1\end{pmatrix}$$

does $(\boldsymbol{A}+\boldsymbol{B})^2=\boldsymbol{A}^2+\boldsymbol{B}^2$?

1.3 Given the two matrices

$$\boldsymbol{A}=\begin{pmatrix}1 & 4 & -3\\ 2 & 5 & 4\end{pmatrix}\quad \text{and}\quad \boldsymbol{B}=\begin{pmatrix}4 & 1\\ 2 & 6\\ 0 & 3\end{pmatrix}$$

Find the matrix products $\boldsymbol{AB}$ and $\boldsymbol{BA}$.

1.4 Given the following matrices and their respective orders: $\boldsymbol{A}$ $(n\times m)$, $\boldsymbol{B}$ $(p\times m)$, and $\boldsymbol{C}$ $(n\times s)$, show one way in which these three matrices can be multiplied. What is the order of the resulting matrix?

1.5 Given

$$\boldsymbol{A}=\begin{pmatrix}ab & b^2\\ -a^2 & -ab\end{pmatrix}$$

Determine $\boldsymbol{A}^2$.

1.6 Given the matrix

$$A = \begin{pmatrix} -4 & -3 & -1 \\ 2 & 1 & 1 \\ 4 & -2 & 4 \end{pmatrix}$$

Determine the value of $4I - 4A - A^2 + A^3$.

Section 1.3

1.7 Given the following matrices:

$$x = \begin{Bmatrix} 1 \\ 2 \end{Bmatrix}, \quad A = \begin{pmatrix} 2 & a \\ 3 & 4 \end{pmatrix}, \quad B = \begin{pmatrix} 6 & 4 \\ 7 & 5 \end{pmatrix}$$

What is the value of a that satisfies the following equation?

$$x^T A x = \det B$$

1.8 Show that

$$\det \begin{pmatrix} a & b+c & 1 \\ b & a+c & 1 \\ c & a+b & 1 \end{pmatrix} = 0$$

1.9 Expand the following determinants and reduce them to their simplest terms.

a) $\det \begin{pmatrix} 1+a & a & a \\ b & 1+b & b \\ b & b & 1+b \end{pmatrix}$

b) $\det \begin{pmatrix} x^3+1 & 1 & 1 \\ 1 & x^3+1 & 1 \\ 1 & 1 & x^3+1 \end{pmatrix}$

1.10 Determine if the following determinant is a function of a

$$\det \begin{pmatrix} e^x & \sin x & \cos x \\ e^x & \cos x & \sin x \\ 1 & 1-a & a \end{pmatrix}$$

1.11 Show that

$$\det \begin{pmatrix} x_1^2 & x_1 & 1 \\ x_2^2 & x_2 & 1 \\ x_3^2 & x_3 & 1 \end{pmatrix} = (x_1 - x_2)(x_1 - x_3)(x_2 - x_3)$$

Section 1.4

1.12 Given

$$A = \begin{pmatrix} 17 & 7 \\ 19 & 9 \end{pmatrix}$$

Determine A^{-1} and verify your result.

Section 1.5

1.13 Given the two matrices

$$A = \begin{pmatrix} 3 & 1 & 4 \\ 2 & 1 & 2 \\ 4 & 2 & 3 \end{pmatrix} \quad \text{and} \quad B = \begin{pmatrix} 2 & 1 & 3 \\ 1 & 2 & 5 \\ 0 & 2 & 1 \end{pmatrix}$$

Show that $(AB)^T = B^T A^T$.

Section 1.7

1.14 Does the following system of equations have a solution?

$$100y_1 + 420y_2 + 486y_3 = 17$$

$$400y_1 + 1050y_2 + 972y_3 = 18$$

$$700y_1 + 1680y_2 + 1458y_3 = -3$$

1.15 Determine whether the following system of equations has a solution.

$$a_1 + 2a_2 + 3a_3 = 1$$

$$4a_1 + 5a_2 + 6a_3 = 0$$

$$7a_1 + 8a_2 + 9a_3 = -7$$

1.16 Given the following system of equations

$$\begin{bmatrix} 7 & 2 \\ 5 & 1 \end{bmatrix} \begin{Bmatrix} x_1 \\ x_2 \end{Bmatrix} - \lambda \begin{bmatrix} 1 & 0 \\ 0 & 1 \end{bmatrix} \begin{Bmatrix} x_1 \\ x_2 \end{Bmatrix} = \begin{Bmatrix} 0 \\ 0 \end{Bmatrix}$$

When $\lambda = 4$, what are the values of x_1 and x_2? Justify your answer.

1.17 Under what conditions does the following system of equations have a non-trivial solution?

$$(1-\lambda)y_1 + 2y_2 + 3y_3 = 0$$

$$4y_1 + (6-\lambda)y_2 + 5y_3 = 0$$

$$7y_1 + 8y_2 + (9-\lambda)y_3 = 0$$

1.18 Given the following system of equations

$$(5-\lambda)x_1 + x_2 = 2$$

$$-2x_1 + (2-\lambda)x_2 = 1$$

Under what specific conditions can one solve for x_1 and x_2?

1.19 Determine the eigenvalues and eigenvectors for the following matrix and verify that the eigenvectors are not orthogonal.

$$\mathbf{A} = \begin{pmatrix} 1 & 0 & -1 \\ -2 & -1 & 2 \\ -1 & 2 & 1 \end{pmatrix}$$

2 Introduction to Complex Variables

2.1 COMPLEX NUMBERS

A complex number z has the form

$$z = a + jb \tag{2.1}$$

where a and b are real numbers, $j = \sqrt{-1}$, and $j^2 = -1$. The quantity a is called the *real part* of z and the quantity b is called the *imaginary part* of z. The notations Re[z] and Im[z] stand for the real and imaginary parts of z, respectively. For example, $w = \text{Re}[z]$ means that $w = a$, and $w = \text{Im}[z]$ means that $w = b$. When $b = 0$, $z = a$ is real, and when $a = 0$, $z = jb$ is imaginary. For $z = 0$, $a = 0$ and $b = 0$.

The quantity

$$\bar{z} = a - jb \tag{2.2}$$

is called the *complex conjugate* of z, that is, the sign of the imaginary part of Eq. (2.1) is changed, in this case from plus to minus. Consider the two complex numbers

$$\begin{aligned} z_1 &= a_1 + jb_1 \\ z_2 &= a_2 + jb_2 \end{aligned} \tag{2.3}$$

The complex numbers are equal if and only if their real parts are equal and their imaginary parts are equal. Thus, for $z_1 = z_2$ in Eq. (2.3),

$$\begin{aligned} a_1 &= a_2 \\ b_1 &= b_2 \end{aligned} \tag{2.4}$$

Next, we consider basic operations with complex numbers and the different characteristics of these numbers. The addition and subtraction of two complex numbers z_1 and z_2 given by Eq. (2.3) are

$$\begin{aligned} z = z_1 \pm z_2 &= (a_1 + jb_1) \pm (a_2 + jb_2) \\ &= (a_1 \pm a_2) + j(b_1 \pm b_2) \end{aligned} \tag{2.5}$$

Furthermore, from Eqs. (2.5) and (2.2),

$$\begin{aligned} \overline{z_1 \pm z_2} &= \overline{(a_1 + jb_1) \pm (a_2 + jb_2)} = \overline{(a_1 \pm a_2) + j(b_1 \pm b_2)} \\ &= (a_1 \pm a_2) - j(b_1 \pm b_2) = a_1 - jb_1 \pm a_2 \mp jb_2 \\ &= a_1 - jb_1 \pm (a_2 \mp jb_2) \\ &= \bar{z}_1 \pm \bar{z}_2 \end{aligned} \tag{2.6}$$

Equations (2.5) and (2.6), respectively, are verified with Mathematica procedure M2.1.

The *multiplication* of two complex numbers z_1 and z_2 given by Eq. (2.3) is

$$\begin{aligned} z_1 z_2 &= (a_1 + jb_1)(a_2 + jb_2) \\ &= a_1 a_2 + j(b_1 a_2 + b_2 a_1) + j^2 b_1 b_2 \\ &= (a_1 a_2 - b_1 b_2) + j(b_1 a_2 + b_2 a_1) \end{aligned} \tag{2.7}$$

where we have used the identity $j^2 = -1$ in arriving at the final expression. Also, we see from Eq. (2.7) that

$$\begin{aligned} \overline{z}_1 \overline{z}_2 &= (a_1 - jb_1)(a_2 - jb_2) \\ &= a_1 a_2 - j(b_1 a_2 + b_2 a_1) + j^2 b_1 b_2 \\ &= (a_1 a_2 - b_1 b_2) - j(b_1 a_2 + b_2 a_1) \end{aligned} \tag{2.8}$$

Therefore, from Eqs. (2.7) and (2.8),

$$\begin{aligned} \overline{z_1 z_2} &= \overline{(a_1 a_2 - b_1 b_2) + j(b_1 a_2 + b_2 a_1)} \\ &= (a_1 a_2 - b_1 b_2) - j(b_1 a_2 + b_2 a_1) \\ &= \overline{z}_1 \overline{z}_2 \end{aligned} \tag{2.9}$$

The case of multiplying a complex number by its complex conjugate results in

$$\begin{aligned} z_1 \overline{z}_1 &= (a_1 + jb_1)(a_1 - jb_1) \\ &= a_1^2 + b_1^2 \end{aligned} \tag{2.10}$$

which is a real quantity.

Equations (2.7), (2.9), and (2.10), respectively, are verified with Mathematica procedure M2.2.

The *division* of two complex numbers z_1 and z_2 is performed in the following manner

$$\begin{aligned} z = \frac{z_1}{z_2} = \frac{z_1 \overline{z}_2}{z_2 \overline{z}_2} &= \frac{(a_1 + jb_1)(a_2 - jb_2)}{(a_2 + jb_2)(a_2 - jb_2)} \\ &= \frac{a_1 a_2 + b_1 b_2}{a_2^2 + b_2^2} + j\frac{b_1 a_2 - b_2 a_1}{a_2^2 + b_2^2} \end{aligned} \tag{2.11}$$

provided that $z_2 \neq 0$.

Equation (2.11) is verified with Mathematica procedure M2.3.

The *absolute value* (modulus, magnitude) of the complex number z_1 is given by

$$|z_1| = \sqrt{a_1^2 + b_1^2} \tag{2.12}$$

Then, from Eqs. (2.12), (2.10), and (2.2), we see that

$$\begin{aligned} |z_1| &= |\bar{z}_1| \\ |z_1|^2 &= z_1\bar{z}_1 \\ |z_1 z_2| &= |z_1||z_2| \end{aligned} \tag{2.13}$$

The last equation of Eq. (2.13) is obtained from

$$\begin{aligned} |z_1 z_2| &= |(a_1a_2 - b_1b_2) + j(b_1a_2 + b_2a_1)| \\ &= \sqrt{(a_1a_2 - b_1b_2)^2 + (b_1a_2 + b_2a_1)^2} \\ &= \sqrt{a_1^2a_2^2 + b_1^2b_2^2 - 2a_1a_2b_1b_2 + b_1^2a_2^2 + a_1^2b_2^2 + 2a_1a_2b_1b_2} \\ &= \sqrt{a_1^2(a_2^2 + b_2^2) + b_1^2(a_2^2 + b_2^2)} \\ &= \sqrt{(a_1^2 + b_1^2)(a_2^2 + b_2^2)} = \sqrt{(a_1^2 + b_1^2)}\sqrt{(a_2^2 + b_2^2)} \\ &= |z_1||z_2| \end{aligned} \tag{2.14}$$

Referring to Figure 2.1, the polar coordinates (r,θ) corresponding to the complex variable $z=x+jy$ can be determined from the equations

$$x = r\cos\theta, \quad y = r\sin\theta \tag{2.15}$$

where

$$r = |z| = \sqrt{x^2 + y^2} \tag{2.16}$$

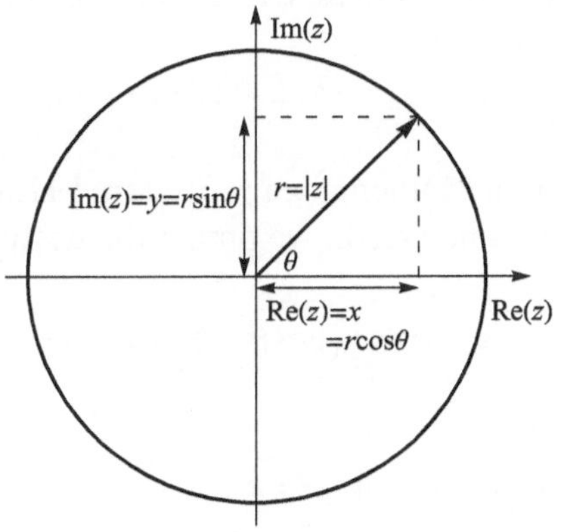

FIGURE 2.1 Relation between Cartesian and polar representations of a complex number.

is called the *magnitude* of z and

$$\theta = \tan^{-1}\frac{y}{x} \tag{2.17}$$

is called the *argument* (or *phase*) of z. To obtain a numerical value of θ, the four-quadrant form of the arctangent function must be used. Thus, the *polar form* of z is

$$z = r(\cos\theta + j\sin\theta) \tag{2.18}$$

From Eq. (2.18), we note that

$$\begin{aligned}\frac{dz}{d\theta} &= r(-\sin\theta + j\cos\theta) = r(j^2\sin\theta + j\cos\theta) = jr(\cos\theta + j\sin\theta)\\ &= jz\end{aligned} \tag{2.19}$$

2.2 COMPLEX EXPONENTIAL FUNCTION: EULER'S FORMULA

Euler's formula is*

$$e^{j\theta} = \cos\theta + j\sin\theta \tag{2.20}$$

which is derived in Eq. (A.4) of Appendix A. It is noted that

$$e^{j(\theta+2m\pi)} = \cos(\theta + 2m\pi) + j\sin(\theta + 2m\pi) = \cos\theta + j\sin\theta \quad m = 0, \pm1, \pm2, \ldots \tag{2.21}$$

From Eq. (2.16), it is seen that

$$\left|e^{j\theta}\right| = \sqrt{\cos^2\theta + \sin^2\theta} = 1 \tag{2.22}$$

Equation (2.22) is verified with Mathematica procedure M2.4.

Equation (2.20) can be used to obtain an expression for a complex exponent. Thus,

$$e^z = e^{x+jy} = e^x e^{jy} = e^x(\cos y + j\sin y) \tag{2.23}$$

Equation (2.23) is obtained with Mathematica procedure M2.5.

Upon combining Eqs. (2.18) and (2.20), we obtain the useful formulas

$$z = x + jy = r(\cos\theta + j\sin\theta) = re^{j\theta} \tag{2.24}$$

---

* Notice the mathematical beauty of the case when $\theta = \pi$

$$e^{j\pi} = -1 \rightarrow e^{j\pi} + 1 = 0$$

which is a relationship of many of the most important quantities in mathematics: e, π, j, 1, and 0.

where

$$r = \sqrt{x^2 + y^2} = \sqrt{\text{Re}(z)^2 + \text{Im}(z)^2}$$
$$\theta = \tan^{-1}(y/x) = \tan^{-1}(\text{Im}(z)/\text{Re}(z)) \tag{2.25}$$

Note that

$$\frac{1}{z} = \frac{1}{re^{j\theta}} = \frac{1}{r}e^{-j\theta} \tag{2.26}$$

and

$$\bar{z} = re^{-j\theta}$$

and, therefore,

$$z\bar{z} = \left(re^{j\theta}\right)\left(re^{-j\theta}\right) = r^2 = x^2 + y^2 \tag{2.27}$$

Equation (2.23) has the following properties:

(a) For all values of $\text{Re}[z] > -\infty$, $e^z \neq 0$.
(b) $|e^z| = e^x$, since $|e^{jy}| = 1$ and $e^x > 0$.
(c) A necessary and sufficient condition for

$$e^z = 1$$

is that $z = 2kj\pi$, where $k = 0, 1, 2, \ldots$ If $e^z = -1$, then $z = (2k+1)j\pi$, where $k = 0, 1, 2 \ldots$
(d) A necessary and sufficient condition that

$$e^{z_1} = e^{z_2} \quad (\text{or } e^{z_1 - z_2} = 1)$$

is that $z_1 - z_2 = 2kj\pi$, where $k = 0, 1, 2, \ldots$
(e) A necessary and sufficient condition that

$$e^z = j$$

is that $z = (4k+1)j\pi/2$, where $k = 0, 1, 2, \ldots$ If $e^z = -j$, then $z = (4k-1)j\pi/2$, where $k = 0, 1, 2, \ldots$
These results are illustrated with the following example.

Example 2.1

We shall use Eqs. (2.23) and (2.24) to solve the equation

$$e^z = 3 + j4$$

It is noted from Eq. (2.25) that

$$3 + j4 = \sqrt{3^2 + 4^2}\,e^{j\varphi} = 5e^{j\varphi}$$

where

$$\varphi = \tan^{-1}\frac{4}{3} = 0.9273 \text{ rad}$$

Then

$$e^z = e^{x+jy} = e^x e^{jy} = 5e^{j\varphi}$$

and, therefore,

$$e^x = 5 \to x = \ln 5 = 1.609$$

$$e^{jy} = e^{j\varphi} \to y = \varphi = 0.9273 \text{ rad}$$

This result is verified with Mathematica procedure M2.6.

Trigonometric and hyperbolic functions with complex arguments are determined as follows. Noting from Eq. (2.20) that

$$\begin{aligned} e^{jy} &= \cos y + j\sin y \\ e^{-jy} &= \cos y - j\sin y \end{aligned} \tag{2.28}$$

we obtain the following relations by respectively adding and subtracting them

$$\begin{aligned} \cos y &= \frac{1}{2}\left(e^{jy} + e^{-jy}\right) \\ \sin y &= \frac{1}{2j}\left(e^{jy} - e^{-jy}\right) \end{aligned} \tag{2.29}$$

Equation (2.29) can be extended to any complex number $z = x + jy$. Thus, for cos(z)

$$\begin{aligned} \cos z &= \frac{1}{2}\left(e^{jz} + e^{-jz}\right) = \frac{1}{2}\left(e^{-y+jx} + e^{y-jx}\right) \\ &= \frac{1}{2}\left(e^{-y}\left[\cos x + j\sin x\right] + e^{y}\left[\cos x - j\sin x\right]\right) \\ &= \cos x\frac{1}{2}\left(e^{y} + e^{-y}\right) - j\sin x\frac{1}{2}\left(e^{y} - e^{-y}\right) \\ &= \cos x\cosh y - j\sin x\sinh y \end{aligned} \tag{2.30}$$

Additionally, it is straightforward to show that

$$\begin{aligned} \left|\cos z\right|^2 &= \cos^2 x + \sinh^2 y \\ \cos jy &= \cosh y \end{aligned} \tag{2.31}$$

In a similar manner, it is found that

$$\begin{aligned} \sin z &= \sin x \cosh y + j \cos x \sinh y \\ |\sin z|^2 &= \sin^2 x + \sinh^2 y \\ \sin jy &= j \sinh y \end{aligned} \tag{2.32}$$

For hyperbolic functions it is found that

$$\begin{aligned} \sinh z &= \cos y \sinh x + j \sin y \cosh x \\ \sinh jz &= j \sin z \\ \cosh z &= \cos y \cosh x + j \sin y \sinh x \\ \cosh jz &= \cos z \end{aligned} \tag{2.33}$$

Equations (2.30) and (2.31) and the last two equations of Eq. (2.33) are verified with Mathematica procedure M2.7.

Furthermore, from Eqs. (2.18) and (2.20), it is seen that

$$\begin{aligned} z^n &= r^n \left(\cos\theta + j \sin\theta\right)^n = r^n e^{jn\theta} \\ &= r^n \left(\cos n\theta + j \sin n\theta\right) \end{aligned} \tag{2.34}$$

which is known as *De Moivre's* theorem. If n is a negative integer, then

$$z^n = \frac{1}{z^{|n|}} \qquad n = -1, -2, \ldots \tag{2.35}$$

and

$$z^{1/n} = \frac{1}{z^{1/|n|}} \qquad n = -1, -2, \ldots \tag{2.36}$$

We now use the results of Eq. (2.33) to determine the solution to

$$\sinh az = 0$$

If we let $az = jg$, then using Eq. (2.33) we obtain

$$\sinh jg = j \sin g = 0 \rightarrow g = \pm n\pi \qquad n = 0, 1, 2, \ldots \tag{2.37}$$

and, therefore, $z = \pm jn\pi/a$. Similarly, for

$$\cosh az = 0$$

we obtain

$$\cosh jg = \cos g = 0 \rightarrow g = \pm(2n+1)\pi/2 \quad n = 0,1,2,\ldots \tag{2.38}$$

and, therefore, $z = \pm j(2n+1)\pi/(2a)$.

Equations (2.37) and (2.38) are verified with Mathematica procedure M2.8.

These relations are used to obtain certain inverse Laplace transforms as shown in Section 8.1.7. See also Example 2.6.

To determine the logarithm of a complex number, we recall that

$$\log_a x = y \rightarrow a^y = x$$

Then

$$\ln e^u = y \rightarrow e^y = e^u$$

Therefore, $y = u$ and

$$\ln e^u = u \tag{2.39}$$

If $u = \ln(z)$, then from Eq. (2.39) $\ln(e^{\ln(z)}) = \ln(z)$ or

$$e^{\ln z} = z \tag{2.40}$$

Thus, for $z \neq 0$, the complex exponent is defined as

$$\begin{aligned} z^n &= e^{n\ln z} \quad n = 0,1,2,\ldots \\ z^{1/n} &= e^{(\ln z)/n} \quad n = 1,2,\ldots \end{aligned} \tag{2.41}$$

The logarithm of a complex quantity $z = x + jy$ is

$$\ln z = \ln(x+jy) = \ln(re^{j(\theta+2m\pi)}) = \ln r + j(\theta + 2m\pi) \quad m = 0,\pm1,\pm2,\ldots \tag{2.42}$$

where r and θ, respectively, are given by Eqs. (2.16) and (2.17). The principal value of Eq. (2.42) is

$$P[\ln z] = \ln r + j\theta \tag{2.43}$$

In addition,

$$\begin{aligned} z^{1/n} &= \exp\left[(\ln z)/n\right] = \exp\left[(\ln r)/n + j(\theta + 2m\pi)/n\right] \\ &= \exp\left[\ln r^{1/n} + j(\theta + 2m\pi)/n\right] \\ &= r^{1/n}\exp\left[j(\theta + 2m\pi)/n\right] \quad m = 0,\pm1,\pm2,\ldots \end{aligned} \tag{2.44}$$

and similarly,

$$z^n = r^n \exp\left[jn\left(\theta + 2m\pi\right)\right] \quad m = 0, \pm 1, \pm 2, \ldots \tag{2.45}$$

Now let us consider the case where $n = w$, where $w = a + jb$ is a complex number and $z = x + jy = re^{j\theta}$, where r and θ, respectively, are given by Eqs. (2.16) and (2.17). Then, from Eqs. (2.41) and (2.42),

$$\begin{aligned} z^w &= e^{w \ln z} = \exp\left[(a + jb)\left(\ln r + j(\theta + 2m\pi)\right)\right] \quad m = 0, \pm 1, \pm 2, \ldots \\ &= \exp\left[a \ln r - b(\theta + 2m\pi)\right]\exp\left[j\left(b \ln r + a(\theta + 2m\pi)\right)\right] \\ &= \exp\left[a \ln r - b(\theta + 2m\pi)\right] \times \\ &\quad \left[\cos\left(b \ln r + a(\theta + 2m\pi)\right) + j \sin\left(b \ln r + a(\theta + 2m\pi)\right)\right] \end{aligned} \tag{2.46}$$

When $b = 0$ and $a = n$, this result simplifies to

$$\begin{aligned} z^n &= \exp\left[n \ln r\right]\exp\left[jn(\theta + 2m\pi)\right] \\ &= r^n \exp\left[jn(\theta + 2m\pi)\right] \quad m = 0, \pm 1, \pm 2, \ldots \end{aligned}$$

which agrees with our previous result. When $a = 0$, the general result simplifies to

$$z^{jb} = \exp\left[-b(\theta + 2m\pi)\right]\exp\left[jb \ln r\right] \quad m = 0, \pm 1, \pm 2, \ldots \tag{2.47}$$

Equation (2.46) is verified with Mathematica procedure M2.9.

These results are illustrated with the following examples.

Example 2.2

We shall evaluate the expression j^j. In this case, $z = 0 + j$ and $w = 0 + j$; therefore, $r = 1$, $\theta = \tan^{-1}(1/0) = \pi/2$, $b = 1$, and $a = 0$. Then, Eq. (2.47) gives

$$\begin{aligned} j^j &= \exp\left[-(\pi/2 + 2m\pi)\right]\exp\left[j \ln 1\right] \\ &= \exp\left[-\pi/2 - 2m\pi\right] \quad m = 0, \pm 1, \pm 2, \ldots \end{aligned}$$

This result is verified with Mathematica procedure M2.10 for the case when $m = 0$.

Example 2.3

We shall evaluate the expression $(1 - j)^{1+j}$. In this case, $z = 1 - j$ and $w = 1 + j$; therefore, $r = \sqrt{1 + 1} = \sqrt{2}$, $\theta = \tan^{-1}(-1/1) = -\pi/4$, $b = 1$, and $a = 1$. Then, from Eq. (2.46)

$$(1-j)^{1+j} = \exp\left[\ln\sqrt{2} - (-\pi/4 + 2m\pi)\right]\left[\cos\left(\ln\sqrt{2} + (-\pi/4 + 2m\pi)\right)\right.$$

$$\left. + j\sin\left(\ln\sqrt{2} + (-\pi/4 + 2m\pi)\right)\right]$$

$$= \sqrt{2}e^{\pi/4 - 2m\pi}\left[\cos\left(\ln\sqrt{2} - \pi/4\right) + j\sin\left(\ln\sqrt{2} - \pi/4\right)\right]$$

$$= \left(2.80788 - 1.31787j\right)e^{-2m\pi} \qquad m = 0, \pm1, \pm2, \ldots$$

This result is verified with Mathematica procedure M2.11 for the case when $m=0$.

2.3 ANALYTIC FUNCTIONS

2.3.1 Cauchy–Riemann Conditions

Consider the function of the complex variable

$$f(z) = u(x,y) + jv(x,y)$$

The function $f(z)$ is said to be continuous at $z=z_0=x_0+jy_0$ if and only if both u and v are continuous at (x_0,y_0). Thus, from the definition of the derivative

$$\frac{df(z)}{dz} = \lim_{\Delta z \to 0} \frac{f(z+\Delta z) - f(z)}{\Delta z} = \lim_{\Delta z \to 0} \frac{f(z + \Delta x + j\Delta y) - f(z)}{\Delta x + j\Delta y}$$

$$= \lim_{\Delta z \to 0} \frac{u(x+\Delta x, y+\Delta y) + jv(x+\Delta x, y+\Delta y) - u(x,y) - jv(x,y)}{\Delta x + j\Delta y}$$

$$= \lim_{\Delta z \to 0} \frac{u(x+\Delta x, y+\Delta y) - u(x,y) + j[v(x+\Delta x, y+\Delta y) - v(x,y)]}{\Delta x + j\Delta y}$$

The value of the limit must be independent of the direction of the vector representing Δz. Thus, when $\Delta y=0$,

$$\frac{df(z)}{dz} = \lim_{\Delta x \to 0} \frac{u(x+\Delta x, y) - u(x,y) + j[v(x+\Delta x, y) - v(x,y)]}{\Delta x}$$

$$= \lim_{\Delta x \to 0} \frac{u(x+\Delta x, y) - u(x,y)}{\Delta x} + j\lim_{\Delta x \to 0} \frac{[v(x+\Delta x, y) - v(x,y)]}{\Delta x}$$

$$= \frac{\partial u}{\partial x} + j\frac{\partial v}{\partial x} = \frac{df}{dx}$$

However, when $\Delta x=0$,

$$\frac{df(z)}{dz} = \lim_{\Delta y \to 0} \frac{u(x, y+\Delta y) - u(x,y) + j[v(x, y+\Delta y) - v(x,y)]}{j\Delta y}$$

$$= \lim_{\Delta y \to 0} \frac{u(x, y+\Delta y) - u(x,y)}{j\Delta y} + j \lim_{\Delta y \to 0} \frac{[v(x, y+\Delta y) - v(x,y)]}{j\Delta y}$$

$$= -j\frac{\partial u}{\partial y} + \frac{\partial v}{\partial y} = -j\frac{df}{dy}$$

Therefore, if the derivatives are to be the same, we must have that

$$\frac{df}{dx} = -j\frac{df}{dy}$$

or

$$\frac{\partial u}{\partial x} + j\frac{\partial v}{\partial x} = -j\frac{\partial u}{\partial y} + \frac{\partial v}{\partial y}$$

In other words,

$$\begin{aligned} \frac{\partial u}{\partial x} &= \frac{\partial v}{\partial y} \\ \frac{\partial v}{\partial x} &= -\frac{\partial u}{\partial y} \end{aligned} \tag{2.48}$$

These are the *Cauchy–Riemann equations*, which state that in order for f to be complex differentiable its real and imaginary parts must satisfy Eq. (2.48).

A function is said to be *analytic* at a point z_0 if its derivatives exist at every point of some neighborhood of z_0. If the function fails to be analytic at z_0 but is analytic at some point in every neighborhood of z_0, then z_0 is said to be a *singular point*. Referring to Figure 2.2, an analytic function on the disk $|z - z_0| < R_0$ that is centered at z_0 with a radius R_0 has a Taylor

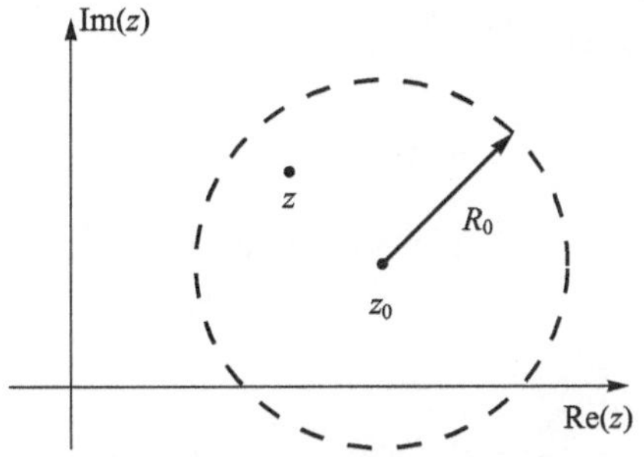

FIGURE 2.2 Disk of radius R_0 centered at z_0.

series representation in a neighborhood of each point in its domain (disk). In other words, the power series converges to $f(z)$ when $|z - z_0| < R_0$. This series is given by

$$f(z) = f(z_0) + \frac{1}{1!} f^{(1)}(z_0)(z - z_0) + \frac{1}{2!} f^{(2)}(z_0)(z - z_0)^2 + \cdots + \frac{1}{(n-1)!} f^{(n-1)}(z_0)(z - z_0)^{n-1} + \cdots \quad |z - z_0| < R_0 \tag{2.49}$$

Satisfaction of the Cauchy–Riemann conditions at a point $z_0 = (x_0, y_0)$ is not sufficient to ensure the existence of the derivative of the function at z_0. However, let

$$f(z) = u(x, y) + jv(x, y)$$

be defined in some neighborhood of z_0 and let the first-order partial derivatives of u and v with respect to x and y exist everywhere in that neighborhood and be continuous at (x_0, y_0). If these derivatives satisfy the Cauchy–Riemann conditions at (x_0, y_0), then the derivative exists at (x_0, y_0) and is given by

$$\frac{df(z_0)}{dz} = \frac{df(z_0)}{dx} = \frac{\partial u(x_0, y_0)}{\partial x} + j\frac{\partial v(x_0, y_0)}{\partial x}$$

or by

$$\frac{df(z_0)}{dz} = -j\frac{df(z_0)}{dy} = \frac{\partial v(x_0, y_0)}{\partial y} - j\frac{\partial u(x_0, y_0)}{\partial y}$$

since the Cauchy–Riemann conditions are satisfied.

Example 2.4

We shall examine two complex functions and determine if they are analytic.

(a) For the first function, we consider

$$f(z) = e^z = e^{x+jy} = e^x \left(\cos y + j \sin y\right)$$

Then,

$$u(x, y) = e^x \cos y$$

$$v(x, y) = e^x \sin y$$

and

$$\frac{\partial u}{\partial x} = e^x \cos y \quad \frac{\partial v}{\partial y} = e^x \cos y$$

$$\frac{\partial v}{\partial x} = e^x \sin y \quad -\frac{\partial u}{\partial y} = e^x \sin y$$

Therefore, the Cauchy–Riemann conditions are satisfied and

$$\frac{df}{dz} = \frac{\partial u}{\partial x} + j\frac{\partial v}{\partial x} \rightarrow \frac{d}{dz}e^z = e^x \cos y + je^x \sin y = e^z$$

(b) For the second function, we consider $f(z) = |z|^2$. Then

$$f(z) = |z|^2 = x^2 + y^2$$

and, therefore,

$$u(x, y) = x^2 + y^2$$

$$v(x, y) = 0$$

Thus, by inspection, it is seen that the Cauchy–Riemann conditions are not satisfied anywhere except at $z = 0$.

These results are verified with Mathematica procedure M2.12.

2.3.2 Cauchy Integral Formula

Consider a curve C that is a set of points $z = (x,y)$ in the complex plane defined by $x = x(t)$, $y = y(t)$, $a \le t \le b$, where $x(t)$ and $y(t)$ are continuous functions of the real parameter t. Then, one may write $z(t) = x(t) + jy(t)$, $a \le t \le b$. The curve is said to be smooth if $z(t)$ has a continuous derivative $dz(t)/dt \neq 0$ at all points along the curve. A contour is defined as a curve consisting of a finite number of smooth curves joined end to end. *A contour is said to be a simple closed contour if the initial and final values of z(t) are the same and the contour does not cross itself.* Let $f(z)$ be any complex function defined in a domain D in the complex plane and let C be any contour contained in D with initial point z_0 and terminal point z. The contour C is divided into n sub-arcs by discrete points $z_0, z_1, z_2, \ldots, z_{n-1}, z_n = z$ arranged consecutively along the direction of increasing t as shown in Figure 2.3. We let ζ_k be an arbitrary point in the sub-arc $z_k z_{k+1}$ and form the sum

$$\sum_{k=0}^{n-1} f(\zeta_k)(z_{k+1} - z_k)$$

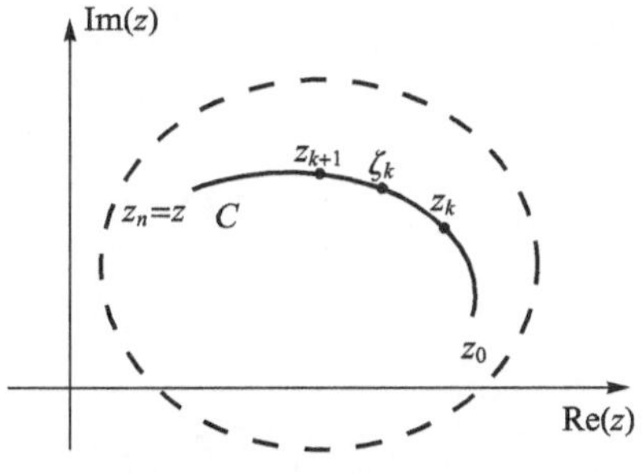

FIGURE 2.3 Nomenclature for n connected arcs of contour C.

If we let $\Delta z_k = z_{k+1} - z_k$ and $\lambda = \max|\Delta z_k|$, then the contour integral is defined as

$$\int_C f(z)dz = \lim_{\substack{\lambda\to 0\\ n\to\infty}} \sum_{k=0}^{n-1} f(\zeta_k)\Delta z_k$$

Thus, if

$$f(z) = u(x,y) + jv(x,y)$$
$$dz = dx + jdy$$

then

$$\int_C f(z)dz = \int_C [u(x,y) + jv(x,y)][dx + jdy]$$
$$= \int_C udx - vdy + j\int_C udy + vdx$$

But, from Green's theorem, which is

$$\iint_R \left(\frac{\partial P}{\partial x} - \frac{\partial Q}{\partial y}\right) dxdy = \int_C (Qdx + Pdy) \tag{2.50}$$

we find that with $Q = u$ and $P = -v$,

$$\int_C udx - vdy = \iint_R \left(-\frac{\partial v}{\partial x} - \frac{\partial u}{\partial y}\right) dxdy \tag{2.51}$$

and with $P = u$ and $Q = v$,

$$\int_C udy + vdx = \iint_R \left(\frac{\partial u}{\partial x} - \frac{\partial v}{\partial y}\right) dxdy \tag{2.52}$$

If $f(z)$ is analytic, that is, the Cauchy–Riemann conditions given by Eq. (2.48) are satisfied, then Eqs. (2.51) and (2.52), respectively, become

$$\int_C udx - vdy = \iint_R \left(-\frac{\partial v}{\partial x} - \frac{\partial u}{\partial y}\right) dxdy = \iint_R \left(\frac{\partial u}{\partial y} - \frac{\partial u}{\partial y}\right) dxdy = 0$$

$$\int_C udy + vdx = \iint_R \left(\frac{\partial u}{\partial x} - \frac{\partial v}{\partial y}\right) dxdy = \iint_R \left(\frac{\partial v}{\partial y} - \frac{\partial v}{\partial y}\right) dxdy = 0$$

and, we have that

$$\int_C f(z)dz = \int_C (udx - vdy) + j\int_C (udy + vdx) = 0 \tag{2.53}$$

This relation is valid if $f(z)$ is analytic and the region D is simply connected, that is, if every closed curve in D encloses only points of D.

The integral of $f(z)$ along a string of contours is equal to the sum of the integrals of $f(z)$ along each of these contours. To show this, consider Figure 2.4 where C is positive counter-clockwise and C_1 and C_2 are clockwise. Then

$$\int_C f(z)dz + \int_{-C_1} f(z)dz + \int_{-C_2} f(z)dz = 0$$

and the paths along the cuts cancel because they are in different directions: Entry versus return. Therefore,

$$\int_C f(z)dz = \int_{C_1} f(z)dz + \int_{C_2} f(z)dz \tag{2.54}$$

and, in general, for K contours

$$\int_C f(z)dz = \sum_{k=1}^{K} \int_{C_k} f(z)dz \tag{2.55}$$

If $z = z(t) = x(t) + jy(t)$, $a \le t \le b$, then

$$\frac{dz}{dt} = \frac{dx}{dt} + j\frac{dy}{dt}$$

and

$$\int_C f(z)dz \to \int_a^b f(z(t))\frac{dz(t)}{dt}dt \tag{2.56}$$

Consider the following important application of this transformation to

$$\int_C \frac{1}{z - z_0}dz \tag{2.57}$$

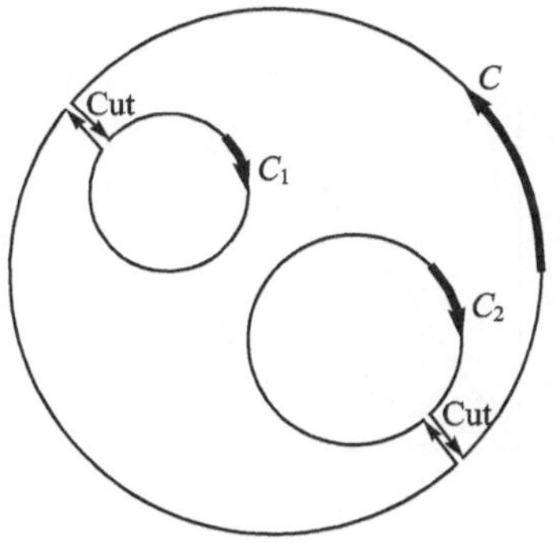

FIGURE 2.4 Three simple closed contours.

where C is a circle centered at z_0 and of radius r, where the quantity z_0 is called a *simple pole*. It is mentioned that $(z - z_0)^{-1}$ is analytic everywhere except at $z=z_0$. Referring to Figure 2.5, the circle can be parameterized as follows

$$z \to z(t) = z_0 + re^{jt} \quad 0 \le t \le 2\pi$$

Therefore,

$$z(t) - z_0 = re^{jt}$$

$$\frac{dz(t)}{dt} = jre^{jt} \quad 0 \le t \le 2\pi$$

and from Eq. (2.56) we obtain

$$\int_C \frac{1}{z - z_0} dz = \int_0^{2\pi} \frac{1}{z(t) - z_0} \frac{dz(t)}{dt} dt = \int_0^{2\pi} \frac{1}{re^{jt}} jre^{jt} dt = j \int_0^{2\pi} dt$$
$$= 2\pi j \tag{2.58}$$

However, notice what happens when we replace $z - z_0$ with $(z - z_0)^n$ in Eq. (2.57) when n is a positive integer different from 1,

$$\int_C \frac{1}{(z - z_0)^n} dz \to \int_0^{2\pi} \frac{1}{(z(t) - z_0)^n} \frac{dz(t)}{dt} dt$$
$$= \int_0^{2\pi} \frac{1}{(re^{jt})^n} jre^{jt} dt = \frac{j}{r^{n-1}} \int_0^{2\pi} e^{-j(n-1)t} dt \tag{2.59}$$
$$= \frac{j}{r^{n-1}} \frac{e^{-j(n-1)t}}{-j(n-1)} \Bigg|_0^{2\pi} = \frac{-1}{(n-1)r^{n-1}} \left(e^{-j2\pi(n-1)} - 1\right) = 0$$

Thus, the only pole with which we need to concern ourselves within the closed contour is a simple pole.

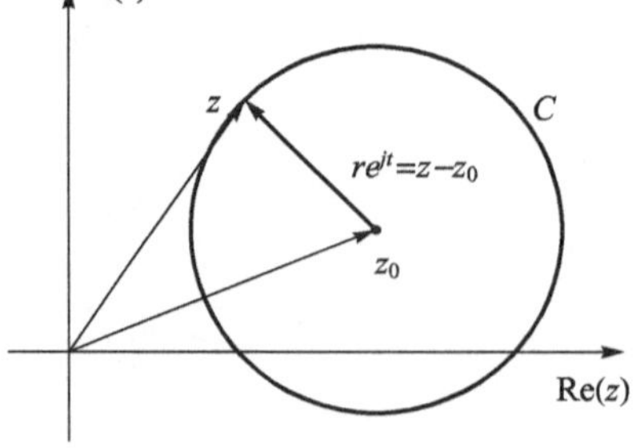

FIGURE 2.5 Circular contour of radius r centered at z_0.

When n is not an integer, the argument is multi-valued and requires what is called a branch cut. This case will not be discussed.

Let $f(z)$ be analytic on and inside a closed simple contour C. Then the *Cauchy integral formula* is

$$\frac{1}{2\pi j}\int_C \frac{f(z)}{z-z_0}dz = f(z_0) \tag{2.60}$$

where the quantity $f(z_0)$ is called the residue of $f(z)$ at the isolated singular point z_0, which we have called a *simple pole.*

To prove this relation, we draw a circle of radius r about z_0 that is small enough to be completely inside C and denote this region C_r as shown in Figure 2.6. Since $f(z)/(z - z_0)$ is analytic in the region between C and C_r, we have that

$$\begin{aligned}\frac{1}{2\pi j}\int_C \frac{f(z)}{z-z_0}dz &= \frac{1}{2\pi j}\int_{C_r} \frac{f(z)}{z-z_0}dz \\ &= \frac{1}{2\pi j}\int_{C_r} \frac{f(z)}{z-z_0}dz + f(z_0)\frac{1}{2\pi j}\int_{C_r} \frac{1}{z-z_0}dz - f(z_0)\frac{1}{2\pi j}\int_{C_r} \frac{1}{z-z_0}dz \\ &= \frac{1}{2\pi j}\int_{C_r} \frac{f(z)-f(z_0)}{z-z_0}dz + f(z_0)\frac{1}{2\pi j}\int_{C_r} \frac{1}{z-z_0}dz \\ &= \frac{1}{2\pi j}\int_{C_r} \frac{f(z)-f(z_0)}{z-z_0}dz + f(z_0)\end{aligned}$$

where we have used Eq. (2.58) and in the second line we have added and subtracted the same quantity.

We now have to show that

$$\frac{1}{2\pi j}\int_{C_r} \frac{f(z)-f(z_0)}{z-z_0}dz = 0 \tag{2.61}$$

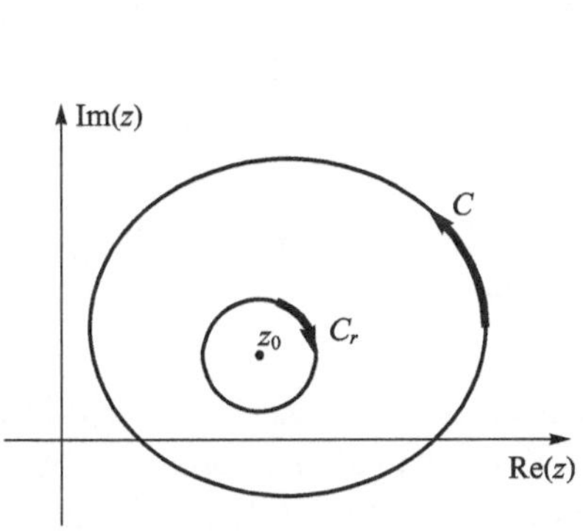

FIGURE 2.6 Circular contour C_r of radius r that centered at z_0 within a simple closed contour C.

Since $f(z)$ is continuous at z_0, we know from calculus that for each $\varepsilon > 0$ there exists a δ such that

$$\left|f(z) - f(z_0)\right| = \left|f(z_0 + re^{j\theta}) - f(z_0)\right| < \varepsilon \quad \text{whenever } r = \left|z - z_0\right| < \delta$$

To guarantee that C_r lies completely inside C, we have chosen $r < \delta$. To proceed, we use the Holder inequality

$$\left|\int h(z)\overline{g(z)}dz\right| \le \int \left|h(z)g(z)\right|\left|dz\right|$$

with $g(z) = 1$. Then Eq. (2.61) yields

$$\left|\frac{1}{2\pi j}\int_{C_r} \frac{f(z) - f(z_0)}{z - z_0} dz\right| \le \frac{1}{2\pi}\int_{C_r} \frac{\left|f(z) - f(z_0)\right|}{\left|z - z_0\right|}\left|dz\right|$$

$$= \frac{1}{2\pi r}\int_{C_r} \left|f(z) - f(z_0)\right|\left|dz\right|$$

$$= \frac{1}{2\pi r}\int_{C_r} \varepsilon\left|dz\right| = \frac{\varepsilon}{2\pi r}\int_0^{2\pi} r d\theta = \frac{\varepsilon}{2\pi r} 2\pi r = \varepsilon$$

where we have used the fact that since $z = re^{j\theta}$, $dz = jre^{j\theta}d\theta$ and, therefore, $|dz| = rd\theta$. Since the magnitude of the above integral is less than any positive value of ε, the value of that integral must be zero. Hence, Eq. (2.61) has been shown to equal zero.

For convenience, we change notation of Eq. (2.60) and let $z = w$ and $z_0 = z$ and investigate the derivative of Eq. (2.60) as follows

$$f(z) = \frac{1}{2\pi j}\int_C \frac{f(w)}{w - z} dw$$

$$\frac{df(z)}{dz} = \frac{1}{2\pi j}\int_C \frac{d}{dz}\left(\frac{f(w)}{w - z}\right) dw = \frac{1}{2\pi j}\int_C \frac{f(w)}{(w - z)^2} dw$$

In general,

$$\frac{d^k f(z)}{dz^k} = \frac{k!}{2\pi j}\int_C \frac{f(w)}{(w - z)^{k+1}} dw \tag{2.62}$$

for any z inside C. This expression is called the *generalized Cauchy integral formula*.

The result expressed by Eq. (2.62) can also be obtained by again considering the integral given in Eq. (2.62) except this time we substitute for $f(z)$ the Taylor series expansion given by Eq. (2.49) to arrive at

$$\begin{aligned}
\int_C \frac{f(z)}{(z-z_0)^n}dz &= \int_C \frac{1}{(z-z_0)^n}\left[f(z_0)+\frac{1}{1!}f^{(1)}(z_0)(z-z_0)+\cdots\right.\\
&\quad +\frac{1}{(n-1)!}f^{(n-1)}(z_0)(z-z_0)^{n-1}\\
&\quad \left.+\sum_{m=n}^{\infty}\frac{1}{m!}f^{(m)}(z_0)(z-z_0)^m\right]dz\\
&= \int_C \left[\frac{f(z_0)}{(z-z_0)^n}+\frac{f^{(1)}(z_0)}{1!(z-z_0)^{n-1}}+\cdots+\frac{f^{(n-1)}(z_0)}{(n-1)!(z-z_0)}\right.\\
&\quad \left.+\sum_{m=n}^{\infty}\frac{1}{m!}f^{(m)}(z_0)(z-z_0)^{m-n}\right]dz\\
&= \frac{2\pi j}{(n-1)!}f^{(n-1)}(z_0)
\end{aligned} \tag{2.63}$$

The final result was obtained by using Eq. (2.59). It is seen that Eqs. (2.62) and (2.63) are the same result when we set $n=k+1$ in Eq. (2.63). In this case, z_0 is called a *pole of order n*.

We shall change our notation used in Eq. (2.60) to the following

$$\int_C g(z)dz = 2\pi j \operatorname{Res}[g(z);z_0] \tag{2.64}$$

where $\operatorname{Res}[g(z);z_0]$ indicates the residue of $g(z)$ at z_0. Then, if $g(z)$ has z_k, $k=1, 2, \ldots, K$, singular points all contained within C, then

$$\int_C g(z)dz = 2\pi j \sum_{k=1}^{K} \operatorname{Res}[g(z);z_k] \tag{2.65}$$

In the case where $g(z)$ has a pole of order n,

$$g(z) = \frac{f(z)}{(z-z_0)^n} \tag{2.66}$$

and its residue is determined from its derivative given by Eq. (2.63) as follows

$$\int_C g(z)dz = \int_C \frac{f(z)}{(z-z_0)^n}dz = \frac{2\pi j}{(n-1)!}\lim_{z\to z_0}\left[\frac{d^{n-1}}{dz^{n-1}}\left((z-z_0)^n g(z)\right)\right] \tag{2.67}$$

When $n=1$, this result simplifies to

$$\int_C \frac{f(z)}{(z-z_0)} dz \to \int_C g(z) dz = 2\pi j \lim_{z\to z_0}\left[(z-z_0)g(z)\right] \tag{2.68}$$

To develop another useful relation, let

$$g(z) = \frac{p(z)}{q(z)}$$

where $p(z_0) \neq 0$, $q(z)$ has a simple pole at z_0 and that $dq/dz \neq 0$ at z_0. Then the residue is given by

$$2\pi j \operatorname{Res}[g(z); z_0] = 2\pi j \lim_{z\to z_0}\left[(z-z_0)\frac{p(z)}{q(z)}\right]$$

Since the numerator and denominator are zero at z_0, we use L'Hôpital's rule to obtain

$$2\pi j \operatorname{Res}[g(z); z_0] = 2\pi j \lim_{z\to z_0}\left[\frac{p(z)+(z-z_0)p'(z)}{q'(z)}\right] = 2\pi j \frac{p(z_0)}{q'(z_0)} \tag{2.69}$$

where the prime indicates the derivative with respect to z.

There are instances where

$$\lim_{z\to z_0}\frac{p(z)}{q(z)} = c \tag{2.70}$$

where c is a finite constant. In this case, z_0 is called a *removable singularity* and the expression is analytic for all values of z including z_0. We illustrate Eq. (2.70) with the following example.

Example 2.5

(i) Consider the following limit

$$\lim_{z\to 0}\frac{z}{\sinh(az)}$$

Using Eq. (A.5) of Appendix A, we find that

$$\lim_{z\to 0}\frac{z}{\sinh(az)} = \lim_{z\to 0} z\left[az + \frac{a^3 z^3}{3!} + \frac{a^5 z^5}{5!} + \cdots\right]^{-1}$$

$$= \lim_{z\to 0}\left[a + \frac{a^3 z^2}{3!} + \frac{a^5 z^4}{5!} + \cdots\right]^{-1} = \frac{1}{a}$$

Thus, we have a removable singularity at $z=0$, and this expression is analytic for all z.

(ii) Consider the following limit

$$\lim_{z\to 0}\left(\frac{\cosh z - 1}{z^2}\right)$$

Using Eq. (A.5) of Appendix A, we find that

$$\lim_{z\to 0}\left(\frac{\cosh z - 1}{z^2}\right) = \lim_{z\to 0}\frac{1}{z^2}\left(\left(1+\frac{z^2}{2!}+\frac{z^4}{4!}+\cdots\right)-1\right)$$

$$= \lim_{z\to 0}\left(\frac{1}{2!}+\frac{z^2}{4!}+\cdots\right) = \frac{1}{2}$$

Thus, we have a removable singularity at $z=0$, and this expression is analytic for all z.

(iii) Consider the following limit

$$\lim_{z\to 0}\left(\frac{\sinh z}{z^3}\right)$$

Using Eq. (A.5) of Appendix A, we find that

$$\lim_{z\to 0}\left(\frac{\sinh z}{z^3}\right) = \lim_{z\to 0}\frac{1}{z^3}\left(z+\frac{z^3}{3!}+\frac{z^5}{5!}+\frac{z^7}{7!}+\cdots\right)$$

$$= \lim_{z\to 0}\left(\frac{1}{z^2}+\frac{1}{3!}+\frac{z^2}{5!}+\frac{z^4}{7!}+\cdots\right)$$

Therefore, we have a second-order pole at $z=0$, and the expression is analytic for all other values of z.

These three examples are verified with Mathematica procedure M2.13.

We now illustrate Eqs. (2.67), (2.68), and (2.69) with the following examples.

Example 2.6

In the following integrals, it is assumed that the contour C encompasses all poles of the argument.

(i) Consider the integral

$$\int_C \frac{e^{bz}}{\sinh(az)}\,dz$$

Using Eq. (2.69) and the results from Eq. (2.37), where it was found that $z_n = \pm jn\pi/a$, $n = 1, 2, \ldots$, we obtain

$$\int_C \frac{e^{bz}}{\sinh(az)} dz = 2\pi j \sum_{n=1}^{\infty} \frac{e^{bz_n}}{a\cosh(az_n)}$$

$$= 2\pi j \sum_{n=1}^{\infty} \left(\frac{e^{jn\pi b/a}}{a\cosh(jn\pi)} + \frac{e^{-jn\pi b/a}}{a\cosh(-jn\pi)} \right)$$

$$= \frac{4\pi j}{a} \sum_{n=1}^{\infty} \frac{\cos(n\pi b/a)}{a\cos(n\pi)} = \frac{4\pi j}{a} \sum_{n=1}^{\infty} (-1)^n \cos(n\pi b/a)$$

(ii) Consider the integral

$$\int_C \frac{e^{-z}}{z - \pi/2} dz$$

Then, since $f(z) = e^{-z}$ and $z_0 = \pi/2$ is a simple pole, we obtain

$$\int_C \frac{e^{-z}}{z - \pi/2} dz = 2\pi j \lim_{z \to \pi/2} \left[\frac{(z - \pi/2)e^{-z}}{z - \pi/2} \right] = 2\pi j e^{-\pi/2}$$

(iii) Consider the integral

$$\int_C \frac{e^{2z}}{(z+1)^4} dz$$

Then, since $f(z) = e^{2z}$ and $z_0 = -1$ in a fourth-order pole, we obtain

$$\int_C \frac{e^{2z}}{(z+1)^4} dz = \frac{2\pi j}{(4-1)!} \lim_{z \to -1} \left[\frac{d^3}{dz^3} \left(\frac{(z+1)^4 e^{2z}}{(z+1)^4} \right) \right] = \frac{8\pi j}{3} \lim_{z \to -1} e^{2z} = \frac{8\pi j}{3} e^{-2}$$

(iv) Consider the integral

$$\int_C \frac{\cos z}{z(z^2+9)} dz$$

Then, since there are simple poles at $z_1 = 0$, $z_2 = 3j$, and $z_3 = -3j$, we obtain

$$\int_C \frac{\cos z}{z(z^2+9)}dz = \int_C \frac{\cos z}{z(z+3j)(z-3j)}dz$$

$$= 2\pi j\left[\lim_{z\to 0}\left(\frac{z\cos z}{z(z+3j)(z-3j)}\right) + \lim_{z\to 3j}\left(\frac{(z-3j)\cos z}{z(z+3j)(z-3j)}\right)\right.$$

$$\left. + \lim_{z\to -3j}\left(\frac{(z+3j)\cos z}{z(z+3j)(z-3j)}\right)\right]$$

$$= 2\pi j\left[\frac{1}{(3j)(-3j)} + \frac{\cos 3j}{(3j)(6j)} + \frac{\cos(-3j)}{(-3j)(-6j)}\right]$$

$$= 2\pi j\left[\frac{1}{9} - \frac{\cosh 3}{18} - \frac{\cosh 3}{18}\right] = \frac{2\pi j}{9}(1-\cosh 3)$$

The results of these four cases are verified with Mathematica procedure M2.14.

MATHEMATICA PROCEDURES

Note: In order to make the Mathematica procedures that follow a little more readable, the `Clear` statements have been omitted.

```
(*M2.1*)
z1=a1+I b1; z2=a2+I b2;
Simplify[z1+z2]
Simplify[Conjugate[z1+z2],Assumptions->{a1,b1,a2,b2}∈Reals]

(*M2.2*)
z1=a1+I b1; z2=a2+I b2;
ComplexExpand[z1 z2]
ComplexExpand[Conjugate[z1 z2]]==ComplexExpand[Conjugate[z1]*
   Conjugate[z2]]
Simplify[ComplexExpand[z1 Conjugate[z1]]]

(*M2.3*)
ComplexExpand[(a1+I b1)/(a2+I b2)]

(*M2.4*)
Simplify[Abs[Exp[I th]],Assumptions->th∈Reals]

(*M2.5*)
Simplify[ComplexExpand[Exp[x+I y]]]

(*M2.6*)
z/.Quiet[NSolve[Exp[z]==3.+I 4,z]][[1]]

(*M2.7*)
ComplexExpand[Cos[x+I y]]
```

```
p=ComplexExpand[Cos[x+I y] Conjugate[Cos[x+I y]]];
Collect[Expand[p/.Cosh[y]^2->1+Sinh[y]^2],Sinh[y]^2]
Cos[I y]
ComplexExpand[Cosh[x+I y]]
Cosh[I y]

(*M2.8*)
Solve[Sinh[a z]==0,z,Complexes]
Solve[Cosh[a z]==0,z,Complexes]

(*M2.9*)
(*Note: x²+y²=r²    th=Arg[x+Iy]*)
Simplify[ComplexExpand[(x+I y)^(a+I b)],
   Assumptions-> {a,b,c,d}∈Reals]

(*M2.10*)
ComplexExpand[I^I]

(*M2.11*)
N[ComplexExpand[(1-I)^(1+I)]]

(*M2.12*)
Cauchy[u_,v_]:={D[u,x]==D[v,y],D[v,x]==-D[u,y]}
Cauchy[Exp[x] Cos[y],Exp[x] Sin[y]]
Cauchy[x^2+y^2,0]

(*M2.13*)
Series[z/Sinh[a z],{z,0,3}]
Series[(Cosh[z]-1)/z^2,{z,0,3}]
Series[Sinh[z]/z^3,{z,0,3}]

(*M2.14*)
(*i*)
Simplify[2 Pi I (1+z)/D[Sinh[a z],z]/.z->I n Pi/a,
   Assumptions->n∈Integers]
(*ii*)
2 Pi I Residue[Exp[-z]/(z-Pi/2),{z,Pi/2}]
(*iii*)
2 Pi I Residue[Exp[2 z]/(z+1)^4,{z,-1}]
(*iv*)
gz=Cos[z]/z/(z^2+9);
2 Pi I (Residue[gz,{z,0}]+Residue[gz,{z,+3 I}]
   +Residue[gz,{z,-3 I}])
```

EXERCISES

Section 2.1

2.1 Show that

$$\left|\frac{a+jb}{b+ja}\right|=1$$

2.2 If $z = x + jy$, then verify the following relations

(a) $\overline{z + 5j} = \overline{z} - 5j$

(b) $\overline{jz} = -j\overline{z}$

(c) $(1 + j)^4 = -4$

(d) $\dfrac{1 + j}{1 - j} = j$

SECTION 2.2

2.3 Show that

(a) $\sqrt{j} = \pm\dfrac{1 + j}{\sqrt{2}}$

(b) $\left(\dfrac{-1 + j\sqrt{3}}{2}\right)^{3/2} = \pm 1$

2.4 Show that

$$(-1 + j)^{10} = -32j$$

2.5 Use Euler's formula to obtain expressions for $\cos(\alpha + \beta)$ and $\sin(\alpha + \beta)$ in terms of $\sin(\alpha)$, $\sin(\beta)$, $\cos(\alpha)$, and $\cos(\beta)$. Hint: Note that $e^{j(\alpha+\beta)} = e^{j\alpha}e^{j\beta}$.

2.6 Show that the four roots of

$$z^4 - 2\cos(2\theta)z^2 + 1 = 0$$

are $\pm e^{j\theta}$ and $\pm e^{-j\theta}$.

2.7 If

$$\sum_{n=0}^{\infty} q^n = \frac{1}{1 - q} \qquad |q| < 1$$

then show that

$$\sum_{n=0}^{\infty} \varepsilon^n \cos(nt) = \frac{1 - \varepsilon\cos t}{1 + \varepsilon^2 - 2\varepsilon\cos t} \qquad |\varepsilon| < 1$$

Hint: Let $q = \varepsilon e^{jt}$.

2.8 Using the appropriate results of Exercise 2.7, show that

$$2\sum_{n=0}^{\infty} e^{-\alpha n}\sin(nt) = \frac{\sin t}{\cosh\alpha - \cos t} \qquad \alpha > 0$$

2.9 If $z=x+jy$, then show that

(a) $\tanh(z/2)=\dfrac{\sinh x + j\sin y}{\cosh x + \cos y}$

(b) $\coth(z/2)=\dfrac{\sinh x - j\sin y}{\cosh x - \cos y}$

Section 2.3

2.10 If $z=x+jy$, which of the following functions are analytic, and for those that are analytic, what is their region of analyticity?

(a) $f(z)=e^{2x}\cos y + je^{2x}\sin y$

(b) $f(z)=\sinh z$

(c) $f(z)=|z|\sin z$

(d) $f(z)=\overline{z}$

(e) $f(z)=jz+|z|$

(f) $f(z)=\dfrac{z-j}{z+j}$

2.11 Express the following expressions in terms of trigonometric and hyperbolic functions and then determine the value(s) of the real quantities a and b that make them analytic:

(a) $\cos(ax+jby)$
(b) $\sin(ax-jby)$

2.12 Given the analytic function

$$f(z)=u(x,y)+jv(x,y)$$

Determine $f(z)$ when:

(a) $u(x,y)=x^3-3xy^2$
(b) $u(x,y)=e^x\cos y$

2.13 Determine the value of the following integrals. Assume that the path C includes all singularities.

(a) $\displaystyle\int_C \frac{e^{-3\pi z}}{2z+j}\,dz$

(b) $\displaystyle\int_C \frac{e^{z}\cos z}{(z-\pi/2)^2}\,dz$

(c) $\int_C \frac{z^n}{(z-a)^{n+1}} dz$

2.14 Evaluate the following integral

$$\int_C \frac{1}{z^3(z+4)} dz$$

for (a) inside the circle $|z|=2$, and (b) inside the circle $|z+2|=3$.

2.15 Determine the value of the following integral when the contour includes all singularities.

$$I = \int_C \frac{e^{tz} dz}{z^2(z^2+2z+2)}$$

2.16 Evaluate the following integral around the unit circle.

$$\oint_{|z|<1} z^2 e^{1/z} dz$$

Hint: Express the exponential function as a series using Eq. (A.3) of Appendix A.

2.17 Determine the order of the pole for the following expressions:

(a) $\frac{e^{tz}}{z \sinh(az)}$

(b) $\frac{e^{z^2}}{z^4}$

(c) $\frac{\sin^2(z)}{z^3}$

3 Fourier Series and Fourier Transforms

3.1 FOURIER SERIES

The Fourier series is a representation of a real* periodic function of time $f(t)$, that is, one for which $f(t+kT)=f(t)$, $-T/2 \leq t \leq T/2$ and $k \pm 1, \pm 2, \pm 3, \ldots$, where $T=1/f_o=2\pi/\omega_o$ is the period and $\omega_o=2\pi/T$ is the fundamental radian frequency. For the Fourier series of $f(t)$ to exist, the following Dirichlet conditions must be satisfied on the interval $-T/2 \leq t \leq T/2$: $f(t)$ is bounded, has at most a finite number of maxima and minima, and is continuous apart from a finite number of discontinuities. Most functions of practical interest satisfy these conditions.

In complex notation, this series is given by†

$$f(t) = \sum_{n=-\infty}^{\infty} c_n e^{j2n\pi t/T} = \sum_{n=-\infty}^{\infty} c_n e^{jn\omega_o t} \tag{3.1}$$

where

$$c_n = \frac{1}{T}\int_{-T/2}^{T/2} f(t)e^{-j2n\pi t/T}dt = \frac{\omega_o}{2\pi}\int_{-\pi/\omega_o}^{\pi/\omega_o} f(t)e^{-jn\omega_o t}dt \quad n = 0, \pm 1, \pm 2, \ldots \tag{3.2}$$

It is seen that c_0 is the average value of $f(t)$ and that the units of c_n are those of $f(t)$. In addition, c_n is only a function of ω_o and its harmonics $n\omega_o$. Therefore, the Fourier series maps the periodic time-domain signal from the time domain to the frequency domain, which is represented by c_n. The mapping has frequency components only at the discrete values of $n\omega_o$, $n = \pm 1, \pm 2, \ldots$ The values of $|c_n|$ as a function of $n\omega_o$ comprise the *frequency spectrum* of the periodic waveform $f(t)$.

Equation (3.2) is obtained in the following manner. We start with Eq. (3.1) and multiply it by $e^{-j2m\pi t/T}$ and then integrate the result over one period to obtain

$$\int_{-T/2}^{T/2} e^{-j2m\pi t/T} f(t)dt = \sum_{n=-\infty}^{\infty} c_n \int_{-T/2}^{T/2} e^{j2(n-m)\pi t/T}dt \tag{3.3}$$

* Limiting $f(t)$ to a real function is not necessary mathematically, but in engineering practice it is rare that one expresses time-varying functions as other than a real quantity.

† As will be seen in Section 3.3.1, the complex form given by Eq. (3.1) has some advantages when the Fourier transform is introduced.

However,

$$\int_{-T/2}^{T/2} e^{j2(n-m)\pi t/T} dt = \frac{T}{\pi}\frac{\sin\left((m-n)\pi\right)}{m-n} = 0 \quad m \neq n \tag{3.4}$$

$$= T \quad m = n$$

We can write Eq. (3.4) in the shorthand notation

$$\int_{-T/2}^{T/2} e^{j2(n-m)\pi t/T} dt = T\delta_{nm} \tag{3.5}$$

where δ_{nm} is the Kronecker delta. It is noted that a shift in the origin of the periodic signal does not alter this result, that is,

$$\int_{0}^{T} e^{j2(n-m)\pi t/T} dt = T\delta_{nm} \tag{3.6}$$

Hence, using Eq. (3.5) in Eq. (3.3), we obtain

$$\int_{-T/2}^{T/2} e^{-j2m\pi t/T} f(t)dt = \sum_{n=-\infty}^{\infty} c_n T\delta_{nm} = Tc_m \tag{3.7}$$

which gives Eq. (3.2).

It is mentioned that Eq. (3.4) satisfies the definition of an orthogonal function on the interval $-T/2 \leq t \leq T/2$ and for Eq. (3.6) on the interval $0 \leq t \leq T$; that is, $-e^{-j2m\pi t/T}$ is an orthogonal function on its respective interval. The generation of orthogonal functions to obtain solutions to differential equations is discussed in Section 6.1.4 and their use illustrated in Section 6.1.6 and Chapters 7 and 8. Let us introduce the notation $\lambda_n = n\omega_o$ and $\varphi_n(t) = e^{-j\lambda_n t}$. As will be seen in the sections cited above, λ_n will be the eigenvalue and $\varphi_n(t)$ the orthogonal function on an interval $a \leq t \leq b$, which satisfies the relation

$$\int_{a}^{b} r(t)\varphi_n(t)\varphi_m(t)dt = \delta_{nm}N_n \tag{3.8}$$

In Eq. (3.8), $r(t)$ is called the weighting function, which in the present case is one,

$$N_n = \left\|\varphi_n(t)\right\|^2 = \int_{a}^{b} r(t)\varphi_n^2(t)dt \tag{3.9}$$

and $\sqrt{N_n}$ is the norm. In this regard, recall Eqs. (1.42), (1.71), and (1.72), respectively, which described the eigenvalues, orthogonal eigenvectors, and the norm for a square symmetric matrix. In general, orthogonal function expansions are not periodic.

Since $f(t)$ is real, Eq. (3.2) can be written as

$$\begin{aligned} c_n &= \frac{\omega_o}{2\pi}\int_{-\pi/\omega_o}^{\pi/\omega_o} f(t)\cos(n\omega_o t)dt - j\frac{\omega_o}{2\pi}\int_{-\pi/\omega_o}^{\pi/\omega_o} f(t)\sin(n\omega_o t)\,dt \\ &= R(n) - jX(n) \\ &= |c_n|e^{-j\varphi_n} \quad n = 0, \pm1, \pm2, \ldots \end{aligned} \tag{3.10}$$

where

$$\begin{aligned} &R(n) = \frac{\omega_o}{2\pi}\int_{-\pi/\omega_o}^{\pi/\omega_o} f(t)\cos(n\omega_o t)dt \quad X(n) = \frac{\omega_o}{2\pi}\int_{-\pi/\omega_o}^{\pi/\omega_o} f(t)\sin(n\omega_o t)dt \\ &|c_n| = \sqrt{R^2(n) + X^2(n)} \qquad \varphi_n = \tan^{-1}\frac{X(n)}{R(n)} \qquad n = 0, \pm1, \pm2, \ldots \end{aligned} \tag{3.11}$$

Note that the definition φ_n is for a negative imaginary part of c_n.

It is also seen that

$$\begin{aligned} c_{-n} &= R(-n) - jX(-n) = R(n) + jX(n) = \overline{c}_n \\ c_{-n}c_n &= \overline{c}_n c_n = |c_n|^2 = R^2(n) + X^2(n) \end{aligned} \tag{3.12}$$

and

$$\varphi_{-n} = \tan^{-1}\frac{X(-n)}{R(-n)} = -\varphi_n \quad n = 0, \pm1, \pm2, \ldots \tag{3.13}$$

where the overbar indicates the complex conjugate (recall Eq. (2.2)). Then, Eq. (3.1) can be written as

$$\begin{aligned} f(t) &= \sum_{n=-\infty}^{-1} |c_n| e^{j(2n\pi t/T - \varphi_n)} + c_0 + \sum_{n=1}^{\infty} |c_n| e^{j(2n\pi t/T - \varphi_n)} \\ &= \sum_{n=1}^{\infty} |c_{-n}| e^{j(-2n\pi t/T - \varphi_{-n})} + c_0 + \sum_{n=1}^{\infty} |c_n| e^{j(2n\pi t/T - \varphi_n)} \\ &= \sum_{n=1}^{\infty} |c_n| e^{j(-2n\pi t/T + \varphi_n)} + c_0 + \sum_{n=1}^{\infty} |c_n| e^{j(2n\pi t/T - \varphi_n)} \\ &= c_0 + \sum_{n=1}^{\infty} |c_n| \left[e^{j(2n\pi t/T - \varphi_n)} + e^{-j(2n\pi t/T - \varphi_n)} \right] \\ &= c_0 + 2\sum_{n=1}^{\infty} |c_n| \cos[2n\pi t/T - \varphi_n] \\ &= c_0 + 2\sum_{n=1}^{\infty} |c_n| \cos[n\omega_o t - \varphi_n] \end{aligned} \tag{3.14}$$

The *root mean square* (rms) value f_{rms} of a real periodic function $f(t)$ is defined as

$$
\begin{aligned}
f_{rms}^2 &= \frac{1}{T}\int_{-T/2}^{T/2} f^2(t)dt \\
&= \frac{1}{T}\int_{-T/2}^{T/2} \left\{ |c_0| + 2\sum_{n=1}^{\infty} |c_n| \cos[2n\pi t/T - \varphi_n] \right\}^2 dt \\
&= \frac{1}{T}\int_{-T/2}^{T/2} |c_0|^2 dt + \frac{4|c_0|}{T}\int_{-T/2}^{T/2} \sum_{n=1}^{\infty} |c_n| \cos[2n\pi t/T - \varphi_n]dt \\
&\quad + \frac{4}{T}\int_{-T/2}^{T/2} \left\{ \sum_{n=1}^{\infty} |c_n| \cos[2n\pi t/T - \varphi_n] \right\}^2 dt \\
&= |c_0|^2 + \frac{4|c_0|}{T}\sum_{n=1}^{\infty} |c_n| \int_{-T/2}^{T/2} \cos[2n\pi t/T - \varphi_n]dt \\
&\quad + \frac{4}{T}\sum_{m=1}^{\infty}\sum_{n=1}^{\infty} |c_n||c_m| \int_{-T/2}^{T/2} \cos[2n\pi t/T - \varphi_n]\cos[2m\pi t/T - \varphi_m]dt
\end{aligned}
\tag{3.15}
$$

However,

$$\int_{-T/2}^{T/2} \cos[2n\pi t/T - \varphi_n]dt = \frac{T}{n\pi}\cos\varphi_n \sin 2n\pi = 0 \quad n = 1, 2, \ldots$$

and

$$\int_{-T/2}^{T/2} \cos[2n\pi t/T - \varphi_n]\cos[2m\pi t/T - \varphi_m]dt = \delta_{nm}\frac{T}{2} \quad n, m = 1, 2, \ldots$$

Then Eq. (3.15) becomes

$$f_{rms} = \sqrt{|c_0|^2 + \frac{4}{T}\frac{T}{2}\sum_{n=1}^{\infty}\sum_{m=1}^{\infty} |c_n||c_m|\delta_{nm}} = \sqrt{|c_0|^2 + 2\sum_{n=1}^{\infty} |c_n|^2} \tag{3.16}$$

and, therefore, we have found that

$$\frac{1}{T}\int_{-T/2}^{T/2} f^2(t)dt = |c_0|^2 + 2\sum_{n=1}^{\infty} |c_n|^2 \tag{3.17}$$

which is called *Parseval's theorem*.

Another form of the Fourier series of a real periodic function is

$$f(t)=\frac{a_0}{2}+\sum_{n=1}^{\infty}\{a_n\cos(n\omega_o t)+b_n\sin(n\omega_o t)\} \tag{3.18}$$

where

$$\begin{aligned} a_n &= \frac{2}{T}\int_0^T f(t)\cos(n\omega_o t)dt=\frac{2}{T}\int_{-T/2}^{T/2} f(t)\cos(n\omega_o t)dt=2R(n) \quad n=0,1,2,\ldots \\ b_n &= \frac{2}{T}\int_0^T f(t)\sin(n\omega_o t)dt=\frac{2}{T}\int_{-T/2}^{T/2} f(t)\sin(n\omega_o t)dt=2X(n) \quad n=1,2,3,\ldots \end{aligned} \tag{3.19}$$

where $R(n)$ and $X(n)$ are given by Eq. (3.11) with $T=2\pi/\omega_o$.

Before showing how the coefficients a_n and b_n are obtained, we note the evaluation of the following integrals when n and m are integers

$$\begin{aligned} &\int_{-T/2}^{T/2}\cos(m\omega_o t)dt=0 \quad m\neq 0 \\ &\int_{-T/2}^{T/2}\cos(n\omega_o t)\cos(m\omega_o t)dt=\frac{T}{2}\delta_{nm} \\ &\int_{-T/2}^{T/2}\sin(n\omega_o t)\sin(m\omega_o t)dt=\frac{T}{2}\delta_{nm} \\ &\int_{-T/2}^{T/2}\sin(n\omega_o t)\cos(m\omega_o t)dt=0 \quad \text{for all } n,m \end{aligned} \tag{3.20}$$

It is again noted from Eq. (3.8) that $\cos(n\omega_o t)$ and $\sin(n\omega_o t)$ satisfy the definition of an orthogonal function on the interval $-T/2 \le t \le T/2$. In other words, from the second and third integral relations of Eq. (3.20) it is seen that the functions cos $(m\omega_o t)$ and $\cos(n\omega_o t)$ and the functions $\sin(m\omega_o t)$ and $\sin(n\omega_o t)$, respectively, are orthogonal to each other on the given interval and from the last equation of Eq. (3.20) that $\cos(m\omega_o t)$ and $\sin(n\omega_o t)$ are always orthogonal to each other irrespective of the values of m and n.

To obtain the expression for a_n given in the first equation of Eq. (3.19), we multiply Eq. (3.18) by $\cos(m\omega_o t)$ $m=1, 2, \ldots$, and integrate over the interval to obtain

$$\begin{aligned} \int_{-T/2}^{T/2} f(t)\cos(m\omega_o t)dt &= \frac{a_0}{2}\int_{-T/2}^{T/2}\cos(m\omega_o t)dt+\sum_{n=1}^{\infty}a_n\int_{-T/2}^{T/2}\cos(n\omega_o t)\cos(m\omega_o t)dt \\ &\quad +\sum_{n=1}^{\infty}b_n\int_{-T/2}^{T/2}\sin(n\omega_o t)\cos(m\omega_o t)dt \end{aligned} \tag{3.21}$$

Using Eq. (3.20), Eq. (3.21) becomes

$$\int_{-T/2}^{T/2} f(t)\cos(m\omega_o t)dt = \frac{T}{2}\sum_{n=1}^{\infty} a_n \delta_{nm} = a_m \frac{T}{2} \tag{3.22}$$

In a similar manner, we multiply Eq. (3.18) by $\sin(m\omega_o t)$ $m = 1, 2, \ldots$, and integrate over the interval to obtain the expression given for b_n in Eq. (3.19).

Equation (3.18) follows directly from Eq. (3.1). We rewrite Eq. (3.1) as

$$f(t) = \sum_{n=-\infty}^{-1} c_n e^{j2n\pi t/T} + c_0 + \sum_{n=1}^{\infty} c_n e^{j2n\pi t/T} = \sum_{n=1}^{\infty} c_{-n} e^{-j2n\pi t/T} + c_0 + \sum_{n=1}^{\infty} c_n e^{j2n\pi t/T}$$

$$= c_0 + \sum_{n=1}^{\infty}\left(c_{-n}e^{-j2n\pi t/T} + c_n e^{j2n\pi t/T}\right)$$

$$= c_0 + \sum_{n=1}^{\infty}\left\{c_{-n}\left(\cos(2n\pi t/T) - j\sin(2n\pi t/T)\right)\right.$$
$$\left. + c_n\left(\cos(2n\pi t/T) + j\sin(2n\pi t/T)\right)\right\} \tag{3.23}$$

$$= c_0 + \sum_{n=1}^{\infty}\left\{(c_n + c_{-n})\cos(2n\pi t/T) + j(c_n - c_{-n})\sin(2n\pi t/T)\right\}$$

$$= \frac{a_0}{2} + \sum_{n=1}^{\infty}\left\{a_n\cos(2n\pi t/T) + b_n\sin(2n\pi t/T)\right\}$$

where, from Eq. (3.12), we see that

$$\begin{aligned} a_n &= c_n + c_{-n} = R(n) - jX(n) + R(n) + jX(n) = 2R(n) = 2\,\mathrm{Re}[c_n] \\ b_n &= j(c_n - c_{-n}) = j\left(R(n) - jX(n) - R(n) - jX(n)\right) = 2X(n) = 2\,\mathrm{Im}[c_n] \\ a_0 &= 2c_0 \end{aligned} \tag{3.24}$$

We shall place Eq. (3.23) in the form given by Eq. (3.14). We note the following identity

$$a\sin\omega t \pm b\cos\omega t = \sqrt{a^2 + b^2}\,\sin(\omega t \pm \psi) \tag{3.25}$$

where $a > 0$, $b > 0$, and

$$\psi = \tan^{-1}\frac{b}{a} \tag{3.26}$$

Therefore,

$$a_n\cos(n\omega_o t) + b_n\sin(n\omega_o t) = \sqrt{a_n^2 + b_n^2}\,\sin(n\omega_o t + \psi_n) \tag{3.27}$$

where

$$\psi_n = \tan^{-1}\frac{a_n}{b_n}$$

and from Eqs. (3.11) and (3.27)

$$\sqrt{a_n^2 + b_n^2} = 2\sqrt{R^2(n) + X^2(n)} = 2|c_n| \tag{3.28}$$

From Eqs. (3.11) and (3.24)

$$\varphi_n = \tan^{-1}\frac{X(n)}{R(n)} = \tan^{-1}\frac{b_n}{a_n} \tag{3.29}$$

Therefore, as seen in Figure 3.1, $\varphi_n = \pi/2 - \psi_n$ and, therefore,

$$\begin{aligned} a_n\cos(n\omega_o t) + b_n\sin(n\omega_o t) &= 2|c_n|\sin(n\omega_o t + \psi_n) \\ &= 2|c_n|\sin(n\omega_o t + \pi/2 - \varphi_n) \\ &= 2|c_n|\cos(n\omega_o t - \varphi_n) \end{aligned}$$

Therefore, Eq. (3.18) can also be written in the form given by Eq. (3.14), that is,

$$f(t) = \frac{a_0}{2} + \sum_{n=1}^{\infty}\sqrt{a_n^2 + b_n^2}\cos[n\omega_o t - \varphi_n]$$

where φ_n is given by Eq. (3.29).

It is seen from what has been presented that to have a periodic function two things are required: (i) the amplitudes of each harmonic must be constant, that is, independent of time; and (ii) the frequencies of each contributing frequency component must be commensurate with every other contributing frequency component; that is, the ratio of any two frequencies must be a rational number – the ratio of two integers. An example of a function with time-dependent amplitude is

$$f(t) = \sin\omega_1 t + t\sin 2\omega_1 t$$

which has two periodic integer-related components, but one of them does not have constant amplitude.

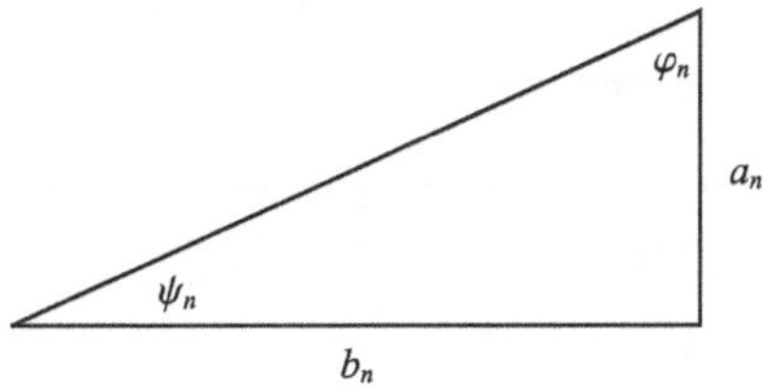

FIGURE 3.1 Definitions of two angles.

To illustrate that each contributing frequency component must be related by an integer number of periods, consider the following function composed of two individual periodic functions

$$f(t) = \sin(\omega_1 t) + \sin(\omega_2 t) \quad \omega_1 < \omega_2$$

The period of each respective term is

$$T_1 = \frac{2\pi}{\omega_1} \text{ and } T_2 = \frac{2\pi}{\omega_2} < T_1$$

Therefore,

$$\frac{T_1}{T_2} = \frac{\omega_2}{\omega_1} > 1$$

It is seen that in order for the two trigonometric functions to have a common period, this ratio must be the ratio of two integers, that is,

$$\frac{T_1}{T_2} = \frac{n_2}{n_1} \quad n_2 > n_1$$

or $n_1T_1 = n_2T_2$. In other words, every n_2T_2 periods of the function at frequency ω_2 must fit exactly within n_1T_1 periods of the function with frequency ω_1. Consequently, selecting T_1 (or T_2), n_1, and n_2 determines the value for T_2 (or T_1). In this case, we see that

$$f(t) = \sin\left(2\pi t / T_1\right) + \sin\left(2\pi t n_2 / (n_1 T_1)\right) \quad n_1 < n_2$$

Each component of this result is plotted in Figure 3.2 for $n_1 = 3$ and $n_2 = 5$.

We shall now illustrate Eqs. (3.1) and (3.2) with several examples.

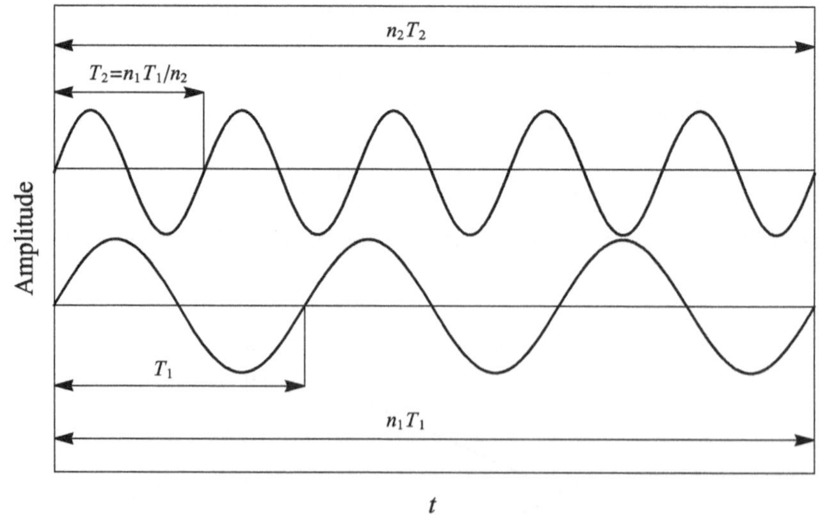

FIGURE 3.2 Periodicity of two sine waves whose frequencies are commensurate. In this case, $n_1 = 3$ and $n_2 = 5$.

Example 3.1

Consider the periodic square wave of magnitude A_o and period T shown in Figure 3.3. The Fourier series for this periodic function is determined as follows. From Eq. (3.2), we find that

$$c_n = \frac{1}{T}\left\{ -A_o \int_{-T/2}^{-T/4} e^{-j2n\pi t/T} dt + A_o \int_{-T/4}^{T/4} e^{-j2n\pi t/T} dt - A_o \int_{T/4}^{T/2} e^{-j2n\pi t/T} dt \right\}$$

$$= \frac{A_o}{n\pi T}\left\{ -Te^{j3n\pi/4}\sin(n\pi/4) + T\sin(n\pi/2) - Te^{-j3n\pi/4}\sin(n\pi/4) \right\}$$

$$= \frac{A_o}{n\pi}\left\{ \sin(n\pi/2) - 2\sin(n\pi/4)\cos(3n\pi/4) \right\}$$

$$= \frac{A_o}{n\pi}\left\{ \sin(n\pi/2) - (\sin n\pi - \sin n\pi/2) \right\} = \frac{2A_o}{n\pi}\sin(n\pi/2)$$

$$= \frac{2A_o}{n\pi}(-1)^{(n-1)/2} \quad n = 1,3,5,\ldots$$

and

$$c_0 = \frac{1}{T}\left\{ -A_o \int_{-T/2}^{-T/4} dt + A_o \int_{-T/4}^{T/4} dt - A_o \int_{T/4}^{T/2} dt \right\}$$

$$= \frac{A_o}{T}\left\{ -\left(-\frac{T}{4} + \frac{T}{2} \right) + \frac{T}{4} + \frac{T}{4} - \left(\frac{T}{2} - \frac{T}{4} \right) \right\} = 0$$

In addition, from Eq. (3.11), we obtain

$$\varphi_n = \tan^{-1}\left[0 / \left(2A_o(-1)^{(n-1)/2}/n\pi \right) \right]$$

$$= \tan^{-1}\frac{0}{(-1)^{(n-1)/2}} = (n-1)\pi/2 \quad n = 1,3,5,\ldots$$

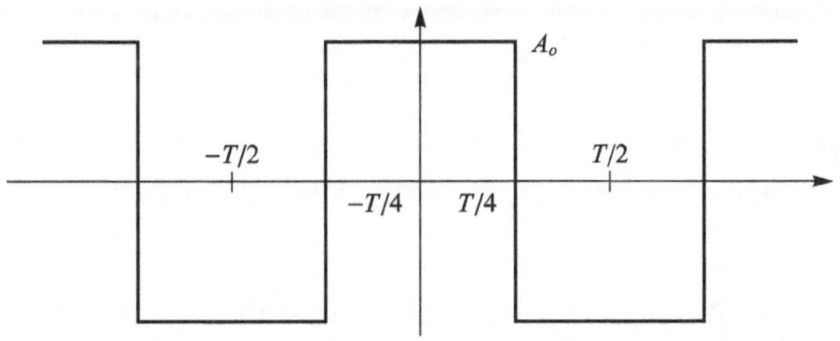

FIGURE 3.3 Periodic square wave.

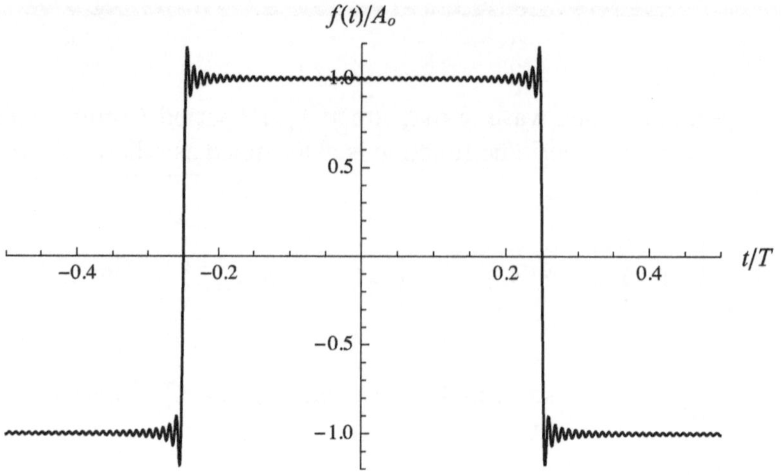

FIGURE 3.4 Evaluation of one period of the Fourier series for a periodic square wave examined in Example 3.1 and shown in Figure 3.3. The oscillation at the discontinuities is called the Gibbs phenomenon.

Then, Eq. (3.14) yields

$$\begin{aligned} f(t) &= \frac{4A_o}{\pi}\sum_{n=1,3,5}^{\infty}\frac{1}{n}\left|(-1)^{(n-1)/2}\right|\cos[2n\pi t/T-(n-1)\pi/2] \\ &= \frac{4A_o}{\pi}\sum_{n=1,3,5}^{\infty}\frac{1}{n}(-1)^{(n-1)/2}\cos[2n\pi t/T] \end{aligned} \tag{a}$$

Equation (a) is verified with Mathematica procedure M3.1 for $T=2\pi$ and is evaluated using 53 terms with Mathematica procedure M3.2. The result is shown in Figure 3.4. It is seen in Figure 3.4 that there is an overshoot at the discontinuity and some rippling of the waveform in the vicinity of the discontinuity. This is called the *Gibbs phenomenon* and is due directly to the nature of the discontinuity. It has been determined numerically in Mathematica procedure M3.3 that the magnitude of the overshoot of the square wave is approximately $1.179A_o$ at $t=0.2498$. The rippling will diminish if enough terms in the series are taken, but the magnitude of the overshoot at the discontinuity will remain very close to this value.

Example 3.2

Consider the periodic sequences of pulses of magnitude A_p, period T, and duration $2t_d<T$ shown in Figure 3.5. The Fourier series for this periodic function is determined as follows

$$c_n = \frac{A_p}{T}\int_{-t_d}^{t_d} e^{-j2n\pi t/T}dt = \frac{A_pT}{T}\frac{\sin(2n\pi t_d/T)}{n\pi} = \alpha A_p\frac{\sin(\alpha n\pi)}{\alpha n\pi}$$

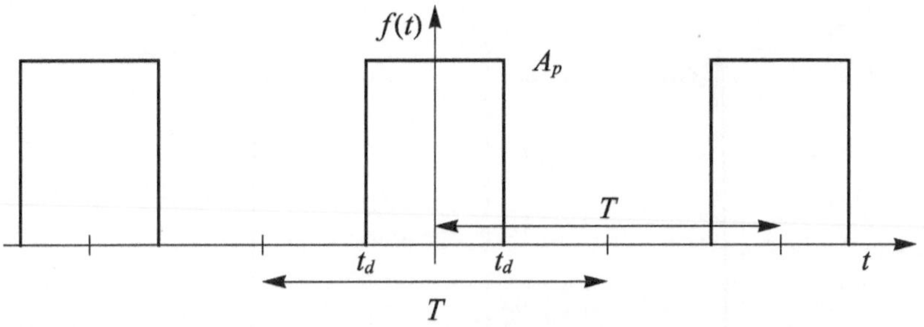

FIGURE 3.5 Periodic pulses of duration $2t_d$ and period T.

where $\alpha = 2t_d/T$ and

$$c_0 = \frac{A_p}{T}\int_{-t_d}^{t_d} dt = \alpha A_p$$

In addition, from Eq. (3.11),

$$\begin{aligned}\varphi_n &= \tan^{-1}\left[0\Big/\left(\alpha A_p \frac{\sin(\alpha n\pi)}{\alpha n\pi}\right)\right]\\ &= \tan^{-1}\frac{0}{\sin(\alpha n\pi)}\\ &= 0 \quad \sin(\alpha n\pi) > 0\\ &= \pi \quad \sin(\alpha n\pi) < 0\end{aligned}$$

Therefore, Eq. (3.14) yields

$$f(t) = \alpha A_p\left\{1 + 2\sum_{n=1}^{\infty}\left|\frac{\sin(\alpha n\pi)}{\alpha n\pi}\right|\left(\cos[2n\pi t/T]\cos\varphi_n + \sin[2n\pi t/T]\sin\varphi_n\right)\right\} \quad \text{(a)}$$

However, since $\varphi_n = 0$ or π depending on the sign of $\sin(\alpha n\pi)$, $\sin\varphi_n = 0$, and $\cos\varphi_n$ = sign of $\sin(\alpha n\pi)$. Therefore, Eq. (a) can be written as

$$f(t) = \alpha A_p\left\{1 + 2\sum_{n=1}^{\infty}\frac{\sin(\alpha n\pi)}{\alpha n\pi}\cos[2n\pi t/T]\right\} \quad \text{(b)}$$

where we have removed the absolute value designation.

Equation (b) is verified with Mathematica procedure M3.4 and is evaluated with Mathematica procedure M3.5 for $\alpha = 0.4$ using 101 terms. The result is shown in Figure 3.6 where it is seen that for this discontinuity the Gibbs phenomenon is again present.

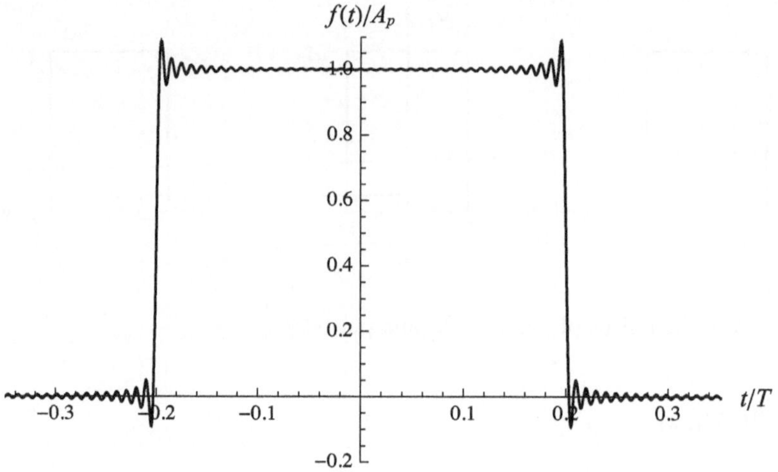

FIGURE 3.6 Evaluation of one period of the Fourier series for periodic pulse given by Eq. (b) of Example 3.2 with $\alpha=0.4$. The oscillation at the discontinuities is called the Gibbs phenomenon.

Example 3.3

Consider the periodic sequence of impulses represented as delta functions (see Section B.1 of Appendix B) of unit magnitude and period T given by

$$\delta_T(t)=\sum_{n=-\infty}^{\infty}\delta(t-nT)$$

Then $f(t)=\delta(t)$, and upon using Eqs. (3.2) and (3.11), we find that

$$c_n=\frac{1}{T}\int_{-T/2}^{T/2}\delta(t)e^{-j2n\pi t/T}dt=\frac{1}{T}$$

$$\varphi_n=\tan^{-1}\frac{0}{1/T}=0$$

Therefore, from Eq. (3.14),

$$\delta_T(t)=\frac{1}{T}\left\{1+2\sum_{n=1}^{\infty}\cos[2n\pi t/T]\right\} \qquad \text{(a)}$$

or from Eq. (3.1)

$$\delta_T(t)=\sum_{n=-\infty}^{\infty}c_n e^{j2n\pi t/T}=\frac{1}{T}\sum_{n=-\infty}^{\infty}e^{j2n\pi t/T} \qquad \text{(b)}$$

It is noted that the units on both sides of Eq. (b) are the same: the reciprocal of T. Equation (a) is verified with Mathematica procedure M3.6.

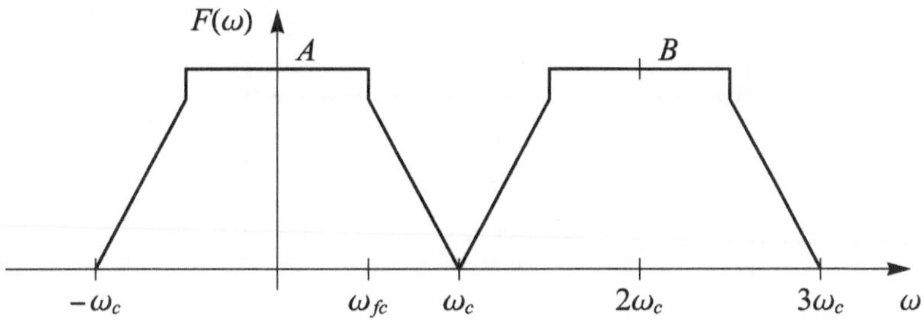

FIGURE 3.7 A representative periodic function in the frequency domain.

3.2 FOURIER SERIES IN THE FREQUENCY DOMAIN

We shall now express the Fourier series representation of a periodic function in the frequency domain $F(\omega)$; that is, one for which $F(\omega + 2k\omega_c) = F(\omega)$, $-\omega_c \le \omega \le \omega_c$, where $k = \pm1, \pm2, \pm3, \ldots$, and $2\omega_c$ is the repetition frequency of $F(\omega)$. A representative periodic function in the frequency domain is shown in Figure 3.7. Then Eq. (3.1) is used to obtain

$$F(\omega) = \sum_{n=-\infty}^{\infty} d_n e^{j2n\pi\omega/(2\omega_c)} \tag{3.30}$$

where

$$d_n = \frac{1}{2\omega_c} \int_{-\omega_c}^{\omega_c} F(\omega) e^{-j2n\pi\omega/(2\omega_c)} d\omega \quad n = 0, \pm1, \pm2, \ldots \tag{3.31}$$

We shall use this relation when we discuss the sampling theorem in Section 3.4.1.

It is pointed out that ω_c is sometimes referred to as the folding frequency; that is, in Figure 3.7 when $F(\omega + 2\omega_c)$, $-\omega_c \le \omega \le \omega_c$, is flipped about ω_c the points A and B will coincide. This is discussed further in Section 3.4.2. In addition, it is important to realize that in order for the signal in the frequency domain to be periodic, the function $F(\omega)$ must be zero outside the region $-\omega_c \le \omega \le \omega_c$. The violation of this assumption introduces a phenomenon called aliasing, which is introduced and discussed in Section 3.4.2. It is also mentioned that the shape of the spectrum shown in Figure 3.7 was chosen to approximate an amplitude response spectrum of a filter. In this case ω_{fc} is called the cutoff frequency of the filter, and the sloping line emanating at this frequency is called the filter's selectivity, usually expressed as dB/octave, the amount of attenuation of the input signal to the filter for each doubling of ω for frequencies greater than ω_{fc}.

To illustrate Eqs. (3.30) and (3.31), we consider the following example, which has application in Section 3.4.2.

Example 3.4

Consider the rectangularly shaped region given by

$$F(\omega) = F_o\left(u(\omega + \alpha\omega_c) - u(\omega - \alpha\omega_c)\right) \tag{a}$$

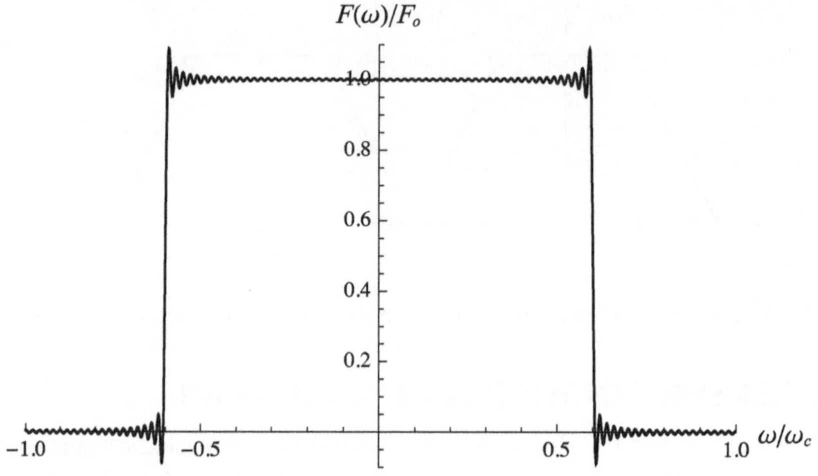

FIGURE 3.8 One period of the periodic function in the frequency domain given by Eq. (a) and computed by Eq. (b) of Example 3.4 for $\alpha = 0.6$.

where $0 < \alpha < 1$ and $u(x)$ is the unit step function. Then Eq. (3.31) gives

$$d_n = \frac{F_o}{2\omega_c} \int_{-\alpha\omega_c}^{\alpha\omega_c} e^{-j2n\pi\omega/(2\omega_c)} d\omega = \frac{F_o}{2\omega_c} \left[\frac{e^{-j2n\pi\omega/(2\omega_c)}}{-j2n\pi/(2\omega_c)} \right]_{-\alpha\omega_c}^{\alpha\omega_c}$$

$$= \frac{F_o}{-j2n\pi}\left(e^{-j\alpha n\pi} - e^{j\alpha n\pi}\right) = \alpha F_o \frac{\sin(\alpha n\pi)}{\alpha n\pi} = d_{-n}$$

Using Eq. (3.30), we find that since $d_n = d_{-n}$

$$F(\omega) = \sum_{n=-\infty}^{\infty} d_n e^{j2n\pi\omega/(2\omega_c)} = d_0 + \sum_{n=1}^{\infty} d_n \left(e^{j2n\pi\omega/(2\omega_c)} + e^{-j2n\pi\omega/(2\omega_c)} \right)$$

$$= \alpha F_o \left(1 + 2\sum_{n=1}^{\infty} \frac{\sin(\alpha n\pi)}{\alpha n\pi} \cos(n\pi\omega/\omega_c) \right) \qquad \text{(b)}$$

Equation (b) is evaluated for $\alpha = 0.6$ and 100 terms and plotted in Figure 3.8 using Mathematica procedure M3.7.

3.3 FOURIER TRANSFORM

3.3.1 An Intuitive Approach

It is clear from the definition of the period T that the spacing between harmonics is

$$\Delta f = (n+1)f_o - nf_o = f_o = 1/T \rightarrow \Delta\omega = 2\pi/T$$

which indicates that as $T \to \infty$, $\Delta\omega \to 0$; that is, the spacing between harmonics approaches zero, and it appears to become a continuous function of ω. To see this, we rewrite Eqs. (3.1) and (3.2), respectively, as

$$\begin{aligned} f(t) &= \sum_{n=-\infty}^{\infty} c_n e^{j2n\pi t/T} = \sum_{n=-\infty}^{\infty} c_n e^{j(n\Delta\omega)t} \\ c_n &= \frac{1}{T}\int_{-T/2}^{T/2} f(t)e^{-j2n\pi t/T}dt = \frac{\Delta\omega}{2\pi}\int_{-\pi/\Delta\omega}^{\pi/\Delta\omega} f(t)e^{-j(n\Delta\omega)t}dt \end{aligned} \tag{3.32}$$

It is seen in Eq. (3.32) that in the limit as $\Delta\omega \to 0$, $c_n \to 0$. However, the ratio $c_n/\Delta\omega$ may exist. Therefore, if one could actually take the limit in Eq. (3.32) as $T \to \infty$, that is, as $\Delta\omega \to 0$, the formalism would look like

$$\begin{aligned} \lim_{\Delta\omega\to 0}\left[\frac{2\pi c_n}{\Delta\omega}\right] &= F(\omega) = \lim_{\Delta\omega\to 0}\left[\int_{-\pi/\Delta\omega}^{\pi/\Delta\omega} f(t)e^{-j(n\Delta\omega)t}dt\right] \overset{?}{\to} \int_{-\infty}^{\infty} f(t)e^{-j\omega t}dt \\ f(t) &= \frac{1}{2\pi}\lim_{\Delta\omega\to 0}\left[\sum_{n=-\infty}^{\infty}\left(\frac{2\pi c_n}{\Delta\omega}\right)e^{j(n\Delta\omega)t}\Delta\omega\right] \overset{?}{\to} \frac{1}{2\pi}\int_{-\infty}^{\infty} F(\omega)e^{j\omega t}d\omega \end{aligned} \tag{3.33}$$

where the rightmost expressions are what the limits may look like. Notice that the units of $F(\omega)$ are the units of $f(t)$ per Hz. In the next section, the definition of the Fourier transform is formally introduced.

3.3.2 Fourier Transform

The Fourier transform pair is defined as

$$\begin{aligned} F(\omega) &= \int_{-\infty}^{\infty} f(t)e^{-j\omega t}dt \to F[f(t)] \\ f(t) &= \frac{1}{2\pi}\int_{-\infty}^{\infty} F(\omega)e^{j\omega t}d\omega \\ &= \frac{1}{2\pi}\int_{-\infty}^{\infty} e^{j\omega t}\left[\int_{-\infty}^{\infty} f(\xi)e^{-j\omega\xi}d\xi\right]d\omega \end{aligned} \tag{3.34}$$

We shall use the notation

$$f(t) \Leftrightarrow F(\omega)$$

to denote a Fourier transform pair: the Fourier transform of $f(t)$ is $F(\omega)$, and the inverse Fourier transform of $F(\omega)$ is $f(t)$.

In Eq. (3.34), $f(t)$ satisfies the Dirichlet conditions and is absolutely convergent, that is,

$$\int_{-\infty}^{\infty} |f(t)|\, dt < \infty \tag{3.35}$$

The Dirichlet conditions are that (i) $f(t)$ has only a finite number of maxima and minima in the interval $-\infty < t < \infty$, (ii) $f(t)$ has only a finite number of finite discontinuities in the interval $-\infty < t < \infty$, and (iii) $f(t)$ has no infinite discontinuities (is bounded).

Since, in general, $f(t)$ can be complex, we note that the complex conjugate of Eq. (3.34) is

$$\overline{f(t)} = \overline{\frac{1}{2\pi}\int_{-\infty}^{\infty} F(\omega) e^{j\omega t}\, d\omega} = \frac{1}{2\pi}\int_{-\infty}^{\infty} \overline{F(\omega)} e^{-j\omega t}\, d\omega = \frac{1}{2\pi}\int_{-\infty}^{\infty} \overline{F(-\omega)} e^{j\omega t}\, d\omega \tag{3.36}$$

and, therefore,

$$\overline{f(t)} \Leftrightarrow \overline{F(-\omega)} \tag{3.37}$$

It is noted that $|F(\omega)|$, which comprises the magnitude of the frequency spectrum, is a continuous function of frequency. This is distinctly different from the frequency spectrum of a periodic function, which has a discrete frequency spectrum comprised of the magnitudes $|c_n|$ defined only at the harmonics $n\omega_o$ of the fundamental frequency. In addition, it is recalled that in the periodic case the units are the units of the periodic function. For the Fourier transformed quantity, the units of $F(\omega)$ are the units of $f(t)$ per Hertz.

3.3.3 Properties of the Fourier Transform

We shall now identify several important properties of the Fourier transform of $f(t)$ and then tabulate them in Section 3.3.5.

Translation: $g(t - t_o)$

Using Eq. (3.34), it is found that

$$F[g(t-t_o)] = \int_{-\infty}^{\infty} g(t-t_o) e^{-j\omega t}\, dt = \int_{-\infty}^{\infty} g(\xi) e^{-j\omega(\xi + t_o)}\, d\xi$$

$$= e^{-j\omega t_o}\int_{-\infty}^{\infty} g(\xi) e^{-j\omega\xi}\, d\xi = e^{-j\omega t_o} G(\omega)$$

Thus,

$$g(t-t_o) \Leftrightarrow e^{-j\omega t_o} G(\omega) \tag{3.38}$$

Since $G(\omega) = |G(\omega)| e^{j\theta(\omega)}$, it is seen that the translation in time does not affect the magnitude of $G(\omega)$, but it does affect its phase.

Modulation: $e^{j\omega_o t} g(t)$

Using Eq. (3.34), it is found that

$$F[e^{j\omega_o t} g(t)] = \int_{-\infty}^{\infty} e^{j\omega_o t} g(t) e^{-j\omega t} dt = \int_{-\infty}^{\infty} g(t) e^{-j(\omega-\omega_o)t} dt = G(\omega - \omega_o)$$

Thus,

$$e^{j\omega_o t} g(t) \Leftrightarrow G(\omega - \omega_o) \quad (3.39)$$

Scaling: $g(at)$

Using Eq. (3.34), we first assume that $a > 0$ and find

$$F[g(at)] = \int_{-\infty}^{\infty} g(at) e^{-j\omega t} dt$$

If we let $a = |a|$ and make the substitution $\xi = |a|t$, then

$$F[g(at)] = \int_{-\infty}^{\infty} g(\xi) e^{-j\omega\xi/a} \frac{d\xi}{|a|} = \frac{1}{|a|} G(\omega / a)$$

For $a < 0$, we set $a = -|a|$ and make the substitution $\xi = -|a|t$, then

$$F[g(at)] = \int_{-\infty}^{\infty} g(-|a|t) e^{-j\omega t} dt = \int_{-\infty}^{\infty} g(-|a|t) e^{-j\omega(-|a|t)/a} dt$$

$$= \int_{\infty}^{-\infty} g(\xi) e^{-j\omega\xi/a} \frac{d\xi}{-|a|} = \frac{1}{|a|} \int_{-\infty}^{\infty} g(\xi) e^{-j\omega\xi/a} d\xi$$

$$= \frac{1}{|a|} G(\omega / a)$$

Therefore,

$$g(at) \Leftrightarrow \frac{1}{|a|} G(\omega / a) \quad (3.40)$$

Time Reversal: $g(-t)$

If, in Eq. (3.40), $a = -1$

$$F[g(-t)] = G(-\omega) \quad (3.41)$$

This result can also be obtained directly from Eq. (3.34) as

$$\int_{-\infty}^{\infty} g(-t)e^{-j\omega t}dt = \int_{\infty}^{-\infty} g(\xi)e^{j\omega\xi}(-d\xi) = \int_{-\infty}^{\infty} g(\xi)e^{j\omega\xi}d\xi = G(-\omega) \tag{3.42}$$

If $g(t)$ is real, then, since $\overline{ab} = \overline{a}\,\overline{b}$ (recall Eq. (2.9)), we obtain from Eq. (3.34)

$$\overline{G(\omega)} = \int_{-\infty}^{\infty} \overline{g(\xi)e^{j\omega\xi}}d\xi = \int_{-\infty}^{\infty} g(\xi)e^{j\omega\xi}d\xi$$

and we see from Eq. (3.42) that

$$G(-\omega) = \overline{G(\omega)} \tag{3.43}$$

Fourier Transform of Derivatives

We now use integration by parts* to determine the Fourier transform of the rth derivative of a function $f(t)$ as follows. From Eq. (3.34)

$$\int_{-\infty}^{\infty} \frac{d^r f(t)}{dt^r}e^{-j\omega t}dt = e^{-j\omega t}\frac{d^{r-1}f(t)}{dt^{r-1}}\bigg|_{-\infty}^{\infty} + j\omega\int_{-\infty}^{\infty} \frac{d^{r-1}f(t)}{dt^{r-1}}e^{-j\omega t}dt$$

If we assume that $f(t)$ and the $r-1$ derivative of $f(t)$ vanish as $t \to \pm\infty$, then

$$\int_{-\infty}^{\infty} \frac{d^r f(t)}{dt^r}e^{-j\omega t}dt = j\omega\int_{-\infty}^{\infty} \frac{d^{r-1}f(t)}{dt^{r-1}}e^{-j\omega t}dt$$

and by repeated application of integration by parts we find that

$$\int_{-\infty}^{\infty} \frac{d^r f(t)}{dt^r}e^{-j\omega t}dt = (j\omega)^r\int_{-\infty}^{\infty} f(t)e^{-j\omega t}dt = (j\omega)^r F(\omega)$$

Thus,

$$\frac{d^r f(t)}{dt^r} \Leftrightarrow (j\omega)^r F(\omega) \tag{3.44}$$

3.3.4 Convolution Integral

Consider two functions $f(t)$ and $g(t)$ that appear in the following integral,

$$f(t) * g(t) = \int_{-\infty}^{\infty} f(t)g(\tau - t)dt = \int_{-\infty}^{\infty} g(t)f(\tau - t)dt \tag{3.45}$$

---

* Recall that

$$\int_a^b f(x)\frac{dg(x)}{dx}dx = f(x)g(x)\Big|_a^b - \int_a^b g(x)\frac{df(x)}{dx}dx$$

Equation (3.45) is called the *convolution integral* and, as indicated in Eq. (3.45), is often denoted as $f(t)*g(t)$. Taking the Fourier transform of Eq. (3.45), we obtain

$$\int_{-\infty}^{\infty} f(t)*g(t)e^{-j\omega\tau}d\tau = \int_{-\infty}^{\infty}\left[\int_{-\infty}^{\infty} f(t)g(\tau-t)dt\right]e^{-j\omega\tau}d\tau = \int_{-\infty}^{\infty}\int_{-\infty}^{\infty} f(t)g(\tau-t)e^{-j\omega\tau}dtd\tau$$

$$= \int_{-\infty}^{\infty} f(t)\left[\int_{-\infty}^{\infty} g(\tau-t)e^{-j\omega\tau}d\tau\right]dt = \int_{-\infty}^{\infty} f(t)e^{-j\omega t}G(\omega)dt$$

$$= G(\omega)\int_{-\infty}^{\infty} f(t)e^{-j\omega t}dt = G(\omega)F(\omega)$$

where we have used Eq. (3.38). Therefore

$$\int_{-\infty}^{\infty} g(t)f(\tau-t)dt \Leftrightarrow G(\omega)F(\omega) \tag{3.46}$$

or using Eq. (3.34), the inverse Fourier transform of the right-hand side of Eq. (3.46) is

$$\int_{-\infty}^{\infty} g(t)f(\tau-t)dt = \frac{1}{2\pi}\int_{-\infty}^{\infty} G(\omega)F(\omega)e^{j\omega\tau}d\omega \tag{3.47}$$

The convolution integral is often used in the following manner. When the inverse transform of, say, $H(\omega)$ is not known, sometimes it is possible to express $H(\omega)$ as the product of two functions such that $H(\omega)=F(\omega)G(\omega)$, where the inverse transforms $f(t)$ and $g(t)$, respectively, of $F(\omega)$ and $G(\omega)$ are known. Then the integration on the left-hand side of Eq. (3.47) provides the inverse transform of $H(\omega)$.

We now introduce the *converse* of Eq. (3.45) with the definition

$$F(\omega)*G(\omega) = \frac{1}{2\pi}\int_{-\infty}^{\infty} F(\omega)G(\Omega-\omega)d\omega \tag{3.48}$$

Using Eq. (3.34), the inverse transform of Eq. (3.48) is

$$\frac{1}{2\pi}\int_{-\infty}^{\infty}\left[F(\omega)*G(\omega)\right]e^{j\Omega t}d\Omega = \frac{1}{(2\pi)^2}\int_{-\infty}^{\infty}\left[\int_{-\infty}^{\infty} F(\omega)G(\Omega-\omega)d\omega\right]e^{j\Omega t}d\Omega$$

$$= \frac{1}{2\pi}\int_{-\infty}^{\infty} F(\omega)\left[\frac{1}{2\pi}\int_{-\infty}^{\infty} G(\Omega-\omega)e^{j\Omega t}d\Omega\right]d\omega$$

$$= \frac{1}{2\pi}\int_{-\infty}^{\infty} F(\omega)e^{j\omega t}g(t)d\omega = \frac{g(t)}{2\pi}\int_{-\infty}^{\infty} F(\omega)e^{j\omega t}d\omega$$

$$= f(t)g(t)$$

where we have used Eq. (3.39). Therefore,

$$\frac{1}{2\pi}\int_{-\infty}^{\infty} F(\omega)G(\Omega-\omega)d\omega \Leftrightarrow f(t)g(t) \tag{3.49}$$

In other words, the Fourier transform of the right-hand side of Eq. (3.49) is

$$\int_{-\infty}^{\infty} f(t)g(t)e^{-j\Omega t}dt = \frac{1}{2\pi}\int_{-\infty}^{\infty} F(\omega)G(\Omega-\omega)d\omega \tag{3.50}$$

Parseval's theorem for the Fourier transform (recall the theorem for the Fourier series given by Eq. (3.17)) is given by

$$\int_{-\infty}^{\infty} |h(t)|^2\, dt = \frac{1}{2\pi}\int_{-\infty}^{\infty} |H(\omega)|^2\, d\omega \tag{3.51}$$

Equation (3.51) is also known as the Rayleigh energy theorem.* We prove Eq. (3.51) as follows

$$\begin{aligned}\int_{-\infty}^{\infty} |h(t)|^2\, dt &= \int_{-\infty}^{\infty} h(t)\overline{h(t)}dt = \int_{-\infty}^{\infty} h(t)\overline{\left[\frac{1}{2\pi}\int_{-\infty}^{\infty} H(\omega)e^{j\omega t}d\omega\right]}dt \\ &= \frac{1}{2\pi}\int_{-\infty}^{\infty} h(t)\left[\int_{-\infty}^{\infty} \overline{H(\omega)}e^{-j\omega t}d\omega\right]dt \\ &= \frac{1}{2\pi}\int_{-\infty}^{\infty} \overline{H(\omega)}\left[\int_{-\infty}^{\infty} h(t)e^{-j\omega t}dt\right]d\omega \\ &= \frac{1}{2\pi}\int_{-\infty}^{\infty} \overline{H(\omega)}H(\omega)d\omega = \frac{1}{2\pi}\int_{-\infty}^{\infty} |H(\omega)|^2\, d\omega\end{aligned} \tag{3.52}$$

Equation (3.51) suggests that the transformation of energy is conserved.

Example 3.5

Consider a pulse that has an amplitude A_p for a duration t_d. The pulse begins at t_1 and ends at t_1+t_d. Then, from Eq. (3.34), the Fourier transform of this pulse is

$$F(\omega) = A_p\int_{t_1}^{t_1+t_d} e^{-j\omega t}dt = t_d A_p\left(\frac{\sin(t_d\omega/2)}{t_d\omega/2}\right)e^{-j(t_1+t_d/2)\omega} \tag{a}$$

---

* If we assume that $f(t) = E(t)$ where $E(t)$ is the instantaneous voltage and P is the instantaneous power in a resistor R, then $P = E^2(t)/R$ W. Thus, the argument of the integral $f^2(t)$ on the left-hand side of Eq. (3.51) can be thought of as being proportional to power, and since it is multiplied by dt, which has the units of time (s), the units of the integral's argument are Ws (= J), which is energy.

We note that

$$|F(\omega)| = t_d A_p \left| \frac{\sin(t_d\omega/2)}{t_d\omega/2} \right| \tag{b}$$

It is seen that $|F(\omega)|_{max} = t_d A_p$, which occurs at $\omega = 0$.

We shall use Eqs. (a) and (b) to confirm Parseval's theorem given by Eq. (3.51), that is,

$$E_t = \int_{-\infty}^{\infty} |f(t)|^2\, dt = \frac{1}{2\pi} \int_{-\infty}^{\infty} |F(\omega)|^2\, d\omega = E_\omega$$

Thus,

$$E_t = \int_{t_1}^{t_1+t_d} A_p^2 dt = t_d A_p^2$$

$$E_\omega = \frac{1}{2\pi} \int_{-\infty}^{\infty} \left| t_d A_p \frac{\sin(t_d\omega/2)}{t_d\omega/2} \right|^2 d\omega = \frac{t_d A_p^2}{\pi} \int_{-\infty}^{\infty} \frac{\sin^2 x}{x^2} dx = t_d A_p^2$$

and, therefore, $E_t = E_\omega$. The integral in the second equation is determined with Mathematica procedure M3.8.

Equation (b) is evaluated with Mathematica procedure M3.9, and the result is shown in Figure 3.9.

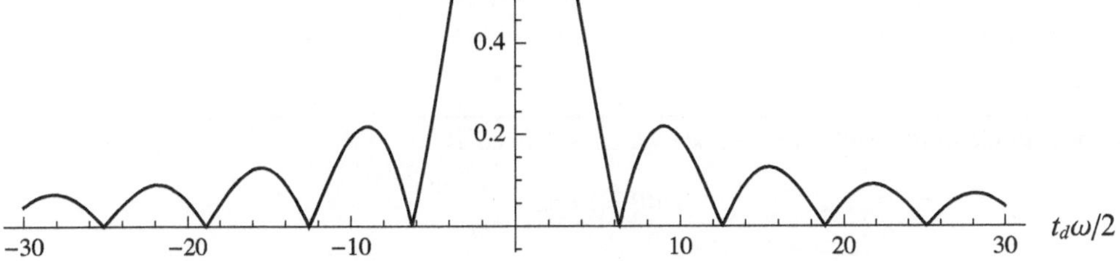

FIGURE 3.9 Normalized Fourier transform of a pulse of duration t_d.

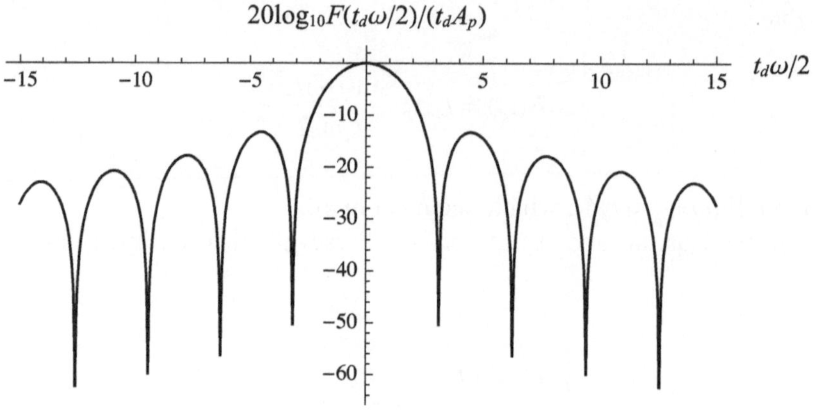

FIGURE 3.10 The absolute value of the Fourier transform of a pulse shown in Figure 3.9 plotted in decibels (dB).

We now re-plot Eq. (b) as*

$$20\log_{10}\left|\frac{F(\omega)}{t_d A_p}\right| = 20\log_{10}\left|\frac{\sin(t_d\omega/2)}{t_d\omega/2}\right| \text{ dB} \tag{c}$$

using Mathematical procedure M3.10. The result is shown in Figure 3.10. The implication of this result will become apparent in Section 3.3.4.

Before leaving this example, it is worthwhile to compare these results to that of the pulse train of Example 3.2, where it was found that

$$\left|\frac{c_n}{\alpha A_p}\right| = \left|\frac{\sin(\alpha n\pi)}{\alpha n\pi}\right| \quad \text{where} \quad \alpha = \frac{t_d}{T} \tag{d}$$

In arriving at Eq. (d), we have noted that $2t_o$ of Example 3.2 equals t_d of this example. As was stated in Section 3.3.1, one should expect that as $T \to \infty$ ($\alpha \to 0$), the envelope of the magnitude of the coefficients of the Fourier series should approach the spectrum of that of the single pulse. We have plotted Eq. (b) and Eq. (d) as functions of α on the same graph in Figure 3.11, which was obtained with Mathematica procedure M3.11. We see that this is indeed the case.

3.3.5 Delta Function

The Fourier transform of the delta function† is

* This logarithmic form is called the decibel, which is defined as

$$\text{dB} = 20\log_{10}\frac{A}{A_{\text{ref}}}$$

where A is the amplitude of interest that is being compared to a reference amplitude A_{ref}.

† See Appendix B.

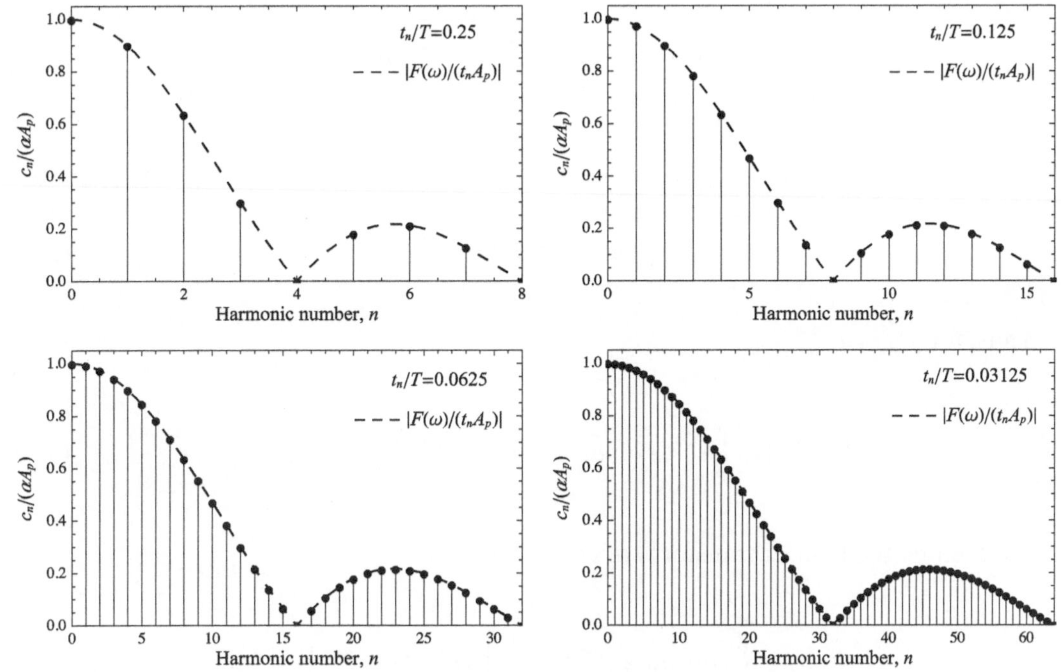

FIGURE 3.11 The coefficients of the Fourier series of a periodic pulse train determined in Example 3.2 for $\alpha = t_d/T$ are plotted along with the Fourier transform of a single pulse determined in Example 3.5. The magnitudes of the Fourier series coefficients are shown with the filled circles and the Fourier transforms by the continuous dashed lines.

$$F(\omega) = \int_{-\infty}^{\infty} \delta(t - t_o) e^{-j\omega t} dt = e^{-j\omega t_o}$$

which implies that

$$\delta(t - t_o) = \frac{1}{2\pi} \int_{-\infty}^{\infty} e^{j\omega(t - t_o)} d\omega$$

Thus,

$$\delta(t - t_o) \Leftrightarrow e^{-j\omega t_o} \tag{3.53}$$

Notice that $|F(\omega)| = 1$, that is, the frequency spectrum of the delta function is constant for $-\infty < \omega < \infty$.

Also,

$$f(t) = \frac{1}{2\pi} \int_{-\infty}^{\infty} \delta(\omega - \omega_o) e^{j\omega t} d\omega = \frac{1}{2\pi} e^{j\omega_o t}$$

which implies that

$$\delta(\omega-\omega_o)=\frac{1}{2\pi}\int_{-\infty}^{\infty}e^{-j(\omega-\omega_o)t}\,dt$$

Hence,

$$\frac{1}{2\pi}e^{j\omega_o t}\Leftrightarrow\delta(\omega-\omega_o) \tag{3.54}$$

Thus, for $f(t)=e^{j\omega_o t}$

$$F[e^{j\omega_o t}]=\int_{-\infty}^{\infty}e^{j\omega_o t}e^{-j\omega t}\,dt=\int_{-\infty}^{\infty}e^{-j(\omega-\omega_o)t}\,dt=2\pi\delta(\omega-\omega_o) \tag{3.55}$$

and, therefore, the Fourier transform of $\cos(\omega_o t)$ is

$$\begin{aligned}F(\omega)&=\int_{-\infty}^{\infty}\cos(\omega_o t)e^{-j\omega t}\,dt=\frac{1}{2}\int_{-\infty}^{\infty}\left(e^{j\omega_o t}+e^{-j\omega_o t}\right)e^{-j\omega t}\,dt\\&=\frac{1}{2}\int_{-\infty}^{\infty}e^{j\omega_o t}e^{-j\omega t}\,dt+\frac{1}{2}\int_{-\infty}^{\infty}e^{-j\omega_o t}e^{-j\omega t}\,dt\\&=\pi\delta(\omega-\omega_o)+\pi\delta(\omega+\omega_o)\end{aligned} \tag{3.56}$$

We are now able to determine the Fourier transform of a Fourier series given by Eq. (3.1) as

$$\begin{aligned}F(\omega)&=\sum_{n=-\infty}^{\infty}c_n\int_{-\infty}^{\infty}e^{jn\omega_o t}e^{-j\omega t}\,dt=\sum_{n=-\infty}^{\infty}c_n\int_{-\infty}^{\infty}e^{-j(\omega-n\omega_o)t}\,dt\\&=2\pi\sum_{n=-\infty}^{\infty}c_n\delta(\omega-n\omega_o)\end{aligned} \tag{3.57}$$

where $\omega_o=2\pi/T$ and we have used Eq. (3.54).

The Fourier transform pairs derived in Sections 3.3.3 to 3.3.5 are listed in Table 3.1.

3.4 FOURIER TRANSFORM AND SIGNAL ANALYSIS

In this section, we shall present the use of the Fourier transform in the context of signal analysis and signal processing. It will be seen that when the independent variable is time and the transformed parameter is frequency, the Fourier series and the Fourier transform lead to some very useful results that have become the basis of digital data acquisition systems and the subsequent manipulation of the acquired data.

TABLE 3.1
Fourier Transform Pairs

Pair	$g(t)$	$G(\omega)$ [$f(t) \Leftrightarrow F(\omega)$]
1	$g(t-t_o)$	$e^{-j\omega t_o}G(\omega)$
2	$e^{j\omega_o t}g(t)$	$G(\omega-\omega_o)$
3	$g(at)$	$\frac{1}{\lvert a\rvert}G(\omega / a)$
4	$\int_{-\infty}^{\infty} g(\tau)f(t-\tau)d\tau$	$G(\omega)F(\omega)$
5	$f(t)g(t)$	$\frac{1}{2\pi}\int_{-\infty}^{\infty} F(\Omega)G(\omega-\Omega)d\Omega$
6	$\delta(t-t_o)$	$e^{-j\omega t_o}$
7	$\frac{1}{2\pi}e^{j\omega_o t}$	$\delta(\omega-\omega_o)$
8	$\frac{d^r f(t)}{dt^r}$	$(j\omega)^r F(\omega)$

Note: Additional Fourier transform pairs, when they exist, can be determined with the Mathematica command `FourierTransform`.

3.4.1 Sampling

Consider a signal whose Fourier transform $F(\omega)$ is such that

$$\begin{aligned} f(t) &\Leftrightarrow F(\omega) \quad &|\omega| \le \omega_c \\ &\Leftrightarrow 0 \quad &|\omega| > \omega_c \end{aligned} \tag{3.58}$$

Thus, $F(\omega)$ essentially has been truncated in the frequency domain by a rectangular windowing function $w(\omega)$ so that $F(\omega)$ is in fact $F(\omega)w(\omega)$ where $w(\omega) = u(\omega+\omega_c) - u(\omega-\omega_c)$ and $u(\omega)$ is the unit step function. Hence, the time-domain signal is determined from

$$f(t) = \frac{1}{2\pi}\int_{-\infty}^{\infty} F(\omega)e^{j\omega t}d\omega = \frac{1}{2\pi}\int_{-\omega_c}^{\omega_c} F(\omega)e^{j\omega t}d\omega \tag{3.59}$$

If we choose to sample $f(t)$ every $t = t_s = n/f_s = 2\pi n/(2\omega_c)$ where $2\pi f_s = 2\omega_c$ and $n = \pm 1, \pm 2, \pm 3, \ldots$, then Eq. (3.59) becomes

$$f(t_s = n\Delta) = \frac{1}{2\pi}\int_{-\omega_c}^{\omega_c} F(\omega)e^{j2n\pi\omega/(2\omega_c)}d\omega \tag{3.60}$$

where $\Delta = 2\pi/(2\omega_c)$ is called the *maximum sampling interval* and, therefore, $f_s = 1/\Delta$ is the *minimum sampling frequency*. We shall now use Eqs. (3.31) and (3.30) as follows. Upon comparing Eq. (3.31) with Eq. (3.60), it is seen that

$$2\omega_c d_{-n} = 2\pi f(n\Delta)$$

or

$$d_n = \frac{\pi}{\omega_c} f(-n\Delta) \tag{3.61}$$

Substituting Eq. (3.30) into Eq. (3.59) and using Eq. (3.61) gives

$$\begin{aligned} f(t) &= \frac{1}{2\pi}\int_{-\omega_c}^{\omega_c}\sum_{n=-\infty}^{\infty} d_n e^{j2n\pi\omega/(2\omega_c)} e^{j\omega t}\, d\omega = \frac{1}{2\omega_c}\sum_{n=-\infty}^{\infty} f(-n\Delta)\int_{-\omega_c}^{\omega_c} e^{j(t+n\pi/\omega_c)\omega}\, d\omega \\ &= \sum_{n=-\infty}^{\infty} f(-n\Delta)\frac{\sin(\omega_c t + n\pi)}{\omega_c t + n\pi} \\ &= \sum_{n=-\infty}^{\infty} f(n\Delta)\frac{\sin(\omega_c t - n\pi)}{\omega_c t - n\pi} \end{aligned} \tag{3.62}$$

Thus, based on Eq. (3.62), $f(t)$ can be reconstructed by its samples spaced no farther apart than Δ. The quantity $f_c = \omega_c/(2\pi)$ is called the *cutoff frequency* of the spectrum.

3.4.2 Aliasing

Consider the sampling of a single-frequency cosine signal of frequency f_o, that is, $f(t) = \cos(2\pi f_o t)$, $0 < f_o < f_c$, where f_c is the cutoff frequency. This signal is sampled every $t = nt_s = n/f_s = n/(2f_c)$. In the previous section, it was noted that $2f_c$ is the minimum sampling frequency. Then

$$f(nt_s) = \cos(2\pi n f_o t_s) = \cos(2\pi n f_o / f_s) \tag{3.63}$$

We vary f_o such that $f_o = mf_s \pm f$, where $0 < f < f_c$, $m = 1, 2, 3, \ldots$, where it is assumed that f_s is fixed and f varies. Then, Eq. (3.63) becomes

$$\begin{aligned} f(nt_s) &= \cos(2\pi n(mf_s \pm f)/f_s) = \cos(2\pi nm \pm 2\pi nf/f_s) \\ &= \cos(2\pi nm)\cos(\pm 2\pi nf/f_s) + \sin(2\pi nm)\sin(\pm 2\pi nf/f_s) \\ &= \cos(2\pi nf/f_s) \end{aligned} \tag{3.64}$$

Thus, from Eqs. (3.63) and (3.64) it is seen that f is indistinguishable from f_o, and we say that f_o has been aliased with f, that is, $f = f_a$. There is no way to remove the aliased signal. For example, if $f_c = 100$ Hz ($f_s = 2f_c$) and $f = 30$ Hz, then the frequencies $f_o = 200 \pm 30$, 400 ± 30, etc. are aliased with $f = 30$ Hz. This phenomenon is shown in Figure 3.12. In practice,

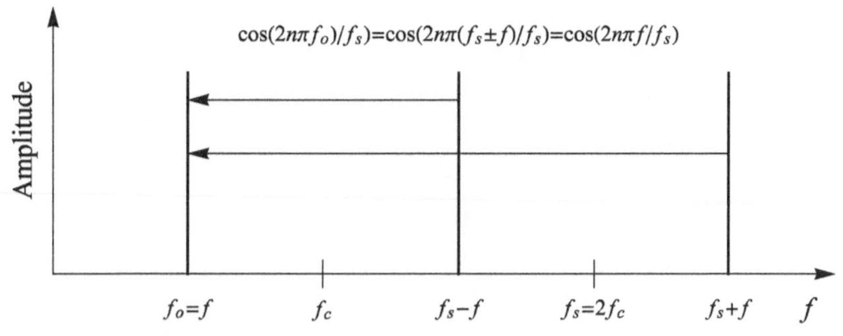

FIGURE 3.12 Representation of aliased frequency components of $f_o = f_s \pm f$.

aliasing is minimized with a combination of filters whose cutoff frequency is f_c and by selecting a sampling frequency somewhat greater than the minimum.

To illustrate aliasing, we select a signal of frequency $f_o = 2$ Hz and greatly under-sample it by using a sampling frequency $f_s = 8f_o/7$. Then, for $m = 1$ and since $f_o = f_s \pm f \to f = f_a = f_s - f_o = 8f_o/7 \pm f_o$, one aliased frequency is $f_o/7$. We use these values and Eq. (3.62) to reconstruct the waveform with Mathematica procedure M3.12. The results are shown in Figure 3.13. A sufficient number of terms to use in Eq. (3.62) was determined empirically to be the smallest integer greater than $3/t_s$.

Another way to look at aliasing is again to consider the spectrum shown in Figure 3.7 except this time we choose the spectrum frequency ω_c to be that shown in Figure 3.14a. Here it is seen that we have improperly selected the spectrum's cutoff frequency because it does not include that portion of the spectrum that lies beyond ω_c. In other words, we have incorrectly chosen the interval that makes $F(\omega)$ periodic. Thus, when the spectrum is folded about ω_c, a portion of the spectrum that should not have been there appears. This portion is shown with the shaded triangle and contains the aliased frequencies. One can prevent these aliased frequencies by choosing a different ω_c, denoted ω'_c, such that $\omega'_c = \alpha\omega_c$ where $\alpha > 1$

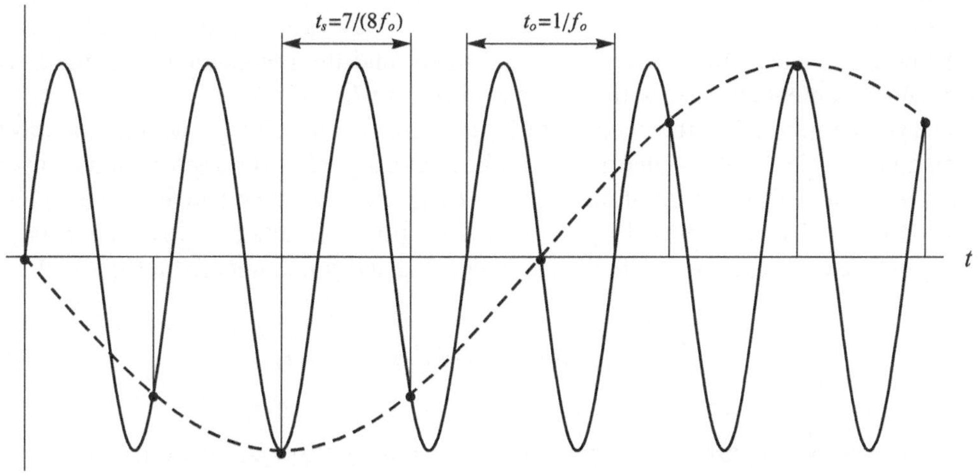

FIGURE 3.13 Effects of aliasing. The filled circles represent the data points that were acquired at a sampling frequency that was very much less than the minimum required. The solid line is the original signal, and the dashed line is the aliased signal.

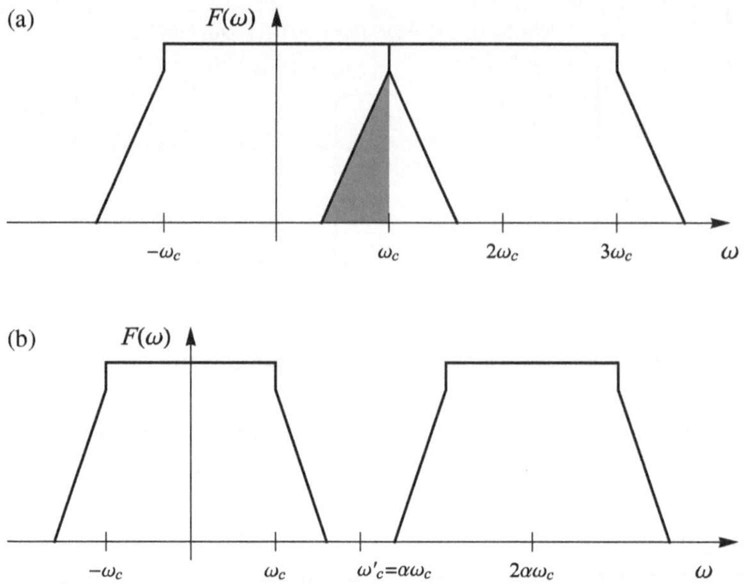

FIGURE 3.14 (a) Aliased spectrum due to insufficient sampling rate as indicated by the folding of the spectra shown as a shaded region; (b) spectrum sampled at a sufficient sampling rate so that aliasing does not occur.

is sufficiently large so that the folded spectrum resides outside (does not overlap) the original spectrum. This is shown in Figure 3.14b. Recall the discussion regarding Figure 3.7.

3.4.3 Short-Time Fourier Transform (STFT)

We shall now examine the practical aspects of the Fourier transform in the context of signal analysis. In signal analysis, the independent variable is time and the dependent variable is amplitude. However, the information that is hidden in the time-domain signal is the frequency content. This frequency content is usually revealed by the Fourier transform. There are two types of signals: stationary signals where on the average the frequency content remains constant each time the signal is examined, and the non-stationary signals where the frequency content changes with time (e.g. chirp, traffic noise).

We start by introducing the short-time Fourier transform (STFT), which is a way that one can analyze pieces of a signal $f(t)$ by looking at a signal of duration t_D and examining adjacent segments of the signal of duration $t_d = t_D/K$, where K is a positive integer. The segment of the signal is determined by a windowing function denoted $w(t)$. The window is typically zero except over a specified time interval, usually t_d. Then, from Eq. (3.49),

$$\frac{1}{2\pi}\int_{-\infty}^{\infty} F(\omega)W(\Omega-\omega)d\omega \Leftrightarrow f(t)w(t) \tag{3.65}$$

is a Fourier transform pair. Let us consider the case of continuous single frequency signal of frequency ω_o expressed in the form $f(t) = e^{j\omega_o t}$. From Eq. (3.54), the Fourier transform of this signal is

$$F(\omega) = 2\pi\delta(\omega - \omega_o) \tag{3.66}$$

Therefore, from Eqs. (3.65) and (3.66), we find that

$$W(\omega - \omega_o) \Leftrightarrow e^{j\omega_o t} w(t) \tag{3.67}$$

where after integration we have set $\Omega = \omega$. From Eq. (3.66), it is seen that the spectrum of a continuous single frequency signal without the window is zero everywhere except at $\omega = \omega_o$. When a window function is used, the frequency spectrum is equal to the frequency spectrum of the window. In other words, the windowing function has introduced frequencies that are not in the original waveform.

The consequence of this is significant. Consider the simplest window function, called the rectangular window, that is given in the time domain by

$$w(t) = 1 - u(t - t_d) \tag{3.68}$$

where $u(t)$ is the unit step function. Using Eq. (3.34), we have that the Fourier transform of $w(t)$ is

$$W(\omega) = \int_0^{t_d} e^{-j\omega t} dt = \frac{1}{\omega}\left(\sin \omega t_d - j\left(1 - \cos \omega t_d\right)\right) \tag{3.69}$$

and, therefore,

$$\left|W(\omega)\right| = \frac{\sqrt{2}}{\omega}\sqrt{1 - \cos \omega t_d} = t_d \left|\frac{\sin\left(\omega t_d / 2\right)}{\omega t_d / 2}\right| \tag{3.70}$$

which agrees with Eq. (b) of Example 3.4 when $A_p = 1$.

We now set the duration of the window $t_d = 2N\pi/\omega_o$, where N is an integer, that is, the duration of the window is N periods. Then, Eq. (3.70) becomes

$$\left|\frac{W(\omega)}{t_d}\right| = \left|\frac{\sin(N\pi\omega/\omega_o)}{N\pi\omega/\omega_o}\right| \tag{3.71}$$

The first maximum of Eq. (3.71) occurs at $\omega/\omega_o = 0$. Using Mathematica procedure M3.14, it is found that the second maximum of Eq. (3.71) is 0.217 which occurs at $N\pi\omega/\omega_o = 4.493$. Thus, irrespective of the value of N, the maximum value remains at 0.217; however, the value of ω/ω_o decreases as N increases. Thus, as the duration of the time domain sample is increased the frequency resolution improves. However, the magnitudes of the local maxima do not change, only their location.

We express Eq. (3.71) in dB and write Eq. (3.71) as

$$20\log_{10}\left|\frac{W(\omega)}{t_d}\right| = 20\log_{10}\left|\frac{\sin(N\pi\omega/\omega_o)}{N\pi\omega/\omega_o}\right| \tag{3.72}$$

It is noted that the second maximum of Eq. (3.71) in dB is $20\log_{10}(0.217) = -13.3$ dB. Equation (3.71) is plotted in Figure 3.15, which was obtained with Mathematica procedure M3.14.

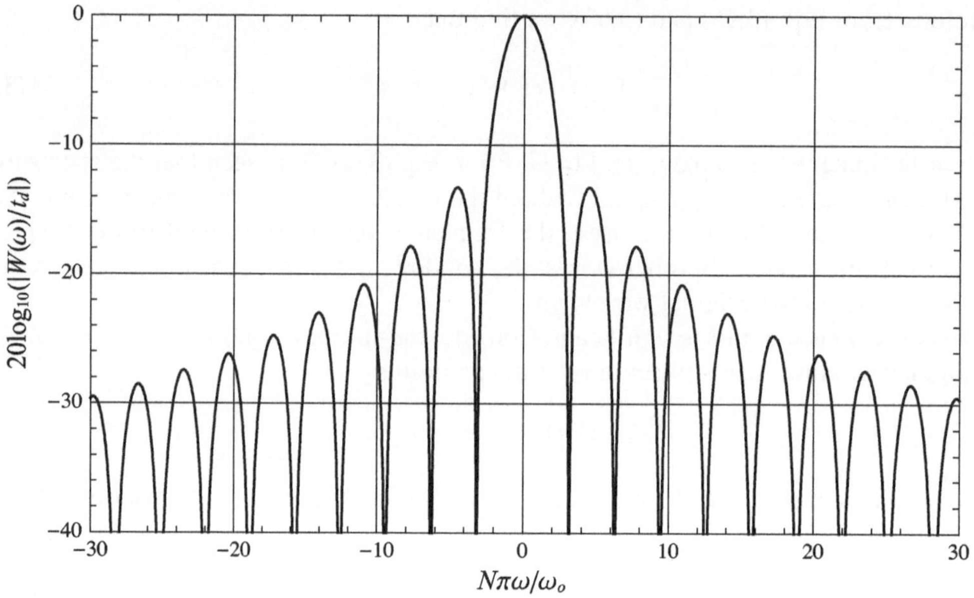

FIGURE 3.15 Fourier transform of N periods of a sine wave of frequency ω_o using a rectangular window function.

3.4.4 Windowing: The Hamming Window

There are numerous windowing functions that have been proposed.* We shall discuss the Hamming window, which is defined as

$$\begin{aligned} w_H(t) &= a - b\cos(2\pi t/T_H) \quad 0 \le t \le T_H \\ &= 0 \qquad \text{otherwise} \end{aligned} \tag{3.73}$$

where $a=0.54$ and $b=0.46$ and T_H is the truncation window duration. Equation (3.73) is shown in Figure 3.16 for $T_H=20\pi$ and was obtained with Mathematica procedure M3.15. The effect of the Hamming window on a sine wave is shown in Figure 3.17 and was obtained with Mathematica procedure M3.16. The maximum value of w_H occurs at a time $t=t_{\max}$ when $2\pi t_{\max}/T_H = \pi$ or $t_{\max}=T_H/2$. At this time, $w_H(t_{\max})=a+b$. The Fourier transform of Eq. (3.73) is

$$\begin{aligned} F_H(\omega) &= \int_0^{T_H} \left[a - b\cos(2\pi t/T_H)\right] e^{-j\omega t}\,dt \\ &= \frac{4a\pi^2 + (b-a)T_H^2\omega^2}{\omega\left(4\pi^2 - T_H^2\omega^2\right)}\left(\sin\omega T_H + j\left(\cos\omega T_H - 1\right)\right) \end{aligned}$$

* See, for example: https://en.wikipedia.org/wiki/Window_function and https://reference.wolfram.com/language/guide/WindowFunctions.html.

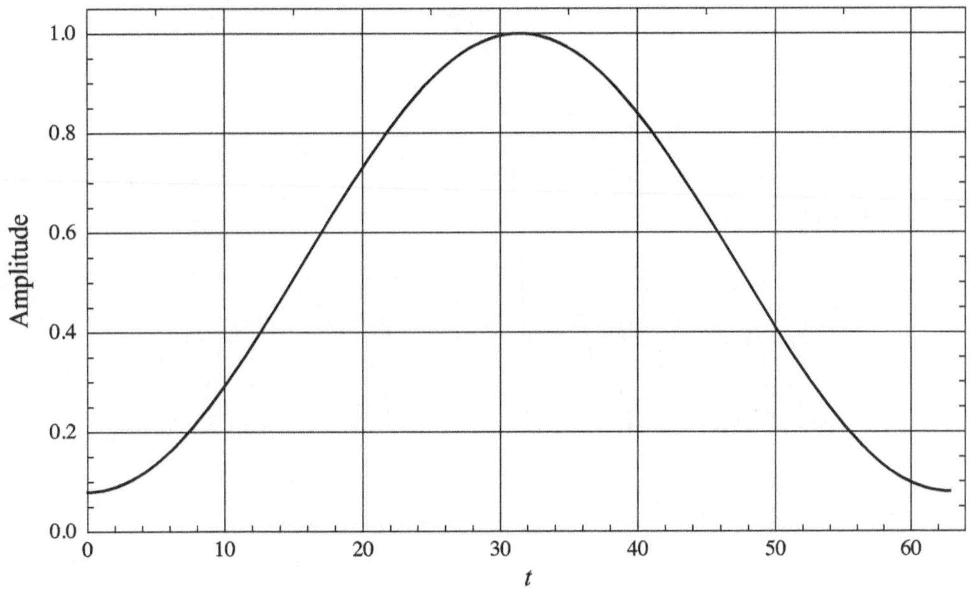

FIGURE 3.16 Hamming window for $T_H = 20\pi$.

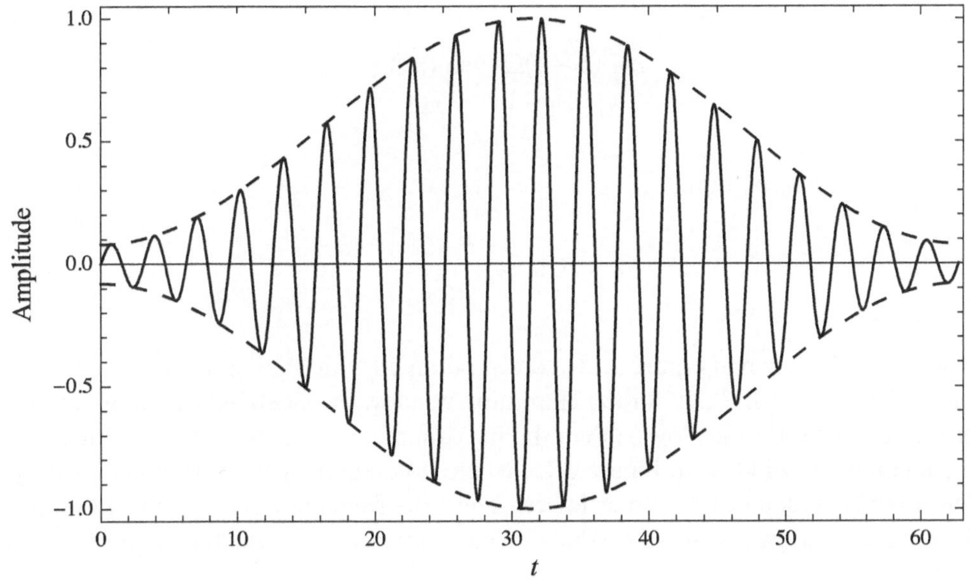

FIGURE 3.17 Sine wave of frequency 2 rad/s modified by a Hamming window for $T_H = 20\pi$.

and, therefore,

$$\left|F_H(\Omega)\right| = \frac{\sqrt{2}}{\omega_H}\left|\frac{a+(b-a)\Omega^2}{\Omega\left(1-\Omega^2\right)}\right|\sqrt{\left(1-\cos 2\pi\Omega\right)} \tag{3.74}$$

where $\Omega = \omega/\omega_H$ and $\omega_H = 2\pi/T_H$. It is noted that $\left|F_H(0)\right| = aT_H$. Thus, we can normalize Eq. (3.74) to obtain

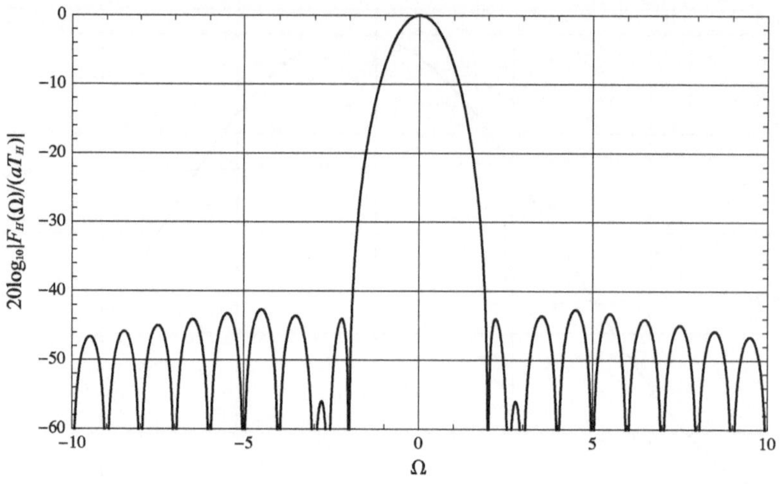

FIGURE 3.18 Normalized Fourier transform of the Hamming window.

$$\frac{|F_H(\Omega)|}{aT_H} = \frac{\sqrt{2}}{a\omega_H T_H}\left|\frac{a+(b-a)\Omega^2}{\Omega\left(1-\Omega^2\right)}\right|\sqrt{\left(1-\cos 2\pi\Omega\right)}$$
$$= \left|\frac{a+(b-a)\Omega^2}{a\left(1-\Omega^2\right)}\right|\left|\frac{\sin(\pi\Omega)}{\pi\Omega}\right| \tag{3.75}$$

which can be expressed in dB as

$$20\log_{10}\frac{|F_H(\Omega)|}{aT_H} \tag{3.76}$$

Equation (3.76) is shown in Figure 3.18 and was obtained with Mathematica procedure M3.17. It is noticed in Figure 3.18 that for the Hamming window the sidebands are more than 40 dB lower (more than 100 times lower) than the maximum value. Compare this with the results for the rectangular window in Figure 3.15, where it is seen that the sidebands are only 13.3 dB lower (about a factor of 4.6 times lower). Thus, the frequency components introduced by the Hamming window are greatly reduced when compared with the rectangular window.

MATHEMATICA PROCEDURES

Note: In order to make the Mathematica procedures that follow a little more readable, the `Clear` statement and many of the graphics-enhancing statements have been omitted.

```
(*M3.1*)
Bb=Sequence[FourierParameters-&gt;{1,-2 Pi}];
Simplify[ComplexExpand[FourierCoefficient[
    Piecewise[{{-1,-1/2<=t<-1/4},{1,-1/4<=t<1/4},{-1,1/4<=t<1/2}}],
    t,n,Bb]],Assumptions->n∈Integers]
```

```
(*M3.2*)
c[t_]:=4/Pi Total[Table[Cos[2 n Pi t-(n-1) Pi/2]/n,{n,1,105,2}]]
Plot[c[t],{t,-0.5,0.5}]

(*M3.3*)
cc[t_]:=4/Pi Total[Table[(-1)^((n-1)/2) Cos[2 n Pi t]/n,
   {n,1,3005,2}]]
FindMaximum[Abs[cc[t]],{t,0.2498,0.2496,0.2501}]
Plot[cc[t],{t,0.249,0.251}]

(*M3.4*)
ComplexExpand[FourierCoefficient[(UnitStep[t+td]-UnitStep[t-td]),
   t,n,FourierParameters->{1,(2 Pi)/T},Assumptions->T>0&&td>0
   &&td<T/2]]/.td->a T/2

(*M3.5*)
cb[t_,a_]:=a (1+2 Total[Table[Sin[a n Pi]*Cos[2 n Pi t]/(a n Pi),
   {n,1,101}]])
Plot[cb[t,0.4],{t,-0.5,0.5}]

(*M3.6*)
PowerExpand[ComplexExpand[FourierCoefficient[DiracDelta[t],t,n,
   FourierParameters->{1,2 Pi/T}]]]

(*M3.7*)
a=0.6;
fg[x_,nn_]:=a (1+2 Total[Table[Sinc[n a Pi] Cos[Pi n x],
   {n,1,nn}]])
Plot[fg[x,100],{x,-1,1}]

(*M3.8*)
Integrate[Sin[x]^2/x^2,{x,-Infinity,Infinity}]

(*M3.9*)
Plot[Abs[Sinc[x/2]],{x,-30,30},PlotRange->All]

(*M3.10*)
Plot[20 Log10[Abs[Sinc[x/2]]],{x,-30,30},PlotRange->All]

(*M3.11*)
ptt[to_]:=Show[ListPlot[Table[{n,Abs[Sinc[n Pi to]]},
   {n,0,2/to}],Filling->Axis],
   Plot[Abs[Sinc[x to Pi]],{x,0,2/to}]]
GraphicsGrid[{{ptt[0.5],ptt[0.25]},{ptt[0.125],ptt[0.0625]}}]

(*M3.12*)
fo=2.;    ts=7/(8 fo);    nmax=Ceiling[3/ts];
sam=Table[{n ts,Sin[2 n Pi fo ts]},{n,0,nmax}];
Show[Plot[Sin[2 Pi fo t],{t,0,nmax ts}],
   ListPlot[sam,Filling->Axis],
   Plot[-Sin[2 Pi fo/7 t],{t,0,nmax ts}]]
```

```
(*M3.13*)
FindMaximum[Abs[Sin[x]/x],{x,4}]

(*M3.14*)
Plot[20. Log10[Abs[Sinc[x]]],{x,-30,30},Frame->True,
   GridLines->Automatic,PlotRange->{{-30,30},{0,-40}}]

(*M3.15*)
a=0.54; b=0.46; to=20 Pi;
Plot[a-b Cos[2 Pi t/to],{t,0,to}]

(*M3.16*)
g[Om_,a_,b_]:=Abs[(a+(-a+b) Om^2) Sin[Pi Om]/(Pi Om (1-Om^2) a)]
a=0.54; b=0.46; r=10; to=20 Pi; n=20;
Plot[{Sin[2 n Pi t/to] (a-b Cos[2 Pi t/to]),(a-b Cos[2 Pi t/to]),
   -(a-b Cos[2 Pi t/to])},{t,0,to}]

(*M3.17*)
g[Om_,a_,b_]:=Abs[(a+(-a+b) Om^2) Sin[Pi Om]/(Pi Om (1-Om^2) a)]
a=0.54;    b=0.46;    to=20 Pi;
Plot[20 Log10[g[x,a,b]],{x,-10,10},PlotRange->{{-10,10},{0,-60}}]
```

EXERCISES

Section 3.1

3.1 Referring to Figure E3.1, consider a sine wave of unit magnitude and frequency ω_b that has a duration $t_b = Nt_p$, where $t_p = 2\pi/\omega_b$ and N is an integer. This sine wave is repeated every period $T = Mt_p = 2\pi/\omega_o$, $M \geq N$, and M is an integer. Determine the Fourier coefficients given by Eq. (3.19) and express them in terms of n, N, and M.

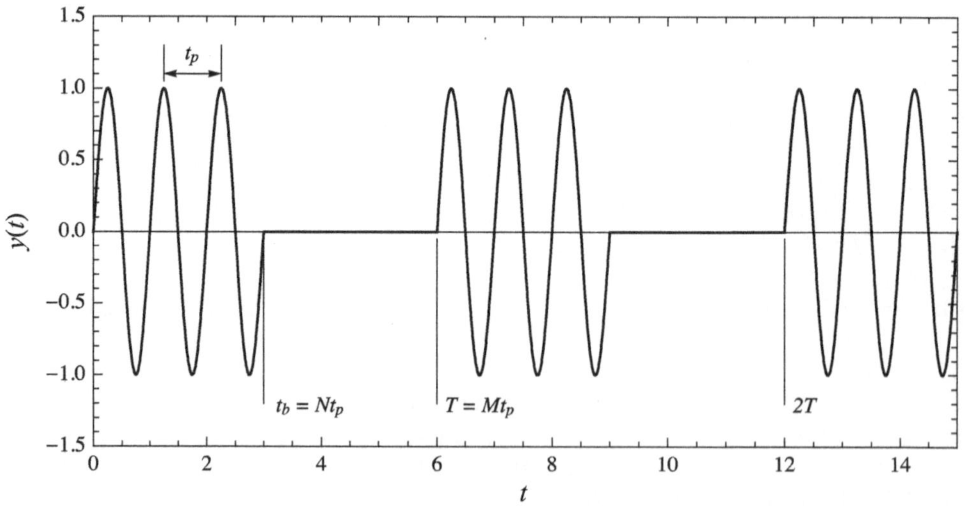

FIGURE E3.1 Periodic finite duration sine wave.

It is suggested that the following quantities be introduced prior to performing the integrations

$$\tau = \omega_o t, \quad \omega_b = M\omega_o, \quad \omega_o t_b = \frac{2\pi N\omega_o}{\omega_b} = \frac{2\pi N}{M}, \quad \text{and} \quad T\omega_o = 2\pi$$

3.2 Expand the following periodic function in complex notation, and use the results to express the series as a real quantity.

$$f(x) = |x| \quad -\pi < x < \pi$$

3.3 (a) Obtain the Fourier series expansion of the following periodic function

$$f(t) = t^2 \quad -\pi < t < \pi$$

(b) Using the results in (a), choose appropriate values for t and show that

$$\sum_{n=1}^{\infty} \frac{(-1)^{n+1}}{n^2} = \frac{\pi^2}{12}$$

$$\sum_{n=1}^{\infty} \frac{1}{n^2} = \frac{\pi^2}{6}$$

(c) Add the results in part (b) and show that (see also Exercise 3.2)

$$\sum_{n=1,3,5}^{\infty} \frac{1}{n^2} = \frac{\pi^2}{8}$$

3.4 Obtain the Fourier series expansion of the following periodic function

$$f(\tau) = t^4 \quad -\pi < t < \pi$$

and choose appropriate values for t to show that

$$\sum_{n=1}^{\infty} \frac{(-1)^{n+1}}{n^4} = \frac{7\pi^4}{720}$$

$$\sum_{n=1}^{\infty} \frac{1}{n^4} = \frac{\pi^4}{90}$$

3.5 Determine the Fourier series expansion for the periodic function $f(t) = 2A_o t/T$, $-T/2 \le t \le T/2$ using Eqs. (3.2) and (3.14) and Eqs. (3.18) and (3.19).

3.6 Determine the Fourier series expansion for $f(t) = A_o e^{-\alpha t/T}$, $\alpha > 0$, over the interval $-T/2 \le t \le T/2$.

3.7 For the wave forms shown in Figure E3.2, use Eq. (3.2) to determine the magnitude and phase of the Fourier coefficients c_n.

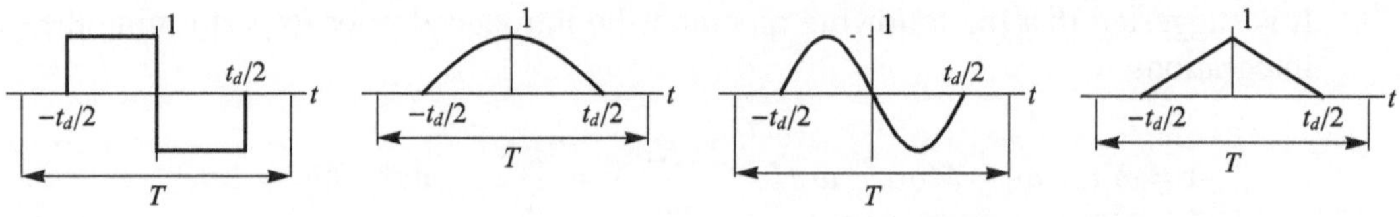

FIGURE E3.2 One period of a periodic waveform: (a) square wave; (b) half sine wave; (c) full sine wave; and (d) triangular wave.

SECTION 3.3

3.8 Find the Fourier transform of the following waveform

$$f(t) = h(1+at) \qquad -1/a \le t \le 0$$
$$= h(1-at) \qquad 0 \le t \le 1/a$$
$$= 0 \qquad |t| > 1/a$$

What happens when $h = a$ and $a \to \infty$? What does this result signify?

3.9 Determine the Fourier transform of the following expression

$$f(\tau) = \frac{\omega_n^2 e^{-\zeta\omega_n\tau}}{\omega_d} \sin(\omega_d\tau) u(\tau)$$

where

$$\omega_d = \omega_n\sqrt{1-\zeta^2}$$

Present your final results in terms of the non-dimensional quantity $\Omega = \omega/\omega_n$.

3.10 Find the Fourier transform of the following waveform

$$f(\tau) = e^{-\alpha|\tau|} \quad \alpha > 0$$

3.11 Find the Fourier transform of the following waveform

$$f(\tau) = \cos\alpha t \qquad -1 < t < 1$$
$$f(\tau) = 0 \qquad \text{otherwise}$$

3.12 For the wave forms shown in Figure E3.3, use Eq. (3.34) to determine the magnitude of their Fourier transforms. Compare these results with the corresponding results obtained in Exercise 3.7.

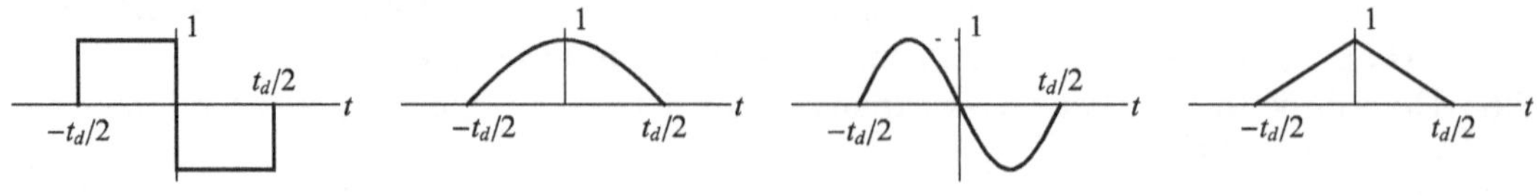

FIGURE E3.3 Pulse waveforms: (a) square wave; (b) half sine wave; (c) full sine wave; and (d) triangular wave.

SECTION 3.4.6

3.13 Consider the pulse

$$f(t) = A_p\left\{\left[u(t) - u(t - t_p)\right] - \left[u(t - t_p) - u(t - 2t_p)\right]\right\}$$

where t_p is the duration of the positive and the negative portions of the pulse and A_p is the magnitude of each portion.

(a) Determine an expression for the magnitude of the Fourier transform of this pulse when a rectangular window function is used, that is, one with no weighting function. The results can be normalized to $2A_pt_p$.

(b) Determine an expression for the magnitude of the Fourier transform of this pulse when a Hamming window function is used.

4 Ordinary Differential Equations Part I

Review of First- and Second-Order Equations

4.1 FIRST-ORDER ORDINARY DIFFERENTIAL EQUATIONS

In this section, we summarize the solution methods for several classes of first-order ordinary differential equations, which play a role in obtaining some general results for the Sturm–Liouville equation discussed Section 6.1.2 and some special nonlinear equations discussed in Section 4.2.10.

4.1.1 Special Cases of First-Order Ordinary Differential Equations

The simplest first-order ordinary differential equation is

$$\frac{dy}{dx} = F(x) \tag{4.1}$$

whose solution is

$$y = \int F(x)dx + C \tag{4.2}$$

for any $F(x)$ that is integrable.

Equation (4.1) can be generalized as one of two equivalent forms. The first form is

$$\frac{dy}{dx} = F(x,y) \tag{4.3}$$

and the second form is

$$N(x,y)\frac{dy}{dx} + M(x,y) = 0 \tag{4.4}$$

or

$$N(x,y)dy + M(x,y)dx = 0 \tag{4.5}$$

We shall now examine several special cases of Eqs. (4.3) to (4.5). When

$$\begin{aligned} N(x,y) &= P_1(x)P_2(y) \\ M(x,y) &= Q_1(x)Q_2(y) \end{aligned} \tag{4.6}$$

the differential equation is said to be *separable* and we can re-write Eq. (4.5) as

$$\frac{P_2(y)}{Q_2(y)}dy+\frac{Q_1(x)}{P_1(x)}dx=0$$

which can be integrated to obtain

$$\int\frac{Q_1(x)}{P_1(x)}dx+\int\frac{P_2(y)}{Q_2(y)}dy=C \tag{4.7}$$

where the constant C is determined from the boundary/initial condition.

We now illustrate this result with the following example.

Example 4.1

Consider the following equation

$$\frac{1}{2x}\frac{dy}{dx}-\frac{y}{x^2-1}=0$$

which can be written as

$$\frac{dy}{y}-\frac{2xdx}{x^2-1}=0$$

Therefore,

$$\int\frac{dy}{y}=\int\frac{2x}{x^2-1}dx$$

$$\ln y=\ln\left(x^2-1\right)+C$$

$$\ln\frac{y}{x^2-1}=C$$

$$y=k\left(x^2-1\right)\quad\left[k=e^C\right]$$

This result is verified with Mathematica procedure M4.1.

In some cases, it may be necessary to use a change of variable to bring an equation to separable form. This technique is described in the next example.

Example 4.2

Consider the equation

$$\frac{dy}{dx}=\frac{x^3y}{x^4+y^4}=\frac{y/x}{1+(y/x)^4} \tag{a}$$

We select the change of variable as $v = y/x$ ($y = xv$). Then

$$\frac{dy}{dx} = x\frac{dv}{dx} + v$$

and Eq. (a) becomes

$$x\frac{dv}{dx} + v = \frac{v}{1+v^4}$$

$$x\frac{dv}{dx} = -\frac{v^5}{1+v^4}$$

$$\frac{\left(1+v^4\right)}{v^5}dv = -\frac{dx}{x}$$

Therefore,

$$\int \frac{1+v^4}{v^5}dv = -\int \frac{dx}{x}$$

$$-\frac{1}{4v^4} + \ln v = -\ln x + C$$

$$4\ln vx = \frac{1}{v^4} + 4C$$

$$\ln(v^4 x^4) = \frac{1}{v^4} + 4C$$

or

$$(vx)^4 = e^{1/v^4 + 4C}$$

$$y^4 = ke^{x^4/y^4}$$

which is an implicit solution. This result is verified with Mathematica procedure M4.2.

The ordinary differential equation given by Eq. (4.5) is said to be *exact* if there exists a function $u(x,y)$ for which

$$\frac{\partial u}{\partial x} = M(x,y)$$
$$\frac{\partial u}{\partial y} = N(x,y) \tag{4.8}$$

Substituting Eq. (4.8) into Eq. (4.5) gives

$$\frac{\partial u}{\partial x}dx + \frac{\partial u}{\partial y}dy = 0 \tag{4.9}$$

which is the definition of the total (exact) differential of u. Thus,

$$du = 0 \text{ or } u = C \tag{4.10}$$

It is noted from Eq. (4.8) that

$$\frac{\partial N}{\partial x} = \frac{\partial^2 u}{\partial x \partial y} = \frac{\partial M}{\partial y}$$

Hence, for the original differential equation to have an exact form, the following must be satisfied

$$\frac{\partial M}{\partial y} = \frac{\partial N}{\partial x} \tag{4.11}$$

It is also seen from the first equation of Eq. (4.8) that

$$u(x, y) = \int M(x, y)dx + f(y) \tag{4.12}$$

where $f(y)$ is a constant with respect to x and is determined from the solution to the second equation of Eq. (4.8), that is,

$$\frac{\partial u}{\partial y} = N(x, y) \tag{4.13}$$

This case is illustrated with the following example.

Example 4.3

Consider the following equation

$$\left(y + 3x^3y^2\right)\frac{dy}{dx} + 3x^2y^3 - 5x^4 = 0 \tag{a}$$

which can be written as

$$\left(y + 3x^3y^2\right)dy + \left(3x^2y^3 - 5x^4\right)dx = 0$$

Therefore,

$$M(x, y) = 3x^2y^3 - 5x^4$$

$$N(x, y) = y + 3x^3y^2$$

and since

$$\frac{\partial M}{\partial y} = 9x^2y^2$$

$$\frac{\partial N}{\partial x} = 9x^2y^2$$

this equation is exact. Hence, from Eq. (4.12),

$$u(x,y) = \int \left(3x^2y^3 - 5x^4\right)dx + f(y)$$

$$= x^3y^3 - x^5 + f(y)$$

From Eq. (4.13), it is found that

$$3x^3y^2 + \frac{df}{dy} = y + 3x^3y^2$$

Therefore,

$$\frac{df}{dy} = y$$

which upon integration gives

$$f = \frac{y^2}{2} + k$$

Then

$$u(x,y) = x^3y^3 - x^5 + \frac{y^2}{2} + k$$

and, from Eq. (4.10),

$$x^3y^3 - x^5 + \frac{y^2}{2} = C_1 \qquad \text{(b)}$$

which is a cubic equation in y.

Equation (b) is verified with Mathematica procedure M4.3.

As a check, we differentiate Eq. (b) with respect to x and obtain

$$3x^2y^3 + 3x^3y^2\frac{dy}{dx} - 5x^4 + y\frac{dy}{dx} = 0$$

$$\left(3x^2y^3 - 5x^4\right)dx + \left(y + 3x^3y^2\right)dy = 0$$

which agrees with Eq. (a).

When the equation is not exact, sometimes it can be made exact by multiplying it by a function $\rho(x,y)$. If $\rho(x,y)$ does indeed make the equation exact, then $\rho(x,y)$ is called an *integrating factor* and

$$\rho(x,y)N(x,y)\frac{dy}{dx} + \rho(x,y)M(x,y) = 0 \qquad (4.14)$$

is an exact equation.

There are three types of general integrating factors that we shall consider.

Type 1: $\rho(x,y) = x^m y^n$

For this form of integrating factor, Eq. (4.14) becomes

$$x^m y^n N(x,y)\frac{dy}{dx} + x^m y^n M(x,y) = 0 \tag{4.15}$$

For this equation to be exact, we use Eq. (4.11) and find that

$$\frac{d}{dx}\left(x^m y^n N(x,y)\right) = \frac{d}{dy}\left(x^m y^n M(x,y)\right) \tag{4.16}$$

We use Eq. (4.16) to determine the values of m and n, as shown in the next example.

Example 4.4

Consider the equation

$$(1 - xy)\frac{dy}{dx} + y^2 + 3xy^3 = 0 \tag{a}$$

Hence

$$\begin{aligned} M(x,y) &= y^2 + 3xy^3 \\ N(x,y) &= 1 - xy \end{aligned} \tag{b}$$

and, therefore,

$$\frac{\partial M}{\partial y} = 2y + 9xy^2$$

$$\frac{\partial N}{\partial x} = -y$$

Thus, the equation is not exact.

In an attempt to obtain an exact solution, we employ Eq. (4.15), which results in

$$(x^m y^n - x^{m+1} y^{n+1})\frac{dy}{dx} + x^m y^{n+2} + 3x^{m+1} y^{n+3} = 0$$

Equation (b) now becomes

$$M(x,y) = x^m y^{n+2} + 3x^{m+1} y^{n+3}$$

$$N(x,y) = x^m y^n - x^{m+1} y^{n+1}$$

Therefore, from Eq. (4.16), it is found that

$$\left(n + m + 3\right)x^m y^{n+1} + 3(n+3)x^{m+1} y^{n+2} - mx^{m-1} y^n = 0$$

In order to satisfy this equation for arbitrary x and y, the coefficients of x and y must equal zero, that is,

$$m+n+3=0$$

$$3(n+3)=0$$

$$m=0$$

Hence, $m=0$ and $n=-3$ and the integrating factor is $\rho(x,y)=y^{-3}$. Therefore, Eq. (a) is made exact by multiplication of this function, and we have that

$$(y^{-3}-xy^{-2})\frac{dy}{dx}+y^{-1}+3x=0 \tag{c}$$

To verify that Eq. (c) is exact, we note that

$$M(x,y)=y^{-1}+3x$$

$$N(x,y)=y^{-3}-xy^{-2}$$

and, therefore,

$$\frac{\partial M}{\partial y}=-\frac{1}{y^2}$$

$$\frac{\partial N}{\partial x}=-\frac{1}{y^2}$$

To obtain a solution to Eq. (c), we use Eq. (4.12) to arrive at

$$u=\int\left(y^{-1}+3x\right)dx+f(y)=\frac{x}{y}+\frac{3}{2}x^2+f(y)$$

From Eq. (4.13), we find that

$$-xy^{-2}+\frac{df}{dy}=y^{-3}-xy^{-2}$$

or

$$\frac{df}{dy}=\frac{1}{y^3}$$

and, therefore,

$$f=-\frac{1}{2y^2}+C_0$$

Then, the solution is

$$u(x,y)=\frac{x}{y}+\frac{3}{2}x^2-\frac{1}{2y^2}+C_0=C_1$$

or

$$\left(3x^2 - 2C_2\right)y^2 + 2xy - 1 = 0 \tag{d}$$

which is a quadratic equation in y.

Equation (d) is verified with Mathematica procedure M4.4.

As a check, we divide Eq. (d) by y^2 and differentiate the result with respect to x to obtain

$$6x + 2y^{-1} - 2xy^{-2}\frac{dy}{dx} + 2y^{-3}\frac{dy}{dx} = 0$$

$$\left(3xy^3 + y^2\right)dx + \left(1 - xy\right)dy = 0$$

which agrees with Eq. (a).

Type 2: $\rho(x, y) = \rho(x)$

Using this form of integrating factor, Eq. (4.14) becomes

$$\rho(x)N(x,y)dy + \rho(x)M(x,y)dx = 0 \tag{4.17}$$

For this equation to be exact

$$\frac{\partial}{\partial x}\left(\rho(x)N(x,y)\right) = \frac{\partial}{\partial y}\left(\rho(x)M(x,y)\right)$$

$$\frac{d\rho(x)}{dx}N(x,y) + \rho(x)\frac{\partial N(x,y)}{\partial x} = \rho(x)\frac{\partial M(x,y)}{\partial y} \tag{4.18}$$

$$\frac{1}{\rho(x)}\frac{d\rho(x)}{dx} = g(x)$$

where

$$g(x) = \frac{1}{N(x,y)}\left(\frac{\partial M(x,y)}{\partial y} - \frac{\partial N(x,y)}{\partial x}\right) \tag{4.19}$$

Upon integrating Eq. (4.18), we obtain

$$\rho(x) = \exp\left(\int g(x)dx\right) \tag{4.20}$$

where the constant of integration has been set to zero.

Special Case of Type 2

As a special case of Type 2, consider the equation

$$\frac{dy}{dx} + p(x)y = q(x) \tag{4.21}$$

Therefore, from Eq. (4.4),

$$M(x,y) = p(x)y - q(x)$$

$$N(x,y) = 1$$

Hence,

$$\frac{\partial M}{\partial y} = p(x)$$

$$\frac{\partial N}{\partial x} = 0$$

Since the equation is not exact, we use the integrating factor determined from Eq. (4.18) where it is seen from Eq. (4.19) that $g(x) = p(x)$. Therefore, Eq. (4.20) becomes

$$\rho(x) = \exp\left(\int p(x)dx\right) \tag{4.22}$$

The original equation, Eq. (4.21), now can be written as

$$\begin{aligned}
\rho(x)\frac{dy}{dx} + \rho(x)p(x)y &= \rho(x)q(x) \\
\rho(x)\frac{dy}{dx} + \frac{d\rho}{dx}y &= \rho(x)q(x) \\
\frac{d}{dx}\left(\rho(x)y\right) &= \rho(x)q(x) \\
y &= \frac{1}{\rho(x)}\left[\int \rho(x)q(x)dx + C\right]
\end{aligned} \tag{4.23}$$

where in the first equation of Eq. (4.23) we have used Eq. (4.18), that is,

$$\frac{1}{\rho(x)}\frac{d\rho(x)}{dx} = g(x) = p(x)$$

Upon substituting Eq. (4.22) into Eq. (4.23), we obtain

$$y = \exp\left[-\int p(x)dx\right]\left[\int\left\{\exp\left[\int p(x)dx\right]q(x)\right\}dx + C\right] \tag{4.24}$$

Equation (4.24) is verified with Mathematica procedure M4.5.
Note that when $q(x) = 0$, Eq. (4.23) simplifies to

$$y = \frac{C}{\rho(x)}$$

which is the solution to the homogeneous differential equation. The particular solution is

$$y = \frac{1}{\rho(x)}\int \rho(x)q(x)dx$$

For the case where $p(x)=b_o$, a constant, we find that $g(x)=b_o$. Then Eq. (4.24) becomes

$$y=\exp\left[-\int b_o dx\right]\left[\int\left\{\exp\left[\int b_o dx\right]q(x)\right\}dx+C\right] \tag{4.25}$$

$$=e^{-b_o x}\left[\int e^{b_o x}q(x)dx+C\right]=Ce^{-b_o x}+e^{-b_o x}\int e^{b_o x}q(x)dx$$

This result will be used in Example 7.12.

This case is illustrated with the following example.

Example 4.5

Consider the equation

$$2xe^{2y}\frac{dy}{dx}=3x^4+e^{2y} \tag{a}$$

To convert this equation into a more amenable form, we use the transformation $v=e^{2y}$. Hence,

$$\frac{dv}{dx}=\frac{dv}{dy}\frac{dy}{dx}=2e^{2y}\frac{dy}{dx}$$

and, therefore,

$$\frac{dy}{dx}=\frac{e^{-2y}}{2}\frac{dv}{dx}$$

Then, Eq. (a) becomes

$$2xe^{2y}\left(\frac{e^{-2y}}{2}\frac{dv}{dx}\right)=3x^4+v$$

$$\frac{dv}{dx}-\frac{v}{x}=3x^3 \tag{b}$$

From Eq. (b), we see that

$$N(x,v)=1$$

$$M(x,v)=-\frac{v}{x}-3x^3$$

and, therefore,

$$\frac{\partial M(x,v)}{\partial v}=-\frac{1}{x}$$

$$\frac{\partial N(x,v)}{\partial x}=0$$

This result indicates that the equation is not exact.

Hence, we seek an integrating factor, which in this case is obtained from Eq. (4.18). Thus,

$$\int \frac{d\rho(x)}{\rho(x)} = -\int \frac{dx}{x}$$

$$\ln \rho = -\ln x$$

$$\rho = \frac{1}{x}$$

and Eq. (b) becomes

$$\frac{1}{x}\frac{dv}{dx} - \frac{v}{x^2} = 3x^2 \tag{c}$$

Equation (c) is of the form given by Eq. (4.21) with

$$p(x) = -\frac{1}{x}$$

$$q(x) = 3x^3$$

Thus, Eq. (4.23) applies giving

$$v = x\int \frac{3x^3}{x}dx + Cx = x^4 + Cx$$

Hence,

$$e^{2y} = x^4 + Cx$$

or

$$y = \frac{1}{2}\ln\left(x^4 + Cx\right)$$

This result is verified with Mathematica procedure M4.6.

Type 3: $\rho(x, y) = \rho(y)$

In this case, we use the procedure to arrive at Eq. (4.18) and obtain

$$\frac{1}{\rho(y)}\frac{d\rho(y)}{dy} = \frac{1}{M(x,y)}\left(\frac{\partial N(x,y)}{\partial x} - \frac{\partial M(x,y)}{\partial y}\right) \tag{4.26}$$

The results of this section are summarized in Table 4.1.

4.1.2 Bernoulli Equation

The Bernoulli equation is a first-order nonlinear differential equation of the form

$$\frac{dy}{dx} + p(x)y = q(x)y^n \quad n \neq 1 \tag{4.27}$$

TABLE 4.1
Summary of Solutions to Classes of First-Order Differential Equations

Case	Type	Differential Equation	Solution
1	Separable	$P_1(x)P_2(y)dy + Q_1(x)Q_2(y)dx = 0$	$\int \frac{Q_1(x)}{P_1(x)}dx + \int \frac{P_2(y)}{Q_2(y)}dy = C$
2	Exact: $\frac{\partial M}{\partial y} = \frac{\partial N}{\partial x}$	$N(x,y)dy + M(x,y)dx = 0$	$u(x,y) = \int M(x,y)dx + f(y)$ where $f(y)$ is determined from $\frac{\partial u}{\partial y} = N(x,y)$
3	Exact: determine m and n so that $\frac{d}{dx}\left(x^m y^n N(x,y)\right) = \frac{d}{dy}\left(x^m y^n M(x,y)\right)$	$x^m y^n N(x,y)dy + x^m y^n M(x,y)dx = 0$	$u(x,y) = y^n \int x^m M(x,y)dx + f(y)$ where $f(y)$ is determined from $\frac{\partial u}{\partial y} = x^m y^n N(x,y)$
4	Exact: determine $\rho(x)$, the integrating factor	$\rho(x)N(x,y)dy + \rho(x)M(x,y)dx = 0$ where $\rho(x) = \exp\left(\int g(x)dx\right)$ $g(x) = \frac{1}{N(x,y)}\left(\frac{\partial M(x,y)}{\partial y} - \frac{\partial N(x,y)}{\partial x}\right)$	$u(x,y) = \int \rho(x)M(x,y)dx + f(y)$ where $f(y)$ is determined from $\frac{\partial u}{\partial y} = \rho(x)N(x,y)$
5	Exact: determine $\rho(x)$, the integrating factor	$\rho(x)\frac{dy}{dx} + \rho(x)p(x)y = \rho(x)q(x)$ where $\rho(x) = \exp\left[\int p(x)dx\right]$	$y = \exp\left[-\int p(x)dx\right]\left[\int\left\{\exp\left[\int p(x)dx\right]q(x)\right\}dx + C\right]$

where n need not be an integer. We have already discussed the solutions for $n=0$, which is given in Eq. (4.21) and $n=1$, which is a particular case of Eq. (4.7) with $P_1=P_2=Q_2=1$ and $Q_1=p(x)-q(x)$.

We change the dependent variable to

$$u = y^{1-n} \tag{4.28}$$

Then,

$$\frac{du}{dx} = \frac{du}{dy}\frac{dy}{dx} = (1-n)y^{-n}\frac{dy}{dx}$$

and

$$\frac{dy}{dx} = \frac{1}{1-n}y^n\frac{du}{dx}$$

Substitution of this result into Eq. (4.27) yields

$$\frac{1}{1-n}\frac{du}{dx} + p(x)u = q(x)$$

where we have used Eq. (4.28). If we let

$$P(x) = (1-n)p(x)$$

$$Q(x) = (1-n)q(x)$$

then Eq. (4.27) becomes

$$\frac{du}{dx} + P(x)u = Q(x)$$

the solution of which is given by Eqs. (4.22) and (4.23), that is,

$$\begin{aligned} \rho(x) &= \exp\left[\int P(x)dx\right] = \exp\left[(1-n)\int p(x)dx\right] \\ u(x) &= \frac{1}{\rho(x)}\left[\int \rho(x)Q(x)dx + C\right] = \frac{1}{\rho(x)}\left[(1-n)\int \rho(x)q(x)dx + C\right] \end{aligned} \tag{4.29}$$

and, from Eq. (4.28), we obtain

$$y = u^{1/(1-n)} \tag{4.30}$$

The solution to the Bernoulli equation given by Eqs. (4.29) and (4.30) is verified with Mathematica procedure M4.7.

For the special case where $q(x)=q_o$ and $p(x)=p_o$, Eq. (4.29) simplifies to

$$\rho(x) = \exp\left[(1-n)\int p_o dx\right] = e^{(1-n)p_o x}$$

and

$$u = e^{(n-1)p_o x}\left[(1-n)q_o \int e^{(1-n)p_o x}dx + C\right] = e^{(n-1)p_o x}\left[\frac{(1-n)q_o e^{(1-n)p_o x}}{p_o(1-n)} + C\right]$$

$$= \frac{q_o}{p_o} + Ce^{(n-1)p_o x}$$

Therefore, Eq. (4.30) becomes

$$y = \left[\frac{q_o}{p_o} + Ce^{(n-1)p_o x}\right]^{1/(1-n)}$$

4.1.3 Direction Fields

To investigate the behavior of first-order differential equations of the form given by Eq. (4.3), we note that a geometric interpretation of Eq. (4.3) is that $m = dy/dx = F(x,y)$ is the slope at the point (x,y). One can then say that the solution to Eq. (4.3) is a graph of a curve having a slope m at each (x,y); that is, at each (x,y) there is a line that is tangent to the curve with a slope m. If, at each (x,y), one constructs a short line segment with a slope m, this line segment is an approximate solution to Eq. (4.3) in the vicinity of this point. Using an appropriate procedure, one can connect these short tangent line segments to obtain a direction field, which represents the solutions to Eq. (4.3). The details of the procedure are omitted since these are standard capabilities of such programs as Mathematica.

To illustrate the direction field, we plot Eq. (a) of Example 4.3 in Figure 4.1 and superimpose upon it the solution given by Eq. (b) of this example for $C_1 = 500$. The superimposed solution is shown with the thick line. This figure was created with Mathematica procedure M4.8.

4.2 SECOND- AND HIGHER-ORDER ORDINARY DIFFERENTIAL EQUATIONS

4.2.1 Introduction

In this section, we shall introduce methods to obtain the solutions to several special cases of the following system of linear ordinary differential equations

$$a_{j,0}(x)\frac{d^n y_j}{dx^n} + a_{j,1}(x)\frac{d^{n-1}y_j}{dx^{n-1}} + \cdots + a_{j,n-1}(x)\frac{dy_j}{dx} + a_{j,n}(x)y_j = Q_j(x) \quad j = 1,2,\ldots,K \tag{4.31}$$

Equation (4.31) is called a nonhomogeneous ordinary differential equation with non-constant coefficients. We shall be concerned primarily with the case when $n = 2$ and $K = 1$. When $a_j(x) = b_j$, a constant, Eq. (4.31) is called an ordinary differential with constant coefficients. When $Q_j(x) = 0$, Eq. (4.31) is called a homogeneous ordinary differential equation. The solution to Eq. (4.31) consists of the sum of two solutions. One solution is the solution

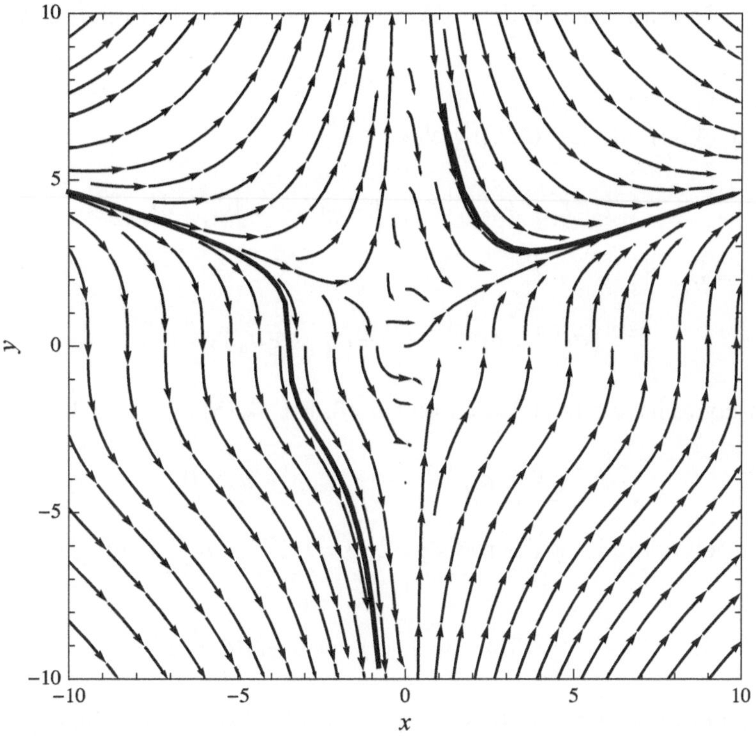

FIGURE 4.1 Direction fields for Eq. (a) of Example 4.3 and the solution given by Eq. (b) for $C_1 = 500$, which is represented by the thick line.

to the homogeneous equation and will often be denoted $y_h(x)$. The other solution, called the particular solution, is the solution when $Q_j(x) \neq 0$ and will often be denoted $y_p(x)$.

When $n = 2$ and $a_j(x)$ takes on certain forms, the resulting differential equation are given names such as Cauchy–Euler (Section 4.2.4), Bessel (Section 5.1.4), Legendre (Section 5.1.9), hypergeometric (Section 5.1.10), and Sturm–Liouville (Section 6.1.2). Obtaining a solution to these equations requires the introduction of several solution techniques. In this chapter, we shall present the variation of parameters and matrix methods. In Chapter 5, we shall introduce power series and the method of Frobenius, in Chapter 6 we shall introduce the Sturm–Liouville system and the generation and application of orthogonal functions, and in Chapter 8 we shall introduce the Laplace transform method.

4.2.2 Homogeneous Differential Equations with Constant Coefficients

Consider the following nth-order homogeneous differential equation with constant coefficients

$$a_0 \frac{d^n y}{dx^n} + a_1 \frac{d^{n-1} y}{dx^{n-1}} + \cdots + a_{n-1} \frac{dy}{dx} + a_n y = 0 \tag{4.32}$$

Equation (4.32) also can be written as

$$a_0 \left(\frac{d}{dx} - \lambda_1 \right) \left(\frac{d}{dx} - \lambda_2 \right) \cdots \left(\frac{d}{dx} - \lambda_{n-1} \right) \left(\frac{d}{dx} - \lambda_n \right) y = 0 \tag{4.33}$$

where the λ_k are determined as follows. We assume a solution to Eq. (4.32) of the form

$$y = e^{\gamma x} \tag{4.34}$$

Substituting Eq. (4.34) into Eq. (4.32) yields

$$\left(a_0\gamma^n + a_1\gamma^{n-1} + \cdots + a_{n-1}\gamma + a_n\right)e^{\gamma x} = 0 \tag{4.35}$$

or, using Eq. (4.34),

$$\left(a_0\gamma^n + a_1\gamma^{n-1} + \cdots + a_{n-1}\gamma + a_n\right)y = 0 \tag{4.36}$$

Thus, in order not to have a trivial solution ($y=0$), Eq. (4.35) or Eq. (4.36) is satisfied when

$$a_0\gamma^n + a_1\gamma^{n-1} + \cdots + a_{n-1}\gamma + a_n = 0 \tag{4.37}$$

This nth-order polynomial has the roots $\gamma=\lambda_k$, $k=1, 2, \ldots, n$. Then Eq. (4.37) can be written as

$$a_0(\gamma - \lambda_1)(\gamma - \lambda_2)\ldots(\gamma - \lambda_{n-1})(\gamma - \lambda_n) = 0 \tag{4.38}$$

In essence, Eq. (4.37) or (4.38) is the characteristic equation with λ_j the eigenvalue and the corresponding solution e^{λ_j} the eigenfunction. Notice that when Eq. (4.34) is substituted into Eq. (4.33) we arrive at Eq. (4.38).

If all the λ_k are distinct (they can be real, imaginary, or complex), then the solution to Eq. (4.32) is

$$y = \sum_{k=1}^{n} A_k e^{\lambda_k x} \tag{4.39}$$

where the A_k are determined from the boundary/initial conditions. Implied in Eq. (4.39) is that the linear combination of each independent solution to Eq. (4.32) is also a solution to Eq. (4.32). This linear combination of solutions to a linear homogeneous equation is referred to as the *superposition principle.* We shall use this superposition principle extensively in Chapter 7.

If the jth root is repeated k times, then Eq. (4.33) becomes

$$\begin{aligned} a_0&\left(\frac{d}{dx} - \lambda_1\right)\left(\frac{d}{dx} - \lambda_2\right)\cdots\left(\frac{d}{dx} - \lambda_{j-1}\right)\left(\frac{d}{dx} - \lambda_j\right)^k \\ &\times\left(\frac{d}{dx} - \lambda_{j+k+1}\right)\cdots\left(\frac{d}{dx} - \lambda_n\right)y = 0 \end{aligned} \tag{4.40}$$

To obtain the "missing solutions" in Eq. (4.40), we have to find the solution to

$$\left(\frac{d}{dx} - \lambda_j\right)^k y = 0 \tag{4.41}$$

We assume a trial solution of the form

$$y_j = x^r e^{\lambda_j x} \tag{4.42}$$

where r is to be determined. Substituting Eq. (4.42) into Eq. (4.41) and noting that

$$\left(\frac{d}{dx} - \lambda_j\right) x^r e^{\lambda_j x} = r x^{r-1} e^{\lambda_j x} + x^r \lambda_j e^{\lambda_j x} - x^r \lambda_j e^{\lambda_j x} = r x^{r-1} e^{\lambda_j x}$$

and

$$\left(\frac{d}{dx} - \lambda_j\right)^2 x^r e^{\lambda_j x} = r\left(\frac{d}{dx} - \lambda_j\right) x^{r-1} e^{\lambda_j x}$$

$$= r\left((r-1)x^{r-2} e^{\lambda_j x} + x^{r-1}\lambda_j e^{\lambda_j x} - x^{r-1}\lambda_j e^{\lambda_j x}\right)$$

$$= r(r-1)x^{r-2} e^{\lambda_j x}$$

we arrive at

$$\left(\frac{d}{dx} - \lambda_j\right)^k x^r e^{\lambda_j x} = r(r-1)(r-2)\ldots(r-k+2)(r-k+1)x^{r-k} e^{\lambda_j x} = 0$$

This equation is satisfied for $r = 0, 1, \ldots, k-1$. Thus, from Eq. (4.42)

$$y_{j+r} = x^r e^{\lambda_j x} \quad r = 0, 1, 2\ldots, k-1$$

and the solution for an equation with repeated roots is

$$y = \sum_{m=1}^{j-1} A_m e^{\lambda_m x} + \left(\sum_{p=0}^{k-1} A_{j+p} x^p\right) e^{\lambda_j x} + \sum_{m=j+k}^{n} A_m e^{\lambda_m x} \tag{4.43}$$

where j refers to the index of the repeated root.

We now illustrate these results with several examples.

Example 4.6

Consider the equation

$$\frac{d^3 y}{dx^3} - 3\frac{d^2 y}{dx^2} + 4y = 0 \tag{a}$$

We assume a solution of the form $y = e^{\lambda x}$ and substitute it into Eq. (a) to find

$$\lambda^3 - 3\lambda^2 + 4 = 0$$

The roots of this polynomial are $\lambda_1 = -1$, $\lambda_2 = 2$, and $\lambda_3 = 2$. Therefore, we have repeated roots with $j = 2$ and $k = 2$ and the solution is

$$y = \sum_{m=1}^{1} A_m e^{\lambda_m x} + \left(\sum_{p=0}^{1} A_{2+p} x^p \right) e^{\lambda_2 x} = A_1 e^{\lambda_1 x} + \left(A_2 + A_3 x\right) e^{\lambda_2 x}$$

$$= A_1 e^{-x} + \left(A_2 + A_3 x\right) e^{2x}$$

This result is verified with Mathematica procedure M4.9.

Example 4.7

Consider the equation

$$\frac{d^2 y}{dx^2} - \Omega^2 y = 0 \quad \Omega \neq 0 \tag{a}$$

We assume a solution of the form $y = e^{\lambda x}$ and substitute it into Eq. (a) to find

$$\lambda^2 - \Omega^2 = 0$$

The roots of this polynomial are $\lambda_1 = \Omega$ and $\lambda_2 = -\Omega$. Then the homogeneous solution is

$$y(x) = A_1 e^{\Omega x} + A_2 e^{-\Omega x} \tag{b}$$

This result is verified with Mathematica procedure M4.10.

We now convert Eq. (b) into another form by letting

$$A_1 = \frac{1}{2}\left(a + b\right) \text{ and } A_2 = \frac{1}{2}\left(a - b\right)$$

Then Eq. (b) becomes

$$\begin{aligned} y(x) &= \frac{1}{2}\left(a + b\right) e^{\Omega x} + \frac{1}{2}\left(a - b\right) e^{-\Omega x} \\ &= \frac{a}{2}\left(e^{\Omega x} + e^{-\Omega x}\right) + \frac{b}{2}\left(e^{s\Omega x} - e^{-\Omega x}\right) \\ &= C_1 \cosh(\Omega x) + C_2 \sinh(\Omega x) \end{aligned} \tag{c}$$

This result is verified with Mathematica procedure M4.11.

The solution to Eq. (a), when $\Omega = 0$, is

$$y_0 = a_1 + a_2 x \tag{d}$$

Notice that the solution given by Eq. (d) is not obtained when Ω is set to zero in Eq. (c).

When $\Omega^2 = -\omega^2$, Eq. (a) becomes

$$\frac{d^2y}{dx^2} + \omega^2 y = 0 \tag{e}$$

and the solution to Eq. (e) is given by Eq. (c) with Ω replaced by $j\omega$. Thus, from Eq. (b)

$$y(x) = D_1 e^{j\omega x} + D_2 e^{-j\omega x} \tag{f}$$

or from Eq. (c)

$$y(x) = D_3 \cos(\omega x) + D_4 \sin(\omega x) \tag{g}$$

where we have used Eq. (2.33). Equation (g) is verified with Mathematica procedure M4.12.

Example 4.8

Consider the fourth-order equation

$$\frac{d^4y}{d\eta^4} - \Omega^4 y = 0 \quad \Omega^4 > 0 \tag{a}$$

We assume a solution of the form $y = e^{\lambda\eta}$ and substitute it into Eq. (a) to find

$$\lambda^4 = \Omega^4$$

or

$$\lambda_{1,2} = \pm\Omega \quad \lambda_{3,4} = \pm j\Omega$$

Hence the solution to Eq. (a) is

$$y = Ae^{\Omega\eta} + Be^{-\Omega\eta} + Ce^{j\Omega\eta} + De^{-j\Omega\eta} \tag{b}$$

Equation (b) can be written in several different forms. To obtain a second form, we let

$$\begin{aligned} A &= \frac{1}{2}(a+b), \quad B = \frac{1}{2}(a-b) \\ C &= \frac{1}{2}(c+d), \quad \frac{1}{2}D = \frac{1}{2}(c-d) \end{aligned} \tag{c}$$

Then, using Eq. (c) in Eq. (b), we obtain

$$\begin{aligned} y &= \frac{1}{2}(a+b)e^{\Omega\eta} + \frac{1}{2}(a-b)e^{-\Omega\eta} + \frac{1}{2}(c+d)e^{j\Omega\eta} + \frac{1}{2}(c-d)e^{-j\Omega\eta} \\ &= \frac{a}{2}\left(e^{\Omega\eta} + e^{-\Omega\eta}\right) + \frac{b}{2}\left(e^{\Omega\eta} - e^{-\Omega\eta}\right) + \frac{c}{2}\left(e^{j\Omega\eta} + e^{-j\Omega\eta}\right) + \frac{d}{2}\left(e^{j\Omega\eta} - e^{-j\Omega\eta}\right) \\ &= a\cosh(\Omega\eta) + b\sinh(\Omega\eta) + c\cos(\Omega\eta) + jd\sin(\Omega\eta) \\ y &= A_1\cosh(\Omega\eta) + A_2\sinh(\Omega\eta) + A_3\cos(\Omega\eta) + A_4\sin(\Omega\eta) \end{aligned} \tag{d}$$

Equation (d) is verified with Mathematica procedure M4.13.

To obtain a third form, we note in Eq. (d) that cosh and cosine are even functions and sinh and sine are odd functions. Thus, we further rearrange Eq. (d) by letting

$$\begin{aligned} A_1 &= \frac{1}{2}(a_1 + a_3), \quad A_3 = \frac{1}{2}(a_1 - a_3) \\ A_2 &= \frac{1}{2}(a_2 + a_4), \quad A_4 = \frac{1}{2}(a_2 - a_4) \end{aligned} \tag{e}$$

Then using Eq. (e) in Eq. (d), we arrive at

$$\begin{aligned} y &= A_1 \cosh(\Omega\eta) + A_2 \sinh(\Omega\eta) + A_3 \cos(\Omega\eta) + A_4 \sin(\Omega\eta) \\ &= \frac{1}{2}(a_1 + a_3)\cosh(\Omega\eta) + \frac{1}{2}(a_2 + a_4)\sinh(\Omega\eta) \\ &\quad + \frac{1}{2}(a_1 - a_3)\cos(\Omega\eta) + \frac{1}{2}(a_2 - a_4)\sin(\Omega\eta) \\ &= \frac{a_1}{2}\left[\cosh(\Omega\eta) + \cos(\Omega\eta)\right] + \frac{a_3}{2}\left[\cosh(\Omega\eta) - \cos(\Omega\eta)\right] \\ &\quad + \frac{a_2}{2}\left[\sinh(\Omega\eta) + \sin(\Omega\eta)\right] + \frac{a_4}{2}\left[\sinh(\Omega\eta) - \sin(\Omega\eta)\right] \\ &= B_1 Q(\Omega\eta) + B_2 R(\Omega\eta) + B_3 S(\Omega\eta) + B_4 T(\Omega\eta) \end{aligned} \tag{f}$$

where

$$\begin{aligned} Q(\Omega\eta) &= \frac{1}{2}\left[\cosh(\Omega\eta) + \cos(\Omega\eta)\right] \\ R(\Omega\eta) &= \frac{1}{2}\left[\sinh(\Omega\eta) + \sin(\Omega\eta)\right] \\ S(\Omega\eta) &= \frac{1}{2}\left[\cosh(\Omega\eta) - \cos(\Omega\eta)\right] \\ T(\Omega\eta) &= \frac{1}{2}\left[\sinh(\Omega\eta) - \sin(\Omega\eta)\right] \end{aligned} \tag{g}$$

The functions given in Eq. (g) will also appear when we use the Laplace transform method to obtain a solution to Eq. (a). See Example 8.19. It is noted from Eq. (g) that $Q(0)=1$ and $R(0)=S(0)=T(0)=0$.

For the case when $\Omega=0$, Eq. (a) becomes

$$\frac{d^4 y}{d\eta^4} = 0 \tag{h}$$

The solution to Eq. (h) is

$$y = C_1 + C_2\eta + C_3\eta^2 + C_4\eta^3$$

Example 4.9

Consider the fourth-order equation

$$\frac{d^4y}{d\eta^4}+\Omega^4 y=0 \quad \Omega^4>0 \tag{a}$$

We assume a solution of the form $y=e^{\lambda\eta}$ and substitute it into Eq. (a) to find

$$\lambda^4=-\Omega^4 \tag{b}$$

We note that

$$\sqrt[4]{-1}=e^{i\pi(2k+1)/4} \quad k=0,1,2,3$$

or, upon expanding,

$$e^{i\pi/4}=\frac{1}{\sqrt{2}}(1+j), \quad e^{i\pi 3/4}=\frac{1}{\sqrt{2}}(-1+j)$$

$$e^{i\pi 5/4}=\frac{1}{\sqrt{2}}(-1-j), \quad e^{i\pi 7/4}=\frac{1}{\sqrt{2}}(1-j)$$

This result can be written more compactly, as

$$\sqrt[4]{-1}=\pm\frac{1}{\sqrt{2}}(1\pm j)$$

Therefore, if we define $q^4=\Omega^4/4$, then Eq. (b) can be written as

$$\lambda=\pm(1\pm j)q \tag{c}$$

and the solution to Eq. (a) is

$$y=e^{q\eta}\left(ae^{jq\eta}+be^{-jq\eta}\right)+e^{-q\eta}\left(ce^{jq\eta}+de^{-jq\eta}\right) \tag{d}$$

To place Eq. (d) into a second form, we make the following substitution

$$a=a_1+a_3, \quad b=a_1-a_3$$

$$c=a_2+a_4, \quad d=a_2-a_4$$

and Eq. (d) becomes

$$y=e^{q\eta}\left(b_1\cos q\eta+b_2\sin q\eta\right)+e^{-q\eta}\left(b_3\cos q\eta+b_4\sin q\eta\right) \tag{e}$$

Equation (e) is verified with Mathematica procedure M4.14.

Lastly, to place Eq. (e) into a third form, we make the substitution

$$b_1=c_1+c_3, \quad b_3=c_1-c_3$$

$$b_2=c_2+c_4, \quad b_4=c_2-c_4$$

and Eq. (e) can be written as

$$y = A\cos q\eta\cosh q\eta + B\cos q\eta\sinh q\eta + C\sin q\eta\cosh q\eta + D\sin q\eta\sinh q\eta \tag{f}$$

Example 4.10

Consider Eq. (4.32) for the case where $n=2$ and $a_0 \neq 0$. Then Eq. (4.32) becomes

$$a_0\frac{d^2y}{dx^2} + a_1\frac{dy}{dx} + a_2y = 0 \tag{a}$$

Anticipating the subsequent results, we re-write Eq. (a) as

$$\frac{d^2y}{dx^2} + 2\zeta\omega\frac{dy}{dx} + \omega^2y = 0 \tag{b}$$

where

$$2\zeta\omega = \frac{a_1}{a_0}, \quad \omega^2 = \frac{a_2}{a_0}$$

Upon substituting Eq. (4.34) into Eq. (b), we obtain the following characteristic equation

$$\lambda^2 + 2\zeta\omega\lambda + \omega^2 = 0$$

Therefore, if we assume that $\zeta < 1$,

$$\lambda_{1,2} = -\omega\zeta \pm j\omega_d$$

where

$$\omega_d = \omega\sqrt{1-\zeta^2}$$

Then, the homogeneous solution is

$$y_h(x) = C_1e^{(-\omega\zeta + j\omega_d)x} + C_2e^{(-\omega\zeta - j\omega_d)x} \tag{c}$$

Equation (c) is verified with Mathematica procedure M4.15.

If we set

$$C_1 = (a+b)/2, \quad C_2 = (a-b)/2$$

where a and b are unknown constants, then Eq. (c) can be written as

$$y_h(x) = D_1e^{-\omega\zeta x}\cos(\omega_dx) + D_2e^{-\omega\zeta x}\sin(\omega_dx) \tag{d}$$

We now examine the four special cases of Eq. (d).

$\zeta=0$: In this case, $\omega_d=\omega$, and Eq. (d) becomes

$$y_h(x) = D_1\cos(\omega x) + D_2\sin(\omega x) \tag{e}$$

$0 \le \zeta < 1$: In this case, Eq. (d) remains as it is.

$\zeta=1$: In this case, $\omega_d=0$, and $\lambda_1=\lambda_2=-\omega$; that is, we have repeating roots. Using Eq. (4.43), the solution is

$$y_h(x) = D_1 e^{-\omega x} + D_2 x e^{-\omega x} \tag{f}$$

$\zeta > 1$: In this case,

$$y_h(x) = E_1 e^{-\omega\zeta x}\cosh(\hat{\omega}_d x) + E_2 e^{-\omega\zeta x}\sinh(\hat{\omega}_d x) \tag{g}$$

where

$$\hat{\omega}_d = \omega\sqrt{\zeta^2 - 1}$$

and we have used Eqs. (2.31) and (2.32).

We shall have frequent use of these results in the subsequent sections.

Example 4.11

We continue with Example 4.10 and determine the solution for the case where $0 \le \zeta < 1$ when the following initial conditions are specified: $y(0)=y_0$ and $y'(0)=v_0$ and the prime denotes the derivative with respect to x. Then, from Eq. (d) of Example 4.10, we see that

$$y_h(x) = D_1 e^{-\omega\zeta x}\cos(\omega_d x) + D_2 e^{-\omega\zeta x}\sin(\omega_d x)$$

$$\begin{aligned}\frac{dy_h(x)}{dx} &= D_1\left(-\omega\zeta e^{-\omega\zeta x}\cos(\omega_d x) - \omega_d e^{-\omega\zeta x}\sin(\omega_d x)\right) \\ &\quad + D_2\left(-\omega\zeta e^{-\omega\zeta x}\sin(\omega_d x) + \omega_d e^{-\omega\zeta x}\cos(\omega_d x)\right)\end{aligned} \tag{a}$$

where

$$\omega_d = \omega\sqrt{1-\zeta^2}$$

Then, using the initial conditions, we find that

$$y_h(0) = D_1 = y_0$$

and

$$\frac{dy_h(0)}{dx} = -\omega\zeta D_1 + \omega_d D_2 = v_0$$

Therefore,

$$D_2 = \frac{v_0 + \omega\zeta}{\omega_d}$$

and the solution is

$$y_h(x) = y_0 e^{-\omega\zeta x}\cos(\omega_d x) + \left(\frac{v_0 + \omega\zeta}{\omega_d}\right)e^{-\omega\zeta x}\sin(\omega_d x) \tag{b}$$

Equation (b) is verified with Mathematica procedure M4.16.
When $\zeta = 0$, Eq. (b) becomes

$$y_h(x) = y_0\cos(\omega x) + \frac{v_0}{\omega}\sin(\omega x) \tag{c}$$

Example 4.12

Consider the fourth-order equation

$$\frac{d^4 y}{d\eta^4} - 2\alpha^2\frac{d^2 y}{d\eta^2} + \alpha^4 y = 0 \qquad \alpha^2 > 0 \tag{a}$$

We note that Eq. (a) can be written as

$$\left(\frac{d^2}{d\eta^2} - \alpha^2\right)^2 y = 0$$

or as

$$\left(\frac{d}{d\eta} - \alpha\right)^2\left(\frac{d}{d\eta} + \alpha\right)^2 y = 0 \tag{b}$$

Equation (b) is of the form given by Eq. (4.41). Thus, the solution to Eq. (b) is obtained from Eq. (4.43) with $k = 2$ as

$$y = \left(\sum_{p=0}^{1} A_{1+p}x^p\right)e^{\alpha x} + \left(\sum_{p=0}^{1} A_{3+p}x^p\right)e^{-\alpha x} \tag{c}$$

which upon expansion gives

$$\begin{aligned} y &= A_1 e^{\alpha x} + A_1 x e^{\alpha x} + A_3 e^{-\alpha x} + A_4 x e^{-\alpha x} \\ &= A_1 e^{\alpha x} + A_3 e^{-\alpha x} + x\left(A_1 e^{\alpha x} + A_4 e^{-\alpha x}\right) \end{aligned}$$

Upon using the techniques introduced in Examples 4.8 and 4.9, this result can be written as

$$y = C_1\cosh\alpha x + C_2\sinh\alpha x + x(C_3\cosh\alpha x + C_4\sinh\alpha x) \tag{d}$$

We shall have use for this result in Section 7.2.7.
Equations (c) and (d) are verified with Mathematica procedure M4.17.

The results of Examples 4.7 to 4.10, 4.12, and subsequent Example 4.14 are summarized in Table 4.2.

4.2.3 Reduction of Order

Suppose that a solution $y_1(x)$ has been found to a homogeneous differential equation of the form

$$a_2(x)\frac{d^2y}{dx^2}+a_1(x)\frac{dy}{dx}+a_0(x)y=0 \tag{4.44}$$

TABLE 4.2
Summary of the Solutions to the Homogeneous Differential Equations of Examples 4.7 to 4.10, 4.12, and 4.14

Example	Differential Equation	Solution
4.7	$\frac{d^2y}{dx^2}-\Omega^2y=0 \quad \Omega^2>0$	$y(x)=C_1\cosh(\Omega x)+C_2\sinh(\Omega x)$ or $y(x)=C_1e^{\Omega x}+C_2e^{-\Omega x}$
4.7	$\frac{d^2y}{dx^2}+\Omega^2y=0 \quad \Omega^2>0$	$y(x)=D_1\cos(\Omega x)+D_2\sin(\Omega x)$
4.8	$\frac{d^4y}{d\eta^4}-\Omega^4y=0 \quad \Omega^4>0$	$y(\eta)=A_1\cosh(\Omega\eta)+A_2\sinh(\Omega\eta)+A_3\cos(\Omega\eta)+A_4\sin(\Omega\eta)$ or $y(\eta)=B_1Q(\Omega\eta)+B_2R(\Omega\eta)+B_3S(\Omega\eta)+B_4T(\Omega\eta)$ where Q, R, S, and T are defined in Eq. (g) of Example 4.8
4.9	$\frac{d^4y}{d\eta^4}+\Omega^4y=0 \quad \Omega^4>0$	$y(\eta)=A\cos q\eta\cosh q\eta+B\cos q\eta\sinh q\eta+C\sin q\eta\cosh q\eta+D\sin q\eta\sinh q\eta$ where $q=\Omega/\sqrt{2}$.
4.10	$\frac{d^2y}{dx^2}+2\zeta\omega\frac{dy}{dx}+\omega^2y=0 \quad 0\le\zeta<1$	$y(x)=D_1e^{-\omega\zeta x}\cos(\omega_d x)+D_2e^{-\omega\zeta x}\sin(\omega_d x)$ where $\omega_d=\omega\sqrt{1-\zeta^2}$.
4.12	$\frac{d^4y}{d\eta^4}-2\alpha^2\frac{d^2y}{d\eta^2}+\alpha^4y=0 \quad \alpha^2>0$	$y(x)=C_1\cosh\alpha x+C_2\sinh\alpha x+x(C_3\cosh\alpha x+C_4\sinh\alpha x)$
4.14	$\left(\frac{d^2}{dr^2}+\frac{d}{rdr}-\frac{m^2}{r^2}\right)\left(\frac{d^2R_m}{dr^2}+\frac{dR_m}{rdr}-\frac{m^2}{r^2}R_m\right)=0$	$R_0=\alpha_{10}+\alpha_{20}\ln r+\alpha_{30}r^2+\alpha_{40}r^2\ln r \quad m=0$ $R_1=\alpha_{11}r+\alpha_{21}r\ln r+\alpha_{31}r^{-1}+\alpha_{41}r^3 \quad m=1$ $R_m=\alpha_{1m}r^m+\alpha_{2m}r^{-m}+\alpha_{3m}r^{2-m}+\alpha_{4m}r^{2+m} \quad m>1$

In this case, the second independent solution to Eq. (4.44) can be found using a method that reduces the second-order equation to a first-order equation. The method is called the *reduction of order.*

Since $y_1(x)$ satisfies Eq. (4.44), we make a change of dependent variable to the form

$$y(x) = v(x)y_1(x) \tag{4.45}$$

where $v(x)$ is to be determined. Noting that

$$\frac{dy}{dx} = \frac{dv}{dx}y_1 + v\frac{dy_1}{dx}$$

$$\frac{d^2y}{dx^2} = \frac{d^2v}{dx^2}y_1 + 2\frac{dv}{dx}\frac{dy_1}{dx} + v\frac{d^2y_1}{dx^2}$$

Eq. (4.44) becomes

$$a_2(x)\left(\frac{d^2v}{dx^2}y_1 + 2\frac{dv}{dx}\frac{dy_1}{dx} + v\frac{d^2y_1}{dx^2}\right) + a_1(x)\left(\frac{dv}{dx}y_1 + v\frac{dy_1}{dx}\right) + a_0(x)vy_1 = 0$$

$$\begin{aligned} y_1\left(a_2(x)\frac{d^2v}{dx^2} + a_1(x)\frac{dv}{dx}\right) + v\overbrace{\left(a_2(x)\frac{d^2y_1}{dx^2} + a_1(x)\frac{dy_1}{dx} + a_0(x)y_1\right)}^{=0} \\ + 2a_2(x)\frac{dv}{dx}\frac{dy_1}{dx} = 0 \\ y_1\left(a_2(x)\frac{d^2v}{dx^2} + a_1(x)\frac{dv}{dx}\right) + 2a_2(x)\frac{dv}{dx}\frac{dy_1}{dx} = 0 \end{aligned} \tag{4.46}$$

where we have used Eq. (4.44) with $y = y_1$.

We now make another change of variable

$$w(x) = \frac{dv}{dx} \tag{4.47}$$

in Eq. (4.46) to arrive at

$$\begin{aligned} a_2(x)y_1\frac{dw}{dx} + w\left(2a_2(x)\frac{dy_1}{dx} + a_1(x)y_1\right) = 0 \\ \frac{1}{w}\frac{dw}{dx} + \frac{a_1(x)}{a_2(x)} + \frac{2}{y_1}\frac{dy_1}{dx} = 0 \end{aligned} \tag{4.48}$$

We see that Eq. (4.48) is a separable first-order equation. Therefore, multiplying Eq. (4.48) by dx and integrating, we arrive at

$$\int \frac{dw}{w} + \int \frac{a_1(x)}{a_2(x)} dx + 2\int \frac{dy_1}{y_1} = 0$$

$$\ln w + \int \frac{a_1(x)}{a_2(x)} dx + 2\ln y_1 = \ln C_1$$

$$\ln \frac{w}{\left(C_1 / y_1^2\right)} = -\int \frac{a_1(x)}{a_2(x)} dx \tag{4.49}$$

$$w = \frac{dv}{dx} = \frac{C_1}{y_1^2} \exp\left[-\int \frac{a_1(x)}{a_2(x)} dx\right]$$

where we have used Eq. (4.47) and expressed the constant as $\ln(C_1)$ in anticipation of the final result. Performing the integration with respect to x gives

$$v = C_1 \int y_1^{-2}(x) \exp\left[-\int \frac{a_1(x)}{a_2(x)} dx\right] dx \tag{4.50}$$

and we have set the constants of integration to zero. The final result is obtained from Eq. (4.45) as

$$y_2(x) = C_1 y_1(x) \int y_1^{-2}(x) \exp\left[-\int \frac{a_1(x)}{a_2(x)} dx\right] dx \tag{4.51}$$

Equation (4.51) will be useful in finding the second independent solution to the Legendre equation as discussed in Section 5.1.9.

We now illustrate this result with the following example.

Example 4.13

It is found that one solution to

$$4x^2 \frac{d^2y}{dx^2} + 4x\frac{dy}{dx} + \left(4x^2 - 1\right)y = 0 \tag{a}$$

is

$$y_1(x) = \frac{\sin x}{\sqrt{x}}$$

The second solution is obtained using the reduction of order as follows. It is noted that

$$a_2(x) = 4x^2$$

$$a_1(x) = 4x$$

Then

$$\int \frac{a_1(x)}{a_2(x)} dx = \int \frac{4x}{4x^2} dx = \int \frac{dx}{x} = \ln x$$

and

$$\int y_1^{-2}(x)\exp\left[-\int \frac{a_1(x)}{a_2(x)}dx\right]dx = \int \frac{x}{\sin^2 x}e^{-\ln x}dx = \int \frac{xx^{-1}}{\sin^2 x}dx$$

$$= \int \frac{1}{\sin^2 x}dx = -\cot x$$

where we used the Eq. (2.40), that is, $e^{-\ln x} = e^{\ln(1/x)} = e^{\ln x^{-1}} = x^{-1}$. Finally, Eq. (4.45) yields

$$y_2(x) = C_1 \frac{\sin x}{\sqrt{x}}(-\cot x) = C_3 \frac{\cos x}{\sqrt{x}}$$

and the solution to Eq. (a) is

$$y(x) = C_1 \frac{\sin x}{\sqrt{x}} + C_3 \frac{\cos x}{\sqrt{x}} \tag{b}$$

This result is verified with Mathematica procedure M4.18.

We now assume that the boundary conditions are that $y(0)$ is finite and that $y(a) = y_0$. To have $y(0)$ finite requires that $C_3 = 0$ since $\cos x / \sqrt{x} \to \infty$ as $x \to 0$. Then, Eq. (b) reduces to

$$y(x) = C_1 \frac{\sin x}{\sqrt{x}} \tag{c}$$

Using Eq. (c) in the boundary at $x = a$, we find that

$$y(a) = C_1 \frac{\sin a}{\sqrt{a}} = y_0$$

Thus,

$$C_1 = y_0 \frac{\sqrt{a}}{\sin a}$$

and the solution becomes

$$y(x) = y_0 \sqrt{\frac{a}{x}} \frac{\sin x}{\sin a} \tag{d}$$

Equation (d) is verified with Mathematica procedure M4.19.

4.2.4 Cauchy–Euler Equation

The Cauchy–Euler equation is an nth-order ordinary differential equation with non-constant coefficients of the form

$$a_n x^n \frac{d^n y}{dx^n} + a_{n-1}x^{n-1}\frac{d^{n-1}y}{dx^{n-1}} + \cdots + a_0 y = 0 \tag{4.52}$$

where a_k are constants. A solution is obtained by assuming that

$$y = x^r \tag{4.53}$$

where r is to be determined.

We shall examine the case for $n=2$, that is, the equation

$$a_2 x^2 \frac{d^2 y}{dx^2} + a_1 x \frac{dy}{dx} + a_0 y = 0 \tag{4.54}$$

and for this equation determine the unknown parameter r. The substitution of Eq. (4.53) into Eq. (4.54) gives

$$\begin{aligned} a_2 x^2 \left(r(r-1)x^{r-2} \right) + a_1 x \left(r x^{r-1} \right) + a_0 x^r &= 0 \\ \left(a_2 r(r-1) + a_1 r + a_0 \right) x^r &= 0 \end{aligned} \tag{4.55}$$

If $x > 0$, then Eq. (4.55) is satisfied when r is a solution to

$$a_2 r^2 + \left(a_1 - a_2 \right) r + a_0 = 0 \tag{4.56}$$

Therefore,

$$r_{1,2} = \frac{1}{2a_2} \left[a_2 - a_1 \pm \sqrt{\left(a_1 - a_2 \right)^2 - 4 a_2 a_0} \right] \tag{4.57}$$

Case 1: Real and Distinct Roots

When

$$\left(a_1 - a_2 \right)^2 - 4 a_2 a_0 > 0$$

there are two real and distinct values for r and the solution for $x > 0$ is

$$y = C_1 x^{r_1} + C_2 x^{r_2} \tag{4.58}$$

Case 2: Only One Root

When

$$\left(a_1 - a_2 \right)^2 - 4 a_2 a_0 = 0$$

there is only one value for r. In this case, we use the reduction of order method as given by Eq. (4.51) to obtain the second solution. Hence, $y_1 = x^{r_1}$, where, from Eq. (4.57), $r_1 = (a_2 - a_1)/(2a_2)$ and we proceed as follows.

It is noted from Eq. (4.44) that

$$a_2(x) = a_2 x^2 \qquad a_1(x) = a_1 x$$

Then

$$\int \frac{a_1(x)}{a_2(x)}dx = \int \frac{a_1 x}{a_2 x^2}dx = \frac{a_1}{a_2}\int \frac{dx}{x} = \frac{a_1}{a_2}\ln x = \ln x^{a_1/a_2}$$

and, therefore, Eq. (4.50) gives

$$\int y_1^{-2}(x)\exp\left[-\int \frac{a_1(x)}{a_2(x)}dx\right]dx = \int x^{-2r_1}e^{-\ln x^{a_1/a_2}}\,dx$$

$$= \int x^{-2r_1}x^{-a_1/a_2}dx = \int x^{-(2r_1+a_1/a_2)}dx$$

However,

$$2r_1 + \frac{a_1}{a_2} = 2\left(\frac{a_2 - a_1}{2a_2}\right) + \frac{a_1}{a_2} = 1$$

and, hence,

$$\int x^{-(2r_1+a_1/a_2)}dx = \int \frac{dx}{x} = \ln x$$

Using Eq. (4.51), the second independent solution for $x > 0$ is

$$y_2(x) = C_2 x^{r_1}\ln x = C_2 x^{(a_2-a_1)/(2a_2)}\ln x$$

and, hence,

$$y(x) = C_1 x^{r_1} + C_2 x^{r_1}\ln x = C_1 x^{(a_2-a_1)/(2a_2)} + C_2 x^{(a_2-a_1)/(2a_2)}\ln x \tag{4.59}$$

Case 3: Roots Are Complex Conjugate Pairs

When

$$(a_1 - a_2)^2 - 4a_2 a_0 < 0$$

there are two distinct values for r, which are a complex conjugate pair. If we denote these complex roots as

$$r_1 = b + jd$$

$$r_2 = b - jd$$

then, using the technique used to arrive at Eq. (2.46), the solution for $x > 0$ is

$$\begin{aligned}
y &= C_1 x^{r_1} + C_2 x^{r_2} = C_1 x^{b+jd} + C_2 x^{b-jd} \\
&= C_1 x^b\left(e^{\ln x}\right)^{jd} + C_2 x^b\left(e^{\ln x}\right)^{-jd} \\
&= x^b\left[C_1 e^{jd\ln x} + C_2 e^{-jd\ln x}\right] \\
&= x^b\left[C_1\left(\cos(d\ln x) + j\sin(d\ln x)\right) + C_2\left(\cos(d\ln x) - j\sin(d\ln x)\right)\right] \\
y &= x^b\left[D_1\cos(d\ln x) + D_2\sin(d\ln x)\right]
\end{aligned} \tag{4.60}$$

Cases 1 and 3 are verified with Mathematica procedure M4.20, and Case 2 is verified with Mathematica procedure M4.21.

We now illustrate this result with the following example.

Example 4.14

Consider the following fourth-order differential equation expressed in the form

$$\left(\frac{d^2}{dr^2}+\frac{d}{rdr}-\frac{m^2}{r^2}\right)\left(\frac{d^2R_m}{dr^2}+\frac{dR_m}{rdr}-\frac{m^2}{r^2}R_m\right)=0 \tag{a}$$

where $m=0, 1, 2, \ldots$, and $r>0$. Assuming a solution of the form $R_m=r^k$ and substituting this assumed solution into Eq. (a), we arrive at

$$\left(\frac{d^2}{dr^2}+\frac{d}{rdr}-\frac{m^2}{r^2}\right)\left(k(k-1)r^{k-2}+kr^{k-2}-m^2r^{k-2}\right)=0$$

$$\left(\frac{d^2}{dr^2}+\frac{d}{rdr}-\frac{m^2}{r^2}\right)r^{k-2}\left(k^2-m^2\right)=0$$

$$\left((k-2)(k-3)r^{k-4}+(k-2)r^{k-4}-m^2r^{k-4}\right)\left(k^2-m^2\right)=0$$

$$\left(k^2-4k+4-m^2\right)\left(k^2-m^2\right)r^{k-4}=0$$

$$\left((k-2)^2-m^2\right)\left(k^2-m^2\right)r^{k-4}=0$$

Therefore, the values of k that satisfy this equation for arbitrary $r>0$ are $k=\pm m$ and $k=2\pm m$. Thus, the form of the solution is a function of the value of m.

For $m=0$, $k=0$, 0, 2, and 2, and, therefore, we have two sets of repeating roots. Thus, for $m=0$, Eq. (4.59) applies, where $r_1=0$ and $r_1=2$ and the solution is

$$R_0=\alpha_{10}+\alpha_{20}\ln r+\alpha_{30}r^2+\alpha_{40}r^2\ln r$$

For $m=1$, $k=1, -1, 1$, and 3; hence, there are two distinct roots, $k=-1$ and 3, for which Eq. (4.58) applies and one set of repeating roots $k=1$ for which Eq. (4.59) applies. In this case, the solution is

$$R_1=\alpha_{11}r+\alpha_{21}r\ln r+\alpha_{31}r^{-1}+\alpha_{41}r^3$$

For $m>1$, $k=m, -m, 2+m, 2-m$, which are all distinct, and Eq. (4.58) applies and the solution is

$$R_m=\alpha_{1m}r^m+\alpha_{2m}r^{-m}+\alpha_{3m}r^{2-m}+\alpha_{4m}r^{2+m} \qquad m>1$$

These results are verified with Mathematica procedure M4.22.

4.2.5 Particular Solutions: Method of Undetermined Coefficients

We shall introduce two techniques to obtain the particular solution to ordinary differential equations. The first technique, called the method of undetermined coefficients, is valid only for a differential equation with constant coefficients and for the nonhomogeneous term of a certain form. The second method, called the variation of parameters, will work, in principle, for the form of the differential equation as given by Eq. (4.31).

We consider the nonhomogeneous form of Eq. (4.31) for which $n=2$ and $K=1$, that is,

$$a_0 \frac{d^2y}{dx^2} + a_1 \frac{dy}{dx} + a_2 y = Q(x) \tag{4.61}$$

where a_j are real constants. When *each* term of $Q(x)$ is either of the form

$$Q(x) = P_n(x, a_n) e^{\alpha x} \cos(\beta x) \tag{4.62}$$

or

$$Q(x) = S_n(x, b_n) e^{\alpha x} \sin(\beta x) \tag{4.63}$$

where

$$P_n(x, a_n) = \sum_{n=0}^{n} a_n x^n, \quad S_n(x, b_n) = \sum_{n=0}^{n} b_n x^n \tag{4.64}$$

one can assume for *each* nonhomogeneous term a trial particular solution of the form

$$y_p(x) = P_n(x, c_n) x^s e^{\alpha x} \cos(\beta x) + S_n(x, d_n) x^s e^{\alpha x} \sin(\beta x) \tag{4.65}$$

where c_n and d_n are to be determined and s is the smallest nonnegative integer that will ensure that no term in y_p is a solution to y_h, the homogeneous solution. It is mentioned that Eqs. (4.62) to (4.64) have many special cases such as $\alpha=0$, and/or $\beta=0$ and/or $P_n=0$, and so on.

To show the significance of s, consider the following equation

$$\frac{d^2y}{dx^2} + 4y = c_o \cos(2x) \tag{4.66}$$

The homogeneous solution to Eq. (4.66) is

$$y_h(x) = A\cos(2x) + B\sin(2x) \tag{4.67}$$

Let us (incorrectly) try using a particular solution of the form

$$y_p(x) = a\cos(2x) + b\sin(2x) \tag{4.68}$$

that is, we have assumed that $s=0$. Substitution of Eq. (4.68) into Eq. (4.66) yields

$$-4a\cos(2x) - 4b\sin(2x) + 4a\cos(2x) + 4b\sin(2x) = c_o \cos(2x)$$

$$0 = c_o \cos(2x)$$

Therefore, y_p given by Eq. (6.68) is not a particular solution for there isn't any way to determine a and b. However, when $s=1$, Eq. (4.68) becomes

$$y_p(x) = ax\cos(2x) + bx\sin(2x) \tag{4.69}$$

Substituting Eq. (4.69) into Eq. (6.66) gives

$$(4b - c_o)\cos(2x) - 4a\sin(2x) = 0$$

Therefore, for arbitrary x, we find that $a=0$ and $b=c_o/4$, and the particular solution is

$$y_p(x) = \frac{c_o}{4} x\sin(2x) \tag{4.70}$$

We shall now illustrate this method with several examples.

Example 4.15

We shall find the solution to

$$\frac{d^2y}{dx^2} - 4y = -2x + 8x^2 \tag{a}$$

subject to the boundary conditions $y(0)=y_0$ and $y'(a)=dy(a)/dx=b_0$.

Using Eq. (4.34), we find that the characteristic equation for the homogeneous form of Eq. (a) is

$$\gamma^2 - 4 = 0 \rightarrow \gamma_{1,2} = \pm 2$$

Thus, the homogeneous solution is

$$y_h(x) = C_1e^{2x} + C_2e^{-2x}$$

Based on the form of the nonhomogeneous term in Eq. (a), we select the terms from the trial particular solution of Eq. (4.65) to be

$$y_p(x) = x^s\left(a_0 + a_1x + a_2x^2\right) \tag{b}$$

Since, when $s=0$, none of these terms is related to the terms in y_h, we can set $s=0$. Then, substituting Eq. (b) with $s=0$ into Eq. (a), we obtain

$$2a_2 - 4a_0 + (2 - 4a_1)x - (4a_2 + 8)x^2 = 0$$

For this equation to be satisfied for all x,

$$2a_2 - 4a_0 = 0$$

$$2 - 4a_1 = 0$$

$$4a_2 + 8 = 0$$

Therefore, $a_1 = 1/2$, $a_2 = -2$, and $a_0 = -1$ and the particular solution is

$$y_p(x) = -1 + x/2 - 2x^2$$

and the solution is

$$y(x) = C_3 e^{2x} + C_4 e^{-2x} - 1 + x/2 - 2x^2$$

or, upon using Eq. (c) of Example 4.7

$$y(x) = C_1 \cosh(2x) + C_2 \sinh(2x) - 1 + x/2 - 2x^2 \tag{c}$$

This result is verified with Mathematica procedure M4.23.

Before satisfying the boundary conditions, we note from Eq. (c) that

$$\frac{dy}{dx} = 2C_1 \sinh(2x) + 2C_2 \cosh(2x) + 1/2 - 4x \tag{d}$$

Then, using Eqs. (c) and (d) in the boundary conditions, we find that

$$\begin{aligned} y(0) &= C_1 - 1 = y_0 \\ y'(a) &= 2C_1 \sinh(2a) + 2C_2 \cosh(2a) + 1/2 - 4a = b_0 \end{aligned} \tag{e}$$

Solving Eq. (e) for C_1 and C_2, we obtain

$$\begin{aligned} C_1 &= 1 + y_0 \\ C_2 &= \frac{1}{2}\left(b_0 + 4a - 1/2\right)\text{sech}(2a) - \left(1 + y_0\right)\tanh(2a) \end{aligned} \tag{f}$$

The final solution is obtained by substituting Eq. (f) into Eq. (c) and is verified with Mathematica procedure M4.24.

Example 4.16

We shall find the particular solution to

$$\frac{d^2y}{dx^2} + 6\frac{dy}{dx} + 13y = e^{-3x}\cos(2x) \tag{a}$$

To determine the homogeneous solution, we note from Example 4.10 that

$$\omega^2 = 13 \to \omega = \sqrt{13}$$

$$2\zeta\omega = 6 \to \zeta = \frac{3}{\sqrt{13}}$$

$$\omega_d = \sqrt{13}\sqrt{1 - \left(\frac{3}{\sqrt{13}}\right)^2} = 2$$

Thus, the homogeneous solution is given by Eq. (d) of that example, which becomes

$$y_h(x) = D_1 e^{-3x}\cos(2x) + D_1 e^{-3x}\sin(2x) \tag{b}$$

Based on the form of the nonhomogeneous term, we select the terms from the trial particular solution of Eq. (4.65) to be

$$y_p(x) = a_0 x^s e^{-3x}\cos(2x) + b_0 x^s e^{-3x}\sin(2x)$$

To determine the value of s, we see that when $s=0$, y_p is proportional to y_h and, therefore, a solution to the homogeneous equation. Since this is not permitted, we set $s=1$. Then, substituting Eq. (b) with $s=1$ into Eq. (a), we obtain, after using Mathematica procedure M4.25,

$$e^{-3x}\left((4b_0 - 1)\cos(2x) - 4a_0\sin(2x)\right) = 0$$

For this equation to be satisfied for all x, we require that

$$a_0 = 0, \qquad b_0 = 1/4$$

and the particular solution is

$$y_p(x) = \frac{1}{4} x e^{-3x}\sin(2x)$$

The complete solution is

$$y(x) = D_1 e^{-3x}\cos(2x) + D_1 e^{-3x}\sin(2x) + \frac{1}{4} x e^{-3x}\sin(2x)$$

This result is verified with Mathematica procedure M4.26.

Example 4.17

Consider the nonhomogeneous equation

$$\frac{d^2y}{dt^2} + 2\zeta\omega\frac{dy}{dt} + \omega^2 y = Q(t) \tag{a}$$

where $Q(t)$ is given by the Fourier series as expressed in Eq. (3.18), that is,

$$Q(t) = \frac{a_0}{2} + \sum_{n=1}^{\infty}\{a_n\cos(\Omega_n t) + b_n\sin(\Omega_n t)\}$$

and $\Omega_n = n\omega_o$ and a_j and b_j are known constants that are determined from Eq. (3.19). Then, Eq. (a) becomes

$$\frac{d^2y}{dt^2} + 2\zeta\omega\frac{dy}{dt} + \omega^2 y = \frac{a_0}{2} + \sum_{n=1}^{\infty}\{a_n\cos(\Omega_n t) + b_n\sin(\Omega_n t)\} \tag{b}$$

When $0 \le \zeta < 1$, the homogeneous solution is given by Eq. (d) of Example 4.10, that is,

$$y_h(t) = D_1 e^{-\omega\zeta t}\cos(\omega_d t) + D_1 e^{-\omega\zeta t}\sin(\omega_d t) \tag{c}$$

where

$$\omega_d = \omega\sqrt{1-\zeta^2}$$

In view of Eq. (4.65) and Eq. (c), we assume a particular solution of the form

$$y_p(t) = A_o + \sum_{n=1}^{\infty}\{A_n\cos(\Omega_n t) + B_n\sin(\Omega_n t)\} \tag{d}$$

since the terms in Eq. (d) are not solutions to the homogeneous solution given by Eq. (c) and provided that when $\zeta=0$, $\omega \neq m\omega_o$.

Substituting Eq. (d) into Eq. (b), we obtain

$$\begin{aligned}&\sum_{n=1}^{\infty} A_n\left\{\left(\omega^2-\Omega_n^2\right)\cos(\Omega_n t) - 2\zeta\omega\Omega_n\sin(\Omega_n t)\right\}\\ &+\sum_{n=1}^{\infty} B_n\left\{\left(\omega^2-\Omega_n^2\right)\sin(\Omega_n t) + 2\zeta\omega\Omega_n\cos(\Omega_n t)\right\}\\ &+\omega^2 A_o = \frac{a_0}{2} + \sum_{n=1}^{\infty}\{a_n\cos(\Omega_n t) + b_n\sin(\Omega_n t)\}\end{aligned} \tag{e}$$

Collecting terms in Eq. (e), we arrive at

$$\begin{aligned}&\sum_{n=1}^{\infty}\left\{A_n\left(\omega^2-\Omega_n^2\right) + 2\zeta\omega\Omega_n B_n\sin(\Omega_n t) - a_n\right\}\cos(\Omega_n t)\\ &\sum_{n=1}^{\infty}\left\{B_n\left(\omega^2-\Omega_n^2\right) - 2\zeta\omega\Omega_n A_n\sin(\Omega_n t) - b_n\right\}\sin(\Omega_n t) + \omega^2 A_o - \frac{a_0}{2} = 0\end{aligned} \tag{f}$$

For Eq. (f) to be satisfied for arbitrary t, the following system of equations in matrix notation must be satisfied

$$\begin{pmatrix}\omega^2-\Omega_n^2 & 2\zeta\omega\Omega_n\\ -2\zeta\omega\Omega_n & \omega^2-\Omega_n^2\end{pmatrix}\begin{Bmatrix}A_n\\ B_n\end{Bmatrix} = \begin{Bmatrix}a_n\\ b_n\end{Bmatrix} \tag{g}$$

In addition, we see from Eq. (f) that

$$A_o = \frac{a_0}{2\omega^2} \tag{h}$$

Solving Eq. (g) for A_n and B_n, we obtain

$$A_n = \frac{1}{D_n}\left\{a_n\left(\omega^2 - \Omega_n^2\right) - 2b_n\zeta\omega\Omega_n\right\}$$

$$B_n = \frac{1}{D_n}\left\{b_n\left(\omega^2 - \Omega_n^2\right) + 2a_n\zeta\omega\Omega_n\right\} \tag{i}$$

where

$$D_n = \left(\omega^2 - \Omega_n^2\right)^2 + \left(2\zeta\omega\Omega_n\right)^2 \tag{j}$$

Substituting Eqs. (h) and (i) into Eq. (d), gives

$$y_p(t) = \frac{a_0}{2\omega^2} + \sum_{n=1}^{\infty}\frac{1}{D_n}\left\{\left(a_n\left(\omega^2 - \Omega_n^2\right) - 2b_n\zeta\omega\Omega_n\right)\cos(\Omega_n t)\right.$$

$$\left. + \left(b_n\left(\omega^2 - \Omega_n^2\right) + 2a_n\zeta\omega\Omega_n\right)\sin(\Omega_n t\right\} \tag{k}$$

or, upon re-arrangement,

$$y_p(t) = \frac{a_0}{2\omega^2} + \sum_{n=1}^{\infty}\frac{a_n}{D_n}\left\{\left(\omega^2 - \Omega_n^2\right)\cos(\Omega_n t) + 2b_n\zeta\omega\Omega_n\sin(\Omega_n t\right\}$$

$$+ \sum_{n=1}^{\infty}\frac{b_n}{D_n}\left\{\left(\omega^2 - \Omega_n^2\right)\sin(\Omega_n t) - 2b_n\zeta\omega\Omega_n\cos(\Omega_n t\right\} \tag{l}$$

Equation (k) is verified for the case of $n=1$ with Mathematica procedure M4.27.

Although Eq. (l) is the correct particular solution, it can be placed in a more compact form as follows. Let us define the quantity

$$\tan\theta_n = \frac{2\zeta\omega\Omega_n}{\omega^2 - \Omega_n^2}$$

Therefore,

$$\sin\theta_n = \frac{2\zeta\omega\Omega_n}{\sqrt{D_n}}, \quad \cos\theta_n = \frac{\omega^2 - \Omega_n^2}{\sqrt{D_n}}$$

where D_n is given by Eq. (j). Then Eq. (l) becomes

$$y_p(t) = \frac{a_0}{2\omega^2} + \sum_{n=1}^{\infty}\frac{a_n}{\sqrt{D_n}}\left\{\cos\theta_n\cos(\Omega_n t) + \sin\theta_n\sin(\Omega_n t\right\}$$

$$+ \sum_{n=1}^{\infty}\frac{b_n}{\sqrt{D_n}}\left\{\cos\theta_n\sin(\Omega_n t) - \sin\theta_n\cos(\Omega_n t\right\}$$

or

$$y_p(t) = \frac{a_0}{2\omega^2} + \sum_{n=1}^{\infty} \frac{1}{\sqrt{D_n}} \{a_n \cos(\Omega_n t - \theta_n) + b_n \sin(\Omega_n t - \theta_n)\}$$

Example 4.18

We shall obtain the particular solution to Eq. (a) of Example 4.17 for the case where $Q(x) = f_o e^{j\Omega x}$ and $0 \le \zeta < 1$. Then Eq. (a) of Example 4.17 becomes,

$$\frac{d^2y}{dx^2} + 2\zeta\omega\frac{dy}{dx} + \omega^2 y = f_o e^{j\Omega x} \tag{a}$$

From Eq. (c) of Example 4.10, it was found that the homogeneous solution to Eq. (a) is

$$y_h(x) = C_1 e^{(-\omega\zeta + j\omega_d)x} + C_2 e^{(-\omega\zeta - j\omega_d)x}$$

From Eq. (4.65), we select the trial function as

$$y_p(x) = A_o x^s e^{j\Omega x} \tag{b}$$

Since the argument of the exponential of the trial function is unrelated to the arguments of the exponentials in the homogeneous solution, we can set $s = 0$. Substituting Eq. (b) with $s = 0$ into Eq. (a), we obtain

$$\left(-\Omega^2 + j2\zeta\omega\Omega + \omega^2\right) A_o e^{j\Omega x} = f_o e^{j\Omega x}$$

which yields

$$A_o = \frac{f_o}{\omega^2 - \Omega^2 + j2\zeta\omega\Omega}$$

However, from Eqs. (2.24) and (2.25),

$$\omega^2 - \Omega^2 + j2\zeta\omega\Omega = re^{j\theta}$$

where

$$r = \sqrt{\left(\omega^2 - \Omega^2\right)^2 + \left(2\zeta\omega\Omega\right)^2}$$

$$\theta = \tan^{-1}\frac{2\zeta\omega\Omega}{\omega^2 - \Omega^2}$$

Therefore,

$$A_o = \frac{f_o}{r} e^{-j\theta}$$

and the particular solution given by Eq. (b) becomes

$$y_p(x) = \frac{f_o}{r} e^{j(\Omega x - \theta)}$$

Compare this solution to that obtained in Example 4.17.

4.2.6 Particular Solutions: Variation of Parameters

Consider the second-order equation

$$L[y] = a_2(x)\frac{d^2y}{dx^2} + a_1(x)\frac{dy}{dx} + a_0(x)y = Q(x) \tag{4.71}$$

Let $y_1(x)$ and $y_2(x)$ be the two linearly independent solutions to the homogeneous equation, that is,

$$L[y_h] = 0 \tag{4.72}$$

where

$$y_h(x) = C_1 y_1(x) + C_2 y_2(x) \tag{4.73}$$

In other words,

$$\begin{aligned} L[y_1] &= 0 \\ L[y_2] &= 0 \end{aligned} \tag{4.74}$$

We assume a particular solution of the form

$$y_p(x) = v_1(x)y_1(x) + v_2(x)y_2(x) \tag{4.75}$$

such that the functions $v_j(x)$ satisfy Eq. (4.71). The total solution is

$$y(x) = y_h(x) + y_p(x) \tag{4.76}$$

Substituting Eq. (4.76) into Eq. (6.71) and noting that

$$\frac{dy_p}{dx} = v_1\frac{dy_1}{dx} + v_2\frac{dy_2}{dx} + \frac{dv_1}{dx}y_1 + \frac{dv_2}{dx}y_2$$

$$\frac{d^2y_p}{dx^2} = v_1\frac{d^2y_1}{dx^2} + \frac{dv_1}{dx}\frac{dy_1}{dx} + v_2\frac{d^2y_2}{dx^2} + \frac{dv_2}{dx}\frac{dy_2}{dx} + \frac{d}{dx}\left(\frac{dv_1}{dx}y_1 + \frac{dv_2}{dx}y_2\right)$$

Eq. (4.71) becomes, with $a_j = a_j(x)$, $j = 0, 1, 2$,

$$a_2\frac{d^2}{dx^2}\left(y_h+y_p\right)+a_1\frac{d}{dx}\left(y_h+y_p\right)+a_0\left(y_h+y_p\right)=Q(x)$$

$$\overbrace{a_2\frac{d^2y_h}{dx^2}+a_1\frac{dy_h}{dx}+a_0y_h}^{L[y_h]=0}+a_2\frac{d^2y_p}{dx^2}+a_1\frac{dy_p}{dx}+a_0y_p=Q(x)$$

$$\begin{gathered}a_2\left[v_1\frac{d^2y_1}{dx^2}+\frac{dv_1}{dx}\frac{dy_1}{dx}+v_2\frac{d^2y_2}{dx^2}+\frac{dv_2}{dx}\frac{dy_2}{dx}+\frac{d}{dx}\left(\frac{dv_1}{dx}y_1+\frac{dv_2}{dx}y_2\right)\right]\\+a_1\left[v_1\frac{dy_1}{dx}+v_2\frac{dy_2}{dx}+\frac{dv_1}{dx}y_1+\frac{dv_2}{dx}y_2\right]+a_0\left[v_1y_1+v_2y_2\right]=Q(x)\end{gathered}\tag{4.77}$$

$$\begin{gathered}\overbrace{v_1\left[a_2\frac{d^2y_1}{dx^2}+a_1\frac{dy_1}{dx}+a_0y_1\right]}^{L[y_1]=0}+\overbrace{v_2\left[a_2\frac{d^2y_2}{dx^2}+a_1\frac{dy_2}{dx}+a_0y_2\right]}^{L[y_2]=0}\\+a_2\left[\frac{dv_1}{dx}\frac{dy_1}{dx}+\frac{dv_2}{dx}\frac{dy_2}{dx}+\frac{d}{dx}\left(\frac{dv_1}{dx}y_1+\frac{dv_2}{dx}y_2\right)\right]\\+a_1\left[\frac{dv_1}{dx}y_1+\frac{dv_2}{dx}y_2\right]=Q(x)\end{gathered}$$

As indicated, the first two terms in the last equation are zero since $y_1(x)$ and $y_2(x)$ satisfy Eq. (4.74). Then, the remaining terms in the last equation of Eq. (4.77) can be written as

$$a_2\left[\frac{dv_1}{dx}\frac{dy_1}{dx}+\frac{dv_2}{dx}\frac{dy_2}{dx}\right]+\left(a_2\frac{d}{dx}+a_1\right)\left[\frac{dv_1}{dx}y_1+\frac{dv_2}{dx}y_2\right]=Q(x)\tag{4.78}$$

We have one equation with two unknowns: dv_j/dx, $j=1, 2$. Hence, if we choose

$$\frac{dv_1}{dx}y_1+\frac{dv_2}{dx}y_2=0\tag{4.79}$$

then Eq. (4.78) becomes

$$\frac{dv_1}{dx}\frac{dy_1}{dx}+\frac{dv_2}{dx}\frac{dy_2}{dx}=\frac{Q(x)}{a_2}\tag{4.80}$$

Writing Eqs. (4.79) and (4.80) in matrix form yields

$$\begin{pmatrix}y_1 & y_2\\ \dfrac{dy_1}{dx} & \dfrac{dy_2}{dx}\end{pmatrix}\begin{Bmatrix}\dfrac{dv_1}{dx}\\ \dfrac{dv_2}{dx}\end{Bmatrix}=\begin{Bmatrix}0\\ \dfrac{Q(x)}{a_2}\end{Bmatrix}\tag{4.81}$$

Solving Eq. (4.81) for the derivatives of v_j, $j=1, 2$, results in [recall Eq. (1.46)]

$$\frac{dv_1}{dx} = -\frac{y_2(x)Q(x)}{a_2(x)W[y_1, y_2]}$$
$$\frac{dv_2}{dx} = \frac{y_1(x)Q(x)}{a_2(x)W[y_1, y_2]} \tag{4.82}$$

where

$$W[y_1, y_2] = \begin{vmatrix} y_1 & y_2 \\ \dfrac{dy_1}{dx} & \dfrac{dy_2}{dx} \end{vmatrix} \tag{4.83}$$

is called the *Wronskian*. If the Wronskian is not equal to zero, the solutions are linearly independent. [The Wronskian is the determinant of the coefficients; hence, if it is not zero, the solutions are linearly independent. Recall Eq. (1.27ff).] Integrating Eq. (4.82) gives

$$v_1 = -\int \frac{y_2(x)Q(x)}{a_2(x)W[y_1, y_2]} dx$$
$$v_2 = \int \frac{y_1(x)Q(x)}{a_2(x)W[y_1, y_2]} dx \tag{4.84}$$

where we have set the constants of integration to zero. Note that Eq. (4.82) satisfies Eq. (4.79). Then, from Eqs. (4.75) and (4.84), the particular solution is

$$y_p(x) = -y_1(x)\int \frac{y_2(x)Q(x)}{a_2(x)W[y_1, y_2]} dx + y_2(x)\int \frac{y_1(x)Q(x)}{a_2(x)W[y_1, y_2]} dx \tag{4.85}$$

Upon using Eqs. (4.73), (4.76), and (4.85), the complete solution can be written as

$$y(x) = C_1 y_1(x) + C_2 y_2(x) + \int \frac{[y_2(x)y_1(\xi) - y_1(x)y_2(\xi)]Q(\xi)}{a_2(\xi)W[y_1(\xi), y_2(\xi)]} d\xi \tag{4.86}$$

After $Q(x)$ has been specified, one uses Eq. (4.86) in the boundary/initial conditions to obtain C_1 and C_2.

We now illustrate this technique with two examples.

Example 4.19

Consider the differential equation

$$\frac{d^2y}{dx^2} - 3\frac{dy}{dx} + 2y = -\frac{e^{2x}}{1+e^x} \tag{a}$$

The homogeneous equation is obtained by using Eq. (4.34), that is, $y_h = e^{\lambda x}$. Then substituting y_h into the homogeneous portion of Eq. (a) results in

$$\lambda^2 - 3\lambda + 2 = 0$$

and, therefore, $\lambda_1 = 2$ and $\lambda_2 = 1$. Then the homogeneous solution becomes

$$y_h(x) = C_1 e^x + C_2 e^{2x}$$

and we find that

$$y_1(x) = e^x \quad \text{and} \quad \frac{dy_1}{dx} = e^x$$

$$y_2(x) = e^{2x} \quad \text{and} \quad \frac{dy_2}{dx} = 2e^{2x}$$

The Wronskian is

$$W[y_1, y_2] = \begin{vmatrix} e^x & e^{2x} \\ e^x & 2e^{2x} \end{vmatrix} = 2e^{3x} - e^{3x} = e^{3x} \neq 0$$

The particular solution is obtained from Eq. (4.85), thus,

$$y_p(x) = -e^x \int \frac{e^{2x}}{e^{3x}} \left(-\frac{e^{2x}}{1+e^x} \right) dx + e^{2x} \int \frac{e^x}{e^{3x}} \left(-\frac{e^{2x}}{1+e^x} \right) dx$$

$$= e^x \int \frac{e^x}{1+e^x} dx - e^{2x} \int \frac{1}{1+e^x} dx$$

$$= e^x \ln\left(1+e^x\right) - e^{2x}\left(x - \ln\left(1+e^x\right)\right)$$

where the fact that $a_2(x) = 1$ was used. The total solution is

$$y(x) = C_1 e^x + C_2 e^{2x} + e^x \ln\left(1+e^x\right) - e^{2x}\left(x - \ln\left(1+e^x\right)\right)$$

The boundary/initial conditions are used to determine C_1 and C_2.

This result is verified with Mathematica procedure M4.28.

Example 4.20

We extend Example 4.10 and consider the nonhomogeneous equation

$$\frac{d^2y}{dx^2} + 2\zeta\omega \frac{dy}{dx} + \omega^2 y = Q(x) \tag{a}$$

When $0 \leq \zeta < 1$, the homogeneous solution is given by Eq. (d) of that example. Thus,

$$y_h(x) = D_1 e^{-\omega\zeta x} \cos\left(\omega_d x\right) + D_1 e^{-\omega\zeta x} \sin\left(\omega_d x\right)$$

where

$$\omega_d = \omega\sqrt{1-\zeta^2}$$

Therefore,

$$y_1(x) = e^{-\omega\zeta x}\cos(\omega_d x)$$

$$y_2(x) = e^{-\omega\zeta x}\sin(\omega_d x)$$

and the Wronskian is

$$W[y_1, y_2] = \begin{vmatrix} e^{-\omega\zeta x}\cos(\omega_d x) & e^{-\omega\zeta x}\sin(\omega_d x) \\ -e^{-\omega\zeta x}\left(\omega\cos(\omega_d x) + \omega_d\sin(\omega_d x)\right) & e^{-\omega\zeta x}\left(-\omega\sin(\omega_d x) + \omega_d\cos(\omega_d x)\right) \end{vmatrix}$$

$$= \omega_d e^{-2\omega\zeta x}$$

which does not equal zero provided that $\zeta \neq 1$ or $\omega = 0$.

Using the particular form of the solution given in Eq. (4.86) with $a_2(x) = 1$, we obtain

$$\begin{aligned} y_p(x) &= \int \frac{e^{2\omega\zeta\xi}}{\omega_d}\Big[e^{-\omega\zeta x}\sin(\omega_d x)e^{-\omega\zeta\xi}\cos(\omega_d\xi) \\ &\quad - e^{-\omega\zeta x}\cos(\omega_d x)e^{-\omega\zeta\xi}\sin(\omega_d\xi)\Big]Q(\xi)d\xi \\ &= \frac{1}{\omega_d}\int e^{\omega\zeta(\xi - x)}\left[\sin(\omega_d x)\cos(\omega_d\xi) - \cos(\omega_d x)\sin(\omega_d\xi)\right]Q(\xi)d\xi \\ &= \frac{1}{\omega_d}\int e^{-\omega\zeta(x-\xi)}\sin(\omega_d(x-\xi))Q(\xi)d\xi \end{aligned}$$

Then the complete solution is

$$\begin{aligned} y(x) &= D_1 e^{-\omega\zeta x}\cos(\omega_d x) + D_1 e^{-\omega\zeta x}\sin(\omega_d x) \\ &\quad + \frac{1}{\omega_d}\int e^{-\omega\zeta(x-\xi)}\sin(\omega_d(x-\xi))Q(\xi)d\xi \end{aligned} \tag{b}$$

Equation (b) is verified with Mathematica procedure M4.29. When $\zeta = 0$, Eq. (b) reduces to

$$y(x) = D_1\cos(\omega x) + D_1\sin(\omega x) + \frac{1}{\omega}\int \sin(\omega(x-\xi))Q(\xi)d\xi \tag{c}$$

We shall have use for Eqs. (b) and (c) subsequently.

The results from Examples 4.17, 4.18, and 4.20 are summarized in Table 4.3.

4.2.7 Conversion to a System of First-Order Differential Equations

An nth-order differential equation or a system of differential equations can be converted to a system of first-order equations by making a series of substitutions as follows. Consider the differential equation

$$a_0(x)\frac{d^n y}{dx^n} + a_1(x)\frac{d^{n-1}y}{dx^{n-1}} + \cdots + a_{n-1}(x)\frac{dy}{dx} + a_n(x)y = Q(x) \tag{4.87}$$

We introduce a set of functions y_k that satisfy the following equations

$$y_{k+1} = \frac{d^k y}{dx^k} \quad k = 0,1,\ldots,n-1 \tag{4.88}$$

Then,

$$\begin{aligned}
y_1 &= y \\
y_2 &= \frac{dy}{dx} = \frac{dy_1}{dx} \to \frac{dy_1}{dx} = y_2 \\
y_3 &= \frac{d^2y}{dx^2} = \frac{d}{dx}\left(\frac{dy_1}{dx}\right) = \frac{dy_2}{dx} \to \frac{dy_2}{dx} = y_3 \\
&\vdots \\
y_n &= \frac{d^{n-1}y}{dx^{n-1}} = \frac{dy_{n-1}}{dx} \to \frac{dy_{n-1}}{dx} = y_n
\end{aligned} \tag{4.89}$$

TABLE 4.3
Summary of the Solutions to the Nonhomogeneous Differential Equations of Examples 4.17, 4.18, and 4.20

Example	Differential Equation	Solution
4.17	$\frac{d^2y}{dt^2} + 2\zeta\omega\frac{dy}{dt} + \omega^2 y = \frac{a_0}{2} + \sum_{n=1}^{\infty}\{a_n\cos(\Omega_n t) + b_n\sin(\Omega_n t)\}$	$y(t) = D_1 e^{-\omega\zeta t}\cos(\omega_d t) + D_1 e^{-\omega\zeta t}\sin(\omega_d t) + \frac{a_0}{2\omega^2} + \sum_{n=1}^{\infty}\frac{1}{\sqrt{D_n}}\begin{Bmatrix} a_n\cos(\Omega_n t - \theta_n) \\ +b_n\sin(\Omega_n t - \theta_n)\end{Bmatrix}$ where $0 \le \zeta < 1$, $\omega_d = \omega\sqrt{1-\zeta^2}$, and $\theta_n = \tan^{-1}\frac{2\zeta\omega\Omega_n}{\omega^2 - \Omega_n^2}$, $D_n = (\omega^2 - \Omega_n^2)^2 + (2\zeta\omega\Omega_n)^2$
4.18	$\frac{d^2y}{dx^2} + 2\zeta\omega\frac{dy}{dx} + \omega^2 y = f_o e^{j\Omega x}$	$y(x) = C_1 e^{(-\omega\zeta + j\omega_d)x} + C_2 e^{(-\omega\zeta - j\omega_d)x} + \frac{f_o}{r}e^{j(\Omega x - \theta)}$ where $0 \le \zeta < 1$, $\omega_d = \omega\sqrt{1-\zeta^2}$, and $r = \sqrt{(\omega^2 - \Omega^2)^2 + (2\zeta\omega\Omega)^2}$, $\theta = \tan^{-1}\frac{2\zeta\omega\Omega}{\omega^2 - \Omega^2}$
4.20	$\frac{d^2y}{dx^2} + 2\zeta\omega\frac{dy}{dx} + \omega^2 y = Q(x)$	$y(x) = D_1 e^{-\omega\zeta x}\cos(\omega_d x) + D_1 e^{-\omega\zeta x}\sin(\omega_d x) + \frac{1}{\omega_d}\int e^{-\omega\zeta(x-\xi)}\sin(\omega_d(x-\xi))Q(\xi)d\xi$ where $0 \le \zeta < 1$, $\omega > 0$, and $\omega_d = \omega\sqrt{1-\zeta^2}$.

Substituting Eq. (4.89) into Eq. (4.87) yields

$$a_0(x)\frac{d}{dx}\left(\frac{d^{n-1}y}{dx^{n-1}}\right)+a_1(x)\frac{d^{n-1}y}{dx^{n-1}}+\cdots+a_{n-1}(x)\frac{dy}{dx}+a_n(x)y=Q(x)$$

$$a_0(x)\frac{dy_n}{dx}+a_1(x)y_n+\cdots+a_{n-1}(x)y_2+a_n(x)y_1=Q(x)$$

and, therefore,

$$\frac{dy_n}{dx}=\frac{1}{a_0(x)}\left[-a_1(x)y_n-\cdots-a_{n-1}(x)y_2-a_n(x)y_1+Q(x)\right] \tag{4.90}$$

It is important to realize that the boundary/initial conditions are also transformed by using Eq. (4.89).

Equations (4.89) and (4.90) can be placed in matrix form as follows

$$\begin{Bmatrix}\frac{dy_1}{dx}\\ \frac{dy_2}{dx}\\ \vdots\\ \frac{dy_n}{dx}\end{Bmatrix}=\begin{pmatrix}0 & 1 & 0 & \cdots & 0\\ 0 & 0 & 1 & \cdots & 0\\ \vdots & \vdots & & & \vdots\\ \frac{-a_n(x)}{a_0(x)} & \frac{-a_{n-1}(x)}{a_0(x)} & & \cdots & \frac{-a_1(x)}{a_0(x)}\end{pmatrix}\begin{Bmatrix}y_1\\ y_2\\ \vdots\\ y_n\end{Bmatrix}+\begin{Bmatrix}0\\ 0\\ \vdots\\ \frac{Q(x)}{a_0(x)}\end{Bmatrix} \tag{4.91}$$

In control theory, Eq. (4.91) is known as the *state-space* form.

For a system of differential equations, the same procedure is followed; however, it is cumbersome to do so for a general case. Instead, this case will be illustrated in Example 4.22.

Example 4.21

Consider the following non-dimensional second-order differential equation describing the motions of a single-degree-of-freedom system given by

$$\frac{d^2w}{d\tau^2}+2\zeta\frac{dw}{d\tau}+w=f(\tau)$$

subject to the initial conditions $w(0)=x_o$ and $\dot{w}(0)=v_o$ and the over dot indicates the derivative with respect to τ. Then, using Eqs. (4.89) and (4.90)

$$w_1=w,\qquad \frac{dw}{d\tau}=\frac{dw_1}{d\tau}=w_2$$

$$\frac{d}{d\tau}\left(\frac{dw}{d\tau}\right)=\frac{dw_2}{d\tau}=-2\zeta\frac{dw}{d\tau}-w+f(\tau)=-2\zeta w_2-w_1+f(\tau)$$

and the initial conditions become $w_1(0)=x_o$ and $w_2(0)=v_o$. Using Eq. (4.91), the matrix form of these equations is

$$\begin{Bmatrix} \dfrac{dw_1}{d\tau} \\ \dfrac{dw_2}{d\tau} \end{Bmatrix} = \begin{pmatrix} 0 & 1 \\ -1 & -2\zeta \end{pmatrix} \begin{Bmatrix} w_1 \\ w_2 \end{Bmatrix} + \begin{Bmatrix} 0 \\ f(\tau) \end{Bmatrix}$$

Example 4.22

Consider the following set of coupled ordinary differential equations

$$\begin{aligned} m_1 \frac{d^2x_1}{dt^2} + (c_1+c_2)\frac{dx_1}{dt} + (k_1+k_2)x_1 - c_2\frac{dx_2}{dt} - k_2x_2 = F_1 f_1(t) \\ m_2 \frac{d^2x_2}{dt^2} + c_2\frac{dx_2}{dt} + k_2x_2 - c_2\frac{dx_1}{dt} - k_2x_1 = F_2 f_2(t) \end{aligned} \tag{a}$$

For this case, we assume the following

$$\begin{aligned} x_1 = y_1 \quad \frac{dx_1}{dt} = y_2 \rightarrow \frac{dy_1}{dt} = y_2 \\ x_2 = y_3 \quad \frac{dx_2}{dt} = y_4 \rightarrow \frac{dy_3}{dt} = y_4 \end{aligned} \tag{b}$$

Substituting Eq. (b) into Eq. (a), results in

$$\begin{aligned} m_1\frac{dy_2}{dt} + (c_1+c_2)y_2 - c_2y_4 + (k_1+k_2)y_1 - k_2y_3 = f_1(t) \\ m_2\frac{dy_4}{dt} + c_2y_4 - c_2y_2 + k_2y_3 - k_2y_1 = f_2(t) \end{aligned} \tag{c}$$

Placing Eqs. (b) and (c) into matrix form gives

$$\begin{Bmatrix} \dfrac{dy_1}{dt} \\ \dfrac{dy_2}{dt} \\ \dfrac{dy_3}{dt} \\ \dfrac{dy_4}{dt} \end{Bmatrix} = \begin{pmatrix} 0 & 1 & 0 & 0 \\ -\dfrac{k_1+k_2}{m_1} & -\dfrac{c_1+c_2}{m_1} & \dfrac{k_2}{m_1} & \dfrac{c_2}{m_1} \\ 0 & 0 & 0 & 1 \\ \dfrac{k_2}{m_2} & \dfrac{c_2}{m_2} & -\dfrac{k_2}{m_2} & -\dfrac{c_2}{m_2} \end{pmatrix} \begin{Bmatrix} y_1 \\ y_2 \\ y_3 \\ y_4 \end{Bmatrix} + \begin{Bmatrix} 0 \\ \dfrac{f_1(t)}{m_1} \\ 0 \\ \dfrac{f_2(t)}{m_2} \end{Bmatrix}$$

4.2.8 Orthogonal Functions and the Solutions to a System of Second-Order Equations

Using the material of Section 1.7 as the basis for what is to follow, we consider the system of N coupled second-order equations

$$\boldsymbol{M\ddot{y}} + \boldsymbol{Ky} = \boldsymbol{f} \tag{4.92}$$

where $\boldsymbol{y}=\boldsymbol{y}(t)$ is an $(N\times 1)$ column matrix, $\boldsymbol{f}=\boldsymbol{f}(t)$ is an $(N\times 1)$ column matrix, $\boldsymbol{M}$ and $\boldsymbol{K}$ are symmetric $(N\times N)$ matrices, and the over dot indicates the derivative with respect to time t. This system of equations is subjected to a set of initial conditions $\boldsymbol{y}(0)$ and $\dot{\boldsymbol{y}}(0)$.

We start by setting $\boldsymbol{f}=0$ and assuming that $\boldsymbol{y}=\boldsymbol{x}e^{j\omega t}$, where

$$\boldsymbol{x} = \begin{Bmatrix} x_1 \\ x_2 \\ \vdots \\ x_N \end{Bmatrix} \tag{4.93}$$

Then, Eq. (4.92) becomes

$$\left(\boldsymbol{K} - \omega^2\boldsymbol{M}\right)\boldsymbol{x} = 0 \tag{4.94}$$

Using the procedure given in Section 1.7, we assume that we have obtained the solution to Eq. (4.94) which is composed of the eigenvalues ω_n^2, the corresponding eigenvectors $\boldsymbol{x}^{(n)}$, $n=1, 2, \ldots, N$, and the matrix of eigenvectors $\boldsymbol{\Phi}$ [recall Eq. (1.59)], which is given by

$$\boldsymbol{\Phi} = \left\{\boldsymbol{x}^{(1)} \quad \boldsymbol{x}^{(2)} \quad \ldots \quad \boldsymbol{x}^{(N)}\right\}$$

Since $\boldsymbol{M}$ and $\boldsymbol{K}$ are symmetric matrices, the eigenvectors are orthogonal. Each $\boldsymbol{x}^{(j)}$ is a solution to

$$\left(\boldsymbol{K} - \omega_j^2\boldsymbol{M}\right)\boldsymbol{x}^{(j)} = 0 \quad j=1,2,\ldots,N \tag{4.95}$$

Upon pre-multiplying Eq. (4.95) by the transpose of $\boldsymbol{x}^{(j)}$, we arrive at

$$\left(\boldsymbol{x}^{(j)}\right)^T\left[\boldsymbol{K} - \omega_j^2\boldsymbol{M}\right]\boldsymbol{x}^{(j)} = 0$$

$$\left(\boldsymbol{x}^{(j)}\right)^T\boldsymbol{K}\boldsymbol{x}^{(j)} - \omega_j^2\left(\boldsymbol{x}^{(j)}\right)^T\boldsymbol{M}\boldsymbol{x}^{(j)} = 0$$

$$\hat{K}_{jj} - \hat{M}_{jj}\omega_j^2 = 0$$

$$\omega_j^2 = \frac{\hat{K}_{jj}}{\hat{M}_{jj}} \quad j=1,2,\ldots,N$$

where

$$\begin{aligned}\hat{K}_{jj} &= \left(\boldsymbol{x}^{(j)}\right)^T\boldsymbol{K}\boldsymbol{x}^{(j)} \\ \hat{M}_{jj} &= \left(\boldsymbol{x}^{(j)}\right)^T\boldsymbol{M}\boldsymbol{x}^{(j)}\end{aligned} \tag{4.96}$$

and it was shown in Eq. (1.38) that expressions of the form $(\boldsymbol{x}^{(j)})^T\boldsymbol{M}\boldsymbol{x}^{(j)}$ and $(\boldsymbol{x}^{(j)})^T\boldsymbol{K}\boldsymbol{x}^{(j)}$ are scalars.

To solve the system of differential equations given by Eq. (4.92), we assume a solution of the form

$$\boldsymbol{y} = \boldsymbol{\Phi}\boldsymbol{\eta} \tag{4.97}$$

where $\boldsymbol{\eta}=\boldsymbol{\eta}(t)$ is an $(N\times 1)$ column matrix given by

$$\eta = \begin{Bmatrix} \eta_1(t) \\ \eta_2(t) \\ \vdots \\ \eta_N(t) \end{Bmatrix} \tag{4.98}$$

Substituting Eq. (4.97) into Eq. (4.92) and then pre-multiplying the result by $\boldsymbol{\Phi}^T$ yields

$$\Phi^T \boldsymbol{M}\Phi\ddot{\eta} + \Phi^T \boldsymbol{K}\Phi\eta = \Phi^T \boldsymbol{f} \tag{4.99}$$

Since the eigenfunctions are orthogonal, Eq. (4.99) becomes

$$\begin{aligned} &\boldsymbol{M}_D\ddot{\eta} + \boldsymbol{K}_D\eta = \Phi^T \boldsymbol{f} \\ &\ddot{\eta} + \left(\boldsymbol{M}_D\right)^{-1}\boldsymbol{K}_D\eta = \left(\boldsymbol{M}_D\right)^{-1}\Phi^T \boldsymbol{f} \\ &\ddot{\eta} + \omega_D^2\eta = \boldsymbol{q} \end{aligned} \tag{4.100}$$

where

$$\boldsymbol{q} = \left(\boldsymbol{M}_D\right)^{-1}\Phi^T \boldsymbol{f}$$

$$\boldsymbol{M}_D = \begin{pmatrix} \hat{M}_{11} & & 0 \\ & \ddots & \\ 0 & & \hat{M}_{NN} \end{pmatrix} \qquad \boldsymbol{K}_D = \begin{pmatrix} \hat{K}_{11} & & 0 \\ & \ddots & \\ 0 & & \hat{K}_{NN} \end{pmatrix}$$

and

$$\omega_D^2 = \left(\boldsymbol{M}_D\right)^{-1}\boldsymbol{K}_D = \begin{pmatrix} \hat{K}_{11}/\hat{M}_{11} & & 0 \\ & \ddots & \\ 0 & & \hat{K}_{NN}/\hat{M}_{NN} \end{pmatrix}$$

The matrix $\boldsymbol{q}$ is an $(N\times 1)$ column matrix.

Therefore, Eq. (4.100) is the system of uncoupled equations

$$\ddot{\eta}_j + \omega_j^2\eta_j = q_j(t) \quad j = 1,2,\ldots,N \tag{4.101}$$

subject to the initial conditions $\eta_j(0)$ and $\dot{\eta}_j(0)$. The solution to Eq. (4.101) is given by Eq. (c) of Example 4.20. Thus, with the appropriate change in notation

$$\eta_j(t) = C_{j,1}\cos\left(\omega_j t\right) + C_{j,2}\sin\left(\omega_j t\right) + \frac{1}{\omega_j}\int_0^t \sin\left(\omega_j(t-\xi)\right)q_j(\xi)d\xi \tag{4.102}$$

If the initial conditions are $\eta_j(0)=\eta_{o,j}$ and $\dot{\eta}_j(0)=v_{o,j}$, then from Eq. (4.102) it is found that

$$\begin{aligned}\eta_j(0)&=\eta_{o,j}=C_{j,1}\\ \dot{\eta}_j(0)&=v_{o,j}=\omega_j C_{j,2}\end{aligned} \tag{4.103}$$

and Eq. (4.102) becomes

$$\eta_j(t)=\eta_{o,j}\cos\left(\omega_j t\right)+\frac{v_{o,j}}{\omega_j}\sin\left(\omega_j t\right)+\frac{1}{\omega_j}\int_0^t \sin\left(\omega_j(t-\xi)\right)q_j(\xi)d\xi \tag{4.104}$$

Then, using Eq. (4.97), the solution is

$$\begin{Bmatrix} y_1(t)\\ y_2(t)\\ \vdots\\ y_N(t)\end{Bmatrix}=\begin{pmatrix} x_1^{(1)} & x_1^{(2)} & \cdots & x_1^{(N)}\\ x_2^{(1)} & x_2^{(2)} & \cdots & x_2^{(N)}\\ \vdots & \vdots & \ddots & \vdots\\ x_N^{(1)} & x_N^{(2)} & \cdots & x_N^{(N)}\end{pmatrix}\begin{Bmatrix} \eta_1(t)\\ \eta_2(t)\\ \vdots\\ \eta_N(t)\end{Bmatrix}$$

or

$$\boldsymbol{y}(t)=\sum_{j=1}^{N}\boldsymbol{x}^{(j)}\eta_j(t) \tag{4.105}$$

or

$$y_i(t)=\sum_{j=1}^{N}X_{ij}\eta_j(t) \tag{4.106}$$

where $X_{ij}=x_i^{(j)}$. To determine the expressions for the initial conditions $\eta_j(0)$ and $\dot{\eta}_j(0)$, we set $t=0$ in Eq. (4.105) and multiply it by $\left(\boldsymbol{x}^{(i)}\right)^T \boldsymbol{M}$ to obtain

$$\left(\boldsymbol{x}^{(i)}\right)^T \boldsymbol{M}\boldsymbol{y}(0)=\sum_{j=1}^{N}\left(\boldsymbol{x}^{(i)}\right)^T \boldsymbol{M}\boldsymbol{x}^{(j)}\eta_j(0)=\hat{M}_{ii}\eta_i(0)$$

or

$$\eta_i(0)=\left(\boldsymbol{x}^{(i)}\right)^T \boldsymbol{M}\boldsymbol{y}(0)\Big/\hat{M}_{ii} \tag{4.107}$$

and we have used Eq. (4.96). In a similar manner, it can be shown that

$$\dot{\eta}_i(0)=\left(\boldsymbol{x}^{(i)}\right)^T \boldsymbol{M}\dot{\boldsymbol{y}}(0)\Big/\hat{M}_{ii} \tag{4.108}$$

We now extend these results and consider the following system of differential equations

$$\boldsymbol{M}\ddot{\boldsymbol{y}}+\boldsymbol{C}\dot{\boldsymbol{y}}+\boldsymbol{K}\boldsymbol{y}=\boldsymbol{f} \tag{4.109}$$

where $\boldsymbol{C}$ is an $(N \times N)$ matrix. We start by substituting Eq. (4.97) into Eq. (4.109) and then pre-multiply the result by $\boldsymbol{\Phi}^T$ to obtain

$$\boldsymbol{\Phi}^T \boldsymbol{M}\boldsymbol{\Phi}\ddot{\eta} + \boldsymbol{\Phi}^T \boldsymbol{C}\boldsymbol{\Phi}\dot{\eta} + \boldsymbol{\Phi}^T \boldsymbol{K}\boldsymbol{\Phi}\eta = \boldsymbol{\Phi}^T \boldsymbol{f} \tag{4.110}$$

Since the matrices $\boldsymbol{M}$ and $\boldsymbol{K}$ are symmetric, the eigenvectors are orthogonal and we can use the previous results in Eq. (4.110) to arrive at

$$\begin{aligned} &\boldsymbol{M}_D\ddot{\eta} + \boldsymbol{\Phi}^T \boldsymbol{C}\boldsymbol{\Phi}\dot{\eta} + \boldsymbol{K}_D\eta = \boldsymbol{\Phi}^T \boldsymbol{f} \\ &\ddot{\eta} + (\boldsymbol{M}_D)^{-1} \boldsymbol{\Phi}^T \boldsymbol{C}\boldsymbol{\Phi}\dot{\eta} + (\boldsymbol{M}_D)^{-1} \boldsymbol{K}_D\eta = (\boldsymbol{M}_D)^{-1} \boldsymbol{\Phi}^T \boldsymbol{f} \\ &\ddot{\eta} + (\boldsymbol{M}_D)^{-1} \boldsymbol{\Phi}^T \boldsymbol{C}\boldsymbol{\Phi}\dot{\eta} + \omega_D^2\eta = \boldsymbol{q} \end{aligned} \tag{4.111}$$

where, again,

$$\boldsymbol{q} = (\boldsymbol{M}_D)^{-1} \boldsymbol{\Phi}^T \boldsymbol{f} \tag{4.112}$$

For Eq. (4.111) to uncouple, we require that

$$(\boldsymbol{M}_D)^{-1} \boldsymbol{\Phi}^T \boldsymbol{C}\boldsymbol{\Phi} \to \boldsymbol{C}_D \tag{4.113}$$

where $\boldsymbol{C}_D$ is a diagonal matrix. The diagonalization can be accomplished by assuming that

$$\boldsymbol{C} = \alpha\boldsymbol{M} + \beta\boldsymbol{K} \tag{4.114}$$

where α and β are real-valued constants, and because $\boldsymbol{M}$ and $\boldsymbol{K}$ are symmetric matrices, $\boldsymbol{C}$ is symmetric. In this case, substitution of Eq. (4.114) into Eq. (4.113) yields

$$\begin{aligned} \boldsymbol{C}_D &= (\boldsymbol{M}_D)^{-1} \boldsymbol{\Phi}^T (\alpha\boldsymbol{M} + \beta\boldsymbol{K})\boldsymbol{\Phi} \\ &= \alpha(\boldsymbol{M}_D)^{-1} \boldsymbol{\Phi}^T \boldsymbol{M}\boldsymbol{\Phi} + \beta(\boldsymbol{M}_D)^{-1} \boldsymbol{\Phi}^T \boldsymbol{K}\boldsymbol{\Phi} \\ &= \alpha(\boldsymbol{M}_D)^{-1} \boldsymbol{M}_D + \beta(\boldsymbol{M}_D)^{-1} \boldsymbol{K}_D \\ &= \alpha\boldsymbol{I} + \beta\omega_D^2 \end{aligned} \tag{4.115}$$

where $\boldsymbol{C}_D$ can be written as

$$\boldsymbol{C}_D = \begin{pmatrix} 2\zeta_1\omega_1 & & 0 \\ & \ddots & \\ 0 & & 2\zeta_N\omega_N \end{pmatrix} = \begin{pmatrix} \alpha + \beta\omega_1^2 & & 0 \\ & \ddots & \\ 0 & & \alpha + \beta\omega_N^2 \end{pmatrix}$$

Then, Eq. (4.111) becomes

$$\ddot{\eta} + \boldsymbol{C}_D\dot{\eta} + \omega_D^2\eta = \boldsymbol{q} \tag{4.116}$$

and we have the following system of uncoupled equations

$$\ddot{\eta}_j + 2\zeta_j\omega_j\dot{\eta}_j + \omega_j^2\eta_j = q_j(t) \quad j = 1,2,\ldots,N \tag{4.117}$$

It is noted that the assumption given by Eq. (4.114) is only necessary if this solution method is chosen. If a direct numerical approach is selected (as shown in Mathematica procedure M4.30 used in Example 4.23) or if the Laplace transform method is used (as shown in Example 8.13), this assumption is unnecessary.

The solution to Eq. (4.117) is given by Eq. (b) of Example 4.20. Thus, with an appropriate change in notation and assuming that $0 \le \zeta_j < 1$,

$$\begin{aligned}\eta_j(t) &= D_{j,1}e^{-\omega_j\zeta_j t}\cos\left(\omega_{j,d}t\right) + D_{j,1}e^{-\omega_j\zeta_j t}\sin\left(\omega_{j,d}t\right)\\ &+ \frac{1}{\omega_{j,d}}\int_0^t e^{-\omega_j\zeta_j(t-\xi)}\sin\left(\omega_{j,d}(t-\xi)\right)q_j(\xi)d\xi\end{aligned} \tag{4.118}$$

where

$$\omega_{j,d} = \omega_j\sqrt{1-\zeta_j^2} \tag{4.119}$$

If the initial conditions are $\eta_j(0) = \eta_{o,j}$ and $\dot{\eta}_j(0) = v_{o,j}$, then from Eq. (4.118) it is found that

$$\begin{aligned}D_{j,1} &= \eta_{o,j}\\ D_{j,2} &= \frac{1}{\omega_{j,d}}\left(v_{o,j} + \eta_{o,j}\omega_j\zeta_j\right)\end{aligned} \tag{4.120}$$

and Eq. (4.118) can be written as

$$\begin{aligned}\eta_j(t) &= \eta_{o,j}e^{-\omega_j\zeta_j t}\cos\left(\omega_{j,d}t\right) + \frac{1}{\omega_{j,d}}\left(v_{o,j} + \eta_{o,j}\omega_j\zeta_j\right)e^{-\omega_j\zeta_j t}\sin\left(\omega_{j,d}t\right)\\ &+ \frac{1}{\omega_{j,d}}\int_0^t e^{-\omega_j\zeta_j(t-\xi)}\sin\left(\omega_{j,d}(t-\xi)\right)q_j(\xi)d\xi\end{aligned} \tag{4.121}$$

We shall now illustrate this result.

Example 4.23

Consider the following system of differential equations

$$\boldsymbol{M}\ddot{\boldsymbol{y}} + \boldsymbol{C}\dot{\boldsymbol{y}} + \boldsymbol{K}\boldsymbol{y} = 0 \tag{a}$$

where

$$\boldsymbol{M} = \begin{pmatrix} 1.5 & 0 \\ 0 & 2.4 \end{pmatrix}, \quad \boldsymbol{K} = \begin{pmatrix} 40 & -10 \\ -10 & 45 \end{pmatrix}, \quad \boldsymbol{C} = 0.2\boldsymbol{M}, \quad \boldsymbol{y} = \begin{Bmatrix} x_1(t) \\ x_2(t) \end{Bmatrix} \tag{b}$$

Upon comparing the expression of C in Eq. (b) with Eq. (4.114), we see that $\alpha = 0.2$ and $\beta = 0$. The system given by Eq. (a) is subject to the following initial conditions

$$\begin{Bmatrix} y_1(0) \\ y_2(0) \end{Bmatrix} = \begin{Bmatrix} 0.05 \\ -0.05 \end{Bmatrix} \text{ and } \begin{Bmatrix} \dot{y}_1(0) \\ \dot{y}_2(0) \end{Bmatrix} = \begin{Bmatrix} 0 \\ 0 \end{Bmatrix} \tag{c}$$

The over dot in Eq. (c) indicates the derivative with respect to time.

The first step is to determine the eigenvalues and eigenvectors. Since the matrices in Eq. (b) are symmetric, the eigenvectors will be orthogonal. We assume that $\boldsymbol{y} = \boldsymbol{x}e^{j\omega t}$ where

$$\boldsymbol{x} = \begin{Bmatrix} x_1 \\ x_2 \end{Bmatrix} \tag{d}$$

To determine the eigenvalues and eigenvectors, we set $C = 0$ and Eq. (a) becomes

$$\left(\boldsymbol{A} - \omega^2 \boldsymbol{I}\right)\boldsymbol{x} = 0 \tag{e}$$

where

$$\boldsymbol{A} = \boldsymbol{M}^{-1}\boldsymbol{K} = \begin{pmatrix} 1.5 & 0 \\ 0 & 2.4 \end{pmatrix}^{-1} \begin{pmatrix} 40 & -10 \\ -10 & 45 \end{pmatrix} = \begin{pmatrix} 26.67 & -6.667 \\ -4.167 & 18.75 \end{pmatrix} \tag{f}$$

With Eq. (e) in this form, we can use Eq. (1.42) to determine the eigenvalues and Eq. (1.50) to determine the eigenvectors. Then, since

$$\text{tr}(\boldsymbol{A}) = 26.67 + 18.75 = 45.417$$

$$\det(\boldsymbol{A}) = 26.67 \times 18.75 - 6.667 \times 4.167 = 472.22$$

the eigenvalues are

$$\omega_1 = \sqrt{\frac{45.417}{2} - \sqrt{\left(\frac{45.417}{2}\right)^2 - 472.22}} = 4.015$$

$$\omega_2 = \sqrt{\frac{45.417}{2} + \sqrt{\left(\frac{45.417}{2}\right)^2 - 472.22}} = 5.413$$

and the corresponding eigenvectors are

$$\boldsymbol{\Phi} = \left\{\boldsymbol{x}^{(1)} \quad \boldsymbol{x}^{(2)}\right\} = \begin{pmatrix} \dfrac{(4.015)^2 - 18.75}{-4.167} & \dfrac{(5.413)^2 - 18.75}{-4.167} \\ 1 & 1 \end{pmatrix}$$

$$= \begin{pmatrix} 0.632 & -2.532 \\ 1 & 1 \end{pmatrix} \tag{g}$$

We now return the Eq. (a) and assume a solution of the form given by Eq. (4.97), which upon substitution into Eq. (a) results in Eq. (4.116) where

$$\boldsymbol{C}_D = \begin{pmatrix} 2\zeta_1\omega_1 & 0 \\ 0 & 2\zeta_2\omega_2 \end{pmatrix} = \begin{pmatrix} 0.2 & 0 \\ 0 & 0.2 \end{pmatrix} \tag{h}$$

Thus,

$$\begin{aligned} \zeta_1 &= \frac{0.2}{2\omega_1} = 0.0249 \\ \zeta_2 &= \frac{0.2}{2\omega_2} = 0.0185 \end{aligned} \tag{h}$$

The solution to Eq. (a) is given by Eq. (4.121), which requires $\eta_j(0) = \eta_{o,j}$ and $\dot{\eta}_j(0) = v_{o,j}$. To determine these quantities, we use Eqs. (4.107) and (4.108). From Eqs. (4.108) and (b), we see that $\dot{\eta}_j(0) = v_{o,j} = 0$. From Eq. (4.107), we find that

$$\begin{aligned} \eta_1(0) &= \left(\boldsymbol{x}^{(1)}\right)^T \boldsymbol{My}(0) \Big/ \hat{M}_{11} \\ &= \frac{1}{2.999}\{0.632 \quad 1\}\begin{pmatrix} 1.5 & 0 \\ 0 & 2.4 \end{pmatrix}\begin{Bmatrix} 0.05 \\ -0.05 \end{Bmatrix} \\ &= -0.0242 \\ \eta_2(0) &= \left(\boldsymbol{x}^{(2)}\right)^T \boldsymbol{My}(0) \Big/ \hat{M}_{22} \\ &= \frac{1}{12.017}\{-2.532 \quad 1\}\begin{pmatrix} 1.5 & 0 \\ 0 & 2.4 \end{pmatrix}\begin{Bmatrix} 0.05 \\ -0.05 \end{Bmatrix} \\ &= -0.02579 \end{aligned} \tag{i}$$

where we have used Eq. (4.96) to obtain

$$\hat{M}_{11} = \left(\boldsymbol{x}^{(1)}\right)^T \boldsymbol{Mx}^{(1)} = \{0.632 \quad 1\}\begin{pmatrix} 1.5 & 0 \\ 0 & 2.4 \end{pmatrix}\begin{Bmatrix} 0.632 \\ 1 \end{Bmatrix} = 2.999$$

$$\hat{M}_{22} = \left(\boldsymbol{x}^{(2)}\right)^T \boldsymbol{Mx}^{(2)} = \{-2.532 \quad 1\}\begin{pmatrix} 1.5 & 0 \\ 0 & 2.4 \end{pmatrix}\begin{Bmatrix} -2.532 \\ 1 \end{Bmatrix} = 12.017$$

Since $\boldsymbol{f} = 0$ and $v_{o,j} = 0$, Eq. (4.121) gives

$$\begin{aligned} \eta_1(t) &= -0.0242e^{-0.1t}\cos(4.015t) - 0.000602e^{-0.1t}\sin(4.015t) \\ \eta_2(t) &= -0.02579e^{-0.1t}\cos(5.413t) - 0.000477e^{-0.1t}\sin(5.413t) \end{aligned} \tag{j}$$

where we have use Eq. (4.119) and Eq. (h) to find that

$$\begin{aligned} \omega_{1,d} &= \omega_1\sqrt{1-\zeta_1^2} \approx \omega_1 = 4.015 \\ \omega_{2,d} &= \omega_2\sqrt{1-\zeta_2^2} \approx \omega_2 = 5.413 \end{aligned}$$

Then, from Eqs. (4.106) and (g), we obtain

$$y_1(t) = 0.632\eta_1(t) - 2.532\eta_2(t)$$

$$= e^{-0.1t}\left\{\begin{matrix}-0.01529\cos(4.015t) - 0.00038\sin(4.015t)\\ +0.0653\cos(5.413t) + 0.0012\sin(5.413t)\end{matrix}\right\} \quad \text{(k)}$$

and

$$y_2(t) = \eta_1(t) + \eta_2(t)$$

$$= -e^{-0.1t}\left\{\begin{matrix}0.0242\cos(4.015t) + 0.000602\sin(4.015t)\\ +0.02579\cos(5.413t) + 0.000477\sin(5.413t)\end{matrix}\right\} \quad \text{(l)}$$

It is easily shown that Eqs. (k) and (l) satisfy the initial conditions given by Eq. (c). Equations (k) and (l) are verified with Mathematica procedure M4.30.

4.2.9 Making Differential Equations Non-Dimensional

In this section, we shall examine several ordinary differential equations that are encountered in engineering and show how they can be converted to non-dimensional form. We shall give additional examples for partial differential equations in Section 7.3 and in Examples 8.12 and 8.19. It is mentioned that the way in which one makes an equation non-dimensional is not unique. There is often more than one way to do it; however, one goal of nondimensionalizing an equation is to have a minimum number of independent non-dimensional parameters appearing in the final result.

Example 4.24: Displacement of a Single-Degree-of-Freedom System

Consider a single-degree-of-freedom system with mass m_o (kg), stiffness k (N/m), and viscous damping c (Ns/m). The mass is subjected to a force $F_o f(t)$, where F_o (N) is the magnitude of the force, $f(t)$ is a non-dimensional shape factor, and t (s) is time. If the displacement about the static equilibrium position is $y=y(t)$ (m), then the governing equation of motion of the mass is

$$m_o\frac{d^2y}{dt^2} + c\frac{dy}{dt} + ky = F_o f(t) \quad \text{N} \quad \text{(a)}$$

To make the equation non-dimensional, we first divide Eq. (a) by m_o to obtain

$$\frac{d^2y}{dt^2} + \frac{c}{m_o}\frac{dy}{dt} + \frac{k}{m_o}y = \frac{F_o}{m_o}f(t) \quad \text{(b)}$$

We notice that the units of k/m_o are

$$\frac{k}{m_o} \to \frac{\text{N}}{\text{m}}\frac{1}{\text{kg}} = \frac{\text{kg}\cdot\text{m}}{\text{s}^2}\frac{1}{\text{m}\cdot\text{kg}} = \frac{1}{\text{s}^2}$$

where we have used Newton's law, that is, N = kg·m/s². Therefore, we define

$$\omega_n^2 = \frac{k}{m_o}\ \mathrm{s}^{-2}$$

and Eq. (b) becomes

$$\frac{d^2y}{dt^2} + \frac{c}{m_o}\frac{dy}{dt} + \omega_n^2 y = \frac{F_o}{m_o} f(t) \tag{c}$$

We next define the non-dimensional time τ as $\tau = t\omega_n$ and Eq. (c) becomes

$$\frac{\omega_n^2}{\omega_n^2}\frac{d^2y}{dt^2} + \frac{\omega_n}{\omega_n}\frac{c}{m_o}\frac{dy}{dt} + \omega_n^2 y = \frac{F_o}{m_o} f(t)$$

$$\omega_n^2 \frac{d^2y}{d\tau^2} + \frac{c\omega_n}{m_o}\frac{dy}{d\tau} + \omega_n^2 y = \frac{F_o}{m_o} g(\tau) \tag{d}$$

$$\frac{d^2y}{d\tau^2} + \frac{c}{m_o\omega_n}\frac{dy}{d\tau} + y = \frac{F_o}{\omega_n^2 m_o} g(\tau) = \frac{F_o}{k} g(\tau)$$

It is seen that

$$\frac{c}{m_o\omega_n} \to \frac{\mathrm{Ns}}{\mathrm{m}}\frac{1}{\mathrm{kg}}\frac{1}{1/\mathrm{s}} = \frac{\mathrm{N}}{\mathrm{kg}\cdot\mathrm{m/s}^2} = \frac{\mathrm{N}}{\mathrm{N}} = 1 \quad \text{(no dimensions)}$$

$$\frac{F_o}{k} \to \frac{\mathrm{N}}{\mathrm{N/m}} = \mathrm{m}$$

We define the following non-dimensional quantity

$$2\zeta = \frac{c}{m_o\omega_n}$$

and Eq. (d) becomes

$$\frac{d^2y}{d\tau^2} + 2\zeta\frac{dy}{d\tau} + y = \frac{F_o}{k} g(\tau)$$

If we assume that $F_o \neq 0$, then for our last conversion we set

$$w = \frac{y}{F_o / k} \to \frac{\mathrm{m}}{\mathrm{m}} = 1$$

and our final result is

$$\frac{d^2w}{d\tau^2} + 2\zeta\frac{dw}{d\tau} + w = g(\tau) \tag{e}$$

In a similar manner, we convert the initial conditions. It is assumed that the initial conditions are $y(0) = X_o$ (m) and $dy(0)/dt = V_o$ (m/s). These quantities are converted to non-dimensional quantities as follows

$$y(0) = X_o \rightarrow \frac{y(0)}{F_o/k} = \frac{X_o}{F_o/k} \rightarrow w(0) = x_o$$

where

$$x_o = \frac{X_o}{F_o/k} \rightarrow \frac{\text{m}}{\text{m}} = 1$$

and

$$\frac{dy(0)}{dt} = V_o \rightarrow \frac{dy(0)}{(\omega_n F_o/k)dt} = \frac{V_o}{\omega_n F_o/k} \rightarrow \frac{dw(0)}{d\tau} = v_o$$

where

$$v_o = \frac{V_o}{\omega_n F_o/k} \rightarrow \frac{\text{m/s}}{(1/\text{s})\text{m}} = 1$$

To apply the results to a specific system, one uses the definitions of the various parameters to return to dimensional quantities.

Example 4.25: Simple LRC Series Circuit

Consider the following circuit equation for an inductor, resistance, and capacitance in series

$$L\frac{di(t)}{dt} + Ri(t) + \frac{1}{C}\int_0^t i(\xi)d\xi = V_o v(t) \quad \text{V} \tag{a}$$

In Eq. (a), i is the current (A), V_o is the voltage (V), R is a resistor (ohms), L is an inductor (H), C is a capacitor (F), $v(t)$ is a non-dimensional voltage shape function, and t (s) is time. The dummy variable of integration ξ has the units of t; that is, ξ also has to be made non-dimensional. In addition, J = Ws = Nm, A = C/s, ohm = Js/C^2, V = J/C, F = C/V, and H = Js^2/C^2, where J is joule and C is coulomb. Note: In this example, C is coulomb when not italicized and C italicized denotes capacitance.

We divide Eq. (a) by L, set $\omega_o^2 = 1/(CL)$, and note that

$$\omega_o^2 = 1/(CL) \rightarrow \frac{1}{\text{FH}} = \frac{1}{(\text{C/V})(\text{Js}^2/\text{C}^2)} = \frac{\text{CV}}{\text{Js}^2} = \frac{1}{\text{s}^2}$$

Therefore, we introduce the non-dimensional time $\tau = \omega_o t$. Regarding the integrand, we make the following change of variables: $\omega_o \xi = \eta$, where η is non-dimensional, $d\xi = d\eta/\omega_o$, and the limit of the integral becomes $\omega_o t = \tau$. Then, after dividing Eq. (a) by L and using these definitions, we obtain

$$\frac{di(t)}{dt}+\frac{R}{L}i(t)+\omega_o^2\int_0^t i(\xi)d\xi=\frac{V_o}{L}v(t)$$

$$\omega_o\frac{di(\tau)}{d\tau}+\frac{R}{L}i(\tau)+\omega_o\int_0^\tau i(\eta)d\eta=\frac{V_o}{L}v(\tau) \qquad \text{(b)}$$

$$\frac{di(\tau)}{d\tau}+\frac{R}{L\omega_o}i(\tau)+\int_0^\tau i(\eta)d\eta=\frac{V_o}{\omega_o L}v(\tau)$$

However,

$$\frac{R}{L\omega_o}\to\frac{\text{Js}}{\text{C}^2}\frac{\text{C}^2}{\text{Js}^2}\frac{1}{1/\text{s}}=1$$

$$\frac{V_o}{\omega_o L}\to\frac{\text{J}}{\text{C}}\frac{1}{1/\text{s}}\frac{\text{C}^2}{\text{Js}^2}=\frac{\text{C}}{\text{s}}=\text{A}$$

Therefore, we introduce the following definitions

$$2\zeta=\frac{R}{L\omega_o}=\frac{R}{L}\sqrt{CL}=R\sqrt{\frac{C}{L}}$$

$$i_o=\frac{V_o}{\omega_o L}\quad \text{A}$$

$$\hat{i}(\tau)=\frac{i(\tau)}{i_o}$$

in Eq. (b) and arrive at our final result

$$\frac{di(\tau)}{d\tau}+2\zeta i(\tau)+\int_0^\tau i(\eta)d\eta=i_o v(\tau)$$

$$\frac{d\hat{i}(\tau)}{d\tau}+2\zeta\hat{i}(\tau)+\int_0^\tau \hat{i}(\eta)d\eta=v(\tau)$$

Example 4.26: Static Deflection of a Beam

A thin elastic beam has a length L (m), an area moment of inertia I (m^4), and a Young's modulus E (N/m^2). If the transverse displacement of the beam is $y=y(x)$ (m), and the distance along the beam is x (m), then the governing equation describing the displacement of the beam is

$$EI\frac{d^4y}{dx^4}=Q_o q(x)\quad \text{N/m} \qquad \text{(a)}$$

where Q_o (N/m) is the magnitude of an externally applied force per unit length and $q(x)$ is a non-dimensional shape function.

To make this equation non-dimensional, we start by letting $\eta = x/L$. Then Eq. (a) becomes

$$\frac{EI}{L^4}\frac{d^4y}{d\eta^4} = Q_o q(\eta)$$
$$\frac{d^4y}{d\eta^4} = \frac{Q_o L^4}{EI} q(\eta) \qquad \text{(b)}$$

We introduce a quantity d_o, where

$$d_o = \frac{Q_o L^4}{EI} \rightarrow \frac{\text{N}}{\text{m}}\frac{\text{m}^4}{\text{m}^4\text{N/m}^2} = \text{m}$$

Then Eq. (b) becomes

$$\frac{d^4y}{d\eta^4} = d_o q(\eta) \quad \text{m} \qquad \text{(c)}$$

To obtain our final result, we introduce the non-dimensional quantity

$$w = \frac{y}{d_o} \rightarrow \frac{\text{m}}{\text{m}} = 1$$

into Eq. (c) and obtain

$$\frac{d^4w}{d\eta^4} = q(\eta)$$

Example 4.27: Two-Degree-of-Freedom System

Consider a two-degree-of-freedom system that is composed of springs k_j (N/m), viscous dampers c_j (Ns/m), and masses m_j (kg), where $j = 1, 2$. The displacement of the respective masses is x_j (m) and a force $F_j f_j(t)$ is applied to m_j, where F_j (N) is the magnitude and $f_j(t)$ is a non-dimensional shape function. The governing equations for this system are

$$m_1\frac{d^2x_1}{dt^2} + (c_1 + c_2)\frac{dx_1}{dt} + (k_1 + k_2)x_1 - c_2\frac{dx_2}{dt} - k_2x_2 = F_1 f_1(t) \quad \text{N}$$
$$m_2\frac{d^2x_2}{dt^2} + c_2\frac{dx_2}{dt} + k_2x_2 - c_2\frac{dx_1}{dt} - k_2x_1 = F_2 f_2(t) \quad \text{N} \qquad \text{(a)}$$

where t (s) is time.

To start the process of making the equations non-dimensional, we extend the definition introduced in Example 4.24; that is, we divide both equations by m_1 and introduce the quantities

$$\omega_{nj} = \sqrt{\frac{k_j}{m_j}} \text{ s} \quad j = 1,2, \quad m_r = \frac{m_2}{m_1}$$

Then, Eq. (a) becomes

$$\begin{aligned}
&\frac{d^2x_1}{dt^2} + \left(\frac{c_1}{m_1} + \frac{c_2}{m_1}\right)\frac{dx_1}{dt} + (\omega_{n1}^2 + m_r\omega_{n2}^2)x_1 - \frac{c_2}{m_1}\frac{dx_2}{dt} - m_r\omega_{n2}^2 x_2 = \frac{F_1}{m_1} f_1(t) \\
&m_r \frac{d^2x_2}{dt^2} + \frac{c_2}{m_1}\frac{dx_2}{dt} + m_r\omega_{n2}^2 x_2 - \frac{c_2}{m_1}\frac{dx_1}{dt} - m_r\omega_{n2}^2 x_1 = \frac{F_2}{m_1} f_2(t)
\end{aligned} \tag{b}$$

where we have used the fact that

$$\frac{k_2}{m_1} = \frac{m_2}{m_1}\frac{k_2}{m_2} = m_r\omega_{n2}^2 \text{ s}^{-2}$$

Again, following Example 4.24, we introduce the non-dimensional time $\tau = \omega_{n1}t$ and Eq. (b) becomes

$$\begin{aligned}
&\omega_{n1}^2 \frac{d^2x_1}{d\tau^2} + \omega_{n1}\left(\frac{c_1}{m_1} + \frac{c_2}{m_1}\right)\frac{dx_1}{d\tau} + (\omega_{n1}^2 + m_r\omega_{n2}^2)x_1 - \frac{\omega_{n1}c_2}{m_1}\frac{dx_2}{d\tau} - m_r\omega_{n2}^2 x_2 = \frac{F_1}{m_1} f_1(\tau) \\
&m_r\omega_{n1}^2 \frac{d^2x_2}{d\tau^2} + \frac{\omega_{n1}c_2}{m_1}\frac{dx_2}{d\tau} + m_r\omega_{n2}^2 x_2 - \frac{\omega_{n1}c_2}{m_1}\frac{dx_1}{d\tau} - m_r\omega_{n2}^2 x_1 = \frac{F_2}{m_1} f_2(\tau)
\end{aligned} \tag{c}$$

The first equation of Eq. (c) is divided by ω_{n1}^2, the second equation is divided by $m_r\omega_{n1}^2$, and the following definitions are introduced

$$\omega_r = \frac{\omega_{n2}}{\omega_{n1}} \rightarrow \frac{1/\text{s}}{1/\text{s}} = 1$$

$$2\zeta_j = \frac{c_j}{m_j\omega_{nj}} \rightarrow \frac{\text{Ns}}{\text{m} \cdot \text{kg/s}} = 1$$

where, again, from Newton's law, N = kg·m/s². Then, Eq. (c) becomes

$$\begin{aligned}
&\frac{d^2x_1}{d\tau^2} + \left(2\zeta_1 + 2\zeta_2 m_r\omega_r\right)\frac{dx_1}{d\tau} + \left(1 + m_r\omega_r^2\right)x_1 - 2\zeta_2 m_r\omega_r \frac{dx_2}{d\tau} - m_r\omega_r^2 x_2 = \frac{F_1}{k_1} f_1(\tau) \text{ m} \\
&\frac{d^2x_2}{d\tau^2} + 2\zeta_2\omega_r \frac{dx_2}{d\tau} + \omega_r^2 x_2 - 2\zeta_2\omega_r \frac{dx_1}{d\tau} - \omega_r^2 x_1 = \frac{F_2}{k_1 m_r} f_2(\tau) \text{ m}
\end{aligned} \tag{d}$$

where we have used the fact that

$$\frac{c_2}{m_1\omega_{n1}} = \frac{m_2\omega_{n2}}{m_2\omega_{n2}}\frac{c_2}{m_1\omega_{n1}} = 2\zeta_2 m_r\omega_r$$

In this case, x_j retains its original units (m). Also note that

$$\frac{F_j}{k_1} \rightarrow \frac{\text{N}}{\text{N/m}} = \text{m}$$

4.2.10 Nonlinear Differential Equations: A Few Special Cases

A nonlinear differential equation is one in which the dependent variable or any of its derivatives appear with exponents greater than 1 and in which there are terms that are products of the dependent variable and its derivatives. Obtaining analytical solutions to these types of equations presents great challenges. In this section, we shall give a brief introduction to a small class of nonlinear ordinary differential equations that are amenable to analytical solutions. However, the solution techniques presented here cannot be generalized, and as with all approaches used to obtain solutions to nonlinear equations great caution must be exercised.

The main approach to these examples will be to reduce the second-order differential equation to a first-order differential equation by using the transformation

$$p = \frac{dy}{dx} \tag{4.122}$$

Then, the derivative of p can be described in two ways. When a differential equation is of the form

$$\frac{d^2y}{dx^2} = f(x, dy/dx) \tag{4.123}$$

then one uses the transformation

$$\frac{dp}{dx} = \frac{d^2y}{dx^2} \tag{4.124}$$

When a differential equation is of the form

$$\frac{d^2y}{dx^2} = f(y, dy/dx) \tag{4.125}$$

then one uses the transformation

$$\frac{dp}{dx} = \frac{dp}{dy}\frac{dy}{dx} = p\frac{dp}{dy} = \frac{d^2y}{dx^2} \tag{4.126}$$

We now illustrate the technique with several examples.

Example 4.28

Consider the nonlinear differential equation

$$\left(1+x^2\right)\frac{d^2y}{dx^2} + \left(\frac{dy}{dx}\right)^2 + 1 = 0 \tag{a}$$

Equation (a) is of the form given by Eq. (4.123). Thus, we use Eqs. (4.122) and (4.124) in Eq. (a) and obtain

$$\left(1+x^2\right)\frac{dp}{dx}+p^2+1=0$$

$$\frac{dp}{dx}+\frac{p^2+1}{1+x^2}=0 \tag{b}$$

$$\frac{dp}{p^2+1}+\frac{dx}{1+x^2}=0$$

Integrating Eq. (b), we find that

$$\int\frac{dp}{p^2+1}+\int\frac{dx}{1+x^2}=0$$

$$\tan^{-1}p+\tan^{-1}x=C$$

$$\tan^{-1}\left(\frac{p+x}{1-xp}\right)=C$$

$$\frac{p+x}{1-xp}=\tan C=C_1$$

Solving for p, we arrive at

$$p=\frac{dy}{dx}=\frac{C_1-x}{1+C_1x}$$

$$\int dy=\int\frac{C_1-x}{1+C_1x}dx$$

$$y=-\frac{x}{C_1}+\frac{1+C_1^2}{C_1^2}\ln\left(1+C_1x\right)+C_2$$

This result is verified with Mathematica procedure M4.31.

Example 4.29

Consider the nonlinear differential equation

$$2y\frac{d^2y}{dx^2}=\left(\frac{dy}{dx}\right)^2+1 \tag{a}$$

Equation (a) is of the form given by Eq. (4.125). Thus, we use Eqs. (4.122) and (4.126) in Eq. (a) and obtain

$$2yp\frac{dp}{dy}=p^2+1 \tag{b}$$

Equation (b) can be rewritten as

$$\frac{2pdp}{1+p^2} = \frac{dy}{y} \tag{c}$$

Integrating Eq. (c), we arrive at

$$2\int \frac{pdp}{1+p^2} = \int \frac{dy}{y}$$

$$\ln\left(1+p^2\right) = \ln y + \ln C \tag{d}$$

$$1+p^2 = Cy$$

$$p = \pm\sqrt{Cy-1}$$

Then, using the definition p in Eq. (d), we find that

$$\frac{dy}{dx} = \pm\sqrt{Cy-1}$$

$$\int \frac{dy}{\sqrt{Cy-1}} = \pm\int dx$$

$$\frac{2}{C}\sqrt{Cy-1} = \pm x + C_1$$

Solving for y yields

$$y = \frac{C}{4}\left(\pm x + C_1\right)^2 + \frac{1}{C}$$

This result is verified with Mathematica procedure M4.32.

Example 4.30

Consider the nonlinear differential equation

$$\frac{d^2y}{dx^2} = \frac{dy}{dx}e^y \tag{a}$$

Equation (a) is of the form given by Eq. (4.125). Thus, we use Eqs. (4.122) and (4.126) in Eq. (a) and obtain

$$\frac{dp}{dy} = e^y \tag{b}$$

Integrating Eq. (b), we obtain

$$\int dp = \int e^y dy$$

$$p = \frac{dy}{dx} = e^y + C$$

Again integrating, we find

$$\int \frac{dy}{e^y + C} = \int dx$$

$$\frac{1}{C}\ln\frac{e^y}{e^y + C} = x + C_1$$

$$\frac{e^y}{e^y + C} = e^{Cx + CC_1}$$

Solving for y yields

$$e^{-Cx - CC_1} = 1 + Ce^{-y}$$

$$e^{-y} = \frac{1}{C}\left(e^{-Cx - CC_1} - 1\right)$$

$$y = -\ln\left[\left(e^{-Cx - CC_1} - 1\right)/C\right]$$

This result is verified with Mathematica procedure M4.33.

4.2.11 Phase Plane and Direction Fields

The phase plane plot and the superposition of direction fields are very useful ways to visualize the solutions of second-order linear and nonlinear ordinary differential equations. Consider the second-order differential equation

$$\frac{d^2x}{dt^2} + a_1(t,x,\dot{x})\frac{dx}{dt} + a_2(t,x,\dot{x})x = f(t) \tag{4.127}$$

where the over dot indicates the derivative with respect to t, $a_j(t,x,\dot{x}), j = 1, 2$, are functions of t, x, and $\dot{x}$. When $a_j(t,x,\dot{x})$ is a function of x and $\dot{x}$, the equation is nonlinear. We assume that the equation is subject to the initial conditions $x(0) = x_o$ and $\dot{x}(0) = v_o$. Then the solution to Eq. (4.127) is, in general, a function of the initial conditions and $f(t)$.

To obtain a *phase plane* plot (*phase portrait*), one obtains the solution to Eq. (4.127) for a set of initial conditions and a given $f(t)$ and plots at every instance of time $(x(t), \dot{x}(t))$.

To obtain another form of the phase plane plot, one in which the entire field of the trajectories of the solutions encompassing all combinations of initial conditions are graphed, we use the results of Section 4.2.7 and transform Eq. (4.127) into the following two first-order equations

$$\frac{dx}{dt} = y$$
$$\frac{dy}{dt} = -a_1(t,x,y)y - a_2(t,x,y)x + f(t) \tag{4.128}$$

If Eq. (4.128) is independent of any explicit dependency on t, in other words, $a_j(t,x,y) \to a_j(x,y), j=1, 2$, and either $f(t)=0$ or $f(t)=\text{constant}$, then such a system is called an *autonomous system*.

For an autonomous system, we can write Eq. (4.128) as

$$\frac{dx}{dt} = F(x,y)$$
$$\frac{dy}{dt} = G(x,y) \tag{9.129}$$

in which the independent variable t does not appear explicitly in F and G. It is assumed that F and G are continuously differentiable in a region R in the (x,y)-plane, which is called the *phase plane*. Any curve in the phase plane is called a *trajectory*. At $t=t_0$, the system satisfies the initial conditions

$$x(t_0) = x_0$$
$$y(t_0) = y_0 \tag{4.130}$$

A *critical point* of the system, denoted (x_c, y_c), is a point such that

$$F(x_c, y_c) = G(x_c, y_c) = 0 \tag{4.131}$$

Then, $x(t)=x_c$ and $y(t)=y_c$ satisfy Eq. (4.129) and such a solution is called the *equilibrium solution*.

From Eq. (4.129), it follows that

$$\frac{dy}{dx} = \frac{dy/dt}{dx/dt} = \frac{G(x,y)}{F(x,y)} \tag{4.132}$$

In general, the solution to Eq. (4.132) cannot be expressed such that y is an explicit function of x. However, the solution can always be expressed as

$$P(x,y) = \text{constant} \tag{4.133}$$

If the value of the constant is held fixed, then Eq. (4.133) is a curve in the phase plane, which is called the *integral curve* of Eq. (4.132). In general, each integral curve is composed of one or more trajectories.

We shall now classify differential equations that describe a general linear system by considering the following special case of Eq. (4.129),

$$\frac{dx}{dt} = ax + by$$
$$\frac{dy}{dt} = cx + dy \tag{4.134}$$

A solution to Eq. (4.134) may be obtained by setting

$$x(t) = Ae^{\lambda t} \quad \text{and} \quad y(t) = Be^{\lambda t} \tag{4.135}$$

which results in

$$(a - \lambda)A + bB = 0$$
$$cA + (d - \lambda)B = 0 \tag{4.136}$$

or, in matrix form,

$$(\boldsymbol{A} - \lambda \boldsymbol{I})\boldsymbol{C} = 0$$

where

$$\boldsymbol{A} = \begin{pmatrix} a & b \\ c & d \end{pmatrix} \quad \boldsymbol{C} = \begin{Bmatrix} A \\ B \end{Bmatrix}$$

For Eq. (4.136) to have a non-trivial solution, λ must be a solution to

$$\begin{vmatrix} a - \lambda & b \\ c & d - \lambda \end{vmatrix} = 0 \tag{4.137}$$

From Eq. (1.42), we find that the solution to Eq. (4.137) can be written as

$$\lambda_{1,2} = \frac{\text{tr}}{2} \pm \sqrt{\left(\frac{\text{tr}}{2}\right)^2 - \det} \tag{4.138}$$

where tr is the trace of $\boldsymbol{A}$ and det is the determinant of $\boldsymbol{A}$, which are given by

$$\text{tr} = a + d$$
$$\det = ad - cb \tag{4.139}$$

From Eqs. (4.135) and (4.138), it is seen that when the real part of $\lambda_{1,2} > 0$, the solution becomes unbounded as $t \to \infty$.

We now consider several special cases of Eqs. (4.134) and (4.138).

Case 1

If det < 0, then Eq. (4.138) shows that there are two real eigenvalues, one positive and one negative. This type of linear system is called a *saddle*. An example of a saddle is provided by the equation

$$\frac{d^2x}{dt^2} - x = 0 \rightarrow \begin{array}{l} \dfrac{dx}{dt} = y \\ \dfrac{dy}{dt} = x \end{array} \tag{4.140}$$

Therefore, $a = d = 0$, $b = 1$, $c = 1$, and, consequently, tr = 0, and det = −1. Then from Eq. (4.138), we obtain

$$\lambda_{1,2} = \pm\sqrt{-\det} = \pm 1$$

Case 2

If det > 0 and tr² > 4det, then there are still two real eigenvalues, but both have the same sign as tr. If tr > 0, then both eigenvalues are positive, and the solution becomes unbounded as $t \rightarrow \infty$. This linear system is called an *unstable node*. The general solution is a linear combination of the two solutions $e^{\lambda_{1,2}t}$, and for large time the solution corresponding to the larger eigenvalue dominates. Similarly, if tr < 0, we have a *stable node*. An example of a stable node is provided by

$$\frac{d^2x}{dt^2} + \frac{5}{2}\frac{dx}{dt} + x = 0 \rightarrow \begin{array}{l} \dfrac{dx}{dt} = y \\ \dfrac{dy}{dt} = -x - \dfrac{5}{2}y \end{array} \tag{4.141}$$

Therefore, $a = 0$, $b = 1$, $c = -1$, $d = -5/2$, and, hence, tr = −5/2, and det = 1. Then from Eq. (4.138), we obtain

$$\lambda_{1,2} = -5/4 \pm \sqrt{25/16 - 1} = -5/4 \pm 3/4$$

Case 3

If det > 0 and tr² < 4det, then there are two complex eigenvalues with real part equal to tr/2. If the tr > 0, we have unbounded growth and the linear system is called an *unstable spiral* or *focus*. Similarly, if the tr < 0, we have a *stable spiral* or focus. An example of a stable spiral is provided by

$$\frac{d^2x}{dt^2} + \frac{dx}{dt} + x = 0 \rightarrow \begin{array}{l} \dfrac{dx}{dt} = y \\ \dfrac{dy}{dt} = -x - y \end{array} \tag{4.142}$$

Therefore, $a = 0$, $b = 1$, $c = -1$, $d = -1$, and, therefore, tr = −1, and det = 1. Then from Eq. (4.138), we obtain

$$\lambda_{1,2} = -1/2 \pm \sqrt{1/4 - 1} = -1/2 \pm j\sqrt{3}/2$$

Case 4

An example of an unstable spiral is provided by

$$\frac{d^2x}{dt^2} - \frac{dx}{dt} + x = 0 \rightarrow \begin{array}{l} \dfrac{dx}{dt} = y \\ \dfrac{dy}{dt} = -x + y \end{array} \tag{4.143}$$

Therefore, $a=0$, $b=1$, $c=-1$, $d=1$, and thus tr$=1$, and det$=1$. Then from Eq. (4.138), we obtain

$$\lambda_{1,2} = 1/2 \pm \sqrt{1/4-1} = 1/2 \pm j\sqrt{3}/2$$

Case 5

If det > 0 and tr$=0$, then there are two imaginary eigenvalues. The corresponding linear system is called a *center.* An example of a center is provided by

$$\frac{d^2x}{dt^2} + x = 0 \rightarrow \begin{array}{l} \dfrac{dx}{dt} = y \\ \dfrac{dy}{dt} = -x \end{array} \tag{4.144}$$

Therefore, $a=d=0$, $b=1$, $c=-1$, and, therefore, tr$=0$, and det$=1$. Then from Eq. (4.138), we obtain

$$\lambda_{1,2} = \pm\sqrt{-\text{det}} = \pm j$$

The results represented by Eqs. (4.140) to (4.144) are shown in two ways. In the first way, tr versus det is plotted in Figure 4.2 and the different regions identified. For the second way, the direction fields are plotted and shown in Figure 4.3, which is a plot of the state variables (dx/dt,dy/dt) given by Eq. (4.129) by considering x and y independent variables. In other words, we are plotting ($F(x,y)$,$G(x,y)$) for all combinations of x and y, which are the coordinates of a phase plane plot. Thus, a plot of (dx/dt,dy/dt) indicates the direction in which a point on the trajectory is moving, and as noted in Section 4.1.3, this type of graph is called

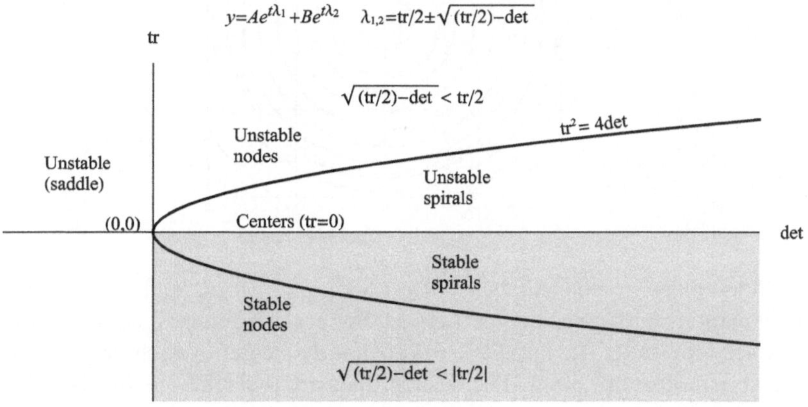

FIGURE 4.2 Stable and unstable regions of a general linear second-order differential equation as a function of tr and det.

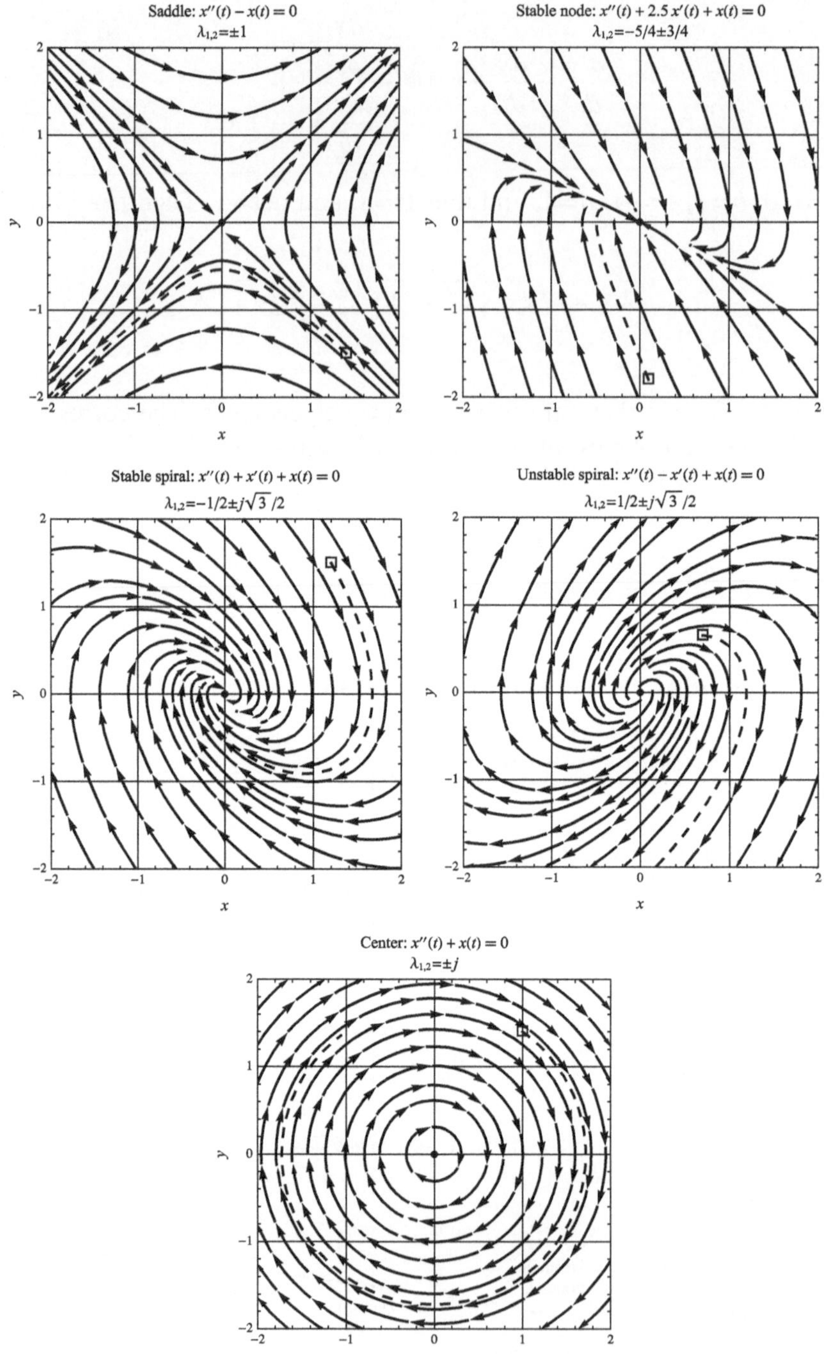

FIGURE 4.3 Phase plane graphs and their corresponding direction fields for various types of linear differential equations with constant coefficients. For each differential equation, the eigenvalues and the stability are indicated. The large dots located at the center of each graph indicate the location of the critical (equilibrium) point. The dashed lines are the solutions for the initial conditions indicated by the squares.

a direction field. These two state representations contain the same information, since they are two different forms of the same equation. Therefore, one can overlay the phase plane plot for specified initial values x_o and v_o on the direction field plot as shown in Figure 4.3.

MATHEMATICA PROCEDURES

Note: In order to make the Mathematica procedures that follow a little more readable, the `Clear` statement and many of the graphics enhancing statements have been omitted.

```
(*M4.1*)
DSolveValue[y'[x]/(2 x)-y[x]/(x^2-1)==0,y[x],x]

(*M4.2*)
DSolveValue[y'[x]==x^3 y[x]/(x^4+y[x]^4),y[x],x]

(*M4.3*)
z=y[x]/.DSolve[(y[x]+3 x^3 y[x]^2)y'[x]+3 x^2 y[x]^3-5 x^4==0,
y,x];
Simplify[Expand[(y-z[[1]]) (y-z[[2]]) (y-z[[3]])]]

(*M4.4*)
z=y[x]/.DSolve[(1-x y[x])y'[x]+y[x]^2+3 x y[x]^3==0,y,x];
Simplify[Expand[(y-z[[1]]) (y-z[[2]])]]

(*M4.5*)
Simplify[DSolveValue[y'[x]+p[x] y[x]==q[x],y[x],x]]

(*M4.6*)
y[x]/.DSolve[2 x Exp[2 y[x]] y'[x]==3 x^4+Exp[2 y[x]],y[x],x][[1]]

(*M4.7*)
Simplify[DSolveValue[y'[x]+p[x] y[x]==q[x] y[x]^n,y[x],x]]

(*M4.8*)
fccv[xx_]:=(fgh=yy/.NSolve[xx^3 yy^3+yy^2/2-c1-xx^5==0,yy];
   Extract[fgh,Position[fgh,_Real]][[1]])
c1=500; r=10;
curve=-(3 x^2 y^3-5 x^4)/(y+3 x^3 y^2);
datam=Table[{xx,fccv[xx]},{xx,Range[-10,-0.8,0.2]}];
datap=Table[{xx,fccv[xx]},{xx,Range[0.2,10,0.5]}];
Show[StreamPlot[{1,curve},{x,-r,r},{y,-r,r},
     StreamPoints->100,FrameLabel->{"x","y"}],
   ListLinePlot[datam,PlotStyle->{Black,Thickness[0.007]}],
   ListLinePlot[datap,PlotStyle->{Black,Thickness[0.007]}]]

(*M4.9*)
Collect[DSolveValue[y'''[x]-3 y''[x]+4 y[x]==0,y[x],x],
   Exp[2 x]]
```

```
(*M4.10*)
DSolveValue[z''[x]-Om^2 z[x]==0,z[x],x]

(*M4.11*)
Simplify[ExpToTrig[DSolveValue[z''[x]-Om^2 z[x]==0, z[x],x]]]

(*M4.12*)
DSolveValue[z''[x]+w^2 z[x]==0,z[x],x]

(*M4.13*)
Simplify[ExpToTrig[DSolveValue[z''''[x]-Om^4 z[x]==0,z[x],x]]]

(*M4.14*)
Collect[ComplexExpand[DSolveValue[z''''[x]+Om^4 z[x]==0,z[x],x]]
   /.Om/Sqrt[2]->q,{Exp[q x],Exp[-q x],Sin[q x],Cos[q x]}]

(*M4.15*)
Collect[DSolveValue[y''[x]+2 w z y'[x]+w^2 y[x]==0,y[x],x]
   /.{Exp[x (-w z-Sqrt[-w^2+w^2 z^2])]->Exp[-z w x] Exp[-I wd x],
   Exp[x (-w z+Sqrt[-w^2+w^2 z^2])]->Exp[-z w x] Exp[+I wd x]},
   Exp[-w x z]]

(*M4.16*)
FullSimplify[DSolveValue[{y''[x]+2 w z y'[x]+w^2 y[x]==0,y[0]==yo,
   y'[0]==vo},y[x],x]]

(*M4.17*)
Collect[DSolveValue[y''''[x]-2 a^2 y''[x]+a^4 y[x]==0,y[x],x],x]
Collect[ExpToTrig[DSolveValue[y''''[x]-2 a^2 y''[x]+a^4 y[x]==0,
   y[x],x]],{x,Cosh[a x],Sinh[a x]}]

(*M4.18*)
Simplify[ExpToTrig[DSolveValue[4 x^2 y''[x]+4 x y'[x]+
   (4 x^2-1) y[x]==0,y[x],x]]]

(*M4.19*)
Simplify[ExpToTrig[DSolveValue[{4 x^2 y''[x]+4 x y'[x]+
   (4 x^2-1) y[x]==0,y[0]==0,y[a]==y0},y[x],x]]]

(*M4.20*)
DSolveValue[a2 x^2 y''[x]+a1 x y'[x]+a0 y[x]==0,y[x],x]

(*M4.21*)
p=(a1-a2)^2/(4 a2);
DSolveValue[a2 x^2 y''[x]+a1 x y'[x]+p y[x]==0,y[x],x]

(*M4.22*)
fcn[w_,m_]:=D[w,r,r]+D[w,r]/r-m^2 w/r^2
p[m_]:=DSolveValue[fcn[fcn[y[r],m],m]==0,y[r],r]
p[0]
p[1]
p[m]
```

```
(*M4.23*)
Simplify[ExpToTrig[DSolveValue[y''[x]-4 y[x]==-2 x+8 x^2,y[x],
   x]]]

(*M4.24*)
FullSimplify[ExpToTrig[DSolveValue[{y''[x]-4 y[x]==-2 x+8 x^2,
   y[0]==y0,y'[a]==b0},y[x],x]]]

(*M4.25*)
yp=a0 x Exp[-3 x] Cos[2 x]+b0 x Exp[-3 x] Sin[2 x];
v=Simplify[D[yp,x,x]+6 D[yp,x]+13 yp-Exp[-3 x] Cos[2 x]]

(*M4.26*)
FullSimplify[DSolveValue[y''[x]+6 y'[x]+13 y[x]==Exp[-3 x]*
   Cos[2 x],y[x],x]]

(*M4.27*)
Simplify[DSolveValue[y''[t]+2 z w y'[t]+w^2 y[t]==
   ao/2+a Cos[Om t]+b Sin[Om t],y[t],t]]

(*M4.28*)
DSolveValue[y''[x]-3 y'[x]+2 y[x]==-Exp[2 x]/(1+Exp[x]),
   y[x],x]

(*M4.29*)
Simplify[DSolveValue[y''[t]+2 z w y'[t]+w^2 y[t]==q[t],y[t],t]]

(*M4.30*)
Chop[Simplify[DSolveValue[{1.5y1''[t]+0.3 y1'[t]+40 y1[t]-
   10 y2[t]==0,2.4y2''[t]+0.48 y2'[t]-10 y1[t]+45 y2[t]==0,
   y1[0]==0.05,y2[0]==-0.05,y1'[0]==0,y2'[0]==0},
   {y1[t],y2[t]},t]]]

(*M4.31*)
DSolveValue[(1+x^2) D[y[x],x,x]+D[y[x],x]^2+1==0,y[x],x]

(*M4.32*)
DSolveValue[2 y[x] D[y[x],x,x]-D[y[x],x]^2-1==0,y[x],x]

(*M4.33*)
DSolveValue[D[y[x],x,x]-D[y[x],x] Exp[y[x]]==0,y[x],x]
```

EXERCISES

Section 4.1.1

4.1 Obtain the solution to the following equations:

(a) $2x^2 \dfrac{dy}{dx} = x^2 + y^2$

(b) $\left(e^x + x\sin y\right)\dfrac{dy}{dx} + ye^x - \cos y = 0$

(c) $\dfrac{dy}{dx} = \dfrac{y}{x} - k\left[1 + \left(\dfrac{y}{x}\right)^2\right]^{1/2}$

(d) $\left(3y^2 + 6xe^y + \sin(2x)\right)dy + \left(4x^3 + 6e^y + 2y\cos(2x)\right)dx = 0$

4.2 For $\varepsilon \neq 0$, find the solution to

$$\frac{d\theta}{d\tau} + \theta^{1+\varepsilon} = 0$$

when the initial condition is $\theta(0) = 1$.

4.3 Given

$$\frac{dw}{d\varphi} - w\cot\varphi = k\left(\cos\varphi - \frac{2}{1+\cos\varphi}\right)$$

with the boundary condition $w(\varphi_o) = 0$. The parameter k is a constant. Show that

$$w(\varphi) = k\left(\ln\frac{1+\cos\varphi}{1+\cos\varphi_o} - \frac{1}{1+\cos\varphi} + \frac{1}{1+\cos\varphi_o}\right)\sin\varphi$$

Section 4.1.2

4.4 Obtain the solution to the following equations:

(a) $y^2\dfrac{dy}{dx} = e^x - y^3$

(b) $\dfrac{dy}{dx} - 2xy = 2x^3y^2$

for the boundary condition $y(0) = 1$.

Section 4.2.2

4.5 Find the general solution to

$$\frac{d^2}{d\eta^2}\left(e^{-\beta\eta}\frac{d^2W}{d\eta^2}\right) - \Omega^4 e^{-\beta\eta}W = 0$$

when $\Omega^2 > \beta^2/4$.

Section 4.2.3

4.6 Find the solution to

$$(x-1)^2\frac{d^2y}{dx^2} + 3(x-1)\frac{dy}{dx} + y = 0$$

Hint: Assume a solution of the form $y = (x-1)^r$ to find the first solution and then use the reduction of order to find the complete solution.

4.7 Find the solution to

$$4x^2\frac{d^2y}{dx^2}-4x\frac{dy}{dx}+\left(3-16x^2\right)y=0$$

when given that $y_1(x)=x^{1/2}e^{2x}$.

4.8 Find the solution to

$$(2x+1)x\frac{d^2y}{dx^2}-2(2x^2-1)\frac{dy}{dx}-4(x+1)y=0$$

when given that $y_1(x)=x^{-1}$.

Section 4.2.4

4.9 Find the solution to

$$x^2\frac{d^2y}{dx^2}-2y=-2$$

with the boundary conditions $y(\varepsilon)=0$ and $y(1)=1$ when $\varepsilon\neq 0$.

4.10 Find the solution to

$$x^2\frac{d^2u}{dx^2}-3x\frac{du}{dx}+13u=0$$

when $u(1)=-1$ and $du(1)/dx=1$.

4.11 Find the solution to

$$\frac{d^2y}{dx^2}+2\frac{dy}{dx}+\left[1+\frac{2}{(1+3x)^2}\right]y=0$$

Hint: Assume a solution of the form $y=ue^{-x}$ and then use the change of variables $1+3x=e^t$.

Section 4.2.5

4.12 Obtain a solution to

$$\frac{d^2w}{d\varphi^2}+w=-k_1-k_2\cos\varphi$$

with the boundary conditions $w'(0)=w'(\pi/2)=0$.

4.13 Find the solution to the following boundary value problem.

$$\frac{d^2y}{dx^2}+2\frac{dy}{dx}+y=x$$

subject to $y(0)=0$ and $y(2)=3$.

4.14 Find the solution to

$$\frac{d^2w}{d\eta^2}+\beta^2 w=-\alpha\left(\eta-\eta^2\right)$$

subject to the boundary conditions $w(0)=w(1)=0$.

4.15 Determine the solution to

$$\frac{d^2y}{dx^2}+6\frac{dy}{dx}+9y=e^{3x}$$

when the initial conditions are $y(0)=0$ and $y'(0)=6$.

4.16 Find the solution to

$$\frac{1}{r}\frac{d}{dr}\left(r\frac{du}{dr}\right)=-r^k \quad 0<r$$

Section 4.2.6

4.17 Use the variation of parameters to find the particular solution to

$$\frac{d^2y}{dx^2}+2\frac{dy}{dx}+y=\frac{2}{x^2}e^{-x}$$

4.18 Use the variation of parameters to find the particular solution to

$$\frac{d^2y}{dx^2}-2\frac{dy}{dx}+y=14x^{3/2}e^x$$

4.19 Use the variation of parameters to find the particular solution to

$$x\frac{d^2y}{dx^2}+2(1-x)\frac{dy}{dx}+(x-2)y=e^{2x}$$

where the independent solutions to the homogeneous equation are

$$y_1(x)=e^x,\quad y_1(x)=x^{-1}e^x$$

4.20 Use the variation of parameters to find the particular solution to

$$4x^2\frac{d^2y}{dx^2}-4x\frac{dy}{dx}+\left(4x^2+3\right)y=x^{7/2}$$

where the independent solutions to the homogeneous equation are

$$y_1(x)=\sqrt{x}\sin x,\quad y_2(x)=\sqrt{x}\cos x$$

Section 4.2.7

4.21 Convert the following system of nonlinear equations and boundary conditions to a system of first-order differential equations.

$$\frac{d^3 f}{d\eta^3} + 3f\frac{d^2 f}{d\eta^2} - 2\left(\frac{df}{d\eta}\right)^2 + T^* = 0$$

$$\frac{d^{2*}T^*}{d\eta^2} + 3\Pr f\frac{dT^*}{d\eta} = 0$$

The initial conditions at $\eta = 0$ are

$$f = 0 \quad \frac{df}{d\eta} = 0 \quad \frac{d^2 f}{d\eta^2} = a$$

$$T^* = 1 \quad \frac{dT^*}{d\eta} = -b$$

where a and b are constants.

SECTION 4.2.10

4.22 Determine a solution to the nonlinear equation

$$\left(x^2 - 1\right)x^2\left(\frac{d^2 y}{dx^2}\right)^2 - 1 = 0$$

5 Ordinary Differential Equations Part II

Power Series Solutions

5.1 POWER SERIES SOLUTIONS TO ORDINARY DIFFERENTIAL EQUATIONS

5.1.1 Classification of Singularities

Consider the second-order differential equation

$$d_0(x)\frac{d^2y}{dx^2}+d_1(x)\frac{dy}{dx}+d_2(x)y=0 \tag{5.1}$$

where $d_k(x)$, $k = 0, 1, 2$, are polynomials. If $d_0(x) \neq 0$, then we define

$$p(x)=\frac{d_1(x)}{d_0(x)}$$
$$q(x)=\frac{d_2(x)}{d_0(x)} \tag{5.2}$$

If either of the ratios $p(x)$ or $q(x)$ is unbounded at $x = x_0$, then the equation has a singularity at this point. In Eq. (5.2), it is assumed that the ratios d_1/d_0 and d_2/d_0 do not contain any common factors.

Whether x_0 is a regular singular point or an irregular singular point is determined from the following limits. A singular point is called a *regular singular point* at x_0 if

$$\lim_{x\to x_0}\left[(x-x_0)p(x)\right]\to \text{finite}$$

and

$$\lim_{x\to x_0}\left[(x-x_0)^2 q(x)\right]\to \text{finite}$$

Otherwise, if either limit is unbounded then the singular point is called an *irregular singular point* at x_0. When x_0 is neither a regular singular point nor an irregular singular point it is called an *ordinary point*.

We now illustrate these definitions with the following example.

Example 5.1

Consider the equation

$$x^3(x-2)^2\frac{d^2y}{dx^2}+5(x+2)(x-2)\frac{dy}{dx}+3x^2y=0$$

To determine if the singularities at $x = 0$ and $x = 2$ are regular singular points we proceed as follows. First, we note that

$$p(x)=\frac{5(x+2)(x-2)}{x^3(x-2)^2}=\frac{5(x+2)}{x^3(x-2)}$$

$$q(x)=\frac{3x^2}{x^3(x-2)^2}=\frac{3}{x(x-2)^2}$$

Therefore, at $x = 0$ we find that

$$\lim_{x\to 0}\left[xp(x)\right]=\lim_{x\to 0}\left[x\frac{5(x+2)}{x^3(x-2)}\right]=\lim_{x\to 0}\left[\frac{5(x+2)}{x^2(x-2)}\right]\to\infty$$

$$\lim_{x\to 0}\left[x^2q(x)\right]=\lim_{x\to 0}\left[x^2\frac{3}{x(x-2)^2}\right]=\lim_{x\to 0}\left[\frac{3x}{(x-2)^2}\right]\to 0$$

and at $x = 2$ we find that

$$\lim_{x\to 2}\left[(x-2)p(x)\right]=\lim_{x\to 2}\left[(x-2)\frac{5(x+2)}{x^3(x-2)}\right]=\lim_{x\to 2}\left[\frac{5(x+2)}{x^3}\right]\to \text{finite}$$

$$\lim_{x\to 2}\left[(x-2)^2q(x)\right]=\lim_{x\to 2}\left[(x-2)^2\frac{3}{x(x-2)^2}\right]=\lim_{x\to 2}\left[\frac{3}{x}\right]\to \text{finite}$$

Therefore, $x = 0$ is an irregular singular point and $x = 2$ is a regular singular point.

5.1.2 Power Series Solution about an Ordinary Point*

In this section and in the next section, we shall introduce a general method for two special cases of Eq. (5.1). In this section, the approach is fairly general but is limited to solutions about the ordinary point $x = 0$ and about the ordinary point $x = x_0$. Using the results of this section, we shall be able to obtain the solution to Legendre's equation, which is discussed in Section 5.1.9.

* The approach used in Sections 5.1.2 and 5.1.3 is similar to that given by Trench WT (2013) *Elementary Differential Equations with Boundary Value Problems*, Free Edition, available at http://ramanujan.math.trinity.edu/wtrench/texts/index.shtml, Chapter 7.

Case 1: We shall consider the following special case of Eq. (5.1) where

$$
\begin{aligned}
d_0(x) &= \alpha_0 + \alpha_1 x + \alpha_2 x^2 \\
d_1(x) &= \beta_0 + \beta_1 x + \beta_2 x^2 \\
d_2(x) &= \gamma_0 + \gamma_1 x + \gamma_2 x^2
\end{aligned}
\tag{5.3}
$$

Then Eq. (5.1) becomes

$$
\begin{aligned}
L[y] = \left(\alpha_0 + \alpha_1 x + \alpha_2 x^2\right)\frac{d^2 y}{dx^2} + \left(\beta_0 + \beta_1 x + \beta_2 x^2\right)\frac{dy}{dx} \\
+\left(\gamma_0 + \gamma_1 x + \gamma_2 x^2\right) y = 0
\end{aligned}
\tag{5.4}
$$

where it is assumed that $\alpha_0 \neq 0$. It is seen that $x = 0$ is an ordinary point. Then a solution to Eq. (5.4) is assumed to be the power series

$$
y(x) = \sum_{n=0}^{\infty} c_n x^n \tag{5.5}
$$

Before substituting Eq. (5.5) into Eq. (5.4), we note the following

$$
\begin{aligned}
y(x) &= \sum_{n=0}^{\infty} c_n \left(x - x_0\right)^n = c_0 + c_1\left(x - x_0\right) + \sum_{n=2}^{\infty} c_n \left(x - x_0\right)^n \\
\frac{dy(x)}{dx} &= c_1 + \sum_{n=2}^{\infty} c_n n\left(x - x_0\right)^{n-1} = \sum_{n=1}^{\infty} c_n n\left(x - x_0\right)^{n-1} = \sum_{n=0}^{\infty} c_n n\left(x - x_0\right)^{n-1} \\
\frac{d^2 y(x)}{dx^2} &= \sum_{n=2}^{\infty} c_n n(n-1)\left(x - x_0\right)^{n-2} = \sum_{n=0}^{\infty} c_{n+2}(n+2)(n+1)\left(x - x_0\right)^n \\
&= \sum_{n=0}^{\infty} c_n n(n-1)\left(x - x_0\right)^{n-2}
\end{aligned}
\tag{5.6}
$$

In the expression for the first derivative, it is seen that changing the summation index from $n = 1$ to $n = 0$ did nothing to the summation since when $n = 0$, that term is zero. Likewise, in the expression for the second derivative, it is seen that changing the summation index from $n = 2$ to $n = 0$ did nothing to the summation since when $n = 0$ and when $n = 1$ those terms are zero. Also, for the second derivative, an alternative way to express the summation starting at $n = 0$ is to set $n = n + 2$ in the left-most summation, which results in the expression for the second summation.

Upon substituting Eq. (5.5) into Eq. (5.4), we obtain

$$
\begin{aligned}
&\left(\alpha_0 + \alpha_1 x + \alpha_2 x^2\right)\sum_{n=2}^{\infty} n(n-1) c_n x^{n-2} \\
&+\left(\beta_0 + \beta_1 x + \beta_2 x^2\right)\sum_{n=1}^{\infty} n c_n x^{n-1} + \left(\gamma_0 + \gamma_1 x + \gamma_2 x^2\right)\sum_{n=0}^{\infty} c_n x^n = 0
\end{aligned}
\tag{5.7}
$$

Expanding Eq. (5.7) gives

$$\begin{aligned}&\sum_{n=2}^{\infty}\alpha_0 n(n-1)c_n x^{n-2}+\sum_{n=2}^{\infty}\alpha_1 n(n-1)c_n x^{n-1}+\sum_{n=2}^{\infty}\alpha_2 n(n-1)c_n x^{n}\\&+\sum_{n=1}^{\infty}\beta_0 n c_n x^{n-1}+\sum_{n=1}^{\infty}\beta_1 n c_n x^{n}+\sum_{n=1}^{\infty}\beta_2 n c_n x^{n+1}\\&+\sum_{n=0}^{\infty}\gamma_0 c_n x^{n}+\sum_{n=0}^{\infty}\gamma_1 c_n x^{n+1}+\sum_{n=0}^{\infty}\gamma_2 c_n x^{n+2}=0\end{aligned} \tag{5.8}$$

Changing all the appropriate summation indices to start from $n = 2$, Eq. (5.8) becomes

$$\begin{aligned}&\sum_{n=2}^{\infty}\left(\alpha_0 n(n-1)c_n+\beta_0(n-1)c_{n-1}+\gamma_0 c_{n-2}\right)x^{n-2}\\&+\sum_{n=2}^{\infty}\left(\alpha_1 n(n-1)c_n+\beta_1(n-1)c_{n-1}+\gamma_1 c_{n-2}\right)x^{n-1}\\&+\sum_{n=2}^{\infty}\left(\alpha_2 n(n-1)c_n+\beta_2(n-1)c_{n-1}+\gamma_2 c_{n-2}\right)x^{n}=0\end{aligned} \tag{5.9}$$

Defining

$$p_j(n)=\alpha_j n(n-1)c_n+\beta_j(n-1)c_{n-1}+\gamma_j c_{n-2} \qquad j=0,1,2 \tag{5.10}$$

Eq. (5.9) can be written as

$$\sum_{n=2}^{\infty}p_0(n)x^{n-2}+\sum_{n=2}^{\infty}p_1(n)x^{n-1}+\sum_{n=2}^{\infty}p_2(n)x^{n}=0 \tag{5.11}$$

Shifting the summation indices in Eq. (5.11) so that in each summation x^n appears, we obtain

$$\sum_{n=0}^{\infty}p_0(n+2)x^{n}+\sum_{n=1}^{\infty}p_1(n+1)x^{n}+\sum_{n=2}^{\infty}p_2(n)x^{n}=0 \tag{5.12}$$

Equation (5.12) can be manipulated further as

$$\begin{aligned}&p_0(2)+p_0(3)x+\sum_{n=2}^{\infty}p_0(n+2)x^{n}+p_1(2)x\\&+\sum_{n=2}^{\infty}p_1(n+1)x^{n}+\sum_{n=2}^{\infty}p_2(n)x^{n}=0\end{aligned} \tag{5.13}$$

$$p_0(2)+\left(p_0(3)+p_1(2)\right)x+\sum_{n=2}^{\infty}b_n x^{n}=0$$

where

$$b_n = p_0(n+2) + p_1(n+1) + p_2(n) \quad n \geq 2 \tag{5.14}$$

Using Eq. (5.10), we find that

$$\begin{aligned} p_0(n+2) &= \alpha_0(n+2)(n+1)c_{n+2} + \beta_0(n+1)c_{n+1} + \gamma_0 c_n \\ p_1(n+1) &= \alpha_1 n(n+1)c_{n+1} + \beta_1 n c_n + \gamma_1 c_{n-1} \\ p_2(n) &= \alpha_2 n(n-1)c_n + \beta_2(n-1)c_{n-1} + \gamma_2 c_{n-2} \end{aligned} \tag{5.15}$$

and Eq. (5.14) becomes

$$\begin{aligned} b_n &= \gamma_2 c_{n-2} + \left(\beta_2(n-1) + \gamma_1\right)c_{n-1} + \left(\alpha_2 n(n-1) + \beta_1 n + \gamma_0\right)c_n \\ &\quad + \left(\alpha_1 n(n+1) + \beta_0(n+1)\right)c_{n+1} + \alpha_0(n+2)(n+1)c_{n+2} \end{aligned} \tag{5.16}$$

In addition, from Eq. (5.15),

$$\begin{aligned} p_0(2) &= 2\alpha_0 c_2 + \beta_0 c_1 + \gamma_0 c_0 \\ p_0(3) &= 6\alpha_0 c_3 + 2\beta_0 c_2 + \gamma_0 c_1 \\ p_1(2) &= 2\alpha_1 c_2 + \beta_1 c_1 + \gamma_1 c_0 \\ p_0(3) + p_1(2) &= 6\alpha_0 c_3 + \left(2\alpha_1 + 2\beta_0\right)c_2 + \left(\beta_1 + \gamma_0\right)c_1 + \gamma_1 c_0 \end{aligned} \tag{5.17}$$

In order for Eq. (5.13) to be satisfied for arbitrary x,

$$\begin{aligned} p_0(2) &= 0 \\ p_0(3) + p_1(2) &= 0 \end{aligned} \tag{5.18}$$

and

$$b_n = 0 \quad n \geq 2 \tag{5.19}$$

Therefore, from Eq. (5.15) to Eq. (5.19)

$$\begin{aligned} c_2 &= -\frac{\beta_0 c_1 + \gamma_0 c_0}{2\alpha_0} \\ c_3 &= -\frac{\left(2\alpha_1 + 2\beta_0\right)c_2 + \left(\beta_1 + \gamma_0\right)c_1 + \gamma_1 c_0}{6\alpha_0} \end{aligned} \tag{5.20}$$

and, for $n \geq 2$,

$$\begin{aligned} c_{n+2} = -\frac{1}{\alpha_0(n+2)(n+1)}\Big[&\left(\alpha_1 n(n+1) + \beta_0(n+1)\right)c_{n+1} \\ &+ \left(\alpha_2 n(n-1) + \beta_1 n + \gamma_0\right)c_n + \left(\beta_2(n-1) + \gamma_1\right)c_{n-1} + \gamma_2 c_{n-2}\Big] \end{aligned} \tag{5.21}$$

Equation (5.21) is called the *recursion relation* (or *recurrence relation*). The constants c_0 and c_1 are usually determined from the specification of $y(0)$ and $dy(0)/dx$. The final solution is obtained by substituting these expressions for c_n into Eq. (5.5).

We shall illustrate these results with the following examples.

Example 5.2

Consider the following equation

$$\left(1+x+2x^2\right)\frac{d^2y}{dx^2}+\left(1+7x\right)\frac{dy}{dx}+2y=0 \tag{a}$$

Comparing Eq. (a) with Eq. (5.3), we see that

$$\begin{aligned} \alpha_0 &= 1, \quad & \beta_0 &= 1, \quad & \gamma_0 &= 2 \\ \alpha_1 &= 1, \quad & \beta_1 &= 7, \quad & \gamma_1 &= 0 \\ \alpha_2 &= 2, \quad & \beta_2 &= 0, \quad & \gamma_2 &= 0 \end{aligned}$$

Therefore, from Eq. (5.20),

$$\begin{aligned} c_2 &= -\frac{c_1}{2} - c_0 \\ c_3 &= -\frac{2}{3}c_2 - \frac{3}{2}c_1 \end{aligned} \tag{b}$$

and from Eq. (5.21)

$$\begin{aligned} c_{n+2} &= -\frac{1}{(n+2)(n+1)}\left[(n+1)^2 c_{n+1} + (2n+1)(n+2)c_n\right] \\ &= -\frac{n+1}{n+2}c_{n+1} - \frac{2n+1}{n+1}c_n \quad n \geq 2 \end{aligned} \tag{c}$$

Then, Eq. (5.5) becomes

$$y(x) = c_0 + c_1 x + c_2 x^2 + c_3 x^3 + \sum_{n=2}^{\infty} c_{n+2} x^{n+2} \tag{d}$$

Using Mathematica procedure M5.1, we find that the first several terms of Eq. (d) can be written as

$$y(x) = c_0\left(1 - x^2 + \frac{2x^3}{3} + \frac{7x^4}{6} - \frac{21x^5}{10}\right) + c_1 x\left(1 - \frac{1}{2}x - \frac{7}{6}x^2 + \frac{41}{24}x^3 + \frac{27}{40}x^4\right) \tag{e}$$

If the initial conditions are $y(0) = y_0$ and $dy(0)/dx = v_0$, then from Eq. (d) we find that

$$y(0) = c_0 = y_0$$

$$dy(0)/dx = c_1 = v_0$$

and Eq. (e) is modified accordingly.

Example 5.3

Consider the following equation, called the Airy equation,

$$\frac{d^2y}{dx^2} - xy = 0 \tag{a}$$

for which all points $-\infty < x < \infty$ are ordinary points. Comparing Eq. (a) with Eq. (5.3), we see that

$$\begin{aligned} \alpha_0 &= 1, & \beta_0 &= 0, & \gamma_0 &= 0 \\ \alpha_1 &= 0, & \beta_1 &= 0, & \gamma_1 &= -1 \\ \alpha_2 &= 0, & \beta_2 &= 0, & \gamma_2 &= 0 \end{aligned}$$

Therefore, from Eq. (5.20),

$$\begin{aligned} c_2 &= 0 \\ c_3 &= \frac{1}{6}c_0 \end{aligned} \tag{b}$$

and from Eq. (5.21)

$$c_{n+2} = \frac{c_{n-1}}{(n+2)(n+1)} \qquad n \geq 2 \tag{c}$$

Then, Eq. (5.5) becomes

$$y(x) = c_0 + c_1x + \frac{c_0}{6}x^3 + \sum_{n=2}^{\infty} c_{n+2}x^{n+2} \tag{d}$$

Using Mathematica procedure M5.2, we find that the first few terms of Eq. (d) can be written as

$$y(x) = c_0\left(1 + \frac{x^3}{6} + \frac{x^6}{180} + \frac{x^9}{12960} + \cdots\right) + c_1x\left(1 + \frac{x^3}{12} + \frac{x^6}{504} + \frac{x^9}{45360} + \cdots\right)$$

Case 2: We shall consider the following special case of Eq. (5.1) with

$$\begin{aligned} d_0(x) &= 1 + \alpha\left(x - x_0\right)^2 = \alpha x^2 - 2\alpha x_0 x + 1 + \alpha x_0^2 \\ d_1(x) &= \beta\left(x - x_0\right) \\ d_2(x) &= \gamma \end{aligned} \tag{5.22}$$

Then Eq. (5.1) becomes

$$L[y] = \left(1 + \alpha\left(x - x_0\right)^2\right)\frac{d^2y}{dx^2} + \beta\left(x - x_0\right)\frac{dy}{dx} + \gamma y = 0 \tag{5.23}$$

It is seen that $x = x_0$ is an ordinary point. The solution to Eq. (5.23) is assumed to be the power series

$$y(x) = \sum_{n=0}^{\infty} c_n (x - x_0)^n \tag{5.24}$$

Substituting Eq. (5.24) into Eq. (5.23) and using Eq. (5.6), we find that

$$\begin{aligned} L[y] &= \left(1 + \alpha (x - x_0)^2\right) \sum_{n=2}^{\infty} c_n n(n-1)(x - x_0)^{n-2} \\ &\quad + \beta (x - x_0) \sum_{n=1}^{\infty} c_n n (x - x_0)^{n-1} + \gamma \sum_{n=0}^{\infty} c_n (x - x_0)^n \\ &= \sum_{n=2}^{\infty} c_n n(n-1)(x - x_0)^{n-2} + \alpha \sum_{n=2}^{\infty} c_n n(n-1)(x - x_0)^n \\ &\quad + \beta \sum_{n=1}^{\infty} c_n n (x - x_0)^n + \gamma \sum_{n=0}^{\infty} c_n (x - x_0)^n \\ &= \sum_{n=2}^{\infty} c_n n(n-1)(x - x_0)^{n-2} + \sum_{n=0}^{\infty} c_n \{\alpha n(n-1) + \beta n + \gamma\}(x - x_0)^n \\ &= \sum_{n=0}^{\infty} \{c_{n+2}(n+2)(n+1) + c_n p(n)\}(x - x_0)^n = 0 \end{aligned} \tag{5.25}$$

where

$$p(n) = \alpha n(n-1) + \beta n + \gamma \tag{5.26}$$

In order for Eq. (5.25) to equal zero for arbitrary x,

$$c_{n+2} = -\frac{c_n p(n)}{(n+2)(n+1)} \quad n = 0,1,2,\ldots \tag{5.27}$$

which is the recursion relation. We re-write Eq. (5.27) as

$$c_{n+2} = -c_n g_n \quad n = 0,1,2,\ldots \tag{5.28}$$

where

$$g_n = \frac{p(n)}{(n+2)(n+1)} \quad n = 0,1,2,\ldots \tag{5.29}$$

Then, writing out the first few terms of Eq. (5.28) gives

$$
\begin{aligned}
c_2 &= -c_0 g_0 \\
c_3 &= -c_1 g_1 \\
c_4 &= -c_2 g_2 = c_0 g_0 g_2 \\
c_5 &= -c_3 g_3 = c_1 g_1 g_3 \\
c_6 &= -c_4 g_4 = -c_0 g_0 g_2 g_4 \\
c_7 &= -c_5 g_5 = -c_1 g_1 g_3 g_5 \\
&\vdots
\end{aligned}
\tag{5.30}
$$

Since c_0 and c_1 are arbitrary constants, the recursion relation for the remaining coefficients c_n can be written in terms of these two constants as

$$c_{2n+2} = -\frac{p(2n)}{(2n+2)(2n+1)} c_{2n} \quad n = 0,1,2,\ldots \tag{5.31}$$

and

$$c_{2n+3} = -\frac{p(2n+1)}{(2n+3)(2n+2)} c_{2n+1} \quad n = 0,1,2,\ldots \tag{5.32}$$

These results will be used to solve Legendre's equation in Section 5.1.9. Legendre's equation often appears as an intermediate equation when the separation of variables method is used to solve partial differential equations in spherical coordinates.

Example 5.4

Consider the following equation

$$\frac{d^2y}{dx^2} + y = 0 \tag{a}$$

Comparing Eq. (a) with Eq. (5.23), we see that $\alpha = x_0 = \beta = 0$ and $\gamma = 1$ and, therefore, from Eq. (5.26), $p(n) = 1$. Using Eq. (5.31), it is found that

$$
\begin{aligned}
c_2 &= -\frac{1}{2\times 1} c_0 \\
c_4 &= -\frac{1}{4\times 3} c_2 = \frac{1}{4\times 3\times 2\times 1} c_0 \\
&\vdots
\end{aligned}
$$

Therefore, in terms of c_0,

$$c_n \to c_{2n} = \frac{(-1)^n}{(2n)!} c_0 \quad n = 1,2,3,\ldots \tag{b}$$

Using Eq. (5.32) it is found that

$$c_3 = -\frac{1}{3\times 2}c_1$$

$$c_5 = -\frac{1}{5\times 4}c_3 = \frac{1}{5\times 4\times 3\times 2}c_1$$

$$\vdots$$

Therefore, in terms of c_1,

$$c_n \to c_{2n+1} = \frac{(-1)^n}{(2n+1)!}c_1 \quad n = 1,2,3,\ldots \tag{c}$$

Then Eq. (5.24) can be written as

$$\begin{aligned} y(x) &= \sum_{n=0}^{\infty} c_{2n}x^{2n} + \sum_{n=0}^{\infty} c_{2n+1}x^{2n+1} \\ &= c_0\sum_{n=0}^{\infty}\frac{c_{2n}}{c_0}x^{2n} + c_1\sum_{n=0}^{\infty}\frac{c_{2n+1}}{c_1}x^{2n+1} \\ &= c_0\sum_{n=0}^{\infty}\frac{(-1)^n}{(2n)!}x^{2n} + c_1\sum_{n=0}^{\infty}\frac{(-1)^n}{(2n+1)!}x^{2n+1} \end{aligned} \tag{d}$$

where we have used Eqs. (b) and (c) and the definition $0! = 1$. However, from Eq. (A.2) of Appendix A, we see that Eq. (d) also can be written as

$$y(x) = c_0\cos(x) + c_1\sin(x) \tag{k}$$

This result, of course, could have been obtained more directly by assuming $y(x) = e^{\lambda x}$ as was demonstrated in Example 4.7.

5.1.3 Power Series about a Regular Singular Point: Method of Frobenius

In this section, we shall introduce a general method for a special case of Eq. (5.1) to obtain solutions about a regular singular point $x = 0$. The approach will be general enough so that it can be used directly to obtain the solutions to the Bessel's equations, which are discussed in Sections 5.1.4, 5.1.6, and 5.1.7 and the hypergeometric equation, which is discussed in Section 5.1.10.

We shall consider the following special case of Eq. (5.1) with

$$\begin{aligned} d_0(x) &= \left(\alpha_0 + \alpha_1 x + \alpha_2 x^2\right)x^2 \\ d_1(x) &= \left(\beta_0 + \beta_1 x + \beta_2 x^2\right)x \\ d_2(x) &= \left(\gamma_0 + \gamma_1 x + \gamma_2 x^2\right) \end{aligned} \tag{5.33}$$

Then Eq. (5.1) becomes

$$L[y] = \left(\alpha_0 + \alpha_1 x + \alpha_2 x^2\right) x^2 \frac{d^2 y}{dx^2} + \left(\beta_0 + \beta_1 x + \beta_2 x^2\right) x \frac{dy}{dx} + \left(\gamma_0 + \gamma_1 x + \gamma_2 x^2\right) y = 0 \tag{5.34}$$

where it is assumed that $\alpha_0 \neq 0$ and $x > 0$. It is seen that $x = 0$ is a regular singular point. The solution to Eq. (5.34) is assumed to be the power series

$$y(x) = \sum_{n=0}^{\infty} c_n x^{n+r} \tag{5.35}$$

where r is to be determined. The form of this solution is known as the *Frobenius method.*

The form of the differential equation was selected to solve a reasonably large class of differential equations with the end result to be able to solve the Bessel equation and the hypergeometric equation. Bessel equations frequently appear as intermediate equations when separable solutions techniques are used to solve partial differential equations in polar cylindrical coordinates.

Upon substituting Eq. (5.35) into Eq. (5.34), we obtain

$$\left(\alpha_0 + \alpha_1 x + \alpha_2 x^2\right) \sum_{n=0}^{\infty} (n+r)(n+r-1) c_n x^{n+r} + \left(\beta_0 + \beta_1 x + \beta_2 x^2\right) \sum_{n=0}^{\infty} (n+r) c_n x^{n+r} + \left(\gamma_0 + \gamma_1 x + \gamma_2 x^2\right) \sum_{n=0}^{\infty} c_n x^{n+r} = 0$$

which, upon re-arrangement, gives

$$\sum_{n=0}^{\infty} \left(\alpha_0 (n+r)(n+r-1) + \beta_0 (n+r) + \gamma_0\right) c_n x^{n+r} + \sum_{n=0}^{\infty} \left(\alpha_1 (n+r)(n+r-1) + \beta_1 (n+r) + \gamma_1\right) c_n x^{n+r+1} + \sum_{n=0}^{\infty} \left(\alpha_2 (n+r)(n+r-1) + \beta_2 (n+r) + \gamma_2\right) c_n x^{n+r+2} = 0 \tag{5.36}$$

Introducing the definition

$$p_j(z) = z(z-1)\alpha_j + z\beta_j + \gamma_j \quad j = 0,1,2 \tag{5.37}$$

Eq. (5.36) can be written as

$$\sum_{n=0}^{\infty} p_0(n+r)c_n x^{n+r} + \sum_{n=0}^{\infty} p_1(n+r)c_n x^{n+r+1} + \sum_{n=0}^{\infty} p_2(n+r)c_n x^{n+r+2} = 0 \tag{5.38}$$

We expand Eq. (5.38) as follows

$$p_0(r)c_0 x^r + \left(p_1(r)c_0 + p_0(1+r)c_1\right)x^{r+1} + \sum_{n=2}^{\infty} p_0(n+r)c_n x^{n+r} + \sum_{n=1}^{\infty} p_1(n+r)c_n x^{n+r+1} + \sum_{n=0}^{\infty} p_2(n+r)c_n x^{n+r+2} = 0 \tag{5.39}$$

Next, we change the summation indices of the second and third summations of Eq. (5.39) so that they both start at $n = 2$. Therefore, in the second summation, we set $n = n - 1$ and in the third summation $n = n - 2$ and arrive at

$$p_0(r)c_0 x^r + \left(p_1(r)c_0 + p_0(1+r)c_1\right)x^{1+r} + \sum_{n=2}^{\infty} \left(p_0(n+r)c_n + p_1(n+r-1)c_{n-1} + p_2(n+r-2)c_{n-2}\right)x^{n+r} = 0 \tag{5.40}$$

which can be written as

$$\sum_{n=0}^{\infty} b_n(r) x^{n+r} = 0 \tag{5.41}$$

where

$$\begin{aligned} b_0(r) &= p_0(r)c_0 \\ b_1(r) &= p_1(r)c_0 + p_0(1+r)c_1 \\ b_n(r) &= p_0(n+r)c_n + p_1(n+r-1)c_{n-1} + p_2(n+r-2)c_{n-2} \quad n \geq 2 \end{aligned} \tag{5.42}$$

In order for Eq. (5.35) to be a solution to Eq. (5.34), in Eq . (5.41) we must set $b_n(r) = 0$, $n \geq 1$. If it is assumed that $p_0(n + r) \neq 0$ for all positive integers n, then setting $b_n(r)$ to zero in the second and third equations of Eq. (5.42) gives

$$\begin{aligned} c_1 &= -\frac{p_1(r)}{p_0(1+r)} c_0 \\ c_n &= -\frac{p_1(n+r-1)c_{n-1} + p_2(n+r-2)c_{n-2}}{p_0(n+r)} \quad n \geq 2 \end{aligned} \tag{5.43}$$

In view of Eq. (5.43), we construct the solution to Eq. (5.34) in the form

$$y(x) = y(x,r) = \sum_{n=0}^{\infty} a_n(r)x^{n+r} \tag{5.44}$$

where

$$\begin{aligned} a_0(r) &= 1 \\ a_1(r) &= -\frac{p_1(r)}{p_0(1+r)} \\ a_n(r) &= -\frac{p_1(n+r-1)a_{n-1}(r) + p_2(n+r-2)a_{n-2}(r)}{p_0(n+r)} \qquad n \geq 2 \end{aligned} \tag{5.45}$$

Substituting Eq. (5.44) into Eq. (5.34) results in Eq. (5.40) with c_n replaced by $a_n(r)$, that is,

$$\begin{aligned} L[y(x,r)] &= p_0(r)a_0(r)x^r + \left(p_1(r)a_0(r) + p_0(1+r)a_1(r)\right)x^{1+r} \\ &+ \sum_{n=2}^{\infty}\begin{pmatrix} p_0(n+r)a_n(r) + p_1(n+r-1)a_{n-1}(r) \\ + p_2(n+r-2)a_{n-2}(r) \end{pmatrix} x^{n+r} \end{aligned} \tag{5.46}$$

Using Eq. (5.45), Eq. (5.46) becomes

$$\begin{aligned} L[y(x,r)] &= p_0(r)x^r + \left(p_1(r) + p_0(1+r)\left\{-\frac{p_1(r)}{p_0(1+r)}\right\}\right)x^{1+r} \\ &+ \sum_{n=2}^{\infty}\left\{\begin{matrix} p_0(n+r)\left(-\dfrac{p_1(n+r-1)a_{n-1}(r) + p_2(n+r-2)a_{n-2}(r)}{p_0(n+r)}\right) \\ + p_1(n+r-1)a_{n-1}(r) + p_2(n+r-2)a_{n-2}(r) \end{matrix}\right\}x^{n+r} \\ &= p_0(r)x^r \end{aligned} \tag{5.47}$$

Since r is still to be determined, we find those values of r, denoted r_1 and r_2, that make the right-hand side of Eq. (5.47) zero; that is, we determine the solutions to

$$p_0(r) = r(r-1)\alpha_0 + r\beta_0 + \gamma_0 = (r-r_1)(r-r_2) = 0 \tag{5.48}$$

where we let $r_1 > r_2$ and for the differential equations that we shall be considering, r_1 and r_2 will be real. The significance of Eq. (5.47) will be apparent subsequently. Equation (5.48) is called the *indicial equation*.

Then the solution to Eq. (5.34) is

$$y(x) = C_1 y(x,r_1) + C_2 y(x,r_2) \tag{5.49}$$

where $y(x,r_1)$ is always given by Eq. (5.44). The specific form of $y(x,r_2)$ depends on the values of r_1 and r_2 and their difference $r_1 - r_2$, which we shall now consider.

Cauchy–Euler equation: Equation (5.34) becomes the Euler equation given by Eq. (4.54) when in Eq. (4.54) $a_2 = \alpha_0$, $a_1 = \beta_0$, $a_0 = \gamma_0$, and when in Eq. (5.34) $\alpha_1 = \alpha_2 = \beta_1 = \beta_2 = \gamma_1 = \gamma_2 = 0$. Then from Eq. (5.37) it is seen that $p_1 = p_2 = 0$ and, therefore, from Eq. (5.45), $a_0(r) = 1$, $a_1(r) = 0$, and $a_n(r) = 0$ for $n \geq 2$. Consequently, the solutions y_j, $j = 1, 2$, each consist of only one term: x^{r_j}.

Case 1: When the roots r_1 and r_2 are real and distinct ($r_1 \neq r_2$) and do not differ by an integer ($r_1 - r_2 \neq$ integer), then the solutions are

$$y_1(x) = y(x, r_1) = x^{r_1} \sum_{n=0}^{\infty} a_n(r_1) x^n \tag{5.50}$$

and

$$y_2(x) = y(x, r_2) = x^{r_2} \sum_{n=0}^{\infty} a_n(r_2) x^n \tag{5.51}$$

Case 2: When the roots r_1 and r_2 are equal ($r_1 = r_2 = r_0$) and, therefore, $r_1 - r_2 = 0$, then to find the second independent solution y_2 we proceed as follows. In this case, Eqs. (5.47) and (4.48) give

$$L[y(x,r)] = (r - r_0)^2 x^r \tag{5.52}$$

Equation (5.52) is now differentiated with respect to r, which results in

$$\begin{aligned} L\left[\frac{\partial y(x,r)}{\partial r}\right] &= \left[2(r - r_0)x^r + (r - r_0)^2 x^r \ln x\right] \\ &= \left[2(r - r_0) + (r - r_0)^2 \ln x\right] x^r \end{aligned}$$

However, at $r = r_0$ the right-hand side equals zero and, therefore,

$$L\left[\frac{\partial y}{\partial r}\right]_{r=r_0} = 0$$

and, hence,

$$y_2 = \left(\frac{\partial y}{\partial r}\right)_{r=r_0}$$

Then, the solutions to Eq. (5.34) are

$$y_1(x) = y(x, r_0) = x^{r_0} \sum_{n=0}^{\infty} a_n(r_0) x^n \tag{5.53}$$

and

$$
\begin{aligned}
y_2(x) &= \left(\frac{\partial y}{\partial r}\right)_{r=r_0} \\
&= \ln x \sum_{n=0}^{\infty} a_n(r_0)x^{n+r_0} + \sum_{n=1}^{\infty} \frac{\partial a_n(r_0)}{\partial r} x^{n+r_0} \\
&= y_1(x)\ln x + x^{r_0} \sum_{n=1}^{\infty} a_n'(r_0)x^n
\end{aligned} \tag{5.54}
$$

where the prime denotes the derivative with respect to r. Recall the solution to the Cauchy–Euler equation for Case 2 given by Eq. (4.59).

Case 3: When the roots r_1 and r_2 ($r_1 > r_2$) differ by a positive integer k, that is, $r_1 - r_2 = k$, to find the second independent solution y_2 we proceed as follows. First, we note that when the roots differ by an integer, we see from Eq. (5.48) that

$$
p_0(r) = (r - r_1)(r - r_2) = (r - r_2 - k)(r - r_2) \tag{5.55}
$$

and, therefore,

$$
p_0(n + r_2) = (r_2 + n - r_2 - k)(r_2 + n - r_2) = (n - k)n \tag{5.56}
$$

Thus, there is a value n such that $p_0(n + r_2) = 0$. Consequently, in this case $a_n(r_2)$ can become unbounded. Recall that the assumption made in arriving at Eq. (5.43) was that $p_0(n + r) \neq 0$.

We obtain the second solution by using a procedure similar to that used for repeated roots. In this case, we substitute Eq. (5.55) into Eq. (5.47), multiply the result by $r - r_2$, differentiate it with respect to r, and then evaluate it at $r = r_2$ to obtain

$$
\begin{aligned}
L\left[\frac{\partial}{\partial r}\{(r - r_2)y\}\right]_{r=r_2} &= \frac{\partial}{\partial r}\left\{(r - r_1)(r - r_2)^2 x^r\right\}\Big|_{r=r_2} \\
&= \left\{(r - r_2)^2 x^r + 2(r - r_1)(r - r_2)x^r \right. \\
&\quad \left. + (r - r_1)(r - r_2)^2 x^r \ln x\right\}_{r=r_2} \\
&= 0
\end{aligned} \tag{5.57}
$$

Therefore,

$$
y_2 = \frac{\partial}{\partial r}\{(r - r_2)y\}_{r=r_2} \tag{5.58}
$$

Obtaining y_2 for the general case given by Eq. (5.33) is complicated; instead, we shall give the final form of Eq. (5.58) for two special cases.

Case 3A: For this case, we assume that $\alpha_2 = \beta_2 = \gamma_2 = 0$ and that $r_1 - r_2 = k$, a positive integer. Then Eq. (5.33) simplifies to

$$
\begin{aligned}
d_0(x) &= (\alpha_0 + \alpha_1 x)x^2 \\
d_1(x) &= (\beta_0 + \beta_1 x)x \\
d_2(x) &= (\gamma_0 + \gamma_1 x)
\end{aligned}
\tag{5.59}
$$

From Eq. (5.37) it is seen that $p_2(r) = 0$ and therefore, $a_n(r)$ given by Eq. (5.45) become

$$
\begin{aligned}
a_0(r) &= 1 \\
a_n(r) &= -\frac{p_1(n+r-1)}{p_0(n+r)} a_{n-1}(r) \quad n \geq 1
\end{aligned}
\tag{5.60}
$$

For these assumptions,

$$
y_1(x) = y(x, r_1) = x^{r_1} \sum_{n=0}^{\infty} a_n(r_1) x^n \tag{5.61}
$$

and*

$$
y_2(x) = x^{r_2} \sum_{n=0}^{k-1} a_n(r_2) x^n + C\left(y_1(x) \ln x + x^{r_1} \sum_{n=1}^{\infty} a_n'(r_1) x^n \right) \tag{5.62}
$$

where

$$
C = -\frac{p_1(r_1 - 1)}{k\alpha_0} a_{k-1}(r_2) \tag{5.63}
$$

Note that in the first equation of Eq. (5.62) the summation stops at $k - 1$ because $a_n(r_2)$ exists only for $0 \leq n \leq k - 1$ since, from Eq. (5.56), $p_0 = 0$ at $n = k$.

Case 3B: For this case, we assume that $\alpha_1 = \beta_1 = \gamma_1 = 0$ and that $r_1 - r_2 = 2k$, where k a positive integer. Then Eq. (5.33) simplifies to

$$
\begin{aligned}
d_0(x) &= (\alpha_0 + \alpha_2 x^2)x^2 \\
d_1(x) &= (\beta_0 + \beta_2 x^2)x \\
d_2(x) &= (\gamma_0 + \gamma_2 x^2)
\end{aligned}
\tag{5.64}
$$

Since in this case, $p_0(n + r_2) = (n - 2k)n$, we consider only even values for n and replace n by $2n$ in what follows. From Eq. (5.37) it is seen that $p_1(r) = 0$ and therefore, $a_n(r)$ given by Eq. (5.45) becomes

$$
\begin{aligned}
a_0(r) &= 1 \\
a_{2n}(r) &= -\frac{p_2(2n+r-2)}{p_0(2n+r)} a_{2n-2}(r) \quad n \geq 1
\end{aligned}
\tag{5.65}
$$

* Trench, *op. cit.*, p. 380.

For these assumptions,

$$y_1(x)=\sum_{n=0}^{\infty}a_{2n}(r_1)x^{2n+r_1} \tag{5.66}$$

and*

$$y_2(x)=x^{r_2}\sum_{n=0}^{k-1}a_{2n}(r_2)x^{2n}+C\left(y_1(x)\ln x+x^{r_1}\sum_{n=1}^{\infty}a'_{2n}(r_1)x^{2n}\right) \tag{5.67}$$

where

$$C=-\frac{p_2(r_1-2)}{2k\alpha_0}a_{2k-2}(r_2) \tag{5.68}$$

Note that in Eq. (5.67) the summation stops at $k-1$ because $a_n(r_2)$ exists only for $0\le n\le k-1$.
We shall now illustrate these different cases with the following examples.

Example 5.5: Distinct roots and $r_1 - r_2 \neq$ integer (Case 1)

Consider the equation

$$2\left(1+x+x^2\right)x^2\frac{d^2y}{dx^2}+\left(9+11x+11x^2\right)x\frac{dy}{dx}+\left(6+10x+7x^2\right)y=0$$

Comparing this equation with Eq. (5.34), it is seen that

$$\begin{aligned}&\alpha_0=2, &&\alpha_1=2, &&\alpha_2=2\\ &\beta_0=9, &&\beta_1=11, &&\beta_2=11\\ &\gamma_0=6, &&\gamma_1=10, &&\gamma_2=7\end{aligned} \tag{a}$$

Then, from Eq. (5.37)

$$\begin{aligned}p_0(r)&=2r(r-1)+9r+6=2r^2+7r+6=(2r+3)(r+2)\\ p_1(r)&=2r(r-1)+11r+10=2r^2+9r+10=(2r+5)(r+2)\\ p_2(r)&=2r(r-1)+11r+7=2r^2+9r+7=(2r+7)(r+1)\end{aligned} \tag{b}$$

From the indicial equation $p_0(r)=0$, we find that $r_1=-3/2$ and $r_2=-2$, which are distinct, and $r_1-r_2=1/2$, which is not a positive integer. Hence, the solutions are given by Eqs. (5.50) and (5.51), that is,

$$\begin{aligned}y_1(x)&=x^{-3/2}\sum_{n=0}^{\infty}a_n(-3/2)x^n\\ y_2(x)&=x^{-2}\sum_{n=0}^{\infty}a_n(-2)x^n\end{aligned} \tag{c}$$

* Trench, *op. cit.*, p. 385.

Using Eq. (5.45), we find that

$$a_0(r) = 1$$

$$a_1(r) = -\frac{(2r+5)(r+2)}{(2r+5)(r+3)} = -\frac{(r+2)}{(r+3)}$$

$$a_n(r) = -\frac{(2n+2r+3)(n+r+1)a_{n-1}(r) + (2n+2r+3)(n+r-1)a_{n-2}(r)}{(2n+2r+3)(n+r+2)} \tag{d}$$

$$= -\frac{(n+r+1)a_{n-1}(r) + (n+r-1)a_{n-2}(r)}{(n+r+2)} \quad n \geq 2$$

Therefore, for $r_1 = -3/2$, Eq. (d) gives

$$a_0(r_1) = 1$$

$$a_1(r_1) = -\frac{(-3/2+2)}{(-3/2+3)} = -\frac{1}{3}$$

$$a_n(r_1) = -\frac{(n-3/2+1)a_{n-1}(r_1) + (n-3/2-1)a_{n-2}(r_1)}{(n-3/2+2)}$$

$$= -\frac{(2n-1)a_{n-1}(r_1) + (2n-5)a_{n-2}(r_1)}{(2n+1)} \quad n \geq 2$$

Then,

$$a_0(r_1) = 1$$

$$a_1(r_1) = -\frac{1}{3}$$

$$a_2(r_1) = -\frac{3a_1(r_1) - a_0(r_1)}{5} = -\frac{-1-1}{5} = \frac{2}{5} \tag{e}$$

$$a_3(r_1) = -\frac{5a_2(r_1) + a_1(r_1)}{7} = -\frac{2-1/3}{7} = -\frac{5}{21}$$

$$a_4(r_1) = -\frac{7a_3(r_1) + 3a_2(r_1)}{9} = -\frac{-5/3+6/5}{9} = \frac{7}{135}$$

For $r_2 = -2$, Eq. (d) gives

$$a_0(r_2) = 1$$

$$a_1(r_2) = 0$$

$$a_n(r_2) = -\frac{(n-1)a_{n-1}(r_2) + (n-3)a_{n-2}(r_2)}{n} \quad n \geq 2$$

Then

$$\begin{aligned}
a_0(r_2) &= 1 \\
a_1(r_2) &= 0 \\
a_2(r_2) &= -\frac{a_1(r_2) - a_0(r_2)}{2} = \frac{1}{2} \\
a_3(r_2) &= -\frac{2a_2(r_2)}{3} = -\frac{1}{3} \\
a_4(r_2) &= -\frac{3a_3(r_2) + a_2(r_2)}{4} = \frac{1}{4} - \frac{1}{8} = \frac{1}{8}
\end{aligned} \tag{f}$$

Substituting Eqs. (e) and (f) into Eq. (c) yields

$$\begin{aligned}
y_1(x) &= x^{-3/2}\left(1 - \frac{1}{3}x + \frac{2}{5}x^2 - \frac{5}{21}x^3 + \frac{7}{135}x^4 \pm \cdots\right) \\
y_2(x) &= x^{-2}\left(1 + \frac{1}{2}x^2 - \frac{1}{3}x^3 + \frac{1}{8}x^4 \pm \cdots\right)
\end{aligned} \tag{g}$$

Equation (g) is verified with Mathematica procedure M5.3, which is valid only for the case where the difference of the roots does not equal to zero or an integer.

Example 5.6: Equal roots (Case 2)

Consider the equation

$$\left(1 - 2x + x^2\right)x^2\frac{d^2y}{dx^2} - (3 + x)x\frac{dy}{dx} + (4 + x)y = 0$$

Comparing this equation with Eq. (5.34), it is seen that

$$\begin{aligned}
&\alpha_0 = 1, \quad &&\alpha_1 = -2, \quad &&\alpha_2 = 1 \\
&\beta_0 = -3, \quad &&\beta_1 = -1, \quad &&\beta_2 = 0 \\
&\gamma_0 = 4, \quad &&\gamma_1 = 1, \quad &&\gamma_2 = 0
\end{aligned} \tag{a}$$

Then, from Eq. (5.37)

$$\begin{aligned}
p_0(r) &= r(r-1) - 3r + 4 = r^2 - 4r + 4 = (r-2)^2 \\
p_1(r) &= -2r(r-1) - r + 1 = -2r^2 + r + 1 = -(2r+1)(r-1) \\
p_2(r) &= r(r-1)
\end{aligned} \tag{b}$$

From the indicial equation $p_0(r) = 0$, we find that $r_1 = r_2 = 2$ and, hence, we have equal roots. Therefore, the solutions are given by Eqs. (5.53) and (5.54), that is,

$$y_1(x) = \sum_{n=0}^{\infty} a_n(2)x^{n+2}$$

$$y_2(x) = y_1(x)\ln x + \sum_{n=0}^{\infty} a_n'(2)x^{n+2} \tag{c}$$

Using Eq. (5.45), it is found that

$$a_0(r) = 1$$

$$a_1(r) = \frac{(2r+1)(r-1)}{(r-1)^2} = \frac{2r+1}{r-1}$$

$$a_n(r) = \frac{(2n+2r-1)(n+r-2)a_{n-1}(r) - (n+r-2)(n+r-3)a_{n-2}(r)}{(n+r-2)^2} \tag{d}$$

$$= \frac{(2n+2r-1)a_{n-1}(r) - (n+r-3)a_{n-2}(r)}{(n+r-2)} \quad n \geq 2$$

It is seen from Eq. (c) that we also need the derivatives of $a_n(r)$. Hence, from Eq. (d)

$$a_0'(r) = 0$$

$$a_1'(r) = -\frac{3}{(r-1)^2}$$

$$a_n'(r) = -\frac{3}{(n+r-2)^2}a_{n-1}(r) + \frac{(2n+2r-1)}{(n+r-2)}a_{n-1}'(r) \tag{e}$$

$$-\frac{1}{(n+r-2)^2}a_{n-2}(r) - \frac{(n+r-3)}{(n+r-2)}a_{n-2}'(r)$$

Therefore, for $r_1 = 2$, Eq. (d) becomes

$$a_0(r_1) = 1$$

$$a_1(r_1) = 5 \tag{f}$$

$$a_n(r_1) = \frac{(2n+3)a_{n-1}(r_1) - (n-1)a_{n-2}(r_1)}{n} \quad n \geq 2$$

and Eq. (e) becomes

$$a_0'(r_1) = 0$$

$$a_1'(r_1) = -3$$

$$a_n'(r_1) = -\frac{3}{n^2}a_{n-1}(r_1) + \frac{2n+3}{n}a_{n-1}'(r_1) \tag{g}$$

$$-\frac{1}{n^2}a_{n-2}(r_1) - \frac{(n-1)}{n}a_{n-2}'(r_1)$$

Then, the first few values of $a_n(r_1)$ are

$$
\begin{aligned}
a_0(r_1) &= 1 \\
a_1(r_1) &= 5 \\
a_2(r_1) &= \frac{7a_1(r_1) - a_0(r_1)}{2} = \frac{7\times 5 - 1}{2} = 17 \\
a_3(r_1) &= \frac{9a_2(r_1) - 2a_1(r_1)}{3} = \frac{9\times 17 - 2\times 5}{3} = \frac{143}{3} \\
a_4(r_1) &= \frac{11a_3(r_1) - 3a_2(r_1)}{4} = \frac{11\times 143/3 - 3\times 17}{4} = \frac{355}{3}
\end{aligned} \tag{h}
$$

and those for $a_n'(r_1)$ are

$$
\begin{aligned}
a_0'(r_1) &= 0 \\
a_1'(r_1) &= -3 \\
a_2'(r_1) &= -\frac{3}{4}a_1(r_1) + \frac{7}{2}a_1'(r_1) - \frac{1}{4}a_0(r_1) - \frac{1}{2}a_0'(r_1) \\
&= -\frac{3}{4}\times 5 + \frac{7}{2}\times(-3) - \frac{1}{4} = -\frac{29}{2} \\
a_3'(r_1) &= -\frac{3}{9}a_2(r_1) + \frac{9}{3}a_2'(r_1) - \frac{1}{9}a_1(r_1) - \frac{2}{3}a_1'(r_1) \\
&= -\frac{3}{9}\times 17 + \frac{9}{3}\times\left(-\frac{29}{2}\right) - \frac{1}{9}\times 5 - \frac{2}{3}\times(-3) = -\frac{859}{18}
\end{aligned} \tag{i}
$$

Substituting Eqs. (h) and (i) into Eq. (c) gives

$$
y_1(x) = x^2\left(1 + 5x + 17x^2 + \frac{143}{3}x^3 + \frac{355}{3}x^4 + \cdots\right)
$$

$$
y_2(x) = y_1(x)\ln x - x^3\left(3 + \frac{29}{2}x + \frac{859}{18}x^2 + \cdots\right)
$$

These results are verified with Mathematica procedure M5.4, which is valid only for equal roots.

Example 5.7: Difference in roots equals an integer (Case 3A)

Consider the equation

$$
2(2+x)x^2\frac{d^2y}{dx^2} + (-4+7x)x\frac{dy}{dx} + (-5+3x)y = 0
$$

Comparing this equation with Eq. (5.34), it is seen that

$$\begin{array}{lll} \alpha_0 = 4, & \alpha_1 = 2, & \alpha_2 = 0 \\ \beta_0 = -4, & \beta_1 = 7, & \beta_2 = 0 \\ \gamma_0 = -5, & \gamma_1 = 3, & \gamma_2 = 0 \end{array} \tag{a}$$

Then, from Eq. (5.37)

$$\begin{aligned} p_0(r) &= 4r(r-1) - 4r - 5 = 4r^2 - 8r - 5 = (2r+1)(2r-5) \\ p_1(r) &= 2r(r-1) + 7r + 3 = 2r^2 + 5r + 3 = (2r+3)(r+1) \\ p_2(r) &= 0 \end{aligned} \tag{b}$$

From the indicial equation $p_0(r) = 0$, we find that $r_1 = 5/2$ and $r_2 = -1/2$ and, hence, $r_1 - r_2 = 3$, a positive integer. Since $\alpha_2 = \beta_2 = \gamma_2 = 0$ and $r_1 - r_2 =$ integer, we have Case 3A. Hence, the solutions are given by Eqs. (5.61) and (5.62), that is,

$$\begin{aligned} y_1(x) &= \sum_{n=0}^{\infty} a_n(r_1) x^{n+r_1} \\ y_2 &= \sum_{n=0}^{2} a_n(r_2) x^{n+r_2} + C\left(y_1(x)\ln x + \sum_{n=1}^{\infty} a_n'(r_1) x^{n+r_1} \right) \end{aligned} \tag{c}$$

where C is given by Eq. (5.63).

Using Eq. (5.60) and Eq. (b), it is found that

$$\begin{aligned} a_0(r) &= 1 \\ a_n(r) &= -\frac{p_1(n+r-1)a_{n-1}(r)}{p_0(n+r)} = -\frac{n+r}{2n+2r-5} a_{n-1}(r) \quad n \geq 1 \end{aligned} \tag{d}$$

Therefore,

$$\begin{aligned} a_0(r) &= 1 \\ a_1(r) &= -\frac{1+r}{2+2r-5} a_0(r) = -\frac{r+1}{2r-3} \\ a_2(r) &= -\frac{2+r}{4+2r-5} a_1(r) = \frac{(r+1)(r+2)}{(2r-3)(2r-1)} \\ &\vdots \\ a_n(r) &= (-1)^n \prod_{j=1}^{n} \frac{(r+j)}{(2r+2j-5)} \quad n \geq 1 \end{aligned} \tag{e}$$

Before proceeding, we determine if $C = 0$ by using Eq. (5.63). If it is equal to zero, then there is no need to obtain $a_n'(r_1)$. Thus,

$$
\begin{aligned}
C &= -\frac{p_1(r_1-1)}{k\alpha_0} a_{k-1}(r_2) = -\frac{(2r_1+1)r_1}{3\times 4} a_2(r_2) \\
&= -\frac{(2r_1+1)r_1}{3\times 4} \frac{(r_2+1)(r_2+2)}{(2r_2-3)(2r_2-1)} \\
&= -\frac{(5+1)(5/2)}{3\times 4} \frac{(-1/2+1)(-1/2+2)}{(-1-3)(-1-1)} \\
&= -\frac{15}{128}
\end{aligned} \tag{f}
$$

For $r_1 = 5/2$, Eq. (e) becomes

$$
a_n(r_1) = (-1)^n \prod_{j=1}^{n} \frac{(5/2+j)}{(5+2j-5)} = \frac{(-1)^n}{4^n n!} \prod_{j=1}^{n} (2j+5) \tag{g}
$$

and the solution corresponding to $r_1 = 5/2$ is given by

$$
y_1(x) = x^{5/2} + \sum_{n=1}^{\infty} \frac{(-1)^n}{4^n n!} \left[\prod_{j=1}^{n} (2j+5) \right] x^{n+5/2}
$$

Expanding this result for a few terms using Mathematica procedure M5.5 gives

$$
y_1(x) = x^{5/2} - \frac{7x^{7/2}}{4} + \frac{63x^{9/2}}{32} - \frac{231x^{11/2}}{128} + \frac{3003x^{13/2}}{2048} - \cdots
$$

To determine the solution given by the second equation of Eq. (c), we note that for $r_2 = -1/2$, Eq. (e) gives

$$
\begin{aligned}
a_0(r_2) &= 1 \\
a_1(r_2) &= -\frac{r_2+1}{2r_2-3} = -\frac{-1/2+1}{-1-3} = \frac{1}{8} \\
a_2(r_2) &= \frac{(r_2+1)(r_2+2)}{(2r_2-3)(2r_2-1)} = \frac{(-1/2+1)(-1/2+2)}{(-1-3)(-1-1)} = \frac{3}{32}
\end{aligned} \tag{h}
$$

Since the coefficients of $a_n(r)$ are expressed as a product, it is easier to employ logarithmic differentiation in the following manner. First, we note that

$$
\begin{aligned}
\ln|a_n(r)| &= \ln\left| (-1)^n \prod_{j=1}^{n} \frac{(r+j)}{(2r+2j-5)} \right| \\
&= \sum_{j=1}^{n} \left(\ln|r+j| - \ln|2r+2j-5| \right)
\end{aligned}
$$

Then,

$$\frac{d}{dr}\left(\ln|a_n(r)|\right)=\sum_{j=1}^{n}\left(\frac{d}{dr}\ln|r+j|-\frac{d}{dr}\ln|2r+2j-5|\right)$$

$$\frac{a_n'(r)}{a_n(r)}=\sum_{j=1}^{n}\left(\frac{1}{r+j}-\frac{2}{2r+2j-5}\right)=-5\sum_{j=1}^{n}\left(\frac{1}{(r+j)(2r+2j-5)}\right) \quad \text{(i)}$$

$$a_n'(r)=-5a_n(r)\sum_{j=1}^{n}\left(\frac{1}{(r+j)(2r+2j-5)}\right)$$

For $r = r_1 = 5/2$, we use Eqs. (g) and (i) to obtain

$$a_n'(5/2)=-5\frac{(-1)^n}{4^n n!}\prod_{j=1}^{n}(2j+5)\sum_{j=1}^{n}\left(\frac{1}{j(2j+5)}\right) \quad \text{(j)}$$

Evaluating Eq. (j) with Mathematica procedure M5.6, we find that

$$a_1'(5/2)=\frac{5}{4}$$

$$a_2'(5/2)=-\frac{125}{64}$$

$$a_3'(5/2)=\frac{1585}{768}$$

Then, the second solution is

$$y_2(x)=x^{-1/2}+\frac{1}{8}x^{1/2}+\frac{3}{32}x^{3/2}$$
$$-\frac{15}{128}\left(y_1(x)\ln x+\frac{5}{4}x^{7/2}-\frac{125}{64}x^{9/2}+\frac{1585}{768}x^{11/2}-\cdots\right)$$

The expressions for $y_1(x)$ and $y_2(x)$ are verified with Mathematica procedure M5.7.

Example 5.8: Difference in roots equals an integer (Case 3A)

Consider the equation

$$x^2\frac{d^2y}{dx^2}+(2+x)x\frac{dy}{dx}-2y=0$$

Comparing this equation with Eq. (5.34), it is seen that

$$\begin{aligned} &\alpha_0 = 1, \quad &\alpha_1 = 0, \quad &\alpha_2 = 0 \\ &\beta_0 = 2, \quad &\beta_1 = 1, \quad &\beta_2 = 0 \\ &\gamma_0 = -2, \quad &\gamma_1 = 0, \quad &\gamma_2 = 0 \end{aligned} \tag{a}$$

Then, from Eq. (5.37)

$$\begin{aligned} p_0(r) &= r(r-1) + 2r - 2 = r^2 + r - 2 = (r-1)(r+2) \\ p_1(r) &= r \\ p_2(r) &= 0 \end{aligned} \tag{b}$$

From the indicial equation $p_0(r) = 0$, we find that $r_1 = 1$ and $r_2 = -2$ and, hence, $r_1 - r_2 = 3$, a positive integer. Since $\alpha_2 = \beta_2 = \gamma_2 = 0$ and $r_1 - r_2 =$ positive integer, we have Case 3A. Hence, the solutions are given by Eqs. (5.61) and (5.62), that is,

$$\begin{aligned} y_1(x) &= \sum_{n=0}^{\infty} a_n(r_1) x^{n+r_1} \\ y_2 &= \sum_{n=0}^{2} a_n(r_2) x^{n+r_2} + C\left(y_1(x)\ln x + \sum_{n=1}^{\infty} a_n'(r_1) x^{n+r_1} \right) \end{aligned} \tag{c}$$

where C is given by Eq. (5.63).

Using Eq. (5.60) and Eq. (b), it is found that

$$\begin{aligned} a_0(r) &= 1 \\ a_n(r) &= -\frac{p_1(n+r-1)a_{n-1}(r)}{p_0(n+r)} \\ &= -\frac{n+r-1}{(n+r-1)(n+r+2)} a_{n-1}(r) = -\frac{a_{n-1}(r)}{(n+r+2)} \qquad n \geq 1 \end{aligned} \tag{d}$$

Therefore,

$$a_n(1) = -\frac{a_{n-1}(1)}{(n+3)} \qquad n \geq 1$$

and the first few terms are

$$\begin{aligned} a_1(1) &= -\frac{a_0(1)}{(1+3)} = -\frac{1}{4} \\ a_2(1) &= -\frac{a_1(1)}{(2+3)} = \left(-\frac{1}{5}\right)\left(-\frac{1}{4}\right) = \frac{1}{5 \times 4} \\ a_3(1) &= -\frac{a_2(1)}{(3+3)} = \left(-\frac{1}{6}\right)\left(\frac{1}{5 \times 4}\right) = -\frac{1}{6 \times 5 \times 4} \end{aligned}$$

Thus, from Eq. (c)

$$y_1(x) = x\sum_{n=0}^{\infty} a_n(1)x^n = x\left(1 - \frac{x}{4} + \frac{x^2}{5\times 4} - \frac{x^3}{6\times 5\times 4}\cdots\right) \quad \text{(c)}$$

To find the second independent solution, we first use Eq. (5.63) to determine C. Thus, since $k = 3$ and using Eq. (b)

$$C = -\frac{p_1(r_1 - 1)}{k\alpha_0}a_{k-1}(r_2) = -\frac{r_1 - 1}{3}a_2(-2) = -\frac{1-1}{3}a_2(-2) = 0$$

Therefore, the second equation of Eq. (c) simplifies to

$$y_2 = x^{-2}\sum_{n=0}^{2} a_n(-2_2)x^n = x^{-2}\left(a_0(-2) + a_1(-2)x + a_3(-2)x^2\right) \quad \text{(d)}$$

where

$$a_n(-2) = -\frac{a_{n-1}(-2)}{n} \qquad n = 1,2$$

Then

$$a_1(-2) = -1$$

$$a_2(-2) = -\frac{1}{2}a_1(-2) = \frac{1}{2}$$

and Eq. (d) becomes

$$y_2(x) = x^{-2}\left(1 - x + \frac{1}{2}x^2\right) = \frac{1}{x^2} - \frac{1}{x} + \frac{1}{2}$$

Example 5.9: Difference in roots equals an integer (Case 3B)

Consider the equation

$$\left(1 + x^2\right)x^2\frac{d^2y}{dx^2} + \left(3 + 10x^2\right)x\frac{dy}{dx} + \left(-15 + 14x^2\right)y = 0$$

Comparing this equation with Eq. (5.34), it is seen that

$$\begin{aligned} &\alpha_0 = 1, \quad &\alpha_1 = 0, \quad &\alpha_2 = 1 \\ &\beta_0 = 3, \quad &\beta_1 = 0, \quad &\beta_2 = 10 \\ &\gamma_0 = -15, \quad &\gamma_1 = 0, \quad &\gamma_2 = 14 \end{aligned} \quad \text{(a)}$$

Then, from Eq. (5.37)

$$p_0(r) = r(r-1) + 3r - 15 = r^2 + 2r - 15 = (r-3)(r+5)$$
$$p_1(r) = 0 \tag{b}$$
$$p_2(r) = r(r-1) + 10r + 14 = r^2 + 9r + 14 = (r+7)(r+2)$$

From the indicial equation $p_0(r) = 0$, we find that $r_1 = 3$ and $r_2 = -5$, $r_1 - r_2 = 8$. Since $\alpha_1 = \beta_1 = \gamma_1 = 0$ and $r_1 - r_2 =$ even integer, we have Case 3B with $k = 4$. Hence, the solutions are given by Eqs. (5.66) and (5.67), that is,

$$y_1(x) = \sum_{n=0}^{\infty} a_{2n}(r_1)x^{2n+r_1}$$
$$y_2(x) = \sum_{n=0}^{k-1} a_{2n}(r_2)x^{2n+r_2} + C\left(y_1(x)\ln x + \sum_{n=1}^{\infty} a'_{2n}(r_1)x^{2n+r_1}\right) \tag{c}$$

where C is given by Eq. (5.68).

Using Eq. (5.65) and Eq. (b), we find that

$$a_0(r) = 1$$
$$a_{2n}(r) = -\frac{2n+r}{2n+r-3}a_{2n-2}(r) \qquad n \geq 1 \tag{d}$$

Therefore, it is straightforward to infer that

$$a_{2n}(r) = (-1)^n \prod_{j=1}^{n} \frac{2j+r}{2j+r-3} \qquad n \geq 1 \tag{e}$$

Before proceeding, we determine if $C = 0$ by using Eq. (5.68). If it is equal to zero, then there is no need to obtain $a'_n(r_1)$. Thus,

$$C = -\frac{p_2(r_1-2)}{2k\alpha_0}a_{2k-2}(r_2) = -\frac{p_2(3-2)}{8}a_6(-5)$$
$$= -\frac{(1+7)(1+2)}{8}(-1)^3 \prod_{j=1}^{3} \frac{2j-5}{2j-8} \tag{f}$$
$$= 3\left(\frac{-3}{-6}\right)\left(\frac{-1}{-4}\right)\left(\frac{1}{-2}\right) = -\frac{3}{16}$$

For $r_1 = 3$, Eq. (e) becomes

$$a_{2n}(r_1) = (-1)^n \prod_{j=1}^{n} \frac{2j+3}{2j} = \frac{(-1)^n}{2^n n!} \prod_{j=1}^{n} (2j+3) \qquad n \geq 1 \tag{g}$$

and from Eq. (c),

$$y_1(x) = x^3 + \sum_{n=1}^{\infty} \frac{(-1)^n}{2^n n!} \prod_{j=1}^{n} (2j+3) x^{2n+3} \tag{h}$$

Using Mathematica procedure M5.8, the first few terms of Eq. (h) are

$$y_1(x) = x^3 - \frac{5x^5}{2} + \frac{35x^7}{8} - \frac{105x^9}{16} + \frac{1155x^{11}}{128} - \cdots \tag{i}$$

To determine the solution given by the second equation of Eq. (c), we note that for $r_2 = -5$, Eq. (e) gives

$$\begin{aligned} a_0(r_2) &= 1 \\ a_{2n}(r_2) &= (-1)^n \prod_{j=1}^{n} \frac{2j-5}{2(j-4)} \qquad n \geq 1 \end{aligned} \tag{j}$$

Then Eq. (j) yields

$$\begin{aligned} a_0(r_2) &= 1 \\ a_2(r_2) &= -\frac{2-5}{2(1-4)} = -\frac{1}{2} \\ a_4(r_2) &= -\frac{4-5}{2(2-4)}\left(-\frac{1}{2}\right) = \frac{1}{8} \\ a_6(r_2) &= -\frac{6-5}{2(3-4)}\left(\frac{1}{8}\right) = \frac{1}{16} \end{aligned} \tag{k}$$

Since the coefficients of $a_{2n}(r)$ are expressed as a product, it is easier to employ logarithmic differentiation in the following manner. Starting with Eq. (e)

$$\begin{aligned} \ln|a_{2n}(r)| &= \ln \prod_{j=1}^{n} \frac{|2j+r|}{|2j+r-3|} \\ &= \sum_{j=1}^{n} \left(\ln|2j+r| - \ln|2j+r-3| \right) \end{aligned}$$

we find that

$$\frac{d}{dr} \ln|a_{2n}(r)| = \sum_{j=1}^{n} \left(\frac{d}{dr} \ln|2j+r| - \frac{d}{dr} \ln|2j+r-3| \right)$$

$$\frac{a_{2n}'(r)}{a_{2n}(r)} = \sum_{j=1}^{n} \left(\frac{1}{2j+r} - \frac{1}{2j+r-3} \right) = -3 \sum_{j=1}^{n} \left(\frac{1}{(2j+r)(2j+r-3)} \right)$$

Then,

$$a'_{2n}(r_1) = -\frac{3}{2}a_{2n}(r_1)\sum_{j=1}^{n}\left(\frac{1}{j(2j+3)}\right)$$

and from the second equation of Eq. (c) and Eqs. (f), (i), (j), and (k)

$$\begin{aligned} y_2(x) &= x^{-5}\left(1 - \frac{1}{2}x^2 + \frac{1}{8}x^4 + \frac{1}{16}x^6\right) - \frac{3}{16}y_1(x)\ln x \\ &+ \frac{9}{32}\sum_{n=1}^{\infty}\frac{(-1)^n}{2^n n!}\prod_{j=1}^{n}(2j+3)\sum_{j=1}^{n}\left(\frac{1}{j(2j+3)}\right)x^{2n+3} \end{aligned} \tag{l}$$

Using Mathematica procedure M5.9 to expand the third expression in Eq. (l) for a few terms gives

$$\begin{aligned} y_2(x) &= x^{-5}\left(1 - \frac{1}{2}x^2 + \frac{1}{8}x^4 + \frac{1}{16}x^6\right) \\ &- \frac{3}{16}\ln x\left(x^3 - \frac{5x^5}{2} + \frac{35x^7}{8} - \frac{105x^9}{16} + \frac{1155x^{11}}{128} - \cdots\right) \\ &- \frac{9}{32}\left(-\frac{x^5}{2} + \frac{19x^7}{16} - \frac{583x^9}{288} + \frac{13771x^{11}}{4608} - \cdots\right) \end{aligned}$$

5.1.4 Bessel's Equation and Bessel Functions

Bessel's equation is of the form

$$x^2\frac{d^2y}{dx^2} + x\frac{dy}{dx} + \left(x^2 - p^2\right)y = 0 \tag{5.69}$$

where p is a real quantity. Comparing Eq. (5.34) with Eq. (5.69), it is seen that

$$\begin{aligned} \alpha_0 &= 1, & \alpha_1 &= 0, & \alpha_2 &= 0 \\ \beta_0 &= 1, & \beta_1 &= 0, & \beta_2 &= 0 \\ \gamma_0 &= -p^2, & \gamma_1 &= 0, & \gamma_2 &= 1 \end{aligned} \tag{5.70}$$

and Eq. (5.37) becomes

$$\begin{aligned} p_0(r) &= r(r-1)1 + r - p^2 = r^2 - p^2 \\ p_1(r) &= 0 \\ p_2(r) &= 1 \end{aligned} \tag{5.71}$$

From the indicial equation, $p_0(r) = 0$, we find that $r_1 = p$ and $r_2 = -p$ and $r_1 - r_2 = 2p$. Hence, depending on the value of p, one of the three cases discussed in Section 5.1.3 will determine the form of $y_2(x)$, the second linearly independent equation. For $r_1 = p$, Eq. (5.45) gives

$$\begin{aligned} a_0(p) &= 1 \\ a_1(p) &= 0 \\ a_n(p) &= -\frac{a_{n-2}(p)}{p_0(n+p)} = -\frac{a_{n-2}(p)}{(n+p)^2 - p^2} = -\frac{a_{n-2}(p)}{n(n+2p)} \quad n \geq 2 \end{aligned} \tag{5.72}$$

Then, the first few terms of Eq. (5.72) yield

$$\begin{aligned} a_0(p) &= 1 \\ a_1(p) &= 0 \\ a_2(p) &= -\frac{a_0(p)}{2(2+p)} = -\frac{1}{2^2(1+p)} \\ a_3(p) &= -\frac{a_1(p)}{3(3+2p)} = 0 \\ a_4(p) &= -\frac{a_2(p)}{4(4+2p)} = \frac{1}{4(4+2p)}\left(\frac{1}{2^2(1+p)}\right) = \frac{1}{2^4 \times 2(1+p)(2+p)} \\ a_5(p) &= -\frac{a_3(p)}{5(5+2p)} = 0 \\ a_6(p) &= -\frac{a_4(p)}{6(6+2p)} = -\frac{1}{6(6+2p)}\left(\frac{1}{2^4 \times 2(1+p)(2+p)}\right) \\ &= -\frac{1}{2^6 \times 1 \times 2 \times 3(1+p)(2+p)(3+p)} \\ &\vdots \end{aligned}$$

Therefore, $a_{2n-1}(p) = 0$, $n = 1, 2, 3, \ldots$ and

$$a_{2n}(p) = \frac{(-1)^n \Gamma(p+1)}{2^{2n} n! \Gamma(p+1+n)} \quad n = 0,1,2,\ldots \tag{5.73}$$

where we have used Eq. (C.6) of Appendix C.

Then the first solution to Eq. (5.69), which is valid for all real values of p, is given by

$$\begin{aligned} y_1(x) &= \sum_{n=0}^{\infty} a_{2n}(p) x^{2n+v} = \sum_{n=0}^{\infty} \frac{(-1)^n}{2^{2n} n! \Gamma(p+1+n)} x^{2n+p} \\ &= 2^p \Gamma(p+1) \sum_{n=0}^{\infty} \frac{(-1)^n (x/2)^{2n+p}}{n! \Gamma(p+1+n)} \\ &= 2^p \Gamma(p+1) J_p(x) \end{aligned} \tag{5.74}$$

where we have changed the indexing to account for the fact that $a_{2n-1}(r_1) = 0$. The quantity $J_p(x)$ is the definition of the *Bessel function of the first kind of order p* and is given by

$$J_p(x)=\sum_{n=0}^{\infty}\frac{(-1)^n(x/2)^{2n+p}}{n!\Gamma(p+1+n)}=\left(\frac{x}{2}\right)^p\sum_{n=0}^{\infty}\frac{(-1)^n(x/2)^{2n}}{n!\Gamma(p+1+n)} \tag{5.75}$$

We shall now determine the expressions for the second independent solution, which is a function of the value of p and the difference $r_1 - r_2 = 2p$.

Case 1: If $p = v$, where v is not an integer and $v \neq 0$, then the solutions to Eq. (5.69) are given by Eqs. (5.50) and (5.51), that is,

$$\begin{aligned} y_1(x) &= 2^{v}\Gamma(v+1)J_v(x) \\ y_2(x) &= 2^{-v}\Gamma(-v+1)J_{-v}(x) \end{aligned} \tag{5.76}$$

Thus, for Case 1, the solution to Eq. (5.69) is

$$y(x)=C_1J_v(x)+C_2J_{-v}(x) \tag{5.77}$$

where the constants $2^{\pm v}\Gamma(\pm v + 1)$ are subsumed in C_j. It is seen from Eq. (C.10) of Appendix C that the gamma function of a negative value exists provided that the value is not an integer.

From Eq. (5.75), it is found that as $x \to 0$, $J_v(x) \to 0$ and $J_{-v}(x) \to \infty$.

Case 2: If $p = 0$, then we have repeated roots, that is, $r_1 = 0$ and $r_2 = 0$. In this case, the solutions to Eq. (5.69) are given by Eqs. (5.53) and (5.54), that is,

$$\begin{aligned} y_1(x) &= \sum_{n=0}^{\infty} a_{2n}(0)x^{2n} \\ y_2(x) &= y_1(x)\ln x + \sum_{n=1}^{\infty} a'_{2n}(0)x^{2n} \end{aligned} \tag{5.78}$$

Then, from Eq. (5.73)

$$a_{2n}(0)=\frac{(-1)^n}{2^{2n}n!\Gamma(1+n)}=\frac{(-1)^n}{2^{2n}(n!)^2} \qquad n=0,1,2,\ldots \tag{5.79}$$

where we have used Eq. (C.5) of Appendix C, and Eq. (5.75) becomes

$$J_0(x)=\sum_{n=0}^{\infty}(-1)^n\frac{(x/2)^{2n}}{(n!)^2} \tag{5.80}$$

This result could have also been obtained by setting $p = 0$ in Eq. (5.75).

From Eq. (5.73), $a'_{2n}(r)$ is determined as follows

$$\frac{d}{dr}a_{2n}(r)=\frac{(-1)^n}{2^{2n}n!}\left(\begin{array}{c}\frac{1}{\Gamma(r+1+n)}\frac{d}{dr}\Gamma(r+1)\\+\Gamma(r+1)\frac{d}{dr}\left[\frac{1}{\Gamma(r+1+n)}\right]\end{array}\right)$$

$$=\frac{(-1)^n}{2^{2n}n!}\left(\begin{array}{c}\frac{1}{\Gamma(r+1+n)}\Gamma(r+1)\psi(1+r)\\-\frac{\Gamma(r+1)}{\Gamma(r+1+n)}\psi(1+r+n)\end{array}\right) \tag{5.81}$$

where have used Eqs. (C.15) and (C.19) of Appendix C and $\psi(x)$ is the digamma function given by Eq. (C.16) of Appendix C. For $r = 0$, Eq. (5.81) becomes

$$a'_{2n}(0)=\frac{(-1)^n}{2^{2n}n!}\left(\frac{\Gamma(1)\psi(1)}{\Gamma(1+n)}-\frac{\Gamma(1)}{\Gamma(1+n)}\psi(1+n)\right)$$

$$=\frac{(-1)^n}{2^{2n}(n!)^2}\left(-\gamma-\psi(1+n)\right)=-\frac{(-1)^n}{2^{2n}(n!)^2}\sum_{k=1}^{n}\frac{1}{k} \tag{5.82}$$

where we have used Eq. (C.18) of Appendix C and $\gamma = 0.5772156649$ is the Euler constant [see Eq. (C.17) in Appendix C].

Then, from the second equation of Eq. (5.78), it is found that

$$y_2(x)=J_0(x)\ln x+\sum_{n=1}^{\infty}\frac{(-1)^{n+1}}{(n!)^2}\left(\sum_{k=1}^{n}\frac{1}{k}\right)\left(\frac{x}{2}\right)^{2n} \tag{5.83}$$

If we scale Eq. (5.83) by multiplying it by $\pi/2$ and then add to it a portion of $y_1(x)$ as

$$\frac{2}{\pi}\left(\gamma-\ln 2\right)J_0(x)$$

Then Eq. (5.83) becomes

$$Y_0(x)=\frac{2}{\pi}y_2(x)+\frac{2}{\pi}\left(\gamma-\ln 2\right)J_0(x)$$

$$=\frac{2}{\pi}\left[\left(\ln(x/2)+\gamma\right)J_0(x)+\sum_{n=1}^{\infty}\left[\frac{(-1)^{n+1}}{(n!)^2}\sum_{k=1}^{n}\frac{1}{k}\right]\left(\frac{x}{2}\right)^{2n}\right] \tag{5.84}$$

When written in this form, $Y_0(x)$ is called the *Bessel function of the second kind of order 0*. The reason for this modification will become apparent when the case for $p = m$ is considered, where m is an integer, and in that result m is set to 0. See Appendix 5.1 at the end of the chapter.

Hence, the solution to Eq. (5.69) when $p = 0$ is

$$y(x) = C_1 J_0(x) + C_2 Y_0(x) \tag{5.85}$$

It is seen from Eq. (5.80) that as $x \to 0$, $J_0(x) \to 1$ and from Eq. (5.84) that $Y_0(x) \to -\infty$.

Case 3: If $p = m$, where $m = 1, 2, 3, \ldots$, then $r_1 = m$, $r_2 = -m$ and $r_1 - r_2 = 2m$. In this case, we set $\nu = m$ in Eq. (5.75) and arrive at

$$J_m(x) = \sum_{n=0}^{\infty} \frac{(-1)^n (x/2)^{2n+m}}{n!\Gamma(m+1+n)} = \left(\frac{x}{2}\right)^m \sum_{n=0}^{\infty} \frac{(-1)^n (x/2)^{2n}}{n!(m+n)!} \tag{5.86}$$

It is seen from Eq. (5.86) that

$$J_m(-x) = (-1)^m J_m(x) \tag{5.87}$$

and as $x \to 0$, $J_m(x) \to 0$, $m \neq 0$. Another relationship, which is stated without proof, is

$$J_{-m}(x) = J_m(-x) = (-1)^m J_m(x) \tag{5.88}$$

This equation indicates that when m is an integer, $J_{-m}(x)$ is not a linearly independent solution.

The determination of the second linearly independent solution is obtained from

$$Y_\nu(x) = \frac{\cos\nu\pi J_\nu(x) - J_{-\nu}(x)}{\sin\nu\pi} \tag{5.89}$$

which is known as the Weber function. When ν is an integer or zero, one takes the limit of Eq. (5.89) with respect to ν as $\nu \to m$, $m = 0,1, 2, 3, \ldots$ This limit exists but obtaining it is somewhat involved and, therefore, has been placed in Appendix 5.1, which appears at the end of the chapter. From this appendix, it is found that

$$\begin{aligned} Y_m(x) = \frac{2}{\pi}\ln\left(\frac{x}{2}\right) J_m(x) - \frac{1}{\pi}\sum_{n=0}^{m-1} \frac{(m-1-n)!}{n!}\left(\frac{x}{2}\right)^{2n-m} \\ - \frac{1}{\pi}\sum_{n=0}^{\infty} \frac{(-1)^n}{n!(m+n)!}\left(\frac{x}{2}\right)^{2n+m} \left[\psi(1+n) + \psi(m+1+n)\right] \end{aligned} \tag{5.90}$$

Equation (5.90) is the conventional form of the *Bessel function of the second kind of order m*. It is seen that as $x \to 0$, $Y_m(x) \to -\infty$. In addition, it can be shown that

$$Y_{-m}(x) = (-1)^m Y_m(x) \tag{5.91}$$

Based on Eqs. (5.89) and (5.90), the Bessel function of the second kind $Y_p(x)$ is valid for any integer or non-integer value of p. Therefore, for any integer or non-integer value of p and when $p = 0$, the solution to Bessel's equation given by Eq. (5.69) is

$$y(x) = C_1 J_p(x) + C_2 Y_p(x) \tag{5.92}$$

The Bessel functions of the first and second kind are plotted in Figure 5.1 for orders p = 0, 1, and 2. Notice that the Bessel functions of the first and second are oscillatory functions that have an infinite number of values for which they equal zero.

The asymptotic forms for Bessel functions of the first and second kind are as follows. For $x \to 0$,

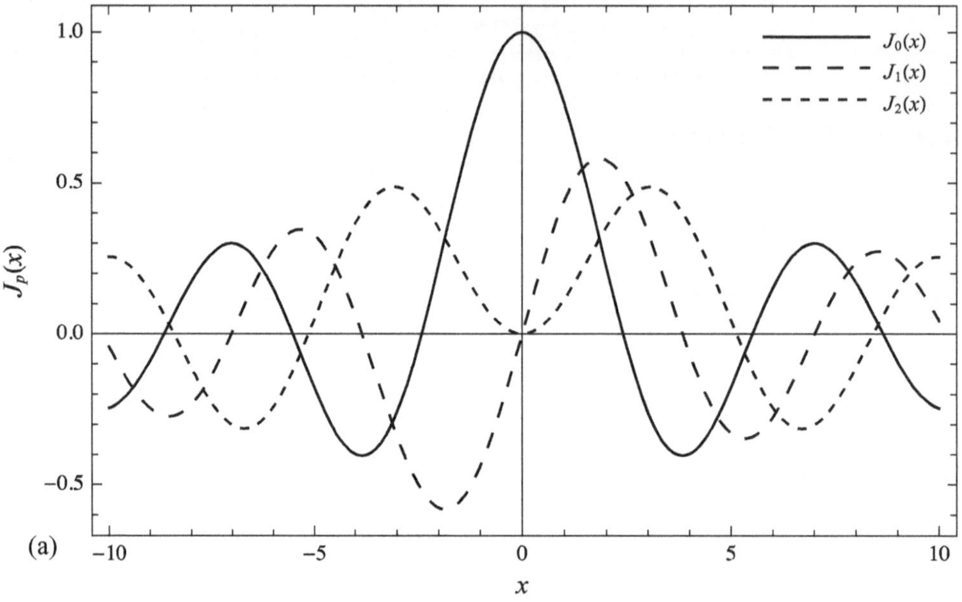

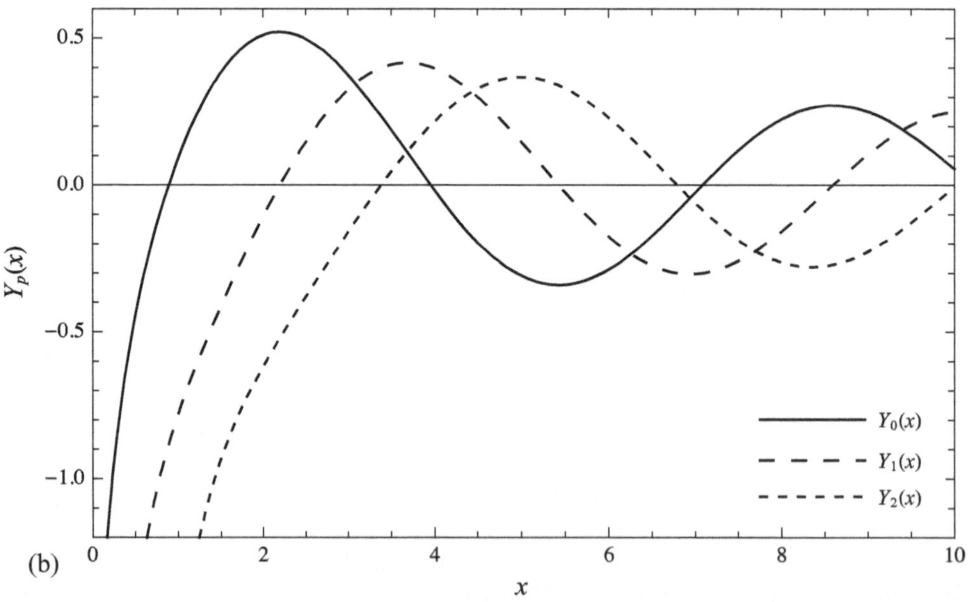

FIGURE 5.1 (a) Bessel functions of the first kind of order p = 0, 1, and 2; (b) Bessel functions of the second kind of order p = 0, 1, and 2.

$$J_0(x) \to 1$$
$$J_p(x) \approx \frac{1}{\Gamma(p+1)}\left(\frac{x}{2}\right)^p \qquad p>0$$
$$Y_0(x) \approx \frac{2}{\pi}\left(\ln\frac{x}{2}+\gamma\right) \tag{5.93}$$
$$Y_p(x) \approx -\frac{\Gamma(p)}{\pi}\left(\frac{2}{x}\right)^p \qquad p>0$$

and for $x \to \infty$,

$$J_p(x) \approx \sqrt{\frac{2}{\pi x}}\cos(x-(2p+1)\pi/4)$$
$$Y_p(x) \approx \sqrt{\frac{2}{\pi x}}\sin(x-(2p+1)\pi/4) \tag{5.94}$$

Equations (5.93) and (5.94), respectively, are verified with Mathematica procedure M5.10.

The *Hankel functions*, sometimes known as the Bessel functions of the third and fourth kind of order p, respectively, are defined as

$$H_p^{(1)}(x) = J_p(x) + jY_p(x)$$
$$H_p^{(2)}(x) = J_p(x) - jY_p(x) \tag{5.95}$$

Therefore, a solution to Eq. (5.69) is

$$y(x) = C_1 H_p^{(1)}(x) + C_2 H_p^{(2)}(x) \tag{5.96}$$

For all values of p, $H_p^{(1)}(x)$ and $H_p^{(2)}(x)$ are linearly independent.

The asymptotic forms for Hankel functions are as follows. For $x \to 0$,

$$H_p^{(1)}(x) \approx -j\frac{\Gamma(p)}{\pi}\left(\frac{2}{x}\right)^p \qquad p \neq -1,-2,\ldots$$
$$H_p^{(2)}(x) \approx j\frac{\Gamma(p)}{\pi}\left(\frac{2}{x}\right)^p \qquad p \neq -1,-2,\ldots \tag{5.97}$$

and for $x \to \infty$,

$$H_p^{(1)}(x) \approx \sqrt{\frac{2}{\pi x}}e^{j(x-(2p+1)\pi/4)}$$
$$H_p^{(2)}(x) \approx \sqrt{\frac{2}{\pi x}}e^{-j(x-(2p+1)\pi/4)} \tag{5.98}$$

5.1.5 Derivatives and Integrals of Bessel Functions of the First and Second Kind

We shall determine the derivative of the Bessel function of the first kind of integer order. To facilitate matters, we shall first obtain the derivative of the product of the Bessel function multiplied by x^{-m}. Thus, from Eq. (5.75)

$$\begin{aligned}\frac{d}{dx}\left(x^{-m}J_m(x)\right) &= \frac{d}{dx}\left(x^{-m}\sum_{n=0}^{\infty}\frac{(-1)^n(x/2)^{2n+m}}{n!(m+n)!}\right) = \frac{d}{dx}\sum_{n=0}^{\infty}\frac{(-1)^n x^{2n}}{2^{2n+m}n!(m+n)!}\\ &= \sum_{n=1}^{\infty}\frac{(-1)^n(2n)x^{2n-1}}{2^{2n+m}n!(m+n)!} = \sum_{n=1}^{\infty}\frac{(-1)^n x^{2n-1}}{2^{2n+m-1}(n-1)!(m+n)!}\\ &= \sum_{n=0}^{\infty}\frac{(-1)^{n+1}x^{2n+1}}{2^{2n+m+1}(n)!(m+1+n)!} = -x^{-m}\sum_{n=0}^{\infty}\frac{(-1)^n(x/2)^{2n+m+1}}{(n)!(m+1+n)!}\\ \frac{d}{dx}\left(x^{-m}J_m(x)\right) &= -x^{-m}J_{m+1}(x)\end{aligned} \tag{5.99}$$

Expanding the left-hand side of Eq. (5.99), we obtain

$$\frac{d}{dx}\left(J_m(x)\right) = \frac{m}{x}J_m(x) - J_{m+1}(x) \tag{5.100}$$

In a similar manner,

$$\begin{aligned}\frac{d}{dx}\left(x^{m}J_m(x)\right) &= \frac{d}{dx}\left(x^{m}\sum_{n=0}^{\infty}\frac{(-1)^n(x/2)^{2n+m}}{n!(m+n)!}\right) = \frac{d}{dx}\sum_{n=0}^{\infty}\frac{(-1)^n x^{2n+2m}}{2^{2n+m}n!(m+n)!}\\ &= \sum_{n=0}^{\infty}\frac{(-1)^n(2n+2m)x^{2n+2m-1}}{2^{2n+m}n!(m+n)!} = \sum_{n=0}^{\infty}\frac{(-1)^n(n+m)x^{2n+2m-1}}{2^{2n+m-1}(n)!(m+n)!}\\ &= \sum_{n=0}^{\infty}\frac{(-1)^n x^{2n+2m-1}}{2^{2n+m-1}(n)!(m-1+n)!} = x^{m}\sum_{n=0}^{\infty}\frac{(-1)^n(x/2)^{2n+m-1}}{(n)!(m-1+n)!}\\ \frac{d}{dx}\left(x^{m}J_m(x)\right) &= x^{m}J_{m-1}(x)\end{aligned} \tag{5.101}$$

Expanding the left-hand side of Eq. (5.101), we obtain

$$\frac{d}{dx}\left(J_m(x)\right) = J_{m-1}(x) - \frac{m}{x}J_m(x) \tag{5.102}$$

Adding Eqs. (5.100) and (5.102) and subtracting Eq. (5.100) from Eq. (5.102), respectively, gives*

$$\frac{d}{dx}\left(J_m(x)\right)=\frac{1}{2}\left[J_{m-1}(x)-J_{m+1}(x)\right]$$
$$\frac{2m}{x}J_m(x)=J_{m+1}(x)+J_{m-1}(x) \tag{5.103}$$

To determine two integrals involving Bessel functions, we use Eq. (5.99), which upon integrating gives

$$\int x^{-m}J_{m+1}(x)dx=-\int d\left(x^{-m}J_m(x)\right)$$

or

$$\int x^{-m}J_{m+1}(x)dx=-x^{-m}J_m(x)+C \tag{5.104}$$

Similarly, integrating Eq. (5.101), we find

$$\int x^{m}J_{m-1}(x)dx=\int d\left(x^{m}J_m(x)\right)$$

or

$$\int x^{m}J_{m-1}(x)dx=x^{m}J_m(x)+C \tag{5.105}$$

For the Bessel function of the second kind, we have

$$\frac{d}{dx}\left(Y_m(x)\right)=\frac{1}{2}\left[Y_{m-1}(x)-Y_{m+1}(x)\right]$$
$$\frac{2m}{x}Y_m(x)=Y_{m+1}(x)+Y_{m-1}(x) \tag{5.106}$$

---

* It is noted that

$$\frac{d}{dx}\left(J_m(ax)\right)=a\frac{d}{dz}\left(J_m(z)\right)=a\left(\frac{m}{z}J_m(z)-J_{m+1}(z)\right)=\frac{m}{x}J_m(ax)-aJ_{m+1}(ax)$$

and, in a similar manner,

$$\frac{d}{dx}\left(J_m(ax)\right)=aJ_{m-1}(x)-\frac{m}{x}J_m(ax)$$

Therefore,

$$\frac{2m}{ax}J_m(ax)=J_{m+1}(ax)+J_{m-1}(ax)$$

and

$$\int x^m Y_{m-1}(x)dx = x^m Y_m(x) + C$$
$$\int x^{-m} Y_{m+1}(x)dx = -x^{-m} Y_m(x) + C \tag{5.107}$$

The respective first equations of Eqs. (5.103) and (5.106) and Eqs. (5.104), (5.105), and (5.107) are verified with Mathematica procedure M5.11.

5.1.6 Spherical Bessel Functions

Spherical Bessel functions are obtained for the case where $p = \pm(k + 1/2)$, $k = 0, 1, 2, \ldots$. Thus, Eq. (5.69) becomes

$$x^2 \frac{d^2y}{dx^2} + x\frac{dy}{dx} + \left(x^2 - \left(k + \frac{1}{2}\right)^2\right)y = 0 \tag{5.108}$$

To obtain a solution to Eq. (5.108), we start by setting $k = 0$. Then using the value $p = 1/2$ in Eq. (5.75) gives

$$\begin{aligned} J_{1/2}(x) &= \sum_{n=0}^{\infty} \frac{(-1)^n (x/2)^{2n+1/2}}{n!\Gamma(1/2+1+n)} = \sqrt{\frac{x}{2}} \sum_{n=0}^{\infty} \frac{(-1)^n x^{2n}}{n! 2^{2n} \Gamma(3/2+n)} \\ &= \sqrt{\frac{x}{2}} \sum_{n=0}^{\infty} \frac{(-1)^n x^{2n} n! 2^{2n+1}}{n! 2^{2n} \sqrt{\pi}(2n+1)!} = \sqrt{\frac{2x}{\pi}} \sum_{n=0}^{\infty} \frac{(-1)^n x^{2n}}{(2n+1)!} \end{aligned} \tag{5.109}$$

where we have used Eq. (C.13) of Appendix C. However, from Eq. (A.2) of Appendix A

$$\begin{aligned} \sin x &= \sum_{n=1,3,5}^{\infty} \frac{(-1)^{(n-1)/2} x^n}{n!} = \sum_{n=1}^{\infty} \frac{(-1)^{n+1} x^{2n-1}}{(2n-1)!} \\ &= \sum_{n=0}^{\infty} \frac{(-1)^{n+2} x^{2n+1}}{(2n+1)!} \\ &= x \sum_{n=0}^{\infty} \frac{(-1)^n x^{2n}}{(2n+1)!} \end{aligned}$$

Then Eq. (5.109) can be written as

$$J_{1/2}(x) = \sqrt{\frac{2x}{\pi}} \frac{\sin x}{x} = \sqrt{\frac{2}{\pi x}} \sin x \tag{5.110}$$

For $p = -1/2$, we use Eqs. (5.77) and (5.75) to find that

$$
\begin{aligned}
J_{-1/2}(x) &= \sum_{n=0}^{\infty} \frac{(-1)^n (x/2)^{2n-1/2}}{n!\Gamma(-1/2+1+n)} = \sqrt{\frac{2}{x}} \sum_{n=0}^{\infty} \frac{(-1)^n x^{2n}}{n! 2^{2n} \Gamma(1/2+n)} \\
&= \sqrt{\frac{2}{x}} \sum_{n=0}^{\infty} \frac{(-1)^n x^{2n} n! 2^{2n}}{\sqrt{\pi} n! 2^{2n} (2n)!} = \sqrt{\frac{2}{\pi x}} \sum_{n=0}^{\infty} \frac{(-1)^n x^{2n}}{(2n)!}
\end{aligned} \tag{5.111}
$$

where we have used Eq. (C.12) of Appendix C. However, from Eq. (A.2) of Appendix A

$$
\cos x = \sum_{n=0,2,4}^{\infty} \frac{(-1)^{n/2} x^n}{n!} = \sum_{n=0}^{\infty} \frac{(-1)^n x^{2n}}{(2n)!}
$$

Then, Eq. (5.111) becomes

$$
J_{-1/2}(x) = \sqrt{\frac{2}{\pi x}} \cos x \tag{5.112}
$$

To obtain the expressions for $k > 0$, we use Eq. (5.99) as follows. First, we set $m = 1/2$ in Eq. (5.99) and obtain

$$
\begin{aligned}
J_{3/2}(x) &= -x^{1/2} \frac{d}{dx}\left(x^{-1/2} J_{1/2}(x)\right) = -x^{1/2} \sqrt{\frac{2}{\pi}} \frac{d}{dx}\left(\frac{\sin x}{x}\right) \\
&= \sqrt{\frac{2}{\pi x}} \left(\frac{\sin x}{x} - \cos x\right)
\end{aligned} \tag{5.113}
$$

Next, we set $m = 3/2$ in Eq. (5.99) to arrive at

$$
\begin{aligned}
J_{5/2}(x) &= -x^{3/2} \frac{d}{dx}\left(x^{-3/2} J_{3/2}(x)\right) \\
&= -x^{3/2} \sqrt{\frac{2}{\pi}} \frac{d}{dx}\left(\frac{\sin x}{x^{3/2}} - \frac{\cos x}{x^{1/2}}\right) \\
&= \sqrt{\frac{2}{\pi x}} \left(\left(\frac{3}{x^2} - 1\right)\sin x - \frac{3\cos x}{x}\right)
\end{aligned} \tag{5.114}
$$

Equations (5.113) and (5.114) are verified with Mathematica procedure M5.12.

To obtain the second linearly independent solution, we set $m = -1/2$ in Eq. (5.101) and arrive at

$$
\begin{aligned}
J_{-3/2}(x) &= x^{1/2} \frac{d}{dx}\left(x^{-1/2} J_{-1/2}(x)\right) = x^{1/2} \sqrt{\frac{2}{\pi}} \frac{d}{dx}\left(\frac{\cos x}{x}\right) \\
&= -\sqrt{\frac{2}{\pi x}} \left(\sin x + \frac{\cos x}{x}\right)
\end{aligned} \tag{5.115}
$$

Next, we set $m = -3/2$ in Eq. (5.101) to arrive at

$$\begin{aligned} J_{-5/2}(x) &= x^{3/2} \frac{d}{dx}\left(x^{-3/2} J_{-3/2}(x)\right) \\ &= -x^{3/2} \sqrt{\frac{2}{\pi}} \frac{d}{dx}\left\{x^{-3/2}\left(\frac{\sin x}{x^{1/2}} + \frac{\cos x}{x^{3/2}}\right)\right\} \\ &= \sqrt{\frac{2}{\pi x}}\left(\left(\frac{3}{x^2} - 1\right)\cos x + \frac{3\sin x}{x}\right) \end{aligned} \tag{5.116}$$

Equations (5.115) and (5.116) are verified with Mathematica procedure M5.13.

The *spherical Bessel function of the first kind* is defined as

$$j_n(x) = \sqrt{\frac{\pi}{2x}} J_{n+1/2}(x) \quad n = 0,1,2,\ldots \tag{5.117}$$

where, from Eqs. (5.110), (5.113), and (5.114), the first few spherical Bessel functions of the first kind are

$$\begin{aligned} j_0(x) &= \frac{\sin x}{x} \\ j_1(x) &= \frac{\sin x}{x^2} - \frac{\cos x}{x} \\ j_2(x) &= \left(\frac{3}{x^3} - \frac{1}{x}\right)\sin x - \frac{3\cos x}{x^2} \end{aligned} \tag{5.118}$$

The *spherical Bessel function of the second kind* is defined as

$$y_n(x) = (-1)^{n+1} \sqrt{\frac{\pi}{2x}} J_{-n-1/2}(x) \quad n = 0,1,2,\ldots \tag{5.119}$$

where, from Eqs. (5.112), (5.115), and (5.116), the first few spherical Bessel functions of the second kind are

$$\begin{aligned} y_0(x) &= -\frac{\cos x}{x} \\ y_1(x) &= -\frac{\cos x}{x^2} - \frac{\sin x}{x} \\ y_2(x) &= \left(-\frac{3}{x^3} + \frac{1}{x}\right)\cos x - \frac{3\cos x}{x^2} \end{aligned} \tag{5.120}$$

The spherical Bessel functions of the first and second kind are plotted in Figure 5.2 for order n = 0, 1, and 2. Their form as given in Eqs. (5.118) and (5.120) is verified with Mathematica procedure M5.14.

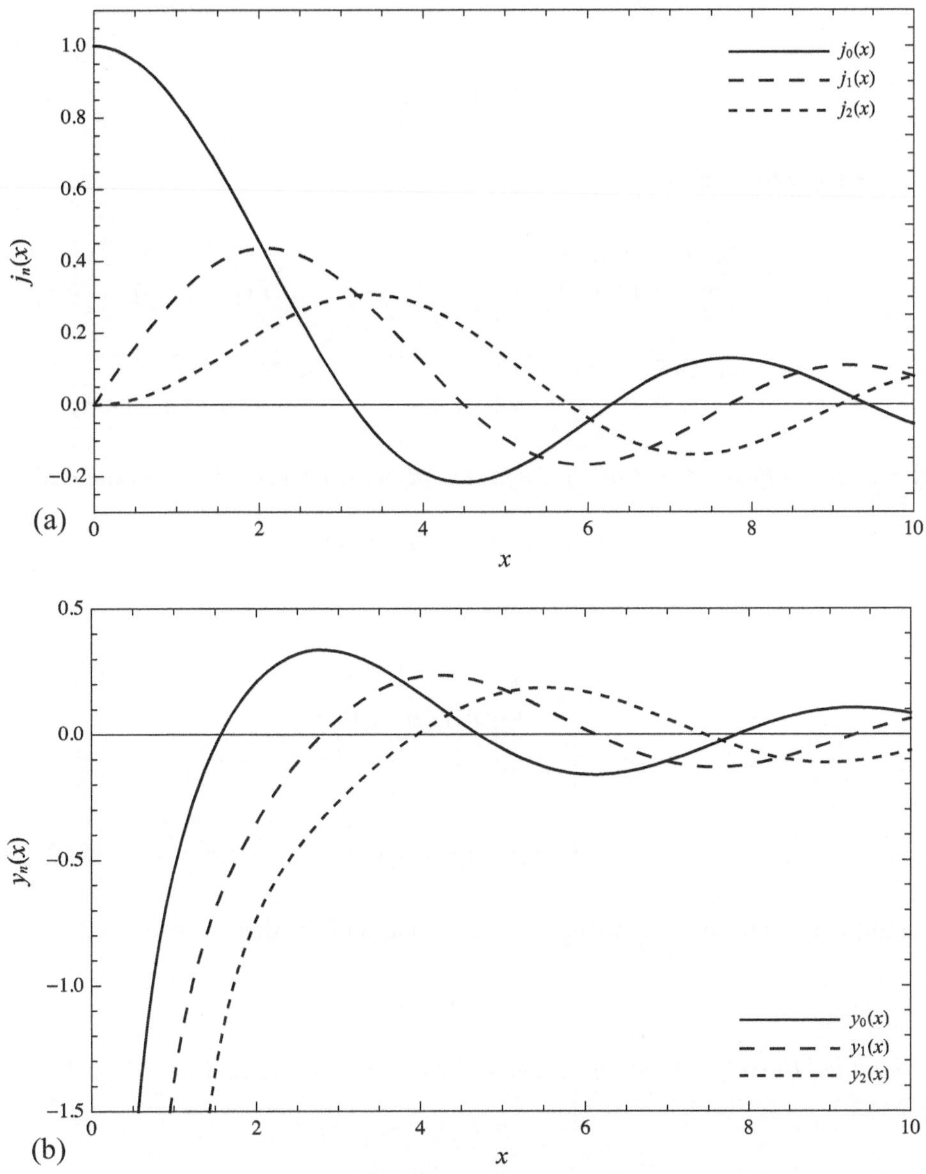

FIGURE 5.2 (a) Spherical Bessel functions of the first kind of order $n = 0$, 1, and 2; (b) spherical Bessel functions of the second kind of order $n = 0$, 1, and 2.

It is mentioned that spherical Bessel functions are considered elementary functions since they are composed of trigonometric functions.

5.1.7 Modified Bessel Functions

The modified Bessel function is the solution to

$$x^2 \frac{d^2y}{dx^2} + x\frac{dy}{dx} + \left(-x^2 - p^2\right)y = 0 \tag{5.121}$$

or

$$x^2 \frac{d^2y}{dx^2} + x\frac{dy}{dx} + \left((\pm jx)^2 - p^2\right)y = 0 \tag{5.122}$$

We use Eq. (5.86) to obtain

$$J_p(\pm jx) = \sum_{n=0}^{\infty} \frac{(-1)^n (\pm jx/2)^{2n+p}}{n!\Gamma(p+1+n)} = (\pm j)^p \sum_{n=0}^{\infty} \frac{(-1)^n (j)^{2n} (x/2)^{2n+p}}{n!\Gamma(p+1+n)}$$

$$= (\pm j)^p \sum_{n=0}^{\infty} \frac{(-1)^n (-1)^n (x/2)^{2n+p}}{n!\Gamma(p+1+n)} = (\pm j)^p \sum_{n=0}^{\infty} \frac{(x/2)^{2n+p}}{n!\Gamma(p+1+n)}$$

Then the *modified Bessel function of the first kind of order p* is defined as the real quantity

$$I_p(\pm x) = (\pm j)^{-p} J_p(\pm jx) \tag{5.123}$$

where

$$I_p(x) = \sum_{n=0}^{\infty} \frac{(x/2)^{2n+p}}{n!\Gamma(p+1+n)} \tag{5.124}$$

and

$$I_p(\pm x) = (\pm 1)^p I_p(x) \tag{5.125}$$

It is straightforward to show by using Eqs. (5.124) and (5.75) that

$$I_p(jx) = (j)^p J_p(x) \tag{5.126}$$

The *modified Bessel function of the second kind* can be obtained from [recall Eq. (5.89)]

$$K_p(x) = \frac{\pi}{2} \frac{I_{-p}(x) - I_p(x)}{\sin p\pi} \tag{5.127}$$

when p is not an integer and in the limit as $p \to m$ (an integer or zero) this limit exists and is given by

$$\begin{aligned} K_m(x) = {} & (-1)^{m+1} \ln\left(\frac{x}{2}\right) I_m(x) \\ & + \frac{1}{2}\left(\frac{x}{2}\right)^{-m} \sum_{n=0}^{m-1} \frac{(m-1-n)!}{n!} \left(-\frac{x}{2}\right)^{2n} \\ & + \frac{(-1)^m}{2}\left(\frac{x}{2}\right)^m \sum_{n=0}^{\infty} \frac{1}{n!(m+n)!} \left(\frac{x}{2}\right)^{2n} \left[\psi(1+n) + \psi(m+1+n)\right] \end{aligned} \tag{5.128}$$

Then the general solution to Eq. (5.121) is

$$y(x) = C_1 I_p(x) + C_2 K_p(x) \tag{5.129}$$

The modified Bessel functions of the first and second kind are plotted in Figure 5.3 for orders $p = 0$, 1, and 2. Notice that the modified Bessel functions of the first and second kind are not oscillatory.

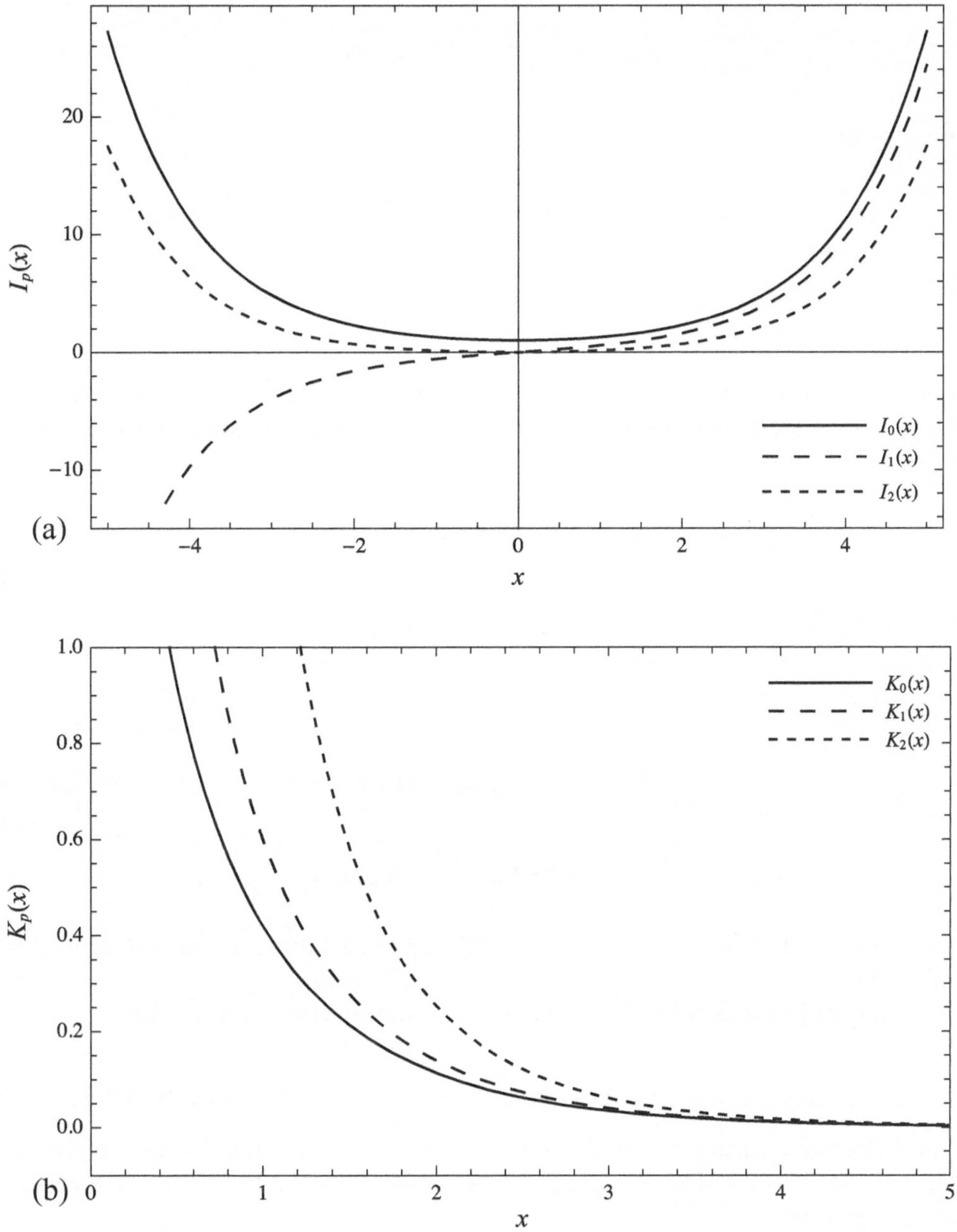

FIGURE 5.3 (a) Modified Bessel functions of the first kind of order $p = 0$, 1, and 2; (b) modified Bessel functions of the second kind of order $p = 0$, 1, and 2.

The asymptotic forms for modified Bessel functions of the first and second kind are as follows. For $x \to 0$,

$$
\begin{aligned}
I_p(x) &\approx \frac{1}{\Gamma(p+1)}\left(\frac{x}{2}\right)^p \quad p>0 \\
K_0(x) &\approx -\ln\frac{x}{2} \\
K_p(x) &\approx \frac{\Gamma(p)}{2}\left(\frac{2}{x}\right)^p \quad p>0
\end{aligned}
\tag{5.130}
$$

and for $x \to \infty$,

$$
\begin{aligned}
I_p(x) &\approx \frac{1}{\sqrt{2\pi x}}e^x \\
K_p(x) &\approx \sqrt{\frac{\pi}{2x}}e^{-x}
\end{aligned}
\tag{5.131}
$$

Equations (5.130) and (5.131), respectively, are verified with Mathematica procedure M5.15.

The following recurrence relations for the modified Bessel equations for integers m are as follows:

$$
\begin{aligned}
&\frac{d}{dx}\left(I_m(x)\right) = \frac{1}{2}\left[I_{m-1}(x) + I_{m+1}(x)\right] \\
&\frac{2m}{x}I_m(x) = I_{m-1}(x) - I_{m+1}(x)
\end{aligned}
\tag{5.132}
$$

and

$$
\begin{aligned}
&\frac{d}{dx}\left(K_m(x)\right) = -\frac{1}{2}\left[K_{m-1}(x) + K_{m+1}(x)\right] \\
&\frac{2m}{x}K_m(x) = -K_{m-1}(x) + K_{m+1}(x)
\end{aligned}
\tag{5.133}
$$

The respective first equations of Eqs. (5.132) and (5.133) are verified with Mathematica procedure M5.16.

The results of Sections 5.1.4, 5.1.5, and 5.1.7 are summarized in Table 5.1.

5.1.8 Differential Equations Whose Solutions Are in Terms of Bessel Functions

Certain differential equations can be converted into a Bessel equation by suitable transformations. We shall consider two classes of equations. For the first type, we consider the following equation

$$
z^2\frac{d^2y}{dz^2} + (1-2a)z\frac{dy}{dz} + \left(z^2 - r^2\right)y = 0 \tag{5.134}
$$

TABLE 5.1
Summary of Bessel Function Relations: $m = 0$ or a Positive Integer

$$w(x) = C_1 w_1(x) + C_2 w_2(x)$$

	Bessel's Equation $x^2 \dfrac{d^2w}{dx^2} + x\dfrac{dw}{dx} + \left(x^2 - p^2\right)w = 0$		**Modified Bessel's Equation** $x^2 \dfrac{d^2w}{dx^2} + x\dfrac{dw}{dx} - \left(x^2 + p^2\right)w = 0$	
	$w_1(x) = J_p(x)$	$w_2(x) = Y_p(x)$	$w_1(x) = I_p(x)$	$w_2(x) = K_p(x)$
Limit $x \to 0$	$J_p(x) \approx \dfrac{1}{\Gamma(p+1)}\left(\dfrac{x}{2}\right)^p \quad p \geq 0$	$Y_0(x) \approx \dfrac{2}{\pi}\left(\ln\dfrac{x}{2} + \gamma\right)$ $Y_p(x) \approx -\dfrac{\Gamma(p)}{\pi}\left(\dfrac{2}{x}\right)^p \quad p > 0$	$I_p(x) \approx \dfrac{1}{\Gamma(p+1)}\left(\dfrac{x}{2}\right)^p \quad p \geq 0$	$K_0(x) \approx -\ln\dfrac{x}{2}$ $K_p(x) \approx \dfrac{\Gamma(p)}{2}\left(\dfrac{2}{x}\right)^p \quad p > 0$
Limit $x \to \infty$	$J_p(x) \approx \sqrt{\dfrac{2}{\pi x}}\cos(x - \varphi)$ $\varphi = (2p+1)\pi / 4$	$Y_p(x) \approx \sqrt{\dfrac{2}{\pi x}}\sin(x - \varphi)$ $\varphi = (2p+1)\pi / 4$	$I_p(x) \approx \dfrac{1}{\sqrt{2\pi x}}e^x$	$K_p(x) \approx \sqrt{\dfrac{\pi}{2x}}e^{-x}$
Derivatives	$\dfrac{d}{dx}\left(J_p(x)\right) = \dfrac{1}{2}\left[J_{p-1}(x) - J_{p+1}(x)\right]$ $\dfrac{2p}{x}J_p(x) = J_{p+1}(x) + J_{p-1}(x)$	$\dfrac{d}{dx}\left(Y_p(x)\right) = \dfrac{1}{2}\left[Y_{p-1}(x) - Y_{p+1}(x)\right]$ $\dfrac{2p}{x}Y_p(x) = Y_{p+1}(x) + Y_{p-1}(x)$	$\dfrac{d}{dx}\left(I_p(x)\right) = \dfrac{1}{2}\left[I_{p-1}(x) + I_{p+1}(x)\right]$ $\dfrac{2p}{x}I_p(x) = I_{p-1}(x) - I_{p+1}(x)$	$\dfrac{d}{dx}\left(K_p(x)\right) = -\dfrac{1}{2}\left[K_{p-1}(x) + K_{p+1}(x)\right]$ $\dfrac{2p}{x}K_p(x) = -K_{p-1}(x) + K_{p+1}(x)$
Integrals	$\int x^m J_{m-1}(x)dx = x^m J_m(x)$ $\int x^{-m} J_{m+1}(x)dx = -x^{-m} J_m(x)$	$\int x^m Y_{m-1}(x)dx = x^m Y_m(x)$ $\int x^{-m} Y_{m+1}(x)dx = -x^{-m} Y_m(x)$	$\int x^m I_{m-1}(x)dx = x^m I_m(x)$ $\int x^{-m} I_{m+1}(x)dx = x^{-m} I_m(x)$	$\int x^m K_{m-1}(x)dx = -x^m K_m(x)$ $\int x^{-m} K_{m+1}(x)dx = -x^{-m} K_m(x)$
Negative orders and arguments	$J_m(-x) = J_{-m}(x)$ $= (-1)^m J_m(x)$	$Y_{-m}(x) = (-1)^m Y_m(x)$	$I_{-m}(x) = I_m(x)$ $= (-1)^m I_m(-x)$	$K_{-m}(x) = K_m(x)$

where a and r are known real constants. Notice that when $a = 0$ in Eq. (5.134), Eq. (5.134) is Bessel's equation given by Eq. (5.69).

Let us assume a solution to Eq. (5.134) of the form $y(z) = z^\nu u(z)$. Then, noting that

$$\frac{dy}{dz} = \nu z^{\nu-1}u(z) + z^\nu \frac{du}{dz}$$
$$\frac{d^2y}{dz^2} = \nu(\nu-1)z^{\nu-2}u(z) + 2\nu z^{\nu-1}\frac{du}{dz} + z^\nu \frac{d^2u}{dz^2} \tag{5.135}$$

we find, upon substituting Eq. (5.135) into Eq. (5.134) and canceling x^ν, that Eq. (5.134) becomes

$$z^2\frac{d^2u}{dz^2} + (2\nu + 1 - 2a)z\frac{du}{dz}$$
$$+\left(\nu(\nu-1) + (1-2a)\nu + z^2 - r^2\right)u = 0$$

For this equation to have the form of a Bessel equation given by Eq. (5.69), we set $a = \nu$ and arrive at

$$z^2\frac{d^2u}{dz^2} + z\frac{du}{dz} + \left(z^2 - p^2\right)u = 0 \tag{5.136}$$

where $p^2 = r^2 + a^2$. Converting Eq. (5.134) to Eq. (5.136) is verified with Mathematica procedure M5.17.

Based on the results of Section 5.1.4, the solution to Eq. (5.136) is

$$u(z) = C_1J_p(z) + C_2Y_p(z)$$

Therefore, since we have set $a = \nu$ and since $y(z) = z^\nu u(z)$

$$y(z) = z^a\left[C_1J_p(z) + C_2Y_p(z)\right] \tag{5.137}$$

We now ask the question: For a given transformation $z = f(x)$, what is the form of the differential equation to which Eq. (5.137) is the solution? To answer this, we note that

$$\frac{dy}{dz} = \frac{dy}{dx}\frac{dx}{dz}$$
$$\frac{d^2y}{dz^2} = \frac{d^2y}{dx^2}\left(\frac{dx}{dz}\right)^2 + \frac{dy}{dx}\frac{d^2x}{dz^2} \tag{5.138}$$

But, from calculus,

$$\frac{dx}{dz} = \frac{1}{f'} \quad \text{and} \quad \frac{d^2x}{dz^2} = -\frac{f''}{f'^3} \tag{5.139}$$

where $f' = df/dx$ and $f'' = d^2f/dx^2$. Then, using Eq. (5.139) in Eq. (5.138), we arrive at

$$\frac{dy}{dz} = \frac{dy}{dx}\left(\frac{1}{f'}\right)$$
$$\frac{d^2y}{dz^2} = \frac{d^2y}{dx^2}\left(\frac{1}{f'}\right)^2 - \frac{dy}{dx}\left(\frac{f''}{f'^3}\right) \tag{5.140}$$

Upon substituting Eq. (5.140) into Eq. (5.134), we obtain

$$f^2\left[\frac{d^2y}{dx^2}\left(\frac{1}{f'}\right)^2 - \frac{dy}{dx}\left(\frac{f''}{f'^3}\right)\right]$$
$$+(1-2a)f\left[\frac{dy}{dx}\left(\frac{1}{f'}\right)\right] + \left(f^2 - r^2\right)y = 0 \tag{5.141}$$
$$\frac{d^2y}{dx^2} + \left[(1-2a)\frac{f'}{f} - \frac{f''}{f'}\right]\frac{dy}{dx} + \frac{f'^2}{f^2}\left(f^2 - r^2\right)y = 0$$

which will result in another form of Bessel's equation after $f(x)$ is specified.

We shall now extend the results given by Eq. (5.141) and consider the transformation

$$y(x) = \frac{u(x)}{g(x)} \tag{5.142}$$

Noting that

$$\frac{dy}{dx} = \frac{u'}{g} - \frac{ug'}{g^2}$$
$$\frac{d^2y}{dx^2} = \frac{u''}{g} + u\left(2\frac{g'^2}{g^3} - \frac{g''}{g^2}\right) - 2\frac{u'g'}{g^2} \tag{5.143}$$

the substitution of Eq. (5.143) into Eq. (5.141) yields

$$\frac{d^2u}{dx^2} + \left((1-2a)\frac{f'}{f} - \frac{f''}{f'} - 2\frac{g'}{g}\right)\frac{du}{dx}$$
$$+\left(\frac{f'^2}{f^2}\left(f^2 - r^2\right) - \frac{g''}{g} - \frac{g'}{g}\left\{(1-2a)\frac{f'}{f} - \frac{f''}{f'} - 2\frac{g'}{g}\right\}\right)u = 0 \tag{5.144}$$

Equation (5.144) will be another form of Bessel's equation after $f(x)$ and $g(x)$ are specified. When g is a constant, it is seen that Eq. (5.144) reduces to Eq. (5.141). Then, from Eqs. (5.142) and Eq. (5.137), the solution to Eq. (5.144) is

$$y(x) = g(x)\left(f(x)\right)^a\left[C_1J_p\left(f(x)\right) + C_2Y_p\left(f(x)\right)\right] \tag{5.145}$$

We shall now examine several cases of $f(x)$ and several combinations of $f(x)$ and $g(x)$ and determine the equivalent Bessel equation.

Case I: $f(x) = kx^b$. For this function, $g(x) = 1$ and Eq. (5.145) gives

$$y(x) = k^a x^{ab}\left[C_1 J_p(kx^b) + C_2 Y_p(kx^b)\right] \tag{5.146}$$

To determine the equation to which Eq. (5.146) is a solution, we note that

$$f = kx^b \qquad f' = kbx^{b-1} \qquad f'' = kb(b-1)x^{b-2}$$

and, therefore,

$$\frac{f'}{f} = bx^{-1} \quad \frac{f''}{f'} = (b-1)x^{-1} \tag{5.147}$$

The derivatives of $g(x)$ are zero. Substituting Eq. (5.147) into Eq. (5.141), it is found that

$$x^2\frac{d^2y}{dx^2} + (1-2ab)x\frac{dy}{dx} + \left(b^2k^2x^{2b} - b^2r^2\right)y = 0 \tag{5.148}$$

Thus, the solution to Eq. (5.148) is given by Eq. (5.146), where $p^2 = r^2 + a^2$.

The verification that Eq. (5.146) is a solution to Eq. (5.148) is determined with Mathematica procedure M5.18.

Case II: $f(x) = ke^{bx}$. For this function, $g(x) = 1$ and Eq. (5.145) gives

$$y(x) = e^{abx}\left[C_1 J_p(ke^{bx}) + C_2 Y_p(ke^{bx})\right] \tag{5.149}$$

To determine the equation to which Eq. (5.149) is a solution, we note that

$$f = ke^{bx} \qquad f' = kbe^{bx} \qquad f'' = kb^2e^{bx}$$

and, therefore,

$$\frac{f'}{f} = b \quad \frac{f''}{f'} = b \tag{5.150}$$

The derivatives of $g(x)$ are zero. Substituting Eq. (5.150) into Eq. (5.141), we arrive at

$$\frac{d^2y}{dx^2} - 2ab\frac{dy}{dx} + b^2\left(k^2e^{2bx} - r^2\right)y = 0 \tag{5.151}$$

Thus, the solution to Eq. (5.151) is given by Eq. (5.149) with $p^2 = r^2 + a^2$.

The verification that Eq. (5.149) is a solution to Eq. (5.151) is determined with Mathematica procedure M5.19.

Case III: $g(x) = e^{cx}$ and $f(x) = kx^b$. For this case,

$$\begin{aligned} f' &= kbx^{b-1} \qquad & f'' &= kb(b-1)x^{b-2} \\ g' &= ce^{cx} \qquad & g'' &= c^2e^{cx} \end{aligned} \tag{5.152}$$

and

$$\begin{aligned} \frac{f'}{f} &= \frac{b}{x} \qquad \frac{f''}{f'} = \frac{b-1}{x} \\ \frac{g'}{g} &= c \qquad \frac{g''}{g} = c^2 \end{aligned} \tag{5.153}$$

Then, Eq. (5.144) becomes

$$\begin{aligned} &x^2 \frac{d^2u}{dx^2} + \left(1 - 2ab - 2cx\right) x \frac{du}{dx} \\ &+ \left(b^2\left(k^2x^{2b} - r^2\right) + c^2x^2 - cx\left(1 - 2ab\right)\right) u = 0 \end{aligned} \tag{5.154}$$

The solution to Eq. (5.154) is obtained from Eq. (5.145) as

$$y(x) = e^{cx} x^{ab} \left[C_1 J_p\left(kx^b\right) + C_2 Y_p\left(kx^b\right) \right] \tag{5.155}$$

Case IV: $g(x) = e^{\beta x^m}$ and $f(x) = kx^b$. For this case,

$$\begin{aligned} f' &= kbx^{b-1} \qquad f'' = kb(b-1)x^{b-2} \\ g' &= m\beta x^{m-1} e^{\beta x^m} \qquad g'' = \left(m(m-1)\beta x^{m-2} + m^2\beta^2 x^{2m-2}\right) e^{\beta x^m} \end{aligned} \tag{5.156}$$

and, therefore,

$$\begin{aligned} \frac{f'}{f} &= \frac{b}{x} \qquad \frac{f''}{f'} = \frac{b-1}{x} \\ \frac{g'}{g} &= m\beta x^{m-1} \qquad \frac{g''}{g} = m(m-1)\beta x^{m-2} + m^2\beta^2 x^{2m-2} \end{aligned} \tag{5.157}$$

Then, Eq. (5.144) becomes

$$\begin{aligned} &x^2 \frac{d^2u}{dx^2} + \left(1 - 2ab - 2m\beta x^m\right) x \frac{du}{dx} \\ &+ \left(b^2\left(k^2x^{2b} - r^2\right) - m\beta\left(m - 2ab\right)x^m + m^2\beta^2 x^{2m}\right) u = 0 \end{aligned} \tag{5.158}$$

The solution to Eq. (5.158) is

$$y(x) = e^{\beta x^m} x^{ab} \left[C_1 J_p\left(kx^b\right) + C_2 Y_p\left(kx^b\right) \right] \tag{5.159}$$

where $p^2 = a^2 + r^2$.

The results of this section are summarized in Table 5.2.

We shall give several examples that use these results.

TABLE 5.2
Differential Equations Whose Solutions Are in Terms of Bessel Functions

Case	Differential Equation	Solution [$p^2 = r^2 + a^2$ Unless Indicated Otherwise]
1	$x^2\frac{d^2y}{dx^2}+(1-2a)x\frac{dy}{dx}+\left(x^2-r^2\right)y=0$	$y(x)=x^a\left[C_1J_p(x)+C_2Y_p(x)\right]$
2	$x^2\frac{d^2y}{dx^2}+(1-2ab)x\frac{dy}{dx}+b^2\left(k^2x^{2b}-r^2\right)y=0$ When $b = 1$ and $k = 1$, we get Case 1.	$y(x)=k^ax^{ab}\left[C_1J_p(kx^b)+C_2Y_p(kx^b)\right]$
3	$\frac{d^2y}{dx^2}-2ab\frac{dy}{dx}+b^2\left(k^2e^{2bx}-r^2\right)y=0$	$y(x)=e^{abx}\left[C_1J_p(ke^{bx})+C_2Y_p(ke^{bx})\right]$
4	$x^2\frac{d^2y}{dx^2}+\left(1-2ab-2cx\right)x\frac{dy}{dx}$ $+\left(b^2\left(k^2x^{2b}-r^2\right)+c^2x^2-cx\left(1-2ab\right)\right)y=0$ When $c = 0$, we get Case 2.	$y(x)=e^{cx}x^{ab}\left[C_1J_p\left(kx^b\right)+C_2Y_p\left(kx^b\right)\right]$
5	$x^2\frac{d^2y}{dx^2}+\left(1-2ab-2m\beta x^m\right)x\frac{dy}{dx}$ $+\left(b^2\left(k^2x^{2b}-r^2\right)-m\beta\left(m-2ab\right)x^m\right.$ $\left.+m^2\beta^2x^{2m}\right)y=0$ When $m = 0$, we get Case 2; when $m = 1$ and $\beta = c$, we get Case 4.	$y(x)=e^{\beta x^m}x^{ab}\left[C_1J_p\left(kx^b\right)+C_2Y_p\left(kx^b\right)\right]$
6*	$x^2\frac{d^2y}{dx^2}+\left(a+2bx^\alpha\right)x\frac{dy}{dx}$ $+\left(c+\gamma^2x^{2\delta}+b(a+\alpha-1)x^\alpha+b^2x^{2\alpha}\right)y=0$ When $\alpha = m$, $c \to -b^2r^2$, $\gamma^2 \to k^2b^2$, $b \to -m\beta$, $\delta \to b$, and $a \to 1-2ab$, we get Case 5.	$y(x)=x^{(1-a)/2}e^{-bx^\alpha/\alpha}\left[AJ_p(z)+AY_p(z)\right]$ $z=\gamma x^\delta/\delta$, $p^2=\left((a-1)^2/4-c\right)/\delta^2$
7	$\frac{d}{dz}\left(z^a\frac{dy}{dz}\right)+bz^cy=0$ or $z\frac{d^2y}{dz^2}+a\frac{dy}{dz}+bz^{c-a+1}y=0$	$y(z)=z^\gamma\left[C_1J_p\left(\sqrt{b}z^g/g\right)+C_2Y_p\left(\sqrt{b}z^g/g\right)\right]$ $p=\gamma/g$ $\gamma=(1-a)/2$ $g=(c-a+2)/2$

*Details have been left to Exercise 5.11.

Example 5.10

Consider the equation

$$\frac{d^2y}{dx^2}-\left(\frac{2\sqrt{3}-1}{x}\right)\frac{dy}{dx}+\left(784x^6-\frac{61}{x^2}\right)y=0$$

We re-write this equation as

$$x^2\frac{d^2y}{dx^2}+\left(1-2\sqrt{3}\right)x\frac{dy}{dx}+\left(784x^8-61\right)y=0 \qquad \text{(a)}$$

This is Case 2 of Table 5.1. Therefore, upon comparing Eq. (a) to the differential equation given in Case 2 of Table 5.1, it is found that

$$2b=8\to b=4$$

$$ab=\sqrt{3}\to a=\frac{\sqrt{3}}{4}$$

$$b^2k^2=784\to bk=28\to k=7$$

$$b^2r^2=61\to r^2=61/16$$

$$p=\sqrt{r^2+a^2}=\sqrt{61/16+3/16}=2$$

Then, from Case 2 of Table 5.1, we obtain

$$y(x)=x^{\sqrt{3}}\left[C_1J_2(7x^4)+C_2Y_2(7x^4)\right]$$

This result is verified with Mathematica procedure M5.20.

Example 5.11

Consider the equation

$$x\frac{d^2y}{dx^2}+\frac{dy}{dx}-\frac{s^2}{g}y=0 \qquad \text{(a)}$$

To get Eq. (a) in the appropriate form, we multiply it by x to obtain

$$x^2\frac{d^2y}{dx^2}+x\frac{dy}{dx}-\frac{s^2}{g}xy=0 \qquad \text{(b)}$$

Equation (b) is a special case of Case 2 of Table 5.1 with

$$2b = 1 \rightarrow b = 1/2$$

$$a = 0$$

$$b^2k^2 = -\frac{s^2}{g} \rightarrow k = j2s\sqrt{\frac{1}{g}}$$

$$r^2 = 0 \rightarrow p = \sqrt{r^2 + a^2} = 0$$

Therefore, from Case 2 of Table 5.1, we find

$$y(x) = C_1 J_0\left(2js\sqrt{x/g}\right) + C_2 Y_0\left(2js\sqrt{x/g}\right)$$
$$= B_1 I_0\left(2s\sqrt{x/g}\right) + B_2 K_0\left(2s\sqrt{x/g}\right)$$

This result is verified with Mathematica procedure M5.21.

Example 5.12

Consider the equation

$$x^2\frac{d^2y}{dx^2} + 2x\frac{dy}{dx} + \left(x^2 - n(n+1)\right)y = 0 \tag{a}$$

where n is an integer. This is Case 1 of Table 5.1 with

$$1 - 2b = 2 \rightarrow a = -1/2$$

$$r^2 = n(n+1) \tag{b}$$

$$p^2 = r^2 + a^2 = n(n+1) + 1/4 = (n+1/2)^2$$

Therefore, from Case 1 of Table 5.1, we find

$$y(x) = x^{-1/2}\left[C_1 J_{n+1/2}(x) + C_2 J_{-n-1/2}(x)\right] \tag{c}$$

It is noted that Eq. (c) is a solution to Eq. (5.136), which in this case becomes

$$z^2\frac{d^2y}{dz^2} + z\frac{dy}{dz} + \left(z^2 - \left(n + \frac{1}{2}\right)^2\right)y = 0 \tag{d}$$

Comparing Eq. (d) with Eq. (5.108), we see that Eq. (d) is the spherical Bessel equation the solution to which is given by Eq. (c). Recall Eqs. (5.117) and (5.119).

These results are verified with Mathematica procedure M5.22.

Equation (a) is the form of the spherical Bessel equation that will result when the separation of variables method in the spherical coordinate system is employed as shown in Tables 7.5, 7.7, and 7.9.

Example 5.13

Consider the equation

$$\frac{d^2u}{dx^2}+x^2\frac{du}{dx}+xu=0 \tag{a}$$

Because of the x^2 in the middle term of Eq. (a), a candidate solution is that given by Case 5 of Table 5.1, which, for convenience, is repeated below

$$\begin{aligned}&x^2\frac{d^2u}{dx^2}+\left(1-2ab-2m\beta x^m\right)x\frac{du}{dx}\\&+\left\{b^2\left(k^2x^{2b}-r^2\right)-m\beta\left(m-2ab\right)x^m+m^2\beta^2x^{2m}\right\}=0\end{aligned} \tag{b}$$

Examining the coefficient of the first derivative term of Eq. (b), we see that upon comparing it with the coefficient of the first derivative term in Eq. (a) the following must be satisfied

$$\left(1-2ab-2m\beta x^m\right)x^{-1}=x^2 \tag{c}$$

For Eq. (c) to be true, we see that

$$\begin{aligned}1-2ab=0 \quad &\rightarrow \quad ab=1/2\\ m=3 \quad &\\ -2m\beta=1 \quad &\rightarrow \quad \beta=-1/6\end{aligned} \tag{d}$$

Then Eq. (b) reduces to

$$\frac{d^2u}{dx^2}+x^2\frac{du}{dx}+\left\{b^2\left(k^2x^{2b-2}-r^2x^{-2}\right)+x+\frac{x^4}{4}\right\}u=0 \tag{e}$$

Based on the third term of Eq. (a), we see that in Eq. (e) the following must be satisfied

$$r=0$$

$$b^2k^2x^{2b-2}+\frac{x^4}{4}=0$$

From the exponents of x, it is seen by inspection that $b = 3$ and, therefore, $k^2 = -1/36$ or $k = j/6$. Then from Eq. (d), it is found that $a = 1/6$. With these values, Eq. (e) reduces to Eq. (a).

Then, from Case 5 of Table 5.1, the solution to Eq. (a) is

$$\begin{aligned}y(x)&=e^{-x^3/6}\sqrt{x}\left[C_1J_{1/6}\left(jx^3/6\right)+C_2J_{-1/6}\left(jx^3/6\right)\right]\\&=e^{-x^3/6}\sqrt{x}\left[C_1I_{1/6}\left(x^3/6\right)+C_2I_{-1/6}\left(x^3/6\right)\right]\end{aligned} \tag{f}$$

This result is verified with Mathematica procedure M5.23.

The second type of equation that can be transformed into a Bessel equation is

$$\frac{d}{dz}\left(z^a\frac{dy}{dz}\right)+bz^c y=0$$

or, upon expansion,

$$z\frac{d^2y}{dz^2}+a\frac{dy}{dz}+bz^{c-a+1}y=0 \tag{5.160}$$

where a, b, and c are real quantities.

We assume a solution to Eq. (5.160) of the form

$$y(z)=z^{\gamma}u(z) \tag{5.161}$$

and use Eq. (5.135) to obtain

$$z^2\frac{d^2u}{dz^2}+(2\gamma+a)z\frac{du}{dz}+\left(\gamma(\gamma-1)+a\gamma+bz^{c-a+2}\right)u=0 \tag{5.162}$$

To get Eq. (5.162) into the desired form, we must set

$$2\gamma+a=1\rightarrow\gamma=\frac{1-a}{2} \tag{5.163}$$

Then, using Eq. (5.163) in Eq. (5.162), Eq. (5.162) becomes

$$z^2\frac{d^2u}{dz^2}+z\frac{du}{dz}+\left(bz^{c-a+2}-\gamma^2\right)u=0 \tag{5.164}$$

Since Eq. (5.164) is still not in the appropriate form, we use the following transformation

$$x=\beta z^g \tag{5.165}$$

Then, from chain differentiation,

$$\frac{du}{dz}=\frac{du}{dx}\frac{dx}{dz}=\beta g z^{g-1}\frac{du}{dx}$$

$$\frac{d^2u}{dz^2}=\frac{du}{dx}\frac{d^2x}{dz^2}+\frac{d^2u}{dx^2}\left(\frac{dx}{dz}\right)^2 \tag{5.166}$$

$$=\beta g(g-1)z^{g-2}\frac{du}{dx}+\beta^2g^2z^{2g-2}\frac{d^2u}{dx^2}$$

Substituting Eq. (5.166) into Eq. (5.164), we obtain

$$x^2\frac{d^2u}{dx^2}+x\frac{du}{dx}+\left(\frac{bz^{c-a+2}}{g^2}-p^2\right)u=0 \tag{5.167}$$

where

$$p^2 = \frac{\gamma^2}{g^2} = \left(\frac{1-a}{2g}\right)^2 \tag{5.168}$$

and we have used Eq. (5.163).

In order for Eq. (5.167) to be in the form of a Bessel's equation,

$$\frac{b}{g^2} z^{c-a+2} = x^2 \to bz^{c-a+2} = g^2 \beta^2 z^{2g} \tag{5.169}$$

Thus,

$$\beta = \frac{\sqrt{b}}{g} \quad \text{and} \quad g = \frac{c-a+2}{2} \tag{5.170}$$

With these values, the solution to Eq. (5.167) is

$$u(x) = C_1 J_p(x) + C_2 Y_p(x)$$

Upon using Eqs. (5.165) and (5.161), we obtain

$$y(z) = z^{\gamma}\left[C_1 J_p\left(\sqrt{b} z^g / g\right) + C_2 Y_p\left(\sqrt{b} z^g / g\right)\right] \tag{5.171}$$

where

$$\gamma = \frac{1-a}{2} \qquad p = \frac{\gamma}{g} \qquad g = \frac{c-a+2}{2}$$

This result appears as Case 7 in Table 5.2.

These results are verified with Mathematica procedure M5.24.

5.1.9 Legendre's Equation and Legendre Polynomials

Legendre's differential equation is

$$\left(1-x^2\right)\frac{d^2y}{dx^2} - 2x\frac{dy}{dx} + n(n+1)y = 0 \tag{5.172}$$

which can be written as

$$\frac{d}{dx}\left[\left(1-x^2\right)\frac{dy}{dx}\right] + n(n+1)y = 0 \tag{5.173}$$

where n is an integer. It is seen that

$$\lim_{x\to\pm1}\left[(x\pm1)\frac{-2x}{1-x^2}\right] \to \text{finite}$$

$$\lim_{x\to\pm1}\left[(x\pm1)^2\frac{n(n+1)}{1-x^2}\right] \to \text{finite}$$

and therefore $x = \pm 1$ are regular singular points. However, we shall be interested in finding a solution about $x = 0$, which is an ordinary point.

Since, in the range $-1 < x < 1$, x is an ordinary point, the solution to Eq. (5.172) is given by the solution to Eq. (5.23), where in Eq. (5.23) $x_0 = 0$, $\alpha = -1$, $\beta = -2$, and $\gamma = n(n + 1)$. Therefore, the solution to Eq. (5.172) is obtained from Eq. (5.24) as

$$y(x) = \sum_{n=0}^{\infty} c_n x^n \tag{5.174}$$

where, upon using Eqs. (5.26), (5.31), and (5.32), we have for even n that

$$c_{2k+2} = g_k(n)c_{2k} \quad k = 0,1,2,\ldots \tag{5.175}$$

where

$$g_k(n) = \frac{2k(2k+1) - n(n+1)}{(2k+2)(2k+1)} \tag{5.176}$$

and for odd n that

$$c_{2k+3} = h_k(n)c_{2k+1} \quad k = 0,1,2,\ldots \tag{5.177}$$

where

$$h_k(n) = \frac{(2k+1)(2k+2) - n(n+1)}{(2k+3)(2k+2)} \tag{5.178}$$

In these relations, c_0 and c_1 are unknown constants.

It is seen in Eq. (5.176) that when $n = 2k$, an even number, $c_{2k+2} = 0$ and, therefore, $c_{2k+2} = 0$ for $k > 2n$. In other words, the series solution terminates at $n = 2(k - 1)$. Similarly, it is seen in Eq. (5.178) that when $n = 2k + 1$, an odd number, $c_{2k+3} = 0$ and, therefore, $c_{2k+3} = 0$ for $2k - 1 > n$. In other words, the series solution terminates at $n = 2k - 1$.

Then, using Eqs. (5.175) and (5.176) and recalling Eq. (5.30), we obtain the following expression for even n in terms of c_0

$$c_{2k+2} = c_0 \prod_{m=0}^{k<n/2} g_m(n) \quad k = 0,1,2,\ldots \tag{5.179}$$

Expanding Eq. (5.177) for a few terms, we have

$$\begin{aligned} c_2 &= -\frac{1}{2!}n(n+1)c_0 \\ c_4 &= \frac{1}{4!}(n-2)n(n+1)(n+3)c_0 \\ c_6 &= -\frac{1}{6!}(n-4)(n-2)n(n+1)(n+3)(n+5)c_0 \end{aligned} \tag{5.180}$$

Similarly, we use Eqs. (5.177) and (5.178) to obtain the following expression for odd n in terms of c_1

$$c_{2k+3} = c_1 \prod_{m=0}^{k<(n-1)/2} h_m(n) \quad k = 0,1,2,\ldots \tag{5.181}$$

Expanding Eq. (5.181) for a few terms have

$$\begin{aligned} c_3 &= \frac{1}{3!}(n-1)(n+2)c_1 \\ c_5 &= \frac{1}{5!}(n-3)(n-1)(n+2)(n+4)c_1 \\ c_7 &= \frac{1}{7!}(n-5)(n-3)(n-1)(n+2)(n+4)(n+6)c_1 \end{aligned} \tag{5.182}$$

Equations (5.180) and (5.182) were obtained with Mathematica procedure M5.25.

Then Eq. (5.174) can be written as

$$y(x) = c_0 y_1(x) + c_1 y_2(x) \tag{5.183}$$

where

$$y_1(x) = 1 + \sum_{k=1}^{\infty} g_k(n) x^{2k} \tag{5.184}$$

and

$$y_2(x) = x + \sum_{k=1}^{\infty} h_k(n) x^{2k+1} \tag{5.185}$$

The solution given by Eq. (5.183) is valid for $-1 < x < 1$. Since y_1 is a function of x to even powers and y_2 is a function of x to odd powers, y_1 and y_2 cannot be proportional to each other and, therefore, are linearly independent solutions. Notice that when n is an even number, y_1 terminates and y_2 is an infinite series, and when n is an odd number, y_2 terminates and y_1 is an infinite series.

While these are the formal results, they can be further manipulated to obtain the following standard formulas, which are valid for all n, even or odd. This formula terminates with the first term for which a given n makes that term zero. Thus, we use the notation $P_n(x)$ to denote the *Legendre polynomial of the first kind* as

$$\begin{aligned} P_0(x) &= 1 \\ P_n(x) &= \sum_{k=0}^{m} (-1)^k \frac{(2n-2k)!\,x^{n-2k}}{2^n k!(n-k)!(n-2k)!} \end{aligned} \tag{5.186}$$

where $m = n/2$ or $m = (n-1)/2$, depending on which value of n results in m being an integer. The first few expressions of Eq. (5.186) are

$$\begin{aligned} P_1(x) &= x \\ P_2(x) &= \frac{1}{2}\left(3x^2 - 1\right) \\ P_3(x) &= \frac{1}{2}\left(5x^3 - 3x\right) \\ P_4(x) &= \frac{1}{8}\left(35x^4 - 30x^2 + 3\right) \end{aligned} \tag{5.187}$$

We state without proof the following recurrence relations for the Legendre function of the first kind.

$$\begin{aligned} (2n+1)xP_n(x) &= (n+1)P_{n+1}(x) + nP_{n-1}(x) \\ \frac{d}{dx}P_{n+1}(x) &= \frac{d}{dx}P_{n-1}(x) + (2n+1)P_n(x) \end{aligned} \tag{5.188}$$

The form of the *Legendre polynomial of the second kind* is the non-terminating series and is denoted $Q_n(x)$. A closed form of this series, however, can be obtained in the following manner. We start with the case for $n = 0$ for which one solution is $P_0(x)$. The second solution can be obtained by using the results of Section 4.2.3 and in particular Eq. (4.51), that is,

$$Q_0(x) = y_1(x)\int y_1^{-2}(x)\exp\left[-\int \frac{a_1(x)}{a_2(x)}dx\right]dx \tag{5.189}$$

Upon comparing Eq. (4.44) with Eq. (5.172), we see that $a_1(x) = -2x$, $a_2(x) = 1 - x^2$ and in this case $y_1 = P_0(x) = 1$. In Eq. (5.189), we note that

$$\begin{aligned} \exp\left[-\int \frac{a_1(x)}{a_2(x)}dx\right] &= \exp\left[2\int \frac{x}{1-x^2}dx\right] \\ &= \exp\left[-\ln\left(x^2-1\right)\right] = \frac{1}{x^2-1} \end{aligned} \tag{5.190}$$

and Eq. (5.189) becomes

$$Q_0(x) = P_0(x)\int \frac{dx}{P_0^2(x)\left(x^2-1\right)} = \int \frac{dx}{\left(x^2-1\right)} = \frac{1}{2}\ln\frac{1-x}{1+x} \tag{5.191}$$

which is valid for $-1 < x < 1$. To find $Q_1(x)$, we have that

$$\begin{aligned} Q_1(x) &= P_1(x)\int \frac{dx}{P_1^2(x)\left(x^2-1\right)} = x\int \frac{dx}{x^2\left(x^2-1\right)} = 1 + \frac{x}{2}\ln\frac{1-x}{1+x} \\ &= P_1(x)Q_0(x) + 1 \end{aligned} \tag{5.192}$$

The convention is to present the above results slightly differently as follows*

$$Q_0(x) = \frac{1}{2}\ln\frac{1+x}{1-x}$$
$$Q_1(x) = P_1(x)Q_0(x) - 1 \tag{5.193}$$

The general formula to determine the Legendre polynomial of the second kind is

$$Q_n(x) = \frac{1}{2}P_n(x)\ln\left(\frac{1+x}{1-x}\right) + \sum_{m=1}^{n}\frac{1}{m}P_{m-1}(x)P_{n-m}(x) \tag{5.194}$$

which is valid for $|x| < 1$. The first few expressions of Eq. (5.194) are

$$Q_0(x) = \frac{1}{2}\ln\left(\frac{1+x}{1-x}\right)$$

$$Q_1(x) = P_1(x)Q_0(x) - 1$$

$$Q_2(x) = P_2(x)Q_0(x) - \frac{3}{2}x$$

$$Q_3(x) = P_3(x)Q_0(x) - \frac{5}{2}x^2 + \frac{2}{3}$$

Therefore, the solution to Eq. (5.172) is

$$y(x) = C_1P_n(x) + C_2Q_n(x) \tag{5.195}$$

Equation (5.195) is verified with Mathematica procedure M5.26.

The Legendre functions of the first and second kind are plotted in Figure 5.4.

We state without proof† the following integral, which will be required in Example 7.6.

$$\int_{-1}^{1} P_n^2(x)dx = \frac{2}{2n+1} \tag{5.196}$$

5.1.10 Associated Legendre's Equation and Legendre Polynomials

When one solves certain partial differential equations in spherical coordinates using separation of variables to attain a solution, the following equation arises (see Table 7.3)

* Note that

$$\ln\frac{1-x}{1+x} = \ln(1-x) - \ln(1+x) = -\big(\ln(1+x) - \ln(1-x)\big) = -\ln\frac{1+x}{1-x}$$

† This integral can be determined one of several ways: Riley KF, Hobson MP, Bence SJ (1997) *Mathematical Methods for Physics and Engineering: A Comprehensive Guide*, Cambridge University Press, p. 454 and Wiley CR, Barrett LC (1995) *Advanced Engineering Mathematics*, 6th ed. McGraw-Hill, p. 849.

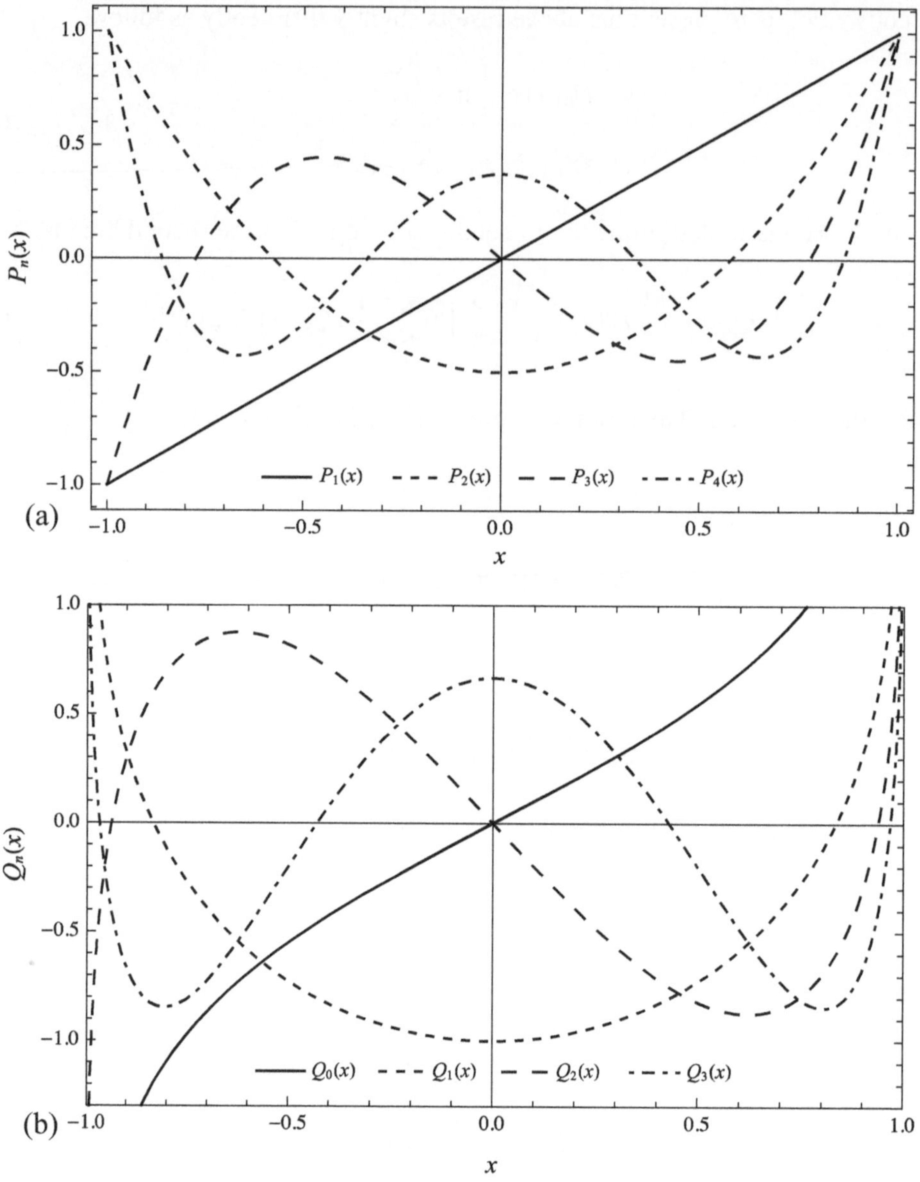

FIGURE 5.4 (a) Legendre functions of the first kind for $n = 1, 2, 3$, and 4; (b) Legendre functions of the second kind for $n = 0, 1, 2$, and 3.

$$\frac{d^2\Phi}{d\varphi^2} + \cot\varphi \frac{d\Phi}{d\varphi} + \left[n(n+1) - \frac{m^2}{\sin^2\varphi} \right] \Phi = 0 \tag{5.197}$$

To get the equation into a more familiar form, we let $x = \cos\varphi$. Then

$$\begin{aligned} \frac{d\Phi}{d\varphi} &= \frac{d\Phi}{dx}\frac{dx}{d\varphi} = -\sin\varphi \frac{d\Phi}{dx} \\ \frac{d^2\Phi}{d\varphi^2} &= \frac{d^2\Phi}{dx^2}\left(\frac{dx}{d\varphi}\right)^2 + \frac{d\Phi}{dx}\frac{d^2x}{d\varphi^2} = \frac{d^2\Phi}{dx^2}\sin^2\varphi - \cos\varphi\frac{d\Phi}{dx} \end{aligned} \tag{5.198}$$

Substituting Eq. (5.198) into Eq. (5.197) gives

$$\sin^2\varphi\frac{d^2\Phi}{dx^2}-2\cos\varphi\frac{d\Phi}{dx}+\left[n(n+1)-\frac{m^2}{\sin^2\varphi}\right]=0 \tag{5.199}$$

However, from trigonometry

$$\sin\varphi=\sqrt{1-x^2}$$

$$\cos\varphi=x$$

and Eq. (5.199) can be written as

$$\left(1-x^2\right)\frac{d^2\Phi}{dx^2}-2x\frac{d\Phi}{dx}+\left[n(n+1)-\frac{m^2}{1-x^2}\right]\Phi=0 \tag{5.200}$$

or

$$\frac{d}{dx}\left[\left(1-x^2\right)\frac{d\Phi}{dx}\right]+\left[n(n+1)-\frac{m^2}{1-x^2}\right]\Phi=0 \tag{5.201}$$

Equations (5.200) and (5.201) are known as the *associated Legendre equations.*

To obtain the solution to Eq. (5.200), we proceed as follows. We let

$$\Phi(x)=\left(x^2-1\right)^{m/2}u(x) \tag{5.202}$$

and, hence,

$$\frac{d\Phi}{dx}=mx\left(x^2-1\right)^{m/2-1}u+\left(x^2-1\right)^{m/2}\frac{du}{dx}$$

$$\frac{d^2\Phi}{dx^2}=\left(x^2-1\right)^{m/2}\frac{d^2u}{dx^2}+2mx\left(x^2-1\right)^{m/2-1}\frac{du}{dx} \tag{5.203}$$

$$+2\left(\frac{m}{2}-1\right)mx^2\left(x^2-1\right)^{m/2-2}u+m\left(x^2-1\right)^{m/2-1}u$$

Then, substituting Eqs. (5.202) and (5.203) into Eq. (5.200), we arrive at

$$\left(1-x^2\right)\frac{d^2u}{dx^2}-2(m+1)x\frac{du}{dx}+(n-m)(n+m+1)u=0 \tag{5.204}$$

To obtain a solution to Eq. (5.204), we first differentiate Legendre's equation, Eq. (5.172), m times. The first differentiation yields

$$\left(1-x^2\right)\frac{d^3y}{dx^3}-4x\frac{d^2y}{dx^2}+\left[n(n+1)-2\right]\frac{dy}{dx}=0$$

The next differentiation yields

$$\left(1-x^2\right)\frac{d^4y}{dx^4}-6x\frac{d^3y}{dx^3}+\left[n(n+1)-6\right]\frac{d^2y}{dx^2}=0$$

The pattern that is emerging is

$$\left(1-x^2\right)\frac{d^{m+2}y}{dx^{m+2}}-2(m+1)x\frac{d^{m+1}y}{dx^{m+1}}+(n-m)(n+m+1)\frac{d^m y}{dx^m}=0 \tag{5.205}$$

Upon comparing Eqs. (5.204) and (5.205), it is seen that

$$\frac{d^2u}{dx^2}=\frac{d^{m+2}y}{dx^{m+2}},\quad \frac{du}{dx}=\frac{d^{m+1}y}{dx^{m+1}},\quad u=\frac{d^m y}{dx^m} \tag{5.206}$$

and from Eqs. (5.195), (5.202), and (5.206), we obtain

$$\Phi(x)=C_1\left(x^2-1\right)^{m/2}\frac{d^m P_n}{dx^m}+C_2\left(x^2-1\right)^{m/2}\frac{d^m Q_n}{dx^m} \tag{5.207}$$

We define the *associated Legendre polynomials* as

$$\begin{aligned} P_n^m(x)&=\left(x^2-1\right)^{m/2}\frac{d^m P_n}{dx^m}\\ Q_n^m(x)&=\left(x^2-1\right)^{m/2}\frac{d^m Q_n}{dx^m}\end{aligned} \tag{5.208}$$

Then the solution given by Eq. (5.207) can be written as

$$\Phi(x)=C_1P_n^m(x)+C_2Q_n^m(x) \tag{5.209}$$

which is the solution to Eq. (5.200). However, since $x=\cos\varphi$,

$$\Phi(\varphi)=C_1P_n^m(\cos\varphi)+C_2Q_n^m(\cos\varphi) \tag{5.210}$$

which is the solution to our original equation, Eq. (5.197). When $m=0$, these results reduce to the Legendre polynomials. Equation (5.209) is verified with Mathematica procedure M5.27.

A plot of the associated Legendre polynomials of the first and second kind is shown in Figure 5.5.

5.1.11 Hypergeometric Equation and Hypergeometric Functions

The hypergeometric equation is given by

$$z(1-z)\frac{d^2y}{dz^2}+[c-(a+b+1)z]\frac{dy}{dz}-aby=0 \tag{5.211}$$

where a, b, and c are real constants. We see that $z=0$ and $z=1$ are regular singular points.

We shall obtain a solution about the regular singular $z=0$. Upon comparing Eq. (5.211) with Eq. (5.34), we see that

$$\begin{aligned} &\alpha_0=1, && \alpha_1=-1, && \alpha_2=0\\ &\beta_0=c, && \beta_1=-(a+b+1), && \beta_2=0\\ &\gamma_0=0, && \gamma_1=-ab, && \gamma_2=0 \end{aligned} \tag{5.212}$$

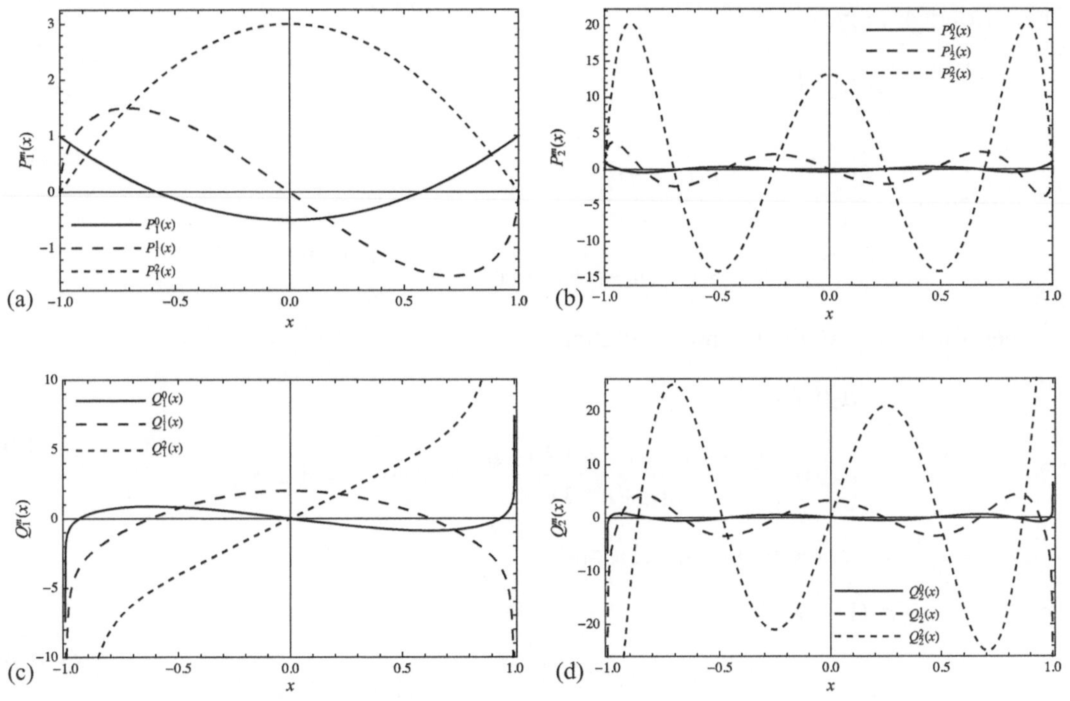

FIGURE 5.5 For the values of $m = 0$, 1, and 2, the associated Legendre polynomials of the first and second kind are plotted for $n = 1$, 2: (a) associated Legendre polynomial of the first kind for $n = 1$; (b) associated Legendre polynomial of the first kind for $n = 2$; (c) associated Legendre polynomial of the second kind for $n = 1$; and (d) associated Legendre polynomial of the second kind for $n = 2$.

From Eq. (5.48), the indicial equation becomes

$$p_0(r) = r(r + c - 1) \tag{5.213}$$

from which we determine that $r_1 = 0$ and $r_2 = 1 - c$. The solutions are a function of the values of c. We shall assume that c does not equal zero or a positive or negative integer. This assumption eliminates the case of equal roots and the case of the roots differing by an integer. With these restrictions, the solutions are given by Eqs. (5.50) and (5.51), that is,

$$y_1(z) = \sum_{n=0}^{\infty} a_n(0) z^n \tag{5.214}$$

and

$$y_2(z) = z^{1-c} \sum_{n=0}^{\infty} a_n(1-c) z^n \tag{5.215}$$

From Eq. (5.37) and (5.212), we find that

$$\begin{aligned} p_1(r) &= -\big(r(r + a + b) + ab\big) \\ p_2(r) &= 0 \end{aligned} \tag{5.216}$$

Then, Eq. (5.45) gives

$$a_0(r)=1$$

$$a_n(r)=-\frac{p_1(r+n-1)a_{n-1}(r)}{p_0(r+n)} \tag{5.217}$$

$$=\frac{(n+r-1)(n+r-1+a+b)+ab}{(n+r)(n+r+c-1)}a_{n-1}(r) \quad n=1,2,3,\ldots$$

Then for $r = r_1 = 0$, the recursion relation is

$$a_0(0)=1$$

$$a_n(0)=\frac{(n-1)(n-1+a+b)+ab}{n(n+c-1)}a_{n-1}(0) \quad n=1,2,3,\ldots \tag{5.218}$$

When $r = r_2 = 1 - c$, the recursion relation is

$$a_0(1-c)=1$$

$$a_n(1-c)=\frac{(n-c)(n-c+a+b)+ab}{n(n+1-c)}a_{n-1}(1-c) \quad n=1,2,3,\ldots \tag{5.219}$$

To determine the first few terms of Eq. (5.218), we use Mathematica procedure M5.28 and obtain

$$a_1(0)=\frac{ab}{c}$$

$$a_2(0)=\frac{a(a+1)b(b+1)}{2c(1+c)}$$

$$a_3(0)=\frac{a(a+1)(a+2)b(b+1)(b+2)}{6c(1+c)(2+c)} \tag{5.220}$$

$$\ldots$$

$$a_n(0)=\frac{a(a+1)(a+2)\cdots(a+n-1)b(b+1)(b+2)\cdots(b+n-1)}{n!c(1+c)(c+2)\cdots(c+n-1)}$$

From Eq. (C.6) in Appendix C with $v = a - 1$, we see that

$$(a)_n=\frac{\Gamma(a+n)}{\Gamma(a)}=a(a+1)(a+2)\cdots(a+n-1) \tag{5.221}$$

where $(a)_n$ is called the *Pochhammer* symbol. Notice that $(1)_n = n!$. Then Eq. (5.220) can be written as

$$a_n(0)=\frac{(a)_n(b)_n}{n!(c)_n} \tag{5.222}$$

Equation (5.222) is verified with Mathematica procedure M5.29.

Using Eq. (5.214), the first solution to the hypergeometric equation is

$$y_1(z) = \sum_{n=0}^{\infty} \frac{(a)_n(b)_n}{n!(c)_n} z^n \tag{5.223}$$

The solution given by Eq. (5.223) is often written as

$${}_2F_1(a,b;c;z) = \sum_{n=0}^{\infty} \frac{(a)_n(b)_n}{n!(c)_n} z^n = \frac{\Gamma(c)}{\Gamma(a)\Gamma(b)} \sum_{n=0}^{\infty} \frac{\Gamma(a+n)\Gamma(b+n)}{\Gamma(c+n)} \frac{z^n}{n!} \tag{5.224}$$

The second solution, which is for $r = r_2 = 1 - c$, is obtained by expanding the first few terms of Eq. (5.219) by using Mathematica procedure M5.30. Thus,

$$\begin{aligned}
a_1(1-c) &= \frac{(a-c+1)(b-c+1)}{(2-c)} \\
a_2(1-c) &= \frac{(a-c+1)(a-c+2)(b-c+1)(b-c+2)}{2(2-c)(3-c)} \\
a_3(1-c) &= \frac{(a-c+1)(a-c+2)(a-c+3)(b-c+1)(b-c+2)(b-c+3)}{6(2-c)(3-c)(4-c)} \\
&\cdots \\
a_n(1-c) &= \frac{(a-c+1)(a-c+2)\cdots(a-c+n)(b-c+1)(b-c+2)\cdots(b-c+n)}{n!(2-c)(3-c)\cdots(n+1-c)}
\end{aligned} \tag{5.225}$$

The last equation of Eq. (5.225) can be obtained directly with the Mathematica procedure M5.31, to obtain

$$a_n(1-c) = \frac{(1+a-c)_n(1+b-c)_n}{n!(2-c)_n} \tag{5.226}$$

Thus, it is seen upon comparing Eq. (5.226) with Eq. (5.222) that Eq. (5.226) can be obtained from Eq. (5.222) by the following substitutions in Eq. (5.222): $a \to a - c + 1$, $b \to b - c + 1$, and $c \to 2 - c$. Then the second solution to Eq. (5.211) is

$$y_2(z) = z^{1-c}\,{}_2F_1(a-c+1, b-c+1; 2-c; z) = z^{1-c} \sum_{n=0}^{\infty} \frac{(a-c+1)_n(b-c+1)_n}{n!(2-c)_n} z^n$$

The complete solution to Eq. (5.211)

$$y(z) = {}_2F_1(a,b;c;z)C_1 + z^{1-c}\,{}_2F_1(a-c+1, b-c+1; 2-c; z)C_2 \tag{5.227}$$

Equation (5.227) is verified with Mathematica procedure M5.32.

This result is illustrated with the following example.

Example 5.14

When determining the natural frequencies of non-uniform thin beams, the following fourth-order equation in non-dimensional form is encountered*

$$\eta^4 \frac{d^4y}{d\eta^4} + 2a\eta^3 \frac{d^3y}{d\eta^3} + a(a-1)\eta^2 \frac{d^2y}{d\eta^2} - \Omega^4\eta^\theta y = 0 \quad 0 \le \eta \le 1 \tag{a}$$

where $\theta = 4 - a + b$, a and b are positive real constants, and Ω is a non-dimensional frequency coefficient. We introduce the change of variable $\xi = \eta\Omega^{(1/\theta)}$ and Eq. (a) becomes

$$\xi^4 \frac{d^4y}{d\xi^4} + 2a\xi^3 \frac{d^3y}{d\xi^3} + a(a-1)\xi^2 \frac{d^2y}{d\xi^2} - \xi^\theta y = 0 \tag{b}$$

Next, we introduce another change of variable $u = \xi^\theta$ into Eq. (b) and use Mathematica procedure M5.33 to obtain

$$\begin{aligned} &\theta^4u^4 \frac{d^4y}{du^4} + 2(a+3\theta-3)\theta^3u^3 \frac{d^3y}{du^3} \\ &+ a(a-1)\left(a^2 + a(6\theta-7) + 7\theta^2 - 18\theta + 11\right)\theta^2u^2 \frac{d^2y}{du^2} \\ &+ (\theta-1)\left(a^2 + a(2\theta-5) + \theta^2 - 5\theta + 6\right)\theta u \frac{dy}{du} - uy = 0 \end{aligned} \tag{c}$$

A solution to Eq. (c) can be obtained with

$$y(u) = \sum_{n=0}^{\infty} c_n(r)u^{n+r} \tag{d}$$

We note that

$$\begin{aligned} \frac{dy}{du} &= \sum_{n=0}^{\infty} c_n(r)(n+r)u^{n+r-1} \\ \frac{d^2y}{du^2} &= \sum_{n=0}^{\infty} c_n(r)(n+r)(n+r-1)u^{n+r-2} \\ \frac{d^3y}{du^3} &= \sum_{n=0}^{\infty} c_n(r)(n+r)(n+r-1)(n+r-2)u^{n+r-3} \\ \frac{d^4y}{du^4} &= \sum_{n=0}^{\infty} c_n(r)(n+r)(n+r-1)(n+r-2)(n+r-3)u^{n+r-4} \end{aligned} \tag{e}$$

* Wang HC (September 1967) Generalized Hypergeometric Function Solutions on the Transverse Vibration of a Class of Nonuniform Beams. ASME J. Applied Mechanics, pp. 702–708; Banks DO, Kurowski GJ (March 1977) The Transverse Vibration of a Doubly Tapered Beam. ASME J. Applied Mechanics, pp. 123–126; and Cazzani A, Rosati L, Ruge R (2017) The contribution of Gustav R. Kirchhoff to the dynamics of tapered beams. Z. Angew. Math. Mech. 97(10): 1174–1203.

Then, substituting Eq. (d) into Eq. (c) and using Eq. (e) gives

$$g_0(r)c_0(r)+\sum_{n=1}^{\infty}\left[g_n(r)c_n(r)-c_{n-1}(r)\right]u^n=0 \tag{f}$$

where

$$g_n(r)=\theta^4(n+r)\left(n+r-\frac{1}{\theta}\right)\left(n+r+\frac{a-2}{\theta}\right)\left(n+r+\frac{a-3}{\theta}\right) \tag{g}$$

and

$$g_0(r)=\theta^4 r\left(r-\frac{1}{\theta}\right)\left(r+\frac{a-2}{\theta}\right)\left(r+\frac{a-3}{\theta}\right) \tag{h}$$

The roots of the indicial equation, which are obtained by setting $g_0(r)$ to zero, are

$$r_1=0,\quad r_2=\frac{1}{\theta},\quad r_3=\frac{2-a}{\theta},\quad r_4=\frac{3-a}{\theta} \tag{i}$$

The general recursion relation is obtained from Eq. (f) as

$$c_n(r)=\frac{c_{n-1}(r)}{g_n(r)}\qquad n=1,2,\ldots \tag{j}$$

The first few terms of Eq. (j) are

$$\begin{aligned}
c_1(r)&=\frac{c_0(r)}{g_1(r)}\\
c_2(r)&=\frac{c_1(r)}{g_2(r)}=\frac{c_0(r)}{g_1(r)g_2(r)}\\
c_3(r)&=\frac{c_2(r)}{g_3(r)}=\frac{c_0(r)}{g_1(r)g_2(r)g_3(r)}\\
&\vdots\\
c_n(r)&=\frac{c_{n-1}(r)}{g_n(r)}=c_0(r)\left(\prod_{m=1}^{n}g_n(r)\right)^{-1}
\end{aligned} \tag{k}$$

We note from Eq. (g) that

$$\prod_{m=1}^{n} g_n(r) = \theta^{4n}(1+r)\left(1+r-\frac{1}{\theta}\right)\left(1+r+\frac{a-2}{\theta}\right)\left(1+r+\frac{a-3}{\theta}\right)$$
$$\times(2+r)\left(2+r-\frac{1}{\theta}\right)\left(2+r+\frac{a-2}{\theta}\right)\left(2+r+\frac{a-3}{\theta}\right)$$
$$\times(3+r)\left(3+r-\frac{1}{\theta}\right)\left(3+r+\frac{a-2}{\theta}\right)\left(3+r+\frac{a-3}{\theta}\right)$$
$$\times\cdots\times(n+r)\left(n+r-\frac{1}{\theta}\right)\left(n+r+\frac{a-2}{\theta}\right)\left(n+r+\frac{a-3}{\theta}\right) \tag{l}$$
$$=\theta^{4n}(1+r)(2+r)(3+r)\cdots(n+r)$$
$$\times\left(1+r-\frac{1}{\theta}\right)\left(2+r-\frac{1}{\theta}\right)\left(3+r-\frac{1}{\theta}\right)\cdots\left(n+r-\frac{1}{\theta}\right)$$
$$\times\left(1+r+\frac{a-2}{\theta}\right)\left(2+r+\frac{a-2}{\theta}\right)\left(3+r+\frac{a-2}{\theta}\right)\cdots\left(n+r+\frac{a-2}{\theta}\right)$$
$$\times\left(1+r+\frac{a-3}{\theta}\right)\left(2+r+\frac{a-3}{\theta}\right)\left(3+r+\frac{a-3}{\theta}\right)\cdots\left(n+r+\frac{a-3}{\theta}\right)$$

But from Eq. (C.6) of Appendix C

$$(1+x)(2+x)(3+x)\cdots(n+x) = \frac{\Gamma(x+1+n)}{\Gamma(x+1)} \tag{m}$$

where $x = r_k$ and r_k is given by Eq. (i). Then, $c_n(r)$ can be written as

$$c_n(r) = \frac{\Gamma(r+1)\Gamma(r+1-1/\theta)\Gamma(r+1+(a-2)/\theta)}{\theta^{4n}\Gamma(r+1+n)\Gamma(n+r+1-1/\theta)\Gamma(n+r+1+(a-2)/\theta)}$$
$$\times\frac{\Gamma(r+1+(a-3)/\theta)}{\Gamma(n+r+1+(a-3)/\theta)} \quad n=1,2,\ldots \tag{n}$$

where we have set $c_0 = 1$. Notice that the numerator of Eq. (n) is independent of n.

If neither the difference in any two roots is an integer nor any two roots are equal, then we have the following four linearly independent solutions*

$$y_k(u) = \sum_{n=0}^{\infty} c_n(r_k)u^{n+r_k} \quad k=1,2,3,4 \tag{o}$$

The solutions given by Eq. (o) are called generalized hypergeometric functions. Then the solution is

* See Wang HC (1967), *loc. cit.* for a discussion of these special cases, which have been excluded in this presentation.

$$y(u)=\sum_{k=1}^{4}C_k y_k(u)$$

However, $u=\xi^{\theta}$ and $\xi=\eta\Omega^{1/\theta}$. Therefore, the solution to Eq. (a) is

$$y(\eta)=\sum_{k=1}^{4}C_k y_k\left(\Omega\eta^{1/\theta}\right) \tag{p}$$

Equation (p) is then used to satisfy the boundary conditions.

Equations (d), (g), (i), and (k) are verified with Mathematica procedure M5.34.

MATHEMATICA PROCEDURES

Note: In order to make the Mathematica procedures that follow a little more readable, the `Clear` statement and many of the graphics enhancing statements have been omitted.

```
(*M5.1*)
(*k>3*)
rec[alp_,bet_,gam_,k_]:=(c=ConstantArray[0,k];
   c[[2]]=-(bet[[1]] c1+gam[[1]] c0)/(2 alp[[1]]);
   c[[3]]=-((2 alp[[2]]+2 bet[[1]]) c[[2]]+(bet[[2]]+gam[[1]]) c1
     +gam[[2]] c0)/(6 alp[[1]]);
   Do[c[[n+2]]=
     -((alp[[2]] n (n+1)+bet[[1]] (n+1)) c[[n+1]]
     +(alp[[3]] n (n-1)+bet[[2]] n+gam[[1]]) c[[n]]
     +(bet[[3]] (n-1)+gam[[2]]) If[n==2,c1,c[[n-1]]]
     +gam[[3]] Which[n==2,c0,n==3,c1,n>3,c[[n-2]]])/
   (alp[[1]] (n+1) (n+2)),{n,2,k-2}])
alp={1,1,2}; bet={1,7,0};     gam={2,0,0};     k=5;
rec[alp,bet,gam,k];
coe=Simplify[c[[2;;k]]];
Collect[c0+c1 x+c[[2]] x^2+c[[3]] x^3
   +Total[Table[c[[n+2]] x^(n+2),{n,2,k-2}]],{c0,c1}]
```

```
(*M5.2*)
(*k>3*)
rec[alp_,bet_,gam_,k_]:=(c=ConstantArray[0,k];
   c[[2]]=-(bet[[1]] c1+gam[[1]] c0)/(2 alp[[1]]);
   c[[3]]=-((2 alp[[2]]+2 bet[[1]]) c[[2]]+(bet[[2]]+gam[[1]]) c1
     +gam[[2]] c0)/(6 alp[[1]]);
   Do[c[[n+2]]=
     -((alp[[2]] n (n+1)+bet[[1]] (n+1)) c[[n+1]]
     +(alp[[3]] n (n-1)+bet[[2]] n+gam[[1]]) c[[n]]
     +(bet[[3]] (n-1)+gam[[2]]) If[n==2,c1,c[[n-1]]]
     +gam[[3]] Which[n==2,c0,n==3,c1,n>3,c[[n-2]]])/
   (alp[[1]] (n+1) (n+2)),
   {n,2,k-2}])
alp={1,0,0}; bet={0,0,0};     gam={0,-1,0};     k=11;
rec[alp,bet,gam,k];
```

```
coe=Simplify[c[[2;;k]]];
Collect[c0+c1 x+c[[2]] x^2+c[[3]] x^3
   +Total[Table[c[[n+2]] x^(n+2),{n,2,k-2}]],{c0,c1}]

(*M5.3*)
(*Valid only for r1 ≠ r2 and r1 - r2 ≠ integer*)
p[j_,r_]:=r (r-1) alp[[j+1]]+r bet[[j+1]]+gam[[j+1]]
c[k_,r_]:=(a=ConstantArray[0,k];
 a[[1]]=1; a[[2]]=-p[1,r]/p[0,r+1];
 Do[a[[n+1]]=-(p[1,n+r-1] a[[n]]+p[2,n+r-2] a[[n-1]])/
   p[0,n+r],{n,2,k-1}]; a)
alp={2,2,2};    bet={9,11,11};    gam={6,10,7};
rj=Sort[r/.Solve[p[0,r]==0,r],Greater];
k=5; (*k term solution*)
xa=Table[x^n,{n,0,k-1}];
x^rj[[1]] Total[c[k,rj[[1]]] xa]
x^rj[[2]] Total[c[k,rj[[2]]] xa]

(*M5.4*)
(*Valid only for r1 = r2*)
p[j_,r_]:=r (r-1) alp[[j+1]]+r bet[[j+1]]+gam[[j+1]]
c[k_,r_]:=(a=ConstantArray[0,k];
a[[1]]=1; a[[2]]=-p[1,r]/p[0,r+1];
Do[a[[n+1]]=-(p[1,n+r-1] a[[n]]+p[2,n+r-2] a[[n-1]])/
   p[0,n+r],{n,2,k-1}];
da=ConstantArray[0,k];
da[[2]]=D[-p[1,v]/p[0,v+1],v]/.v->r;
Do[da[[n+1]]=-(p[1,n+r-1] da[[n]]+p[2,n+r-2] da[[n-1]])/
 p[0,n+r]+ D[-(p[1,n+v-1] a[[n]]+p[2,n+v-2] a[[n-1]])/
 p[0,n+v],v]/.v->r,{n,2,k-1}]; {a,da})
alp={1,-2,1};    bet={-3,-1,0};    gam={4,1,0};
rj=Sort[r/.Solve[p[0,r]==0,r],Greater];
k=5; (*k term solution*)
xa=Table[x^n,{n,0,k-1}];
{a,da}=c[k,rj[[1]]];
x^rj[[1]] Total[a xa]
x^rj[[1]] Total[da xa]

(*M5.5*)
ww[x_,n_]:=(-1)^n/4^n/Factorial[n]Product[(2 k+5),{k,1,n}]*
   x^(n+5/2)
y=x^(5/2)+Total[Table[ww[x,n],{n,1,4}]]

(*M5.6*)
vv[j_]:=1/(j (2 j+5))
app[n_]:=-5 (-1)^n/4^n/Factorial[n]Product[(2 k+5),{k,1,n}]*
   Total[Table[vv[kk],{kk,1,n}]]
Table[Row[{"a'"ToString[n],"(5/2)=",app[n]}],{n,1,3}]//Column

(*M5.7*)
p[j_,z_]:=z (z-1) alp[[j+1]]+z bet[[j+1]]+gam[[j+1]]
```

```
app[k_,r_]:=(a=ConstantArray[0,k]; da=ConstantArray[0,k];a[[1]]=1;
   a[[2]]=-p[1,r]/p[0,r+1];
   Do[a[[n+1]]=-p[1,r+n-1]/p[0,r+n] a[[n]],{n,2,k-1}];
   da[[1]]=0;
   da[[2]]=D[-p[1,v]/p[0,v+1],v]/.v->r;
   Do[da[[n+1]]=-p[1,n+r-1] da[[n]]/p[0,n+r]+
     D[-(p[1,n+v-1] a[[n]])/p[0,n+v],v]/.v->r,{n,2,k-1}];
   {a,da})
alp={4,2,0};   bet={-4,7,0};    gam={-5,3,0};
rj=Sort[r/.Solve[p[0,r]==0,r],Greater];
kk=rj[[1]]-rj[[2]];
k=4; (*k term solution*)
x1=Table[x^n,{n,0,k-1}];
{a1,a2}=app[k,rj[[1]]];
{a11,a21}=app[kk,rj[[2]]];
cee=-p[1,rj[[1]]-1]/(kk alp[[1]])a11[[kk]];
Row[{"y1(x)=",Total[x^rj[[1]]a1 x1]}]
Row[{"y2(x)=",Total[Expand[x^rj[[2]]a11[[1;;kk]] x1[[1;;kk]]]],
   " ",cee,"(y1(x)ln(x)+",Total[Expand[x^rj[[1]]a2 x1]],")"}]

(*M5.8*)
ww[x_,n_]:=(-1)^n/2^n/Factorial[n]Product[(2 k+3),{k,1,n}]*
   x^(2 n+3)
y1=x^3+Total[Table[ww[x,n],{n,1,4}]]

(*M5.9*)
vv[j_]:=1/(j (2 j+3))
ww[x_,n_]:=(-1)^n/2^n/Factorial[n]Product[(2 k+3),{k,1,n}]*
   Total[Table[vv[kk],{kk,1,n}]] x^(2 n+3)
Total[Table[ww[x,n],{n,1,4}]]

(*M5.10*)
Normal[Series[BesselJ[m,x],{x,0,1}]]
Simplify[Normal[Series[BesselY[m,x],{x,0,1}]]]
Simplify[Normal[Series[BesselY[0,x],{x,0,1}]]]
Normal[Series[BesselJ[m,x],{x,Infinity,0}]]
Normal[Series[BesselY[m,x],{x,Infinity,0}]]

(*M5.11*)
D[BesselJ[m,x],x]
D[BesselY[m,x],x]
FullSimplify[Integrate[x^(-m) BesselJ[m+1,x],x]]
FullSimplify[Integrate[x^(m) BesselJ[m-1,x],x]]
FullSimplify[Integrate[x^(m) BesselY[m-1,x],x]]
FullSimplify[Integrate[x^(-m) BesselY[m+1,x],x]]

(*M5.12*)
BesselJ[3/2,x]
BesselJ[5/2,x]
```

```
(*M5.13*)
BesselJ[-3/2,x]
BesselJ[-5/2,x]

(*M5.14*)
Table[FunctionExpand[SphericalBesselJ[n,x]],{n,0,2}]//Column
Table[FunctionExpand[SphericalBesselY[n,x]],{n,0,2}]//Column

(*M5.15*)
Normal[Series[BesselI[m,x],{x,0,1}]]
Normal[Series[BesselK[0,x],{x,0,1}]]
Normal[Series[BesselK[m,x],{x,0,1}]]
Normal[Series[BesselI[m,x],{x,Infinity,0}]]
Normal[Series[BesselK[m,x],{x,Infinity,0}]]

(*M5.16*)
D[BesselI[m,x],x]
D[BesselK[m,x],x]

(*M5.17*)
y=z^v u[z];
ode=Simplify[z^2 D[y,z,z]+(1-2 a) z D[y,z]+(z^2-r^2) y]
ode=ode/.v->a
ode=ode/.-a^2-r^2->p^2

(*M5.18*)
Simplify[DSolveValue[x ^2 y''[x]+x (1-2 a b) y'[x]+
 (b^2 k^2 x^(2 b)-b^2 r^2) y[x]==0,y[x],x]]

(*M5.19*)
Simplify[DSolveValue[y''[x]-2 a b y'[x]+
   b^2 (k^2 Exp[2 b x]-r^2) y[x]==0,y[x],x]]

(*M5.20*)
Simplify[DSolveValue[x^2 y''[x]+(1-2 Sqrt[3]) x y'[x]+
 (784 x^8-61) y[x]==0,y[x],x]]

(*M5.21*)
DSolveValue[x y''[x]+y'[x]-s^2/g y[x]==0,y[x],x]

(*M5.22*)
DSolveValue[x^2 u''[x]+2 x u'[x]+(x^2-n (n+1)) u[x]==0,u[x],x]

(*M5.23*)
DSolveValue[{y''[x]+ x^2 y'[x]+x y[x]==0},y[x],x]

(*M5.24*)
Simplify[DSolveValue[x u''[x]+a u'[x]+b x^(c-a+1) u[x]==0,u[x],x]]

(*M5.25*)
Ceven[k_]:=If[k==0,c0,1]*
   Factor[(2 k (2 k+1)-n (n+1))/((2 k+2) (2 k+1))]
```

```
Codd[k_]:=If[k==0,c1,1]*
   Factor[((2 k +1) (2 k+2)-n (n+1))/((2 k+3) (2 k+2))]
end=2;
Table[Product[Ceven[k],{k,0,mm}],{mm,0,end}]//Column
Table[Product[Codd[k],{k,0,mm}],{mm,0,end}]//Column
```

```
(*M5.26*)
DSolveValue[(1-x^2) y''[x]-2 x y'[x]+n (n+1) y[x]==0,y[x],x]

(*M5.27*)
DSolveValue[(1-x^2) y''[x]-2 x y'[x]+(n (n+1)-m^2/(1-x^2))y[x]==0,
   y[x],x]

(*M5.28*)
hyp[n_,r_]:=((n+r-1) (n+r-1+a+b)+a b)/((n+r) (n+r+c-1))
Table[Simplify[Product[hyp[n,0],{n,1,mm}]],{mm,1,3}]

(*M5.29*)
hyp[n_,r_]:=((n+r-1) (n+r-1+a+b)+a b)/((n+r) (n+r+c-1))
Product[hyp[mm,0],{mm,1,n}]

(*M5.30*)
hyp[n_,r_]:=((n+r-1) (n+r-1+a+b)+a b)/((n+r) (n+r+c-1))
Table[Simplify[Product[hyp[n,1-c],{n,1,mm}]],{mm,1,3}]

(*M5.31*)
hyp[n_,r_]:=((n+r-1) (n+r-1+a+b)+a b)/((n+r) (n+r+c-1))
Product[hyp[mm,1-c],{mm,1,n}]

(*M5.32*)
DSolveValue[z (1-z) y''[z]+(c-(a+b+1) z) y'[z]-a b y[z]==0,y[z],z]

(*M5.33*)
u=x^t;
de=Simplify[x^4 D[y[u],x,x,x,x]+2 a x^3 D[y[u],x,x,x]+a (a-1) x^2
D[y[u],x,x]-x^t y[u]];
Clear[u]
Collect[de/.x^t->u,{y[u],y'[u],y''[u],y'''[u],y''''[u]}]

(*M5.34*)
gnr[n_,r_]:=(n t+r t) (n t+r t-1) (n t+r t+a-2) (n t+r t+a-3)
comp[xx_,rr_]:=(rte=Simplify[Series[xx,{z,0,2}],Assumptions->t>0];
   Simplify[1+z^t/gnr[1,rr]+z^(2 t)/(gnr[1,rr] gnr[2,rr])-rte])
(*First part*)
DSolveValue[z^4 y''''[z]+2 a z^3 y'''[z]+a(a-1) z^2 y''[z]-
   z^t y[z]==0,y[z],z]
(*Second part, which uses the results of the first part*)
z1= HypergeometricPFQ[{},{1-1/t,1-3/t+a/t,1-2/t+a/t},z^t/t^4];
r1=0;
z2=HypergeometricPFQ[{},{1+1/t,1-2/t+a/t,1-1/t+a/t},z^t/t^4];
```

```
r2=1/t;
z3=HypergeometricPFQ[{},{1-1/t,1+1/t-a/t,1+2/t-a/t},z^t/t^4];
r3=(2-a)/t;
z4=HypergeometricPFQ[{},{1+1/t,1+2/t-a/t,1+3/t-a/t},z^t/t^4];
r4=(3-a)/t;
comp[z1,r1]
comp[z3,r3]
comp[z3,r3]
comp[z4,r4]

(*M5.35*)
(*Appendix 5.1*)
Limit[Simplify[D[(x/2)^p/Gamma[p+1+n],p]],p->m]
Limit[Simplify[D[(x/2)^(-p)/Gamma[-p+1+n],p]],p->m]
Simplify[Limit[D[(x/2)^(-p) Gamma[p-1-n] (p-1-n)*
    Sin[Pi(p-1-n)],p],p->m]]
```

EXERCISES

Section 5.1.2

5.1 Use the power series method to obtain a solution to

$$\left(1-x^2\right)\frac{d^2y}{dx^2}-5x\frac{dy}{dx}-4y=0$$

5.2 Use the power series method to obtain a solution to

$$\left(2+4x-2x^2\right)\frac{d^2y}{dx^2}-12(x-1)\frac{dy}{dx}-12y=0$$

Show that the solution can be written as

$$y(x)=c_0\sum_{n=0}^{\infty}\frac{2n+1}{2^n}(x-1)^n+c_1\sum_{n=0}^{\infty}\frac{n+1}{2^n}(x-1)^{n+1}$$

Note that

$$2+4x-2x^2=4-2(x-1)^2$$

Section 5.1.3

5.3 Use the method of Frobenius and the appropriate series relationships to show that the solution to

$$4x\frac{d^2y}{dx^2}+2\frac{dy}{dx}+y=0$$

is

$$y=C_1\cos\left(\sqrt{x}\right)+C_2\sin\left(\sqrt{x}\right)$$

5.4 Use the method of Frobenius to obtain the solution to

$$x^2\frac{d^2y}{dx^2}-3x\frac{dy}{dx}+(4-x)y=0$$

To place your solution a compact form, note the following:

$$\frac{3}{2}=\left(1+\frac{1}{2}\right),\quad \frac{11}{6}=\left(1+\frac{1}{2}+\frac{1}{3}\right),\quad \frac{25}{12}=\left(1+\frac{1}{2}+\frac{1}{3}+\frac{1}{4}\right)\cdots$$
$$g(n)=\left(1+\frac{1}{2}+\frac{1}{3}+\cdots+\frac{1}{n}\right)\text{(define this function as stated)} \tag{5.97}$$

5.5 Use the method of Frobenius to obtain the solution to

$$\left(5+x+10x^2\right)x^2\frac{d^2y}{dx^2}+\left(4+3x+48x^2\right)x\frac{dy}{dx}+\left(x+36x^2\right)y=0$$

5.6 Use the method of Frobenius to obtain the solution to

$$\left(1+2x+x^2\right)x^2\frac{d^2y}{dx^2}+\left(1+3x+4x^2\right)x\frac{dy}{dx}+\left(-x+2x^2\right)y=0$$

5.7 Use the method of Frobenius to obtain the solution to

$$(1+x)x^2\frac{d^2y}{dx^2}-(3+10x)x\frac{dy}{dx}+30xy=0$$

Section 5.1.8

5.8 A fourth-order differential equation can be factored into the following two second-order differential equations

$$\frac{d}{d\varphi}\left(\varphi^{n+1}\frac{dW}{d\varphi}\right)\pm\lambda^2\varphi^nW=0$$

Determine the complete solution, that is, one that contains four independent solution functions.

5.9 Obtain the solution to the following equation that remains finite at $x = 0$.

$$x^2\frac{d^2y}{dx^2}+2x\frac{dy}{dx}+h^2xy=0$$

5.10 Given the equation

$$x^2\frac{d^2y}{dx^2}+x\frac{dy}{dx}-\left(jx^2+n^2\right)y=0$$

where n is an integer, show that one solution to this equation is

$$y=\text{Ber}_nx+j\text{Bei}_nx$$

where

$$\text{Ber}_n x = \sum_{m=0}^{\infty} \frac{(-1)^{m+n}(x/2)^{2m+n}}{m!(m+n)!} \cos\left(\pi(2m+n)/4\right)$$

$$\text{Bei}_n x = \sum_{m=0}^{\infty} \frac{(-1)^{m+n+1}(x/2)^{2m+n}}{m!(m+n)!} \sin\left(\pi(2m+n)/4\right)$$

The quantities Ber and Bei, respectively, are known as the real and imaginary parts of the Kelvin function of the first kind. Hint: The exponential form of $-j$ will prove beneficial.

5.11 Consider the equation

$$x^2 \frac{d^2y}{dx^2} + \left(a + 2bx^\alpha\right) x \frac{dy}{dx} + \left(c + \gamma^2 x^{2\delta} + b(a+\alpha-1)x^\alpha + b^2 x^{2\alpha}\right) y = 0$$

where a, b, c, α, δ, and γ are real constants. Make the following two transformations sequentially

$$y = x^{(1-a)/2} e^{-bx^\alpha/\alpha} u(x) \quad \text{and} \quad z = \frac{\gamma}{\delta} x^\delta$$

and show that the resulting equation is a Bessel equation. It may be easier to use Mathematica to obtain dy/dx and d^2y/dx^2. It will be found that the order p of the Bessel function is

$$p = \frac{1}{\delta} \left\{ \left(\frac{a-1}{2} \right)^2 - c \right\}^{1/2}$$

5.12 Using the appropriate relationship in Table 5.1, determine the solution to

$$\frac{d^2y}{dx^2} + \frac{\beta}{x} \frac{dy}{dx} + k^2 x^{\alpha-\beta} y = 0$$

APPENDIX 5.1

Bessel Function of the Second Kind

To determine the second independent solution to Bessel's equation, we shall use Eq. (5.89). It is seen in Eq. (5.89) that when p is an integer $\cos(p\pi) \longrightarrow (-1)^p$ and $\sin(p\pi) \longrightarrow 0$, the ratio of the numerator and denominator of Eq. (5.89) is indeterminate. However, its limit can be determined by using L'Hôpital's rule. Then, Eq. (5.89) becomes

$$\begin{aligned} Y_m(x) &= \lim_{p\to m} \left[\frac{\cos p\pi J_p(x) - J_{-p}(x)}{\sin p\pi} \right] = \lim_{p\to m} \left[\frac{\frac{d}{dp}\left(\cos p\pi J_p(x) - J_{-p}(x)\right)}{\frac{d}{dp}(\sin p\pi)} \right] \\ &= -\lim_{p\to m} \left[\frac{\pi \sin p\pi J_p(x)}{\pi \cos p\pi} \right] + \lim_{p\to m} \left[\frac{\cos p\pi dJ_p(x) dp}{\pi \cos p\pi} \right] - \lim_{p\to m} \left[\frac{dJ_{-p}(x)/dp}{\pi \cos p\pi} \right] \\ &= \lim_{p\to m} \left[\frac{1}{\pi} \left(\frac{d}{dp} J_p(x) - (-1)^m \frac{d}{dp} J_{-p}(x) \right) \right] \end{aligned} \qquad (5.1.1)$$

For algebraic convenience, we rewrite the expressions for the Bessel functions as follows

$$J_p(x) = \sum_{n=0}^{\infty} \frac{(-1)^n (x/2)^{2n+p}}{n!\Gamma(p+1+n)} = \sum_{n=0}^{\infty} C_n \frac{(x/2)^p}{\Gamma(p+1+n)}$$
$$J_{-p}(x) = \sum_{n=0}^{\infty} \frac{(-1)^n (x/2)^{2n-p}}{n!\Gamma(-p+1+n)} = \sum_{n=0}^{\infty} C_n \frac{(x/2)^{-p}}{\Gamma(-p+1+n)} \tag{5.1.2}$$

where

$$C_n = \frac{(-1)^n (x/2)^{2n}}{n!} \tag{5.1.3}$$

Then Eq. (5.1.1) becomes

$$Y_m(x) = \lim_{p\to m}\left[\frac{1}{\pi}\left\{\sum_{n=0}^{\infty} C_n \frac{d}{dp}\left(\frac{(x/2)^p}{\Gamma(p+1+n)}\right) - (-1)^m \sum_{n=0}^{\infty} C_n \frac{d}{dp}\left(\frac{(x/2)^{-p}}{\Gamma(-p+1+n)}\right)\right\}\right]$$

However, as discussed in Appendix C, $\Gamma(-m + 1 + n)$ is undefined when its argument is a negative integer. In other words, this summation is not valid until $n + 1 > m$ or until $n = m$. Thus, the second summation is broken into two parts to obtain,

$$\begin{aligned} Y_m(x) = \lim_{p\to m}\Bigg[\frac{1}{\pi}\Bigg\{&\sum_{n=0}^{\infty} C_n \frac{d}{dp}\left(\frac{(x/2)^p}{\Gamma(p+1+n)}\right) \\ &- (-1)^m \sum_{n=0}^{m-1} C_n \frac{d}{dp}\left(\frac{(x/2)^{-p}}{\Gamma(-p+1+n)}\right) - (-1)^m \sum_{n=m}^{\infty} C_n \frac{d}{dp}\left(\frac{(x/2)^{-p}}{\Gamma(-p+1+n)}\right)\Bigg\}\Bigg] \\ = \lim_{p\to m}\Bigg[\frac{1}{\pi}\Bigg\{&\sum_{n=0}^{\infty} C_n \frac{d}{dp}\left(\frac{(x/2)^p}{\Gamma(p+1+n)}\right) + (-1)^m \sum_{n=0}^{m-1} C_n \frac{1}{\pi}\frac{d}{dp} g_n(p) \\ &- (-1)^m \sum_{n=m}^{\infty} C_n \frac{d}{dp}\left(\frac{(x/2)^{-p}}{\Gamma(p-1-n)}\right)\Bigg\}\Bigg] \end{aligned} \tag{5.1.4}$$

where

$$g_n(p) = (x/2)^{-p}(p-1-n)\Gamma(p-1-n)\sin[\pi(p-1-n)]$$

and we have used Eq. (C.10) of Appendix C with $-z = -p + 1 + n$ since $-p + 1 + n < 0$, that is, $z = p - 1 - n$.

Using the Mathematica program M5.35 and Eq. (C.19) of Appendix C, it is found that

$$\lim_{p\to m}\left[\frac{d}{dp}\left(\frac{(x/2)^p}{\Gamma(p+1+n)}\right)\right] = \frac{1}{\Gamma(m+1+n)}\left(\frac{x}{2}\right)^m\left[\ln\left(\frac{x}{2}\right) - \psi(m+1+n)\right]$$
$$\lim_{p\to m}\left[\frac{d}{dp}\left(\frac{(x/2)^{-p}}{\Gamma(-p+1+n)}\right)\right] = \frac{1}{\Gamma(-m+1+n)}\left(\frac{x}{2}\right)^{-m}\left[-\ln\left(\frac{x}{2}\right) + \psi(-m+1+n)\right] \tag{5.1.5}$$

and

$$\lim_{p\to m}\left[\frac{d}{dp}g_n(p)\right]=\pi\left(\frac{x}{2}\right)^{-p}(-1)^{n+m+1}\Gamma(m-n) \tag{5.1.6}$$

where $\psi(x)$ is the digamma function given by Eq. (C.16) of Appendix C. Substituting Eqs. (5.1.5) and (5.1.6) into Eq. (5.1.4) yields

$$\begin{aligned}Y_m(x)=\frac{1}{\pi}\Bigg\{&\sum_{n=0}^{\infty}\frac{C_n}{\Gamma(m+1+n)}\left[\left(\frac{x}{2}\right)^m\left\{\ln\left(\frac{x}{2}\right)-\psi(m+1+n)\right\}\right]\\&+(-1)^m\sum_{n=0}^{m-1}C_n\left(\frac{x}{2}\right)^{-m}(-1)^{n+m+1}\Gamma(m-n)\\&-(-1)^m\sum_{n=m}^{\infty}\frac{C_n}{\Gamma(-m+1+n)}\left[\left(\frac{x}{2}\right)^{-m}\left\{-\ln\left(\frac{x}{2}\right)+\psi(-m+1+n)\right\}\right]\Bigg\}\end{aligned} \tag{5.1.7}$$

Changing the summation indices in the last summation of Eq. (5.1.7) results in

$$\begin{aligned}Y_m(x)=\frac{1}{\pi}\Bigg\{&\sum_{n=0}^{\infty}\frac{C_n}{\Gamma(m+1+n)}\left[\left(\frac{x}{2}\right)^m\left\{\ln\left(\frac{x}{2}\right)-\psi(m+1+n)\right\}\right]\\&+(-1)^m\sum_{n=0}^{m-1}C_n\left(\frac{x}{2}\right)^{-m}(-1)^{n+m+1}\Gamma(m-n)\\&-(-1)^m\sum_{n=0}^{\infty}\frac{C_{n+m}}{\Gamma(1+n)}\left[\left(\frac{x}{2}\right)^{-m}\left\{-\ln\left(\frac{x}{2}\right)+\psi(1+n)\right\}\right]\Bigg\}\end{aligned} \tag{5.1.8}$$

Using Eqs. (5.1.3) and (C.5) of Appendix C, we find that Eq. (5.1.8) becomes

$$\begin{aligned}Y_m(x)=\frac{1}{\pi}\Bigg\{&\sum_{n=0}^{\infty}\frac{(-1)^n(x/2)^{2n}}{n!(m+n)!}\left[\left(\frac{x}{2}\right)^m\left\{\ln\left(\frac{x}{2}\right)-\psi(m+1+n)\right\}\right]\\&-\sum_{n=0}^{m-1}\frac{(x/2)^{2n}}{n!}\left(\frac{x}{2}\right)^{-m}\Gamma(m-n)\\&-(-1)^m\sum_{n=0}^{\infty}\frac{(-1)^{n+m}(x/2)^{2n+2m}}{n!(m+n)!}\left[\left(\frac{x}{2}\right)^{-m}\left\{-\ln\left(\frac{x}{2}\right)+\psi(1+n)\right\}\right]\Bigg\}\end{aligned} \tag{5.1.9}$$

Performing the multiplications indicated in Eq. (5.1.9) and simplifying, we arrive at

$$Y_m(x) = \frac{2}{\pi}\ln\left(\frac{x}{2}\right)J_m(x) - \frac{1}{\pi}\sum_{n=0}^{m-1}\frac{(m-1-n)!}{n!}\left(\frac{x}{2}\right)^{2n-m}$$
$$-\frac{1}{\pi}\sum_{n=0}^{\infty}\frac{(-1)^n}{n!(m+n)!}\left(\frac{x}{2}\right)^{2n+m}\left[\psi(1+n)+\psi(m+1+n)\right] \tag{5.1.10}$$

Equation (5.1.10) is the conventional form of the Bessel function of the second kind of order m. It is straightforward to show that when $m = 0$, Eq. (5.1.10) reduces to Eq. (5.84).

It is seen that as $x \to 0$, $Y_m(x) \to -\infty$.

6 Ordinary Differential Equations Part III

Sturm–Liouville Equation

6.1 STURM–LIOUVILLE EQUATION

6.1.1 PRELIMINARIES: ADJOINT EQUATIONS

Consider the following second-order differential operator on the interval $a < x < b$

$$L[y] = a_0(x)\frac{d^2y}{dx^2} + a_1(x)\frac{dy}{dx} + a_4(x)y = 0 \tag{6.1}$$

where

$$a_4(x) = a_2(x) + a_3(x)\lambda_y \tag{6.2}$$

and we have chosen the form of this equation in anticipation of what is to follow. The parameter λ_y is a constant that is linked to the differential equation in $y(x)$. In Eq. (6.1), $a_0(x) \neq 0$, $a_0(a) \neq 0$, $a_0(b) \neq 0$, and $a_j(x)$, $j = 0, 1, \ldots, 3$, are continuous and their first derivatives exist. We now define the following second differential operator on the interval $a < x < b$

$$K[y] = \frac{d^2}{dx^2}\left(a_0(x)y\right) - \frac{d}{dx}\left(a_1(x)y\right) + a_4(x)y \tag{6.3}$$

as the *adjoint operator* to the operator L. The differential equation

$$K[y] = 0$$

is the adjoint differential equation of Eq. (6.1). If the operators $K = L$, then the operators are called *self-adjoint.*

In order for the operators to be self-adjoint, we are stating the following. Expanding Eq. (6.3), it is found that

$$K[y] = a_0(x)\frac{d^2y}{dx^2} + \left(2\frac{da_0}{dx} - a_1(x)\right)\frac{dy}{dx} + \left(\frac{d^2a_0}{dx^2} - \frac{da_1}{dx} + a_4(x)\right)y = 0 \tag{6.4}$$

Therefore, to be self-adjoint, the coefficients of Eqs. (6.1) and (6.4) must be equal, that is,

$$\begin{aligned} a_1 &= 2\frac{da_0}{dx} - a_1 \rightarrow a_1 = \frac{da_0}{dx} \\ a_4 &= \frac{d^2a_0}{dx^2} - \frac{da_1}{dx} + a_4 \rightarrow a_1 = \frac{da_0}{dx} \end{aligned} \tag{6.5}$$

These relationships are not true in general.

However, one can change the second-order operator L by a suitable transformation of the form

$$L_{sa}[y] = z(x)L[y] \tag{6.6}$$

Therefore

$$L_{sa}[y] = z(x)a_0(x)\frac{d^2y}{dx^2} + z(x)a_1(x)\frac{dy}{dx} + z(x)a_4(x)y \tag{6.7}$$

From Eqs. (6.7) and (6.5), it is seen that [$a_1 \to a_1 z$ and $a_0 \to a_0 z$]

$$z(x)a_1(x) = \frac{d}{dx}\left(z(x)a_0(x)\right) = \frac{dz}{dx}a_0(x) + z(x)\frac{da_0}{dx}$$

which can be written as

$$\frac{dz}{z} = \frac{a_1 dx}{a_0} - \frac{da_0}{a_0}$$

Integrating this equation, one obtains

$$\ln z = \int \frac{a_1(x)}{a_0(x)}dx - \ln a_0(x)$$

$$\ln\left(a_0(x)z\right) = \int \frac{a_1(x)}{a_0(x)}dx$$

or

$$z = \frac{p(x)}{a_0(x)} \tag{6.8}$$

where

$$p(x) = \exp\left[\int \frac{a_1(x)}{a_0(x)}dx\right] \tag{6.9}$$

Using Eq. (6.8) in Eq. (6.7), the self-adjoint operator is

$$L_{sa}[y] = p(x)\frac{d^2y}{dx^2} + \frac{p(x)a_1(x)}{a_0(x)}\frac{dy}{dx} + \frac{p(x)a_4(x)}{a_0(x)}y \tag{6.10}$$

Thus, any equation of the form given by Eq. (6.1) with the conditions on a_j as stipulated can be converted to a self-adjoint equation.

We would like to write Eq. (6.10) in the form

$$\begin{aligned} L_{sa}[y] &= \frac{d}{dx}\left(p(x)\frac{dy}{dx}\right) + \frac{p(x)a_4(x)}{a_0(x)}y \\ &= p(x)\frac{d^2y}{dx^2} + \frac{dp}{dx}\frac{dy}{dx} + \frac{p(x)a_4(x)}{a_0(x)}y \end{aligned} \tag{6.11}$$

Comparing Eqs. (6.10) and (6.11), it is seen that in order to put Eq. (6.10) in the form given by Eq. (6.11)

$$\frac{dp}{dx} = \frac{p(x)a_1(x)}{a_0(x)} \tag{6.12}$$

which, upon integration, gives

$$p(x) = \exp\left[\int \frac{a_1(x)}{a_0(x)}dx\right] \tag{6.13}$$

Thus, the same function $p(x)$ that is used to make the operator self-adjoint will also permit one to write the operator in the form given by Eq. (6.11). Then we can write Eq. (6.1) as

$$L_{sa}[y] = \frac{d}{dx}\left(p(x)\frac{dy}{dx}\right) + q(x)y + r(x)\lambda_y y \tag{6.14}$$

where

$$\begin{aligned} q(x) &= \frac{p(x)a_2(x)}{a_0(x)} \\ r(x) &= \frac{p(x)a_3(x)}{a_0(x)} \end{aligned} \tag{6.15}$$

and $p(x)$ is given by Eq. (6.13).

We will have subsequent use for the following integral

$$I_2 = \int_a^b \left(uL_{sa}[v] - vL_{sa}[u]\right)dx \tag{6.16}$$

where

$$\begin{aligned} L_{sa}[u] &= \frac{d}{dx}\left(p(x)\frac{du}{dx}\right) + q(x)u + r(x)\lambda_u u \\ L_{sa}[v] &= \frac{d}{dx}\left(p(x)\frac{dv}{dx}\right) + q(x)v + r(x)\lambda_v v \end{aligned} \tag{6.17}$$

Then the integrand of Eq. (6.16) is

$$\begin{aligned} uL_{sa}[v] - vL_{sa}[u] &= u\frac{d}{dx}\left(p(x)\frac{dv}{dx}\right) + q(x)uv + r(x)\lambda_v uv \\ &\quad - v\frac{d}{dx}\left(p(x)\frac{du}{dx}\right) - q(x)vu - r(x)\lambda_u vu \\ &= u\frac{d}{dx}\left(p(x)\frac{dv}{dx}\right) - v\frac{d}{dx}\left(p(x)\frac{du}{dx}\right) \\ &\quad + r(x)\left(\lambda_v - \lambda_u\right)uv \end{aligned} \tag{6.18}$$

However, it is noted that

$$\begin{aligned}\frac{d}{dx}\left(p(x)u\frac{dv}{dx}\right)&=u\frac{d}{dx}\left(p(x)\frac{dv}{dx}\right)+p(x)\frac{dv}{dx}\frac{du}{dx}\\ \frac{d}{dx}\left(p(x)v\frac{du}{dx}\right)&=v\frac{d}{dx}\left(p(x)\frac{du}{dx}\right)+p(x)\frac{dv}{dx}\frac{du}{dx}\end{aligned} \tag{6.19}$$

Using Eq. (6.19) in Eq. (6.18) results in

$$\begin{aligned}uL_{sa}[v]-vL_{sa}[u]&=\frac{d}{dx}\left(p(x)u\frac{dv}{dx}\right)-p(x)\frac{dv}{dx}\frac{du}{dx}-\frac{d}{dx}\left(p(x)v\frac{du}{dx}\right)\\ &\quad+p(x)\frac{dv}{dx}\frac{du}{dx}+(\lambda_v-\lambda_u)r(x)uv\\ &=\frac{d}{dx}\left(p(x)u\frac{dv}{dx}\right)-\frac{d}{dx}\left(p(x)v\frac{du}{dx}\right)+(\lambda_v-\lambda_u)r(x)uv\\ &=\frac{d}{dx}\left[p(x)\left(u\frac{dv}{dx}-v\frac{du}{dx}\right)\right]+(\lambda_v-\lambda_u)r(x)uv\end{aligned} \tag{6.20}$$

When $r(x) = 0$, Eq. (6.20) is known as Lagrange's identity. Therefore, Eq. (6.16) can be written as

$$\begin{aligned}\int_a^b(uL_{sa}[v]-vL_{sa}[u])dx&=\int_a^b\frac{d}{dx}\left[p(x)\left(u\frac{dv}{dx}-v\frac{du}{dx}\right)\right]dx\\ &\quad+(\lambda_v-\lambda_u)\int_a^b r(x)uvdx\\ &=\left[p(x)\left(u\frac{dv}{dx}-v\frac{du}{dx}\right)\right]_a^b+(\lambda_v-\lambda_u)\int_a^b r(x)uvdx\end{aligned} \tag{6.21}$$

6.1.2 Sturm–Liouville Equation

The Sturm–Liouville equation is represented by the following equation

$$a_0(x)\frac{d^2y}{dx^2}+a_1(x)\frac{dy}{dx}+a_2(x)y+\lambda a_3(x)y=0 \quad a<x<b \tag{6.22}$$

which has already been introduced as Eq. (6.1) and where the restrictions on a_j have been stated. It was shown that Eq. (6.22) is a self-adjoint equation that can be written in the form [recall Eq. (6.14)]

$$\frac{d}{dx}\left(p(x)\frac{dy}{dx}\right)+q(x)y+r(x)\lambda y=0 \quad a<x<b \tag{6.23}$$

where $p(x)$ is given by Eq. (6.13) and $q(x)$ and $r(x)$ are given by Eq. (6.15). In addition, we require that $p(x) > 0$ and that $r(x) > 0$.

With these conditions, Eq. (6.23) is the *Sturm–Liouville equation* when subject to the following homogeneous boundary conditions

$$\begin{aligned} \alpha y(a) + \beta y'(a) = 0 \quad & \alpha^2 + \beta^2 > 0 \\ \delta y(b) + \gamma y'(b) = 0 \quad & \delta^2 + \gamma^2 > 0 \end{aligned} \tag{6.24}$$

where α, β, γ, and δ are real quantities and neither α nor β can simultaneously equal zero and neither γ nor δ can simultaneously equal zero. It is seen that Eqs. (6.23) and (6.24) form a boundary-value problem composed of a homogeneous differential equation with homogeneous boundary conditions and an undefined parameter λ. The unknown parameter λ is selected so that the solution to Eqs. (6.23) and (6.24) is not trivial, that is, $y(x) \neq 0$. These values of λ are denoted λ_n and are the eigenvalues, which will be shown subsequently to be real. In addition, there are an infinite number of λ_n for which there is a smallest one, but there is no largest one. Corresponding to each eigenvalue λ_n is an eigenfunction $\varphi_n(x)$, which has $n - 1$ zeros on the interval (a,b). The determination of λ_n and $\varphi_n(x)$ will be discussed subsequently. It is noted that the solution to Eq. (6.23) will also be a solution to Eq. (6.22) since these two equations are different forms of the same equation.

Historically, boundary conditions have had three different names attached to them and are commonly used to indicate the type of boundary condition specified. The *Dirichlet* (1850) boundary condition specifies that the function (dependent variable) along a boundary is zero. The *Neumann* (1877) boundary condition specifies that the normal derivative of the function (dependent variable) along a boundary is zero. Lastly, the *Robin* (1886) boundary condition specifies that a linear combination of the function and its normal derivative is equal to zero. Notice that Eq. (6.24) represents the Robin boundary conditions. If any one of the constants is zero in each boundary condition, then either the Dirichlet or Neumann boundary conditions will be obtained. It is mentioned that some authors refer to the Dirichlet boundary conditions as the first boundary value problem, the Neumann boundary conditions as the second boundary value problem, and the Robin boundary conditions as the third boundary value problem.

We shall now determine the form of the solution to Eq. (6.23) subject to the boundary conditions given by Eq. (6.24). The solution to Eq. (6.23), a homogeneous differential equation, is assumed to be composed of two linearly independent solutions ψ and θ, that is,

$$y(x) = y_h(x) = C_1\psi(x,\lambda) + C_2\theta(x,\lambda) \tag{6.25}$$

where the dependency on the free parameter λ is explicitly denoted. Substituting Eq. (6.25) into the homogeneous boundary conditions given by Eq. (6.24) yields

$$\begin{pmatrix} g_{11}(a,\lambda) & g_{12}(a,\lambda) \\ g_{21}(b,\lambda) & g_{22}(b,\lambda) \end{pmatrix} \begin{Bmatrix} C_1 \\ C_2 \end{Bmatrix} = 0 \tag{6.26}$$

where

$$\begin{aligned}
g_{11}(a,\lambda) &= \alpha\psi(a,\lambda)+\beta\psi'(a,\lambda)\\
g_{12}(a,\lambda) &= \alpha\theta(a,\lambda)+\beta\theta'(a,\lambda)\\
g_{21}(b,\lambda) &= \delta\psi(b,\lambda)+\gamma\psi'(b,\lambda)\\
g_{22}(b,\lambda) &= \delta\theta(b,\lambda)+\gamma\theta'(b,\lambda)
\end{aligned} \tag{6.27}$$

and the prime (′) denotes the derivative with respect to x. We have a non-trivial solution (recall Section 1.7) only when $\lambda = \lambda_n$, where λ_n are the solutions to

$$\det\begin{pmatrix} g_{11}(a,\lambda_n) & g_{12}(a,\lambda_n) \\ g_{21}(b,\lambda_n) & g_{22}(b,\lambda_n) \end{pmatrix} = 0 \tag{6.28}$$

or

$$g_{11}(a,\lambda_n)g_{22}(b,\lambda_n) - g_{12}(a,\lambda_n)g_{21}(b,\lambda_n) = 0 \tag{6.29}$$

Equations (6.28) and (6.29) are referred to as the *characteristic equation*. It is mentioned that the functions involved in the solution of the homogeneous equation, ψ and θ, must be such that the evaluation of either Eq. (6.28) or (6.29) results in an infinite number of λ_n. The values of λ_n that satisfy this equation are called *eigenvalues*. As shown below, corresponding to the eigenvalues are functions called *eigenfunctions*.

To determine the corresponding eigenfunctions to each λ_n, we start with the first equation of Eq. (6.26)

$$g_{11}(a,\lambda_n)C_1 + g_{12}(a,\lambda_n)C_2 = 0$$

One can select either

$$\frac{C_2}{C_1} \to \frac{C_{2n}}{C_{1n}} = -\frac{g_{11}(a,\lambda_n)}{g_{12}(a,\lambda_n)}$$

or

$$\frac{C_1}{C_2} \to \frac{C_{1n}}{C_{2n}} = -\frac{g_{12}(a,\lambda_n)}{g_{11}(a,\lambda_n)}$$

From the second equation of Eq. (6.26)

$$g_{21}(b,\lambda_n)C_1 + g_{22}(b,\lambda_n)C_2 = 0$$

and one can select either

$$\frac{C_2}{C_1} \to \frac{C_{2n}}{C_{1n}} = -\frac{g_{21}(b,\lambda_n)}{g_{22}(b,\lambda_n)}$$

or

$$\frac{C_1}{C_2} \to \frac{C_{1n}}{C_{2n}} = -\frac{g_{22}(b,\lambda_n)}{g_{21}(b,\lambda_n)}$$

Therefore,

$$\frac{C_{2n}}{C_{1n}}=-\frac{g_{11}(a,\lambda_n)}{g_{12}(a,\lambda_n)}=-\frac{g_{21}(b,\lambda_n)}{g_{22}(b,\lambda_n)} \tag{6.30}$$

or

$$\frac{C_{1n}}{C_{2n}}=-\frac{g_{12}(a,\lambda_n)}{g_{11}(a,\lambda_n)}=-\frac{g_{22}(b,\lambda_n)}{g_{21}(b,\lambda_n)} \tag{6.31}$$

It is noted that the equalities of the two ratios appearing on the right-hand sides of Eqs. (6.30) and (6.31) are simply a rearrangement of Eq. (6.29), the characteristic equation.

Then, using Eq. (6.25), the solution corresponding to λ_n has the following forms, any one of which can be used.

Form 1

$$y(x)\rightarrow y_n(x)=C_{1n}\left(\psi(x,\lambda_n)+\frac{C_{2n}}{C_{1n}}\theta(x,\lambda_n)\right)=C_{1n}\varphi_n(x) \tag{6.32}$$

where, from Eq. (6.30),

$$\begin{aligned}\varphi_n(x)&=\psi(x,\lambda_n)-\frac{g_{11}(a,\lambda_n)}{g_{12}(a,\lambda_n)}\theta(x,\lambda_n)\\&=\psi(x,\lambda_n)-\frac{\alpha\psi(a,\lambda_n)+\beta\psi'(a,\lambda_n)}{\alpha\theta(a,\lambda_n)+\beta\theta'(a,\lambda_n)}\theta(x,\lambda_n)\end{aligned} \tag{6.33}$$

or

$$\begin{aligned}\varphi_n(x)&=\psi(x,\lambda_n)-\frac{g_{21}(b,\lambda_n)}{g_{22}(b,\lambda_n)}\theta(x,\lambda_n)\\&=\psi(x,\lambda_n)-\frac{\delta\psi(b,\lambda_n)+\gamma\psi'(b,\lambda_n)}{\delta\theta(b,\lambda_n)+\gamma\theta'(b,\lambda_n)}\theta(x,\lambda_n)\end{aligned} \tag{6.34}$$

Form 2

$$y(x)\rightarrow y_n(x)=C_{2n}\left(\frac{C_{1n}}{C_{2n}}\psi(x,\lambda_n)+\theta(x,\lambda_n)\right)=C_{2n}\varphi_n(x) \tag{6.35}$$

where, from Eq. (6.31),

$$\begin{aligned}\varphi_n(x)&=-\frac{g_{12}(a,\lambda_n)}{g_{11}(a,\lambda_n)}\psi(x,\lambda_n)+\theta(x,\lambda_n)\\&=-\frac{\alpha\theta(a,\lambda_n)+\beta\theta'(a,\lambda_n)}{\alpha\psi(a,\lambda_n)+\beta\psi'(a,\lambda_n)}\psi(x,\lambda_n)+\theta(x,\lambda_n)\end{aligned} \tag{6.36}$$

or

$$\varphi_n(x) = -\frac{g_{22}(b,\lambda_n)}{g_{21}(b,\lambda_n)}\psi(x,\lambda_n) + \theta(x,\lambda_n)$$
$$= -\frac{\delta\theta(b,\lambda_n) + \gamma\theta'(b,\lambda_n)}{\delta\psi(b,\lambda_n) + \gamma\psi'(b,\lambda_n)}\psi(x,\lambda_n) + \theta(x,\lambda_n) \tag{6.37}$$

The function $\varphi_n(x)$ is the eigenfunction corresponding to the eigenvalue λ_n. Notice that $\varphi_n(x)$ is a shape function that is scaled by the constants C_{1n} or C_{2n}, which at this point are undetermined.

Since $\varphi_n(x)$ and the corresponding λ_n are solutions to Eqs. (6.23) and (6.24) for all n, we use superposition to obtain the solution as

$$y_h(x) = \sum_{n=1}^{\infty} y_n(x) = \sum_{n=1}^{\infty} c_n \varphi_n(x) \tag{6.38}$$

where c_n corresponds to C_{1n} or C_{2n} depending on the form of $\varphi_n(x)$ selected, and $\varphi_n(x)$ is a solution to

$$\frac{d}{dx}\left(p(x)\frac{d\varphi_n}{dx}\right) + q(x)\varphi_n + r(x)\lambda_n\varphi_n = 0 \tag{6.39}$$

and the boundary conditions given by Eq. (6.24), which become

$$\alpha\varphi_n(a) + \beta\varphi_n'(a) = 0$$
$$\delta\varphi_n(b) + \gamma\varphi_n'(b) = 0 \tag{6.40}$$

It is emphasized that the solution given by Eq. (6.38) is a solution to Eqs. (6.23) and (6.24) only when $\lambda = \lambda_n$; otherwise $y_h = 0$, which is a trivial solution.

6.1.3 Examples of Sturm–Liouville Systems*

It will be shown that certain second-order equations with constant coefficients, certain Bessel equations, the Legendre equation, and the hypergeometric equation each with homogeneous boundary conditions given by Eq. (6.24) are Sturm–Liouville systems. It is emphasized that the parameter λ must be chosen so that we obtain nontrivial solutions, that these values must be real, and that there must be an infinite number of them. In addition, one of the requirements to be a Sturm–Liouville system is for $r(x) > 0$. It will be seen in what follows that when $r(x) > 0$ the solutions to the equations are oscillatory functions, a necessary condition to obtain an infinite number of λ.

Equation with Constant Coefficients: Consider the following equation

$$\frac{d^2y}{dx^2} + \mu^2 y = 0 \quad \mu^2 > 0 \tag{6.41}$$

* For a collection of over 50 differential equations that can be classified as being of the Sturm–Liouville type see: Everitt WN (2005) A Catalogue of Sturm-Liouville Differential Equations. In: *Sturm-Liouville Theory*, ed. Amrein WO, Hinz AM, Pearson DP. Birkhäuser, Basel, pp. 271–331.

Upon comparing Eq. (6.41) with Eq. (6.22), we see that $a_0(x) = 1$, $a_1(x) = 0$, $a_2(x) = 0$, $a_3(x) = 1$, and $\lambda = \mu^2$. Then using Eqs. (6.9) and (6.15), we find that

$$p(x) = \exp\left[\int \frac{a_1(x)}{a_0(x)}dx\right] = \exp[0] = 1$$

$$q(x) = \frac{p(x)a_2(x)}{a_0(x)} = 0 \tag{6.42}$$

$$r(x) = \frac{p(x)a_3(x)}{a_0(x)} = 1$$

From Table 4.2, we see that the solutions to Eq. (6.41) are

$$y(x) = A\cos(\mu x) + B\sin(\mu x) \tag{6.43}$$

which are oscillatory functions. Notice, however, that when μ^2 is replaced by $-\mu^2$, that is,

$$\frac{d^2y}{dx^2} - \mu^2 y = 0 \quad \mu^2 > 0 \tag{6.44}$$

$r(x) = -1 < 0$ and this equation is *not* of the Sturm–Liouville type. Recall from Table 4.2 that the solutions to Eq. (6.44) are hyperbolic functions, which are not oscillatory.

Bessel's Equation: Consider the Bessel equation [recall Eq. (5.69)]

$$x^2\frac{d^2y}{dx^2} + x\frac{dy}{dx} + \left(\kappa^2x^2 - m^2\right)y = 0 \quad \kappa^2 > 0 \tag{6.45}$$

Upon comparing Eq. (6.45) with Eq. (6.22), we see that $a_0(x) = x^2$, $a_1(x) = x$, $a_2(x) = -m^2$, $a_3(x) = x^2$, and $\lambda = \kappa^2$. Then using Eqs. (6.9) and (6.15), we find that

$$p(x) = \exp\left[\int \frac{a_1(x)}{a_0(x)}dx\right] = \exp\left[\int \frac{x}{x^2}dx\right] = \exp[\ln x] = x$$

$$q(x) = \frac{p(x)a_2(x)}{a_0(x)} = \frac{x}{x^2}\left(-m^2\right) = -\frac{m^2}{x} \tag{6.46}$$

$$r(x) = \frac{p(x)a_3(x)}{a_0(x)} = \frac{x \times x^2}{x^2} = x$$

Thus, the Sturm–Liouville form is

$$\frac{d}{dx}\left(x\frac{dy}{dx}\right) + \left(-\frac{m^2}{x} + \kappa^2 x\right)y = 0 \tag{6.47}$$

From Table 5.1, we see that the solutions to Eq. (6.45) are

$$y(x) = AJ_m(\kappa x) + BY_m(\kappa x) \tag{6.48}$$

which are oscillatory functions. When κ^2 is replaced by $-\kappa^2$, that is,

$$x^2\frac{d^2y}{dr^2}+x\frac{dy}{dr}-\left(\kappa^2x^2+m^2\right)y=0 \quad \kappa^2>0 \tag{6.49}$$

$r(x) = -x < 0$ and this equation is *not* of the Sturm–Liouville type. From Table 5.1, we see that the solutions to Eq. (6.49) are the modified Bessel functions, which are not oscillatory.

Spherical Bessel's Equation: The spherical Bessel equation that will be most useful for our purposes in Chapter 7 is the form given by Eq. (a) of Example 5.12, that is

$$x^2\frac{d^2y}{dx^2}+2x\frac{dy}{dx}-n(n+1)y+\kappa^2x^2y=0 \quad \kappa\neq 0 \tag{6.50}$$

Upon comparing Eq. (6.50) with Eq. (6.22), we see that $a_0(x) = x^2$, $a_1(x) = x$, $a_2(x) = -n(n+1)$, $a_3(x) = x^2$, and $\lambda = \kappa^2$. Then using Eqs. (6.9) and (6.15), we find that

$$\begin{aligned}
p(x)&=\exp\left[\int\frac{a_1(x)}{a_0(x)}dx\right]=\exp\left[\int\frac{2x}{x^2}dx\right]=\exp(2\ln x)=x^2\\
q(x)&=\frac{p(x)a_2(x)}{a_0(x)}=\frac{x^2}{x^2}\left(-n(n+1)\right)=-n(n+1)\\
r(x)&=\frac{p(x)a_3(x)}{a_0(x)}=\frac{x^2\times x^2}{x^2}=x^2
\end{aligned} \tag{6.51}$$

Thus, the Sturm–Liouville form is

$$\frac{d}{dx}\left(x^2\frac{dy}{dx}\right)+\left(-n(n+1)+\kappa^2x^2\right)y=0 \tag{6.52}$$

The solution to Eq. (6.50) is given by the spherical Bessel functions [recall Eqs. (5.117) and (5.119)] as

$$y(x)=Aj_n(\kappa x)+By_n(\kappa x) \tag{6.53}$$

Legendre's Equation: Consider the Legendre equation [recall Eq. (6.172)]

$$\left(1-x^2\right)\frac{d^2y}{dx^2}-2x\frac{dy}{dx}+n(n+1)y=0 \tag{6.54}$$

Upon comparing Eq. (6.54) with Eq. (6.22), we see that $a_0(x) = 1 - x^2$, $a_1(x) = -2x$, $a_2(x) = 0$, $a_3(x) = 1$, and $\lambda = n(n + 1)$. Then using Eqs. (6.9) and (6.15), we find that

$$\begin{aligned}
p(x)&=\exp\left[\int\frac{a_1(x)}{a_0(x)}dx\right]=\exp\left[\int\frac{-2x}{1-x^2}dx\right]=\exp\left[\ln\left(1-x^2\right)\right]=1-x^2\\
q(x)&=\frac{p(x)a_2(x)}{a_0(x)}=0\\
r(x)&=\frac{p(x)a_3(x)}{a_0(x)}=\frac{\left(1-x^2\right)\times 1}{1-x^2}=1
\end{aligned} \tag{6.55}$$

Thus, the Sturm–Liouville form is

$$\frac{d}{dx}\left(\left(1-x^2\right)\frac{dy}{dx}\right)+n(n+1)y=0 \tag{6.56}$$

The solution to Eq. (6.56) is [recall Eq. (5.195)]

$$y(x)=AP_n(x)+BQ_n(x) \tag{6.57}$$

Associated Legendre's Equation: Consider the associated Legendre equation [recall Eq. (5.200)]

$$\left(1-x^2\right)\frac{d^2y}{dx^2}-2x\frac{dy}{dx}+\left[n(n+1)-\frac{m^2}{1-x^2}\right]y=0 \tag{6.58}$$

The results for the Legendre equation given above apply, except that in this case

$$q(x)=\frac{p(x)a_2(x)}{a_0(x)}=\frac{1-x^2}{1-x^2}\left(\frac{-m^2}{1-x^2}\right)=\frac{-m^2}{1-x^2}$$

Thus, the Sturm–Liouville form is

$$\frac{d}{dx}\left(\left(1-x^2\right)\frac{dy}{dx}\right)+\left[-\frac{m^2}{1-x^2}+n(n+1)\right]y=0 \tag{6.59}$$

From Eq. (5.209), the solution to Eq. (6.59) is

$$y(x)=AP_n^m(x)+BQ_n^m(x) \tag{6.60}$$

Hypergeometric Equation: Consider the hypergeometric equation [recall Eq. (5.211)]

$$x(1-x)\frac{d^2y}{dx^2}+[c-(a+b+1)x]\frac{dy}{dx}-aby=0 \tag{6.61}$$

where a, b, and c are real constants. Upon comparing Eq. (6.61) with Eq. (6.22), we see that $a_0(x)=x(1-x)$, $a_1(x)=c-(a+b+1)x$, $a_2(x)=0$, $a_3(x)=1$, and $\lambda=ab$. Then using Eqs. (6.9) and (6.15), we find that

$$\begin{aligned}
p(x)&=\exp\left[\int\frac{c-(a+b+1)x}{x(1-x)}dx\right]=\exp\left(\ln\left(x^c(1-x)^\gamma\right)\right)=x^c(1-x)^\gamma\\
q(x)&=\frac{p(x)a_2(x)}{a_0(x)}=0\\
r(x)&=\frac{p(x)}{x(1-x)}=x^{c-1}(1-x)^{\gamma-1}
\end{aligned} \tag{6.62}$$

where

$$\gamma=a+b+1-c \tag{6.63}$$

Thus, the Sturm–Liouville form is

$$\frac{d}{dx}\left(x^{c}(1-x)^{\gamma}\frac{dy}{dx}\right)+abx^{c-1}(1-x)^{\gamma-1}y=0 \tag{6.64}$$

From Eq. (5.227), we see that the solutions to Eq. (6.64) are

$$y(x)={}_1F_2(a,b;c;x)C_1+x^{1-c}\,{}_1F_2(a+1-c,b+1-c;2-c;x)C_2 \tag{6.65}$$

The results of this section are summarized in Table 6.1.

6.1.4 Orthogonal Functions: Their Generation and Their Properties

An *orthogonal function* is defined as follows. If a sequence of real functions $\{\varphi_n(x)\}$, $n = 1, 2, \ldots, N$, has the property that over some interval

$$\int_a^b r(x)\varphi_n(x)\varphi_k(x)dx=\delta_{nk}\left\|\varphi_k(x)\right\|^2=\delta_{nk}N_n \tag{6.66}$$

where δ_{nk} is the Kronecker delta, then the functions are said to form an *orthogonal set* with respect to the weighting function $r(x)$ on that interval. The quantity $\sqrt{N_n}=\left\|\varphi_n\right\|$ is called the norm of the function $\varphi_n(x)$. Hence, if k and n are different from each other – that is, we have two different eigenfunctions – the integral given by Eq. (6.66) evaluates to zero. This property is analogous to the orthogonality property associated with vectors; that is, the scalar dot product of two identical vectors is the square of the vector's magnitude, while the scalar dot product of two orthogonal vectors is zero. It can also be thought of as an extension of the definition of the orthogonality of the eigenvectors of a symmetric matrix: $\{\boldsymbol{x}^{(n)}\}^T\{\boldsymbol{x}^{(k)}\} = 0\ n \neq k$. Recall Eq. (1.65). Lastly, this result is similar to the Fourier method given in Section 3.1, where it was seen that the Fourier series was composed of two orthogonal functions each being orthogonal on the interval T.

The solution given by λ_n and the corresponding $\varphi_n(x)$ for the Sturm–Liouville system has several important properties.

Property 1: The first property is that $\varphi_n(x)$ forms a set of orthogonal functions with respect to the weight function $r(x)$.

$\boldsymbol{\lambda_n \neq \lambda_m}$: To show that $\varphi_n(x)$ is an orthogonal function for the Sturm–Liouville system, we consider the two distinct eigenvalues λ_n and λ_m ($\lambda_n \neq \lambda_m$) and their corresponding eigenfunctions $\varphi_n(x)$ and $\varphi_m(x)$, which are each, respectively, a solution to [recall Eq. (6.39)]

$$\begin{aligned}L_{sa}[\varphi_n]&=0\\ L_{sa}[\varphi_m]&=0\end{aligned} \tag{6.67}$$

where

$$\begin{aligned}L_{sa}[\varphi_n]&=\frac{d}{dx}\left(p(x)\frac{d\varphi_n}{dx}\right)+q(x)\varphi_n+r(x)\lambda_n\varphi_n\\ L_{sa}[\varphi_m]&=\frac{d}{dx}\left(p(x)\frac{d\varphi_m}{dx}\right)+q(x)\varphi_m+r(x)\lambda_m\varphi_m\end{aligned} \tag{6.68}$$

TABLE 6.1
Several Sturm–Liouville Equations and Their General Solutions

$$\frac{d}{dx}\left(p(x)\frac{dy}{dx}\right)+q(x)y+r(x)\lambda y=0$$

Case	Type	Sturm–Liouville Form	$p(x)$	$q(x)$	$r(x)$	λ	$y(x)$
1	Constant coefficients	$\frac{d^2y}{dx^2}+\mu^2y=0$	1	0	1	μ^2	$C_1\cos\mu x+C_2\sin\mu x$
2	Bessel	$\frac{d}{dx}\left(x\frac{dy}{dx}\right)+\left(-\frac{m^2}{x}+\kappa^2x\right)y=0$	x	$-m^2/x$	x	κ^2	$C_1J_m(\kappa x)+C_2Y_m(\kappa x)$
3	Spherical Bessel	$\frac{d}{dx}\left(x^2\frac{dy}{dx}\right)$ $+\left(-n(n+1)+\kappa^2x^2\right)y=0$	x^2	$-n(n+1)$	x^2	κ^2	$C_1j_m(\kappa x)+C_2y_m(\kappa x)$
4	Legendre	$\frac{d}{dx}\left(\left(1-x^2\right)\frac{dy}{dx}\right)+n(n+1)y=0$	$1-x^2$	0	1	$n(n+1)$	$C_1P_n(x)+C_2Q_n(x)$
5	Associated Legendre	$\frac{d}{dx}\left(\left(1-x^2\right)\frac{dy}{dx}\right)$ $-\frac{m^2}{1-x^2}y+n(n+1)y=0$	$1-x^2$	$\frac{-m^2}{1-x^2}$	1	$n(n+1)$	$C_1P_n^m(x)+C_2Q_n^m(x)$
6	Hypergeometric	$\frac{d}{dx}\left(x^c(1-x)^\gamma\frac{dy}{dx}\right)$ $+abx^{c-1}(1-x)^{\gamma-1}y=0$ where $\gamma=a+b+1-c$	$x^c(1-x)^\gamma$	0	$x^{c-1}(1-x)^{\gamma-1}$	ab	${}_1F_2(a,b;c;x)C_1$ $+x^d\,{}_1F_2(a+d,b+d;2-c;x)C_2$ where $d=1-c$

The boundary conditions are given by Eq. (6.40), which can be written as

$$\begin{aligned}\alpha\varphi_k(a)+\beta\varphi_k'(a)=0\\ \delta\varphi_k(b)+\gamma\varphi_k'(b)=0 \quad k=m \text{ or } n\end{aligned} \tag{6.69}$$

Multiplying the first equation of Eq. (6.67) by $\varphi_m(x)$ and the second equation of Eq. (6.67) by $\varphi_n(x)$, we obtain

$$\begin{aligned}\varphi_m L_{sa}[\varphi_n]=0\\ \varphi_n L_{sa}[\varphi_m]=0\end{aligned} \tag{6.70}$$

Upon subtracting the equations in Eq. (6.70), we arrive at

$$\varphi_m L_{sa}[\varphi_n]-\varphi_n L_{sa}[\varphi_m]=0 \tag{6.71}$$

We now integrate Eq. (6.71) over its interval and use Eq. (6.21) with $u = \varphi_m$ and $v = \varphi_n$ to obtain

$$\begin{aligned}\int_a^b \left(\varphi_m L_{sa}[\varphi_n]-\varphi_n L_{sa}[\varphi_m]\right)dx &= \left[p(x)\left(\varphi_m\varphi_n'-\varphi_n\varphi_m'\right)\right]_a^b \\ &\quad +\left(\lambda_n-\lambda_m\right)\int_a^b r(x)\varphi_n\varphi_m dx=0\end{aligned} \tag{6.72}$$

where the prime (′) denotes the derivative with respect to x.

Expanding the first term on the right-hand side of Eq. (6.72) gives

$$\begin{aligned}\left[p(x)\left(\varphi_m\varphi_n'-\varphi_n\varphi_m'\right)\right]_a^b &= p(b)\left(\varphi_m(b)\varphi_n'(b)-\varphi_n(b)\varphi_m'(b)\right)\\ &\quad -p(a)\left(\varphi_m(a)\varphi_n'(a)-\varphi_n(a)\varphi_m'(a)\right)\end{aligned} \tag{6.73}$$

Using the boundary conditions given by Eq. (6.69), we find that

$$\begin{aligned}\left[p(x)\left(\varphi_m\varphi_n'-\varphi_n\varphi_m'\right)\right]_a^b &= p(b)\left(-\varphi_m(b)\varphi_n(b)\delta/\gamma+\varphi_n(b)\varphi_m(b)\delta/\gamma\right)\\ &\quad -p(a)\left(-\varphi_m(a)\varphi_n(a)\alpha/\beta+\varphi_n(a)\varphi_m(a)\alpha/\beta\right)\\ &=0\end{aligned} \tag{6.74}$$

Then Eq. (6.72) becomes

$$\int_a^b \left(\varphi_m L_{sa}[\varphi_n]-\varphi_n L_{sa}[\varphi_m]\right)dx=\left(\lambda_n-\lambda_m\right)\int_a^b r(x)\varphi_n\varphi_m dx=0 \tag{6.75}$$

However, since $\lambda_n \neq \lambda_m$, Eq. (6.75) gives

$$\int_a^b r(x)\varphi_m\varphi_n dx = 0 \quad m \neq n \tag{6.76}$$

There are two special cases of Eq. (6.73) that we shall examine.

Special Case I – Periodic Boundary Conditions: A special case of Eq. (6.73) is when $p(a) = p(b)$. In this case, Eq. (6.73) yields

$$\begin{aligned}\left[p(x)\left(\varphi_m\varphi_n' - \varphi_n\varphi_m'\right)\right]_a^b = p(b)\big[&\varphi_m(b)\varphi_n'(b) - \varphi_n(b)\varphi_m'(b)\\ &- \varphi_m(a)\varphi_n'(a) + \varphi_n(a)\varphi_m'(a)\big]\end{aligned} \tag{6.77}$$

Equation (6.77) equals zero when

$$\begin{aligned}\varphi_k(a) &= \varphi_k(b)\\ \varphi_k'(a) &= \varphi_k'(b) \quad k = n,m\end{aligned} \tag{6.78}$$

These conditions are called *periodic boundary conditions*. They too generate orthogonal eigenfunctions and sometimes appear in boundary-value problems in polar cylindrical coordinates.

For example, consider the solution to the Sturm–Liouville equation that is given by Eq. (6.43), that is,

$$y(x) = A\cos(\mu x) + B\sin(\mu x) \tag{6.79}$$

where the boundary conditions are specified at $x = 0$ and $x = 2\pi$. For this system, $p(x) = 1$; therefore, $p(0) = p(2\pi) = 1$. To satisfy the first boundary condition of Eq. (6.78), we have to satisfy $y(x) = y(x+2\pi)$, that is,

$$\begin{aligned}y(x) &= y(x+2\pi)\\ A\cos\mu x + B\sin\mu x &= A\cos(\mu[x+2\pi]) + B\sin(\mu[x+2\pi])\\ &= A\left(\cos\mu x\cos 2\mu\pi - \sin\mu x\sin 2\mu\pi\right)\\ &\quad + B\left(\sin\mu x\cos 2\mu\pi + \cos\mu x\sin 2\mu\pi\right)\end{aligned} \tag{6.80}$$

It is seen from Eq. (6.80) that the equality is only true when $\mu = m$, an integer, since in this case $\sin(2m\pi) = 0$ and $\cos(2m\pi) = 1$ and the right-hand side of Eq. (6.80) equals the left-hand side. In a similar manner, it is straightforward to show that when m is an integer $y'(x) = y'(x+2\pi)$, where the prime denotes the derivative with respect to x. Thus, when m is an integer and the boundary conditions are specified at $x = 0$ and $x = 2\pi$, the definition of periodic boundary conditions is satisfied. We shall use this property frequently in Chapter 7.

Special Case II – Boundness: Another special case arises when a boundary condition is stipulated such that the solution must remain bounded (finite) at $x = a$. In this case, $\varphi_n(a) = c_o$ = constant $\neq 0$. Then, Eq. (6.77) becomes

$$\left[p(x)\left(\varphi_m\varphi_n' - \varphi_n\varphi_m'\right)\right]_a^b = p(b)\left(\varphi_m(b)\varphi_n'(b) - \varphi_n(b)\varphi_m'(b)\right) \\ - p(a)c_o\left(\varphi_n'(a) - \varphi_m'(a)\right) \tag{6.81}$$

Thus, Eq. (6.81) can still be satisfied provided that either $\varphi_k'(a) = 0$, $k = 1, 2, \ldots$, or $p(a) = 0$. The boundary condition at $x = b$, which is given by the second equation of Eq. (6.69), remains the same.

For example, consider the solution to the Sturm–Liouville equation given by Eq. (6.45), which is given by Eq. (6.48), that is,

$$y(x) = AJ_m(\kappa x) + BY_m(\kappa x) \tag{6.82}$$

We know that when $x = a = 0$, $Y_m(0) \to \infty$; therefore, if we require the solution to be bounded at $x = 0$ we must set $B = 0$ so that the solution remains finite at this point. The boundary conditions at $x = b$, which are given by the second equation of Eq. (6.69), make the expression multiplying $p(b)$ equal to zero. In addition, for this system $p(x) = x$ and, consequently, $p(0) = 0$; therefore, the right-hand side of Eq. (6.81) equals zero and the functions are orthogonal.

For another example, consider the case when a boundary condition is stipulated such that it must remain bounded (finite) at $x = b$. In this case, $\varphi_n(b) = c_o = \text{constant} \neq 0$. Then, Eq. (6.73) becomes

$$\left[p(x)\left(\varphi_m\varphi_n' - \varphi_n\varphi_m'\right)\right]_a^b = p(b)c_o\left(\varphi_n'(b) - \varphi_m'(b)\right) \\ - p(a)\left(\varphi_m(a)\varphi_n'(a) - \varphi_n(a)\varphi_m'(a)\right) \tag{6.83}$$

Thus, Eq. (6.83) can still be satisfied provided that either $\varphi_k'(b) = 0$, $k = 1, 2, \ldots$, or $p(b) = 0$. The boundary condition at $x = a$, which is given by the first equation of Eq. (6.69), remains the same.

To illustrate Eq. (6.83), consider the solution to the Sturm–Liouville equation given by Eq. (6.54), which is given by Eq. (6.57), that is,

$$y(x) = AP_n(x) + BQ_n(x) \tag{6.84}$$

We know that when $x = a = 1$, $Q_n(1) \to \infty$; therefore, if we require the solution to be bounded at $x = 1$ we must set $B = 0$ so that the solution remains finite at this point. The boundary conditions at $x = a$, which are given by the first equation of Eq. (6.66), make the expression multiplying $p(a)$ equal to zero. In addition, for this system, $p(x) = 1 - x^2$ and, consequently, $p(1) = 0$; therefore, the right-hand side of Eq. (6.83) equals zero and the functions are orthogonal.

$\lambda_n = \lambda_m$: To determine what happens when $\lambda_n = \lambda_m$, we start with Eq. (6.20) and change its notation so that $u(x) = \varphi_n(\lambda_n x)$ and $v(x) = \varphi_m(\lambda_m x)$. In addition, as a practical notational matter, the parameter λ will usually appear in the form $\lambda = \alpha^2$. Therefore, in anticipation of this usage, we shall temporarily change the notation to Eq. (6.20) by replacing λ_n with λ_n^2. Then Eq. (6.20) becomes

$$\frac{d}{dx}\left[p(x)\left(\varphi_m\frac{d\varphi_n}{dx}-\varphi_n\frac{d\varphi_m}{dx}\right)\right]=\left(\lambda_m^2-\lambda_n^2\right)r(x)\varphi_m\varphi_n \tag{6.85}$$

When $\lambda_n = \lambda_m$ and $\varphi_n = \varphi_m$, both sides of Eq. (6.85) equal zero. Therefore, we use L'Hôpital's rule and differentiate both sides of Eq. (6.85) with respect to λ_n and arrive at

$$\frac{d}{dx}\left[p(x)\frac{d}{d\lambda_n}\left(\varphi_m\frac{\lambda_n d\varphi_n}{d(\lambda_n x)}-\varphi_n\frac{\lambda_m d\varphi_m}{d(\lambda_m x)}\right)\right]=\frac{d}{d\lambda_n}\left(\left(\lambda_m^2-\lambda_n^2\right)r(x)\varphi_m\varphi_n\right)$$

$$\frac{d}{dx}\left[p(x)\left(\varphi_m\varphi_n'+\lambda_n x\varphi_m\varphi_n''-\lambda_m x\varphi_n'\varphi_m'\right)\right]=-2\lambda_n r(x)\varphi_m\varphi_n \tag{6.86}$$

$$+\left(\lambda_m^2-\lambda_n^2\right)r(x)x\varphi_m\varphi_n'$$

where the prime denotes the derivative with respect to its argument, either $\lambda_n x$ or $\lambda_m x$. Taking the limit as $\lambda_n \to \lambda_m$, we obtain

$$\frac{d}{dx}\left[p(x)\left(\varphi_n\varphi_n'+\lambda_n x\varphi_n\varphi_n''-\lambda_n x\varphi_n'^2\right)\right]=-2\lambda_n r(x)\varphi_n^2 \tag{6.87}$$

To obtain our final result, we integrate Eq. (6.87) over the interval and obtain

$$N_n=\left\|\varphi_n(x)\right\|^2=\int_a^b r(x)\varphi_n^2dx=\frac{1}{2\lambda_n}\left[p(x)\left(\lambda_n x\varphi_n'^2-\varphi_n\left(\varphi_n'+\lambda_n x\varphi_n''\right)\right)\right]_a^b \tag{6.88}$$

The right-hand side of Eq. (6.88) is evaluated after the boundary conditions have been specified. Equation (6.88) will prove useful when φ_n is a Bessel function. (See Examples 7.5, 7.11, and 7.18.)

Then, from Eqs. (6.76) and (6.88), we have the following orthogonality condition

$$\int_a^b r(x)\varphi_n(x)\varphi_m(x)dx=\delta_{nm}N_n \tag{6.89}$$

Therefore, eigenfunctions of a Sturm–Liouville equation satisfying a specific set of homogeneous boundary conditions are orthogonal.

Sometimes the orthogonal eigenfunctions are made *orthonormal* by setting

$$\hat{\varphi}_n=\frac{\varphi_n}{\sqrt{N_n}}$$

so that the orthogonality condition is

$$\int_a^b r(x)\hat{\varphi}_m\hat{\varphi}_n dx=\delta_{nm}$$

An orthogonal set of functions $\{\varphi_n(x)\}$, $n = 1, 2, \ldots, N$, forms a set of *linearly independent functions*. We say that the set of N functions $\{\varphi_n(x)\}$ is linearly independent if the only way to satisfy the homogeneous relation

$$\sum_{k=1}^{N} c_k \varphi_k(x) = 0 \tag{6.90}$$

for all x in the interval $a \leq x \leq b$ is for all the constant coefficients c_k to equal zero. Since $\varphi_n(x)$ is orthogonal with respect to the weight function $r(x)$, we can multiply Eq. (6.90) by $r(x)\varphi_n(x)$ and integrate over the interval to obtain

$$\sum_{k=1}^{N} c_k \int_a^b r(x)\varphi_n(x)\varphi_k(x)dx = 0$$

$$\sum_{k=1}^{N} c_k N_n \delta_{nk} = 0$$

$$c_n N_n = 0$$

where we have used Eq. (6.89). Therefore, since $N_n > 0$, $c_n = 0$. Hence, there is only one linearly independent eigenfunction corresponding to each eigenvalue.

Property 2: The second property is that the eigenvalues λ_n of a self-adjoint system are real. To show this, we assume that we have two different eigenvalues: $\lambda = a + jb$ and $\bar{\lambda}$, which is the complex conjugate of λ [recall Eq. (2.2)]. Corresponding to the eigenvalues are the eigenfunctions $\varphi(x)$ and their complex conjugates $\bar{\varphi}(x)$. Then, from Eq. (6.75),

$$\left(\lambda - \bar{\lambda}\right) \int_a^b r(x)\varphi\bar{\varphi}dx = 0$$

$$\left(a + jb - a + jb\right) \int_a^b r(x)\left|\varphi\right|^2 dx = 0$$

$$2jb \int_a^b r(x)\left|\varphi\right|^2 dx = 0$$

Since the integral does not equal to zero, $b = 0$. Therefore, the eigenvalues are real.

Property 3: All eigenvalues are positive.

6.1.5 Fourth-Order Sturm–Liouville Differential Equation

Consider the following fourth-order differential equation

$$L_{4sa}[y] = 0 \tag{6.91}$$

where

$$L_{4sa} = \frac{d^2}{dx^2}\left(S(x)\frac{d^2}{dx^2}\right) + \frac{d}{dx}\left(P(x)\frac{d}{dx}\right) + Q_2(x) - R(x)\lambda_y \tag{6.92}$$

and $R(x) > 0$. It has been shown that Eq. (6.92) is self-adjoint* and is considered a generalization of the second-order Sturm–Liouville equation to the fourth order. It will be seen that with an appropriate set of homogeneous boundary conditions it has all the characteristics associated with the second-order Sturm–Liouville equation.

As was the case with Eq. (6.16), for the fourth-order equation we shall have use for the evaluation of the following integral

$$I_4 = \int_a^b \left(uL_{4sa}[v] - vL_{4sa}[u]\right)dx \tag{6.93}$$

where

$$\begin{aligned} L_{4sa}[u] &= \frac{d^2}{dx^2}\left(S(x)\frac{d^2u}{dx^2}\right) + \frac{d}{dx}\left(P(x)\frac{du}{dx}\right) + Q_2(x)u - R(x)\lambda_u u \\ L_{4sa}[v] &= \frac{d^2}{dx^2}\left(S(x)\frac{d^2v}{dx^2}\right) + \frac{d}{dx}\left(P(x)\frac{dv}{dx}\right) + Q_2(x)v - R(x)\lambda_v v \end{aligned} \tag{6.94}$$

Then, the integrand of Eq. (6.93) becomes

$$\begin{aligned} uL_{4sa}[v] - vL_{4sa}[u] &= u\frac{d^2}{dx^2}\left(S(x)\frac{d^2v}{dx^2}\right) + u\frac{d}{dx}\left(P(x)\frac{dv}{dx}\right) \\ &\quad + uQ_2(x)v - uR(x)\lambda_v v - vQ_2(x)u + vR(x)\lambda_u u \\ &\quad - v\frac{d^2}{dx^2}\left(S(x)\frac{d^2u}{dx^2}\right) - v\frac{d}{dx}\left(P(x)\frac{du}{dx}\right) \\ &= u\frac{d^2}{dx^2}\left(S(x)\frac{d^2v}{dx^2}\right) - v\frac{d^2}{dx^2}\left(S(x)\frac{d^2u}{dx^2}\right) \\ &\quad + \frac{d}{dx}\left(P(x)\left(u\frac{dv}{dx} - v\frac{du}{dx}\right)\right) + \left(\lambda_u - \lambda_v\right)R(x)uv \end{aligned} \tag{6.95}$$

where we have used Eq. (6.19). Substituting Eq. (6.95) into Eq. (6.93) yields

$$\begin{aligned} I_4 &= \int_a^b \left[u\frac{d^2}{dx^2}\left(S(x)\frac{d^2v}{dx^2}\right) - v\frac{d^2}{dx^2}\left(S(x)\frac{d^2u}{dx^2}\right)\right]dx \\ &\quad + \left[P(x)\left(u\frac{dv}{dx} - v\frac{du}{dx}\right)\right]_a^b + \left(\lambda_u - \lambda_v\right)\int_a^b R(x)v(x)u(x)dx \end{aligned} \tag{6.96}$$

* Whyburn WM (1930) On Self-Adjoint Ordinary Differential Equations of the Fourth Order. *American Journal of Mathematics* 52(1): 171–196; Keener MS (1971) On Solutions of Certain Self-Adjoint Differential Equations of Fourth Order. *Journal of Mathematical Analysis and Applications* 33: 278–305; and Leighton W, Neharu Z (1958) On the Oscillation of Solutions of Self-Adjoint Linear Differential Equations of the Fourth Order. *Transactions of the American Mathematical Society* 89: 325–377.

We shall use integration by parts to evaluate the first integral on the right-hand side of Eq. (6.96). Introducing the definitions

$$g(v) = S(x)\frac{d^2v}{dx^2}$$
$$h(v) = \frac{dg}{dx} = \frac{d}{dx}\left(S(x)\frac{d^2v}{dx^2}\right) \tag{6.97}$$

the first term of the first integral in Eq. (6.96) becomes

$$\int_a^b u\frac{d^2}{dx^2}\left(S(x)\frac{d^2v}{dx^2}\right)dx = \int_a^b u\frac{dh}{dx}dx = uh\Big|_a^b - \int_a^b h\frac{du}{dx}dx = uh\Big|_a^b - \int_a^b \frac{du}{dx}\frac{dg}{dx}dx$$
$$= uh\Big|_a^b - \left[\frac{du}{dx}g\Big|_a^b - \int_a^b g\frac{d^2u}{dx^2}dx\right] \tag{6.98}$$
$$= u\frac{d}{dx}\left(S(x)\frac{d^2v}{dx^2}\right)\Bigg|_a^b - S(x)\frac{du}{dx}\frac{d^2v}{dx^2}\Bigg|_a^b + \int_a^b S(x)\frac{d^2v}{dx^2}\frac{d^2u}{dx^2}dx$$

The second term in the first integral in Eq. (6.96) is obtained by interchanging u and v in Eq. (6.98). Thus,

$$\int_a^b v\frac{d^2}{dx^2}\left(S(x)\frac{d^2u}{dx^2}\right)dx = v\frac{d}{dx}\left(S(x)\frac{d^2u}{dx^2}\right)\Bigg|_a^b - S(x)\frac{dv}{dx}\frac{d^2u}{dx^2}\Bigg|_a^b$$
$$+ \int_a^b S(x)\frac{d^2u}{dx^2}\frac{d^2v}{dx^2}dx \tag{6.99}$$

Substituting Eqs. (6.98) and (6.99) into Eq. (6.96) yields

$$\int_a^b \left(uL_{4sa}[v] - vL_{4sa}[u]\right)dx = \left[u(x)\Gamma\left(v(x)\right)\right]_a^b - \left[v(x)\Gamma\left(u(x)\right)\right]_a^b$$
$$-\left[u'(x)\Lambda\left(v(x)\right)\right]_a^b + \left[v'(x)\Lambda\left(u(x)\right)\right]_a^b \tag{6.100}$$
$$+\left(\lambda_u - \lambda_v\right)\int_a^b R(x)v(x)u(x)dx$$

where the prime (′) denotes the derivative with respect to x and

$$\Gamma(\sigma(x)) = h(\sigma) + P(x)\frac{d\sigma}{dx}$$
$$\Lambda(\sigma(x)) = g(\sigma) \tag{6.101}$$

and $h(\sigma)$ and $g(\sigma)$ are given by Eq. (6.97). We shall use this result subsequently when we determine the conditions under which the eigenfunctions are orthogonal functions.

We now consider the fourth-order differential equation given by Eq. (6.92), that is,

$$\frac{d^2}{dx^2}\left(S(x)\frac{d^2y}{dx^2}\right)+\frac{d}{dx}\left(P(x)\frac{dy}{dx}\right)+Q_2(x)y-R(x)\lambda y=0 \tag{6.102}$$

when it is subject to the homogeneous boundary conditions of the general form

$$\begin{aligned}&\Gamma(y(a))+\alpha_a y(a)=0 \quad \text{and} \quad \Lambda(y(a))+\beta_a y'(a)=0\\ &\Gamma(y(b))+\alpha_b y(b)=0 \quad \text{and} \quad \Lambda(y(b))+\beta_b y'(b)=0\end{aligned} \tag{6.103}$$

where Γ and Λ are given by Eq. (6.101) and α_k and β_k, $k = a, b$, are constants that take on any value from zero to infinity. For example, consider the first boundary condition at $x = a$. When $\alpha_a = 0$, the boundary condition becomes $\Gamma(y(a)) = 0$. On the other hand, when $\alpha_a \to \infty$, we first divide by α_a and then take the limit as $\alpha_a \to \infty$. The result is the boundary condition $y(a) = 0$.

It is seen that Eqs. (6.102) and (6.103) form a boundary-value problem composed of a homogeneous differential equation with homogeneous boundary conditions and an undefined parameter λ. The unknown parameter λ is selected so that the solution to Eqs. (6.102) and (6.103) is not trivial, that is, $y(x) \neq 0$. These values of λ are denoted λ_n and are the eigenvalues; it is assumed that there is an infinite number of them. Corresponding to these eigenvalues are the eigenfunctions $\phi_n(x)$. The determination of λ_n and $\phi_n(x)$ is now discussed.

The solution to Eq. (6.102) is assumed to be composed of four linearly independent solutions $h_j(x,\lambda)$, that is,

$$y(x)=\sum_{j=1}^{4}C_j h_j(x,\lambda) \tag{6.104}$$

where the dependency on the free parameter λ is explicitly denoted. Substituting Eq. (6.104) into the boundary conditions given by Eq. (6.103) yields

$$\begin{pmatrix} d_{11}(\lambda) & d_{12}(\lambda) & d_{13}(\lambda) & d_{12}(\lambda)\\ d_{21}(\lambda) & d_{22}(\lambda) & d_{23}(\lambda) & d_{24}(\lambda)\\ d_{31}(\lambda) & d_{32}(\lambda) & d_{33}(\lambda) & d_{34}(\lambda)\\ d_{41}(\lambda) & d_{42}(\lambda) & d_{43}(\lambda) & d_{44}(\lambda)\end{pmatrix}\begin{Bmatrix} C_1\\ C_2\\ C_3\\ C_4\end{Bmatrix}=0 \tag{6.105}$$

where

$$\begin{aligned}d_{1j}(\lambda)&=\Gamma(h_j(a,\lambda))+\alpha_a h_j(a,\lambda)\\ d_{2j}(\lambda)&=\Lambda(h_j(a,\lambda))+\beta_a h_j'(a,\lambda)\\ d_{3j}(\lambda)&=\Gamma(h_j(b,\lambda))+\alpha_b h_j(b,\lambda)\\ d_{4j}(\lambda)&=\Lambda(h_j(b,\lambda))+\beta_b h_j'(b,\lambda)\end{aligned} \tag{6.106}$$

We have a non-trivial solution (recall Section 1.7) only when $\lambda = \lambda_n$, where λ_n are the solutions to

$$\det\begin{pmatrix} d_{11}(\lambda_n) & d_{12}(\lambda_n) & d_{13}(\lambda_n) & d_{12}(\lambda_n) \\ d_{21}(\lambda_n) & d_{22}(\lambda_n) & d_{23}(\lambda_n) & d_{24}(\lambda_n) \\ d_{31}(\lambda_n) & d_{32}(\lambda_n) & d_{33}(\lambda_n) & d_{34}(\lambda_n) \\ d_{41}(\lambda_n) & d_{42}(\lambda_n) & d_{43}(\lambda_n) & d_{44}(\lambda_n) \end{pmatrix} = 0 \tag{6.107}$$

We shall obtain the corresponding eigenfunctions in terms of the ratios $C_2/C_1 \to C_{2n}/C_{1n}$, $C_3/C_1 \to C_{3n}/C_{1n}$, and $C_4/C_1 \to C_{4n}/C_{1n}$. These ratios are determined from the second to fourth equations of Eq. (6.105) as follows

$$\begin{pmatrix} d_{22}(\lambda_n) & d_{23}(\lambda_n) & d_{24}(\lambda_n) \\ d_{32}(\lambda_n) & d_{33}(\lambda_n) & d_{34}(\lambda_n) \\ d_{42}(\lambda_n) & d_{43}(\lambda_n) & d_{44}(\lambda_n) \end{pmatrix} \begin{Bmatrix} C_{2n}/C_{1n} \\ C_{3n}/C_{1n} \\ C_{4n}/C_{1n} \end{Bmatrix} = \begin{Bmatrix} -d_{21}(\lambda_n) \\ -d_{31}(\lambda_n) \\ -d_{41}(\lambda_n) \end{Bmatrix} \tag{6.108}$$

Then, from Eq. (6.104), the solution to Eqs. (6.102) and (6.103) is

$$\begin{aligned} y(x) &= C_{1n}\phi_n(x) \\ &= C_{1n}\left(h_1(x,\lambda_n) + \frac{C_{2n}}{C_{1n}}h_2(x,\lambda_n) + \frac{C_{3n}}{C_{1n}}h_3(x,\lambda_n) + \frac{C_{4n}}{C_{1n}}h_4(x,\lambda_n)\right) \end{aligned} \tag{6.109}$$

where C_{jn}/C_{1n}, $j = 2, 3, 4$ are obtained from Eq. (6.108).

Since $\phi_n(x)$ and the corresponding λ_n are solutions to Eqs. (6.102) and (6.103) for all n, we use superposition to obtain the solution as

$$y(x) = \sum_{n=1}^{\infty} c_n\phi_n(x) \tag{6.110}$$

where we have set $c_n = C_{1n}$ and $\phi_n(x)$ is a solution to

$$\frac{d^2}{dx^2}\left(S(x)\frac{d^2\phi_n}{dx^2}\right) + \frac{d}{dx}\left(P(x)\frac{d\phi_n}{dx}\right) + Q_2(x)\phi_n - R(x)\lambda_n\phi_n = 0 \tag{6.111}$$

and the boundary conditions given by Eq. (6.103), which become

$$\begin{aligned} &\Gamma(\phi_n(a)) + \alpha_a\phi_n(a) = 0 \quad \text{and} \quad \Lambda(\phi_n(a)) + \beta_a\phi_n'(a) = 0 \\ &\Gamma(\phi_n(b)) + \alpha_b\phi_n(b) = 0 \quad \text{and} \quad \Lambda(\phi_n(b)) + \beta_b\phi_n'(b) = 0 \end{aligned} \tag{6.112}$$

Orthogonality: To show that $\phi_n(x)$ is an orthogonal function for the fourth-order equation given by Eq. (6.111) and the boundary conditions given by Eq. (6.112), we start by considering two distinct eigenvalues λ_n and λ_m ($\lambda_n \neq \lambda_m$) and their corresponding eigenfunctions $\phi_n(x)$ and $\phi_m(x)$, which are each, respectively, a solution to

$$\begin{aligned} L_{4sa}[\phi_n] &= 0 \\ L_{4sa}[\phi_m] &= 0 \end{aligned} \tag{6.113}$$

where

$$L_{4sa}[\phi_n] = \frac{d^2}{dx^2}\left(S(x)\frac{d^2\phi_n}{dx^2}\right) + \frac{d}{dx}\left(P(x)\frac{d\phi_n}{dx}\right) + Q_2(x)\phi_n - R(x)\lambda_n\phi_n$$
$$L_{4sa}[\phi_m] = \frac{d^2}{dx^2}\left(S(x)\frac{d^2\phi_m}{dx^2}\right) + \frac{d}{dx}\left(P(x)\frac{d\phi_m}{dx}\right) + Q_2(x)\phi_m - R(x)\lambda_m\phi_m \quad (6.114)$$

Multiplying the first equation of Eq. (6.113) by $\phi_m(x)$ and the second equation of Eq. (6.113) by $\phi_n(x)$, subtracting the result, and integrating over the interval gives

$$I_4 = \int_a^b \left(\phi_m L_{4sa}[\phi_n] - \phi_n L_{4sa}[\phi_m]\right)dx = 0 \quad (6.115)$$

However, from Eq. (6.100), the evaluation of the integral of Eq. (6.115) is

$$I_4 = \left(\lambda_n - \lambda_m\right)\int_a^b R(x)\phi_n(x)\phi_m(x)dx + b_{nm} = 0 \quad (6.116)$$

where

$$\begin{aligned} b_{nm} &= \left[\phi_m(x)\Gamma\left(\phi_n(x)\right)\right]_a^b - \left[\phi_n(x)\Gamma\left(\phi_m(x)\right)\right]_a^b \\ &\quad - \left[\phi_m'(x)\Lambda\left(\phi_n(x)\right)\right]_a^b + \left[\phi_n'(x)\Lambda\left(\phi_m(x)\right)\right]_a^b \\ &= \phi_m(b)\Gamma\left(\phi_n(b)\right) - \phi_m(a)\Gamma\left(\phi_n(a)\right) - \phi_n(b)\Gamma\left(\phi_m(b)\right) \\ &\quad + \phi_n(a)\Gamma\left(\phi_m(a)\right) - \phi_m'(b)\Lambda\left(\phi_n(b)\right) + \phi_m'(a)\Lambda\left(\phi_n(a)\right) \\ &\quad + \phi_n'(b)\Lambda\left(\phi_m(b)\right) - \phi_n'(a)\Lambda\left(\phi_m(a)\right) \end{aligned} \quad (6.117)$$

and Γ and Λ are given by Eq. (6.101). Upon substituting the boundary conditions given by Eq. (6.112) into the right-hand side of Eq. (6.117), we find that

$$\begin{aligned} b_{nm} &= -\alpha_b\phi_m(b)\phi_n(b) + \alpha_a\phi_m(a)\phi_n(a) + \alpha_b\phi_n(b)\phi_m(b) - \alpha_a\phi_n(a)\phi_m(a) \\ &\quad + \beta_b\phi_m'(b)\phi_n'(b) - \beta_a\phi_m'(a)\phi_n'(a) - \beta_b\phi_n'(b)\phi_m'(b) + \beta_a\phi_n'(a)\phi_m'(a) = 0 \end{aligned} \quad (6.118)$$

Since $\lambda_n \neq \lambda_m$, Eq. (6.116) becomes

$$\int_a^b R(x)\phi_n(x)\phi_m(x)dx = 0 \quad n \neq m \quad (6.119)$$

Then, we can state the orthogonality condition as

$$\int_a^b R(x)\phi_m(x)\phi_n(x)dx = N_{4n}\delta_{nm} \tag{6.120}$$

where

$$N_{4n} = \|\phi_n(x)\|^2 = \int_a^b R(x)\phi_n^2(x)dx \tag{6.121}$$

Since $R(x) > 0$, $N_{4n} > 0$.

6.1.6 General Solution to Nonhomogeneous Sturm–Liouville Equations

Second-Order Sturm–Liouville Equations: We shall obtain a solution to the second-order nonhomogeneous differential equation

$$\frac{d}{dx}\left(p(x)\frac{dy}{dx}\right) + q(x)y + r(x)\lambda y = f(x) \tag{6.122}$$

with the boundary conditions given by Eq. (6.24), that is,

$$\begin{aligned} \alpha y(a) + \beta y'(a) = 0 \qquad \alpha^2 + \beta^2 > 0 \\ \delta y(b) + \gamma y'(b) = 0 \qquad \delta^2 + \gamma^2 > 0 \end{aligned} \tag{6.123}$$

We have shown that $y_h(x) = 0$ for arbitrary λ, but when $\lambda = \lambda_n$ the solution is

$$y_h(x) = \sum_{n=1}^{\infty} c_n\varphi_n(x) \tag{6.124}$$

where, from Eq. (6.39), $\varphi_n(x)$ is a solution to

$$\frac{d}{dx}\left(p(x)\frac{d\varphi_n}{dx}\right) + q(x)\varphi_n + r(x)\lambda_n\varphi_n = 0 \tag{6.125}$$

and the boundary conditions given by Eq. (6.40), which are

$$\begin{aligned} \alpha\varphi_n(a) + \beta\varphi_n'(a) = 0 \\ \delta\varphi_n(b) + \gamma\varphi_n'(b) = 0 \end{aligned} \tag{6.126}$$

Since $\varphi_n(x)$ is a solution to the Sturm–Liouville equation and the homogeneous boundary conditions of the appropriate form, that is, Eq. (6.126), $\varphi_n(x)$ is an orthogonal eigenfunction. We now use this fact as follows.

To obtain the particular solution, it seems reasonable to assume a solution of the form given by Eq. (6.124), that is

$$y(x) = \sum_{n=1}^{\infty} c_n\varphi_n(x) \tag{6.127}$$

Since $\varphi_n(x)$ satisfies the boundary conditions, all that remains to satisfy is Eq. (6.122).

Upon substituting Eq. (6.127) into Eq. (6.122), we obtain

$$\sum_{n=1}^{\infty} c_n\left(\frac{d}{dx}\left(p(x)\frac{d\varphi_n}{dx}\right)+q(x)\varphi_n+r(x)\lambda\varphi_n\right)=f(x) \tag{6.128}$$

However, from Eq. (6.125),

$$\frac{d}{dx}\left(p(x)\frac{d\varphi_n}{dx}\right)+q(x)\varphi_n=-r(x)\lambda_n\varphi_n \tag{6.129}$$

Then, substituting Eq. (6.129) into Eq. (6.128), Eq. (6.128) becomes

$$\sum_{n=1}^{\infty} c_n\left(\lambda-\lambda_n\right)r(x)\varphi_n(x)=f(x) \tag{6.130}$$

To determine c_n, we multiply Eq. (6.130) by $\varphi_m(x)$ and integrate over the interval to arrive at

$$\sum_{n=1}^{\infty} c_n\left(\lambda-\lambda_n\right)\int_a^b r(x)\varphi_n(x)\varphi_m(x)dx=\int_a^b f(x)\varphi_m(x)dx \tag{6.131}$$

However, from Eq. (6.89)

$$\int_a^b r(x)\varphi_n(x)\varphi_m(x)dx=\delta_{nm}N_n \tag{6.132}$$

where

$$N_n=\int_a^b r(x)\varphi_n^2(x)dx \tag{6.133}$$

Then, using Eq. (6.132) in Eq. (6.131) yields

$$\begin{aligned} c_n\left(\lambda-\lambda_n\right)N_n &= \int_a^b f(x)\varphi_n(x)dx \\ c_n &= \frac{1}{N_n\left(\lambda-\lambda_n\right)}\int_a^b f(x)\varphi_n(x)dx \end{aligned} \tag{6.134}$$

provided that $\lambda \neq \lambda_n$. Therefore, the solution to Eq. (6.122) and the boundary conditions given by Eq. (6.123) is

$$y(x)=\sum_{n=1}^{\infty}\frac{\varphi_n(x)}{N_n\left(\lambda-\lambda_n\right)}\int_a^b f(x)\varphi_n(x)dx \tag{6.135}$$

It is noted that in engineering and physics problems, the nonhomogeneous differential equation given by Eq. (6.122) can appear in different ways as will be seen in Chapter 7. For the Legendre equation as given by Eq. (6.54), Eq. (6.122) becomes

$$\frac{d}{dx}\left(\left(1-x^2\right)\frac{dy}{dx}\right)+n(n+1)y=f(x) \tag{6.136}$$

which often appears in the form

$$\left(1-x^2\right)\frac{dy}{dx}-2x\frac{dy}{dx}+n(n+1)y=F(x) \tag{6.137}$$

Upon expanding Eq. (6.136) and comparing the result with Eq. (6.137), we see that $f(x) = F(x)$.

For the Bessel equation as given by Eq. (6.52), Eq. (6.122) becomes

$$\frac{d}{dx}\left(x\frac{dy}{dx}\right)+\left(-\frac{m^2}{x}+\kappa^2 x\right)y=f(x) \tag{6.138}$$

Equation (6.138) often appears in the form

$$\frac{d^2y}{dx^2}+\frac{dy}{xdx}+\left(-\frac{m^2}{x^2}+\kappa^2\right)y=F(x) \tag{6.139}$$

Upon expanding Eq. (6.138), we arrive at

$$x\frac{d^2y}{dx^2}+\frac{dy}{dx}+\left(-\frac{m^2}{x}+\kappa^2 x\right)y=f(x) \tag{6.140}$$

To get Eq. (6.139) into the form given by Eq. (6.140), we multiply Eq. (6.139) by x; therefore, $f(x) = xF(x)$. For the case of an equation with constant coefficients as given by Eq. (6.41), there is only one form so that $f(x) = F(x)$. These relations have to be kept in mind when certain nonhomogeneous partial differential equations are solved in Chapter 7.

Fourth-Order Sturm–Liouville equation: We shall obtain a solution to the fourth order nonhomogeneous Sturm–Liouville equation of the form

$$\frac{d^2}{dx^2}\left(S(x)\frac{d^2y}{dx^2}\right)+\frac{d}{dx}\left(P(x)\frac{dy}{dx}\right)+Q_2(x)y-R(x)\lambda y=f(x) \tag{6.141}$$

where $R(x) > 0$. The boundary conditions are assumed to be homogeneous and of the general form given by Eq. (6.103), that is,

$$\begin{aligned} &\Gamma(y(a))+\alpha_a y(a)=0 \quad \text{and} \quad \Lambda(y(a))+\beta_a y'(a)=0 \\ &\Gamma(y(b))+\alpha_b y(b)=0 \quad \text{and} \quad \Lambda(y(b))+\beta_b y'(b)=0 \end{aligned} \tag{6.142}$$

where Γ and Λ are given by Eq. (6.101).

As we did for the second-order Sturm–Liouville equation, it seems reasonable to assume a solution to Eq. (6.141) of the form

$$y(x)=\sum_{n=1}^{\infty}c_n\phi_n(x) \tag{6.143}$$

where the orthogonal function $\phi_n(x)$ is the solution to

$$\frac{d^2}{dx^2}\left(S(x)\frac{d^2\phi_n}{dx^2}\right)+\frac{d}{dx}\left(P(x)\frac{d\phi_n}{dx}\right)+Q_2(x)\phi_n-R(x)\lambda_n\phi_n=0 \tag{6.144}$$

and the boundary conditions

$$\begin{aligned}\Gamma(\phi_n(a))+\alpha_a\phi_n(a)=0 \quad &\text{and} \quad \Lambda(\phi_n(a))+\beta_a\phi_n'(a)=0\\ \Gamma(\phi_n(b))+\alpha_b\phi_n(b)=0 \quad &\text{and} \quad \Lambda(\phi_n(b))+\beta_b\phi_n'(b)=0\end{aligned} \tag{6.145}$$

Since $\phi_n(x)$ is a solution to the fourth-order Sturm–Liouville equation and the homogeneous boundary conditions of the appropriate form, that is, Eq. (6.145), $\phi_n(x)$ is an orthogonal eigenfunction. We now use this fact as follows.

Upon substituting Eq. (6.143) into Eq. (6.141), we obtain

$$\sum_{n=1}^{\infty}c_n\left\{\frac{d^2}{dx^2}\left(S(x)\frac{d^2\phi_n}{dx^2}\right)+\frac{d}{dx}\left(P(x)\frac{d\phi_n}{dx}\right)+Q_2(x)\phi_n-R(x)\lambda\phi_n\right\}=f(x) \tag{6.146}$$

However, from Eq. (6.144),

$$\frac{d^2}{dx^2}\left(S(x)\frac{d^2\phi_n}{dx^2}\right)+\frac{d}{dx}\left(P(x)\frac{d\phi_n}{dx}\right)+Q_2(x)\phi_n=R(x)\lambda_n\phi_n \tag{6.147}$$

and substitution of Eq. (6.147) into Eq. (6.146) yields

$$\sum_{n=1}^{\infty}c_n\left(\lambda_n-\lambda\right)R(x)\phi_n(x)=f(x) \tag{6.148}$$

Multiplying Eq. (6.148) by $\phi_m(x)$ and integrating over the interval results in

$$\sum_{n=1}^{\infty}c_n\left(\lambda_n-\lambda\right)\int_a^b R(x)\phi_n(x)\phi_m(x)dx=\int_a^b f(x)\phi_m(x)dx \tag{6.149}$$

However, from Eq. (6.120)

$$\int_a^b R(x)\phi_m(x)\phi_n(x)dx=N_{4n}\delta_{nm} \tag{6.150}$$

where N_{4n} is given by Eq. (6.121). Then, Eq. (6.149) reduces to

$$\begin{aligned}\sum_{n=1}^{\infty}c_n\left(\lambda_n-\lambda\right)\delta_{nm}N_{4n}&=\int_a^b f(x)\phi_m(x)dx\\ c_n&=\frac{1}{N_{4n}\left(\lambda_n-\lambda\right)}\int_a^b f(x)\phi_n(x)dx\end{aligned} \tag{6.151}$$

provided that $\lambda \neq \lambda_n$. Therefore, the solution to Eq. (6.141) and the boundary conditions given by Eq. (6.142) is

$$y(x) = \sum_{n=1}^{\infty} \frac{\phi_n(x)}{N_{4n}(\lambda_n - \lambda)} \int_a^b f(x)\phi_n(x)dx \tag{6.152}$$

We shall now illustrate these results with the following examples.

Example 6.1

Consider the equation

$$\frac{d^2y}{dx^2} + \alpha^2 y = f(x) \tag{a}$$

with the boundary conditions

$$y(0) = 0 \quad \text{and} \quad y(b) = 0$$

When $f(x) = 0$, this is a special case of the Sturm–Liouville equation given by Eq. (6.41) with $\mu^2 = \alpha^2$. The solution to the homogeneous equation is obtained from Table 4.2 as

$$y_h = A\cos\alpha x + B\sin\alpha x$$

Substituting the homogeneous solution into the boundary conditions gives

$$y(0) = A = 0$$

$$y(b) = B\sin\alpha b = 0$$

Although $B = 0$ is a solution, it results in $y_h = 0$, which is a trivial solution. Therefore, we set $\alpha b = n\pi$ and

$$\alpha = \alpha_n = n\pi / b \tag{b}$$

which is the eigenvalue. Then,

$$y_h = B\sin\alpha_n x$$

and, therefore,

$$\varphi_n(x) = \sin\alpha_n x \tag{c}$$

Because we have a Sturm–Liouville system, φ_n is an orthogonal function on the interval $0 \leq x \leq b$. Since there is an infinite number of solutions, we use superposition and obtain

$$y_h = \sum_{n=1}^{\infty} C_n \sin\alpha_n x$$

Then as indicated by Eq. (6.127), we assume a solution to Eq. (a) of the form

$$y(x) = \sum_{n=1}^{\infty} c_n \varphi_n(x) = \sum_{n=1}^{\infty} c_n \sin \alpha_n x$$

Substituting this equation in the governing equations yields

$$\sum_{n=1}^{\infty} c_n \left(\frac{d^2 \varphi_n(x)}{dx^2} + \alpha^2 \varphi_n(x) \right) = f(x) \tag{d}$$

Since φ_n is a solution to

$$\frac{d^2 \varphi_n(x)}{dx^2} + \alpha_n^2 \varphi_n(x) = 0 \tag{e}$$

we substitute Eq. (e) into Eq. (d) and obtain

$$\begin{aligned} &\sum_{n=1}^{\infty} c_n \left(-\alpha_n^2 \varphi_n(x) + \alpha^2 \varphi_n(x) \right) = f(x) \\ &\sum_{n=1}^{\infty} c_n \left(\alpha^2 - \alpha_n^2 \right) \sin \alpha_n x = f(x) \end{aligned} \tag{f}$$

To determine c_n, we use the fact that φ_n is orthogonal on the interval $0 \le x \le b$. Hence, we multiply Eq. (f) by φ_m and integrate over the interval to obtain

$$\sum_{n=1}^{\infty} c_n \left(\alpha^2 - \alpha_n^2 \right) \int_0^b \sin \alpha_m x \sin \alpha_n x dx = \int_0^b f(x) \sin \alpha_m x dx$$

$$\sum_{n=1}^{\infty} c_n \left(\alpha^2 - \alpha_n^2 \right) N_n \delta_{nm} = \int_0^b f(x) \sin \alpha_m x dx$$

$$c_n = \frac{1}{N_n \left(\alpha^2 - \alpha_n^2 \right)} \int_0^b f(x) \sin \alpha_n x dx$$

where, from Eq. (6.89),

$$N_n = \int_0^b \sin^2(\alpha_n x) dx = \frac{b}{2} \tag{g}$$

Hence, the solution is

$$y(x) = \frac{2}{b} \sum_{n=1}^{\infty} \frac{\sin \alpha_n x}{\alpha^2 - \alpha_n^2} \int_0^T f(x) \sin \alpha_n x dx \tag{h}$$

provided that $\alpha_n \neq \alpha$. This solution satisfies both the homogeneous and particular solutions of the original equation and the boundary conditions.

This result could also have been obtained by using Eq. (6.135) as follows. As was mentioned in the discussion after Eq. (6.140), $F(x) = f(x)$. Then using Eqs.. (b), (c), and (g) in Eq. (6.134) with $\lambda_n = \alpha_n$, we obtain Eq. (h).

Example 6.2

Consider the equation

$$\frac{d^4y}{d\eta^4} - \Omega^4 y = f(\eta) \quad 0 \le \eta \le 1 \tag{a}$$

with the boundary conditions

$$\begin{array}{ll} y(0)=0 & y'(0)=0 \\ y''(1)=0 & y'''(1)=0 \end{array} \tag{b}$$

Equation (a) is a special case of Eq. (6.102), that is, in Eq. (6.102) $S(x) = R(x) = 1$, $P(x) = Q_2(x) = 0$, and $\lambda = \Omega^4$. It is also seen that Eq. (b) is a special case of the boundary conditions given by Eq. (6.103), that is, $\alpha_b = \beta_b = 0$ and $\alpha_a \to \infty$ and $\beta_a \to \infty$. Hence, we shall be able to generate orthogonal functions.

The solution to Eq. (a) when $f(\eta) = 0$ is obtained from Table 4.2 as

$$y_h(\eta) = C_1 Q(\Omega\eta) + C_2 R(\Omega\eta) + C_3 \hat{S}(\Omega\eta) + C_4 T(\Omega\eta) \tag{c}$$

where

$$\begin{aligned} Q(\Omega\eta) &= \frac{1}{2}\left[\cosh(\Omega\eta) + \cos(\Omega\eta)\right] \\ R(\Omega\eta) &= \frac{1}{2}\left[\sinh(\Omega\eta) + \sin(\Omega\eta)\right] \\ \hat{S}(\Omega\eta) &= \frac{1}{2}\left[\cosh(\Omega\eta) - \cos(\Omega\eta)\right] \\ T(\Omega\eta) &= \frac{1}{2}\left[\sinh(\Omega\eta) - \sin(\Omega\eta)\right] \end{aligned} \tag{d}$$

It is noted that the derivatives of Eq. (d) are

$$\begin{array}{lll} Q'(\Omega\eta) = \Omega T(\Omega\eta) & Q''(\Omega\eta) = \Omega^2 S(\Omega\eta) & Q'''(\Omega\eta) = \Omega^3 R(\Omega\eta) \\ R'(\Omega\eta) = \Omega Q(\Omega\eta) & R''(\Omega\eta) = \Omega^2 T(\Omega\eta) & R'''(\Omega\eta) = \Omega^3 S(\Omega\eta) \\ \hat{S}'(\Omega\eta) = \Omega R(\Omega\eta) & \hat{S}''(\Omega\eta) = \Omega^2 Q(\Omega\eta) & \hat{S}'''(\Omega\eta) = \Omega^3 T(\Omega\eta) \\ T'(\Omega\eta) = \Omega S(\Omega\eta) & T''(\Omega\eta) = \Omega^2 R(\Omega\eta) & T'''(\Omega\eta) = \Omega^3 Q(\Omega\eta) \end{array} \tag{e}$$

where the prime denotes the derivative with respect to η. It is also seen that $Q(0) = 1$ and $R(0) = \hat{S}(0) = T(0) = 0$. Therefore, using Eq. (e) in the derivatives of Eq. (c) it is found that

$$
\begin{aligned}
y_h'(\eta) &= \Omega\left(C_1T(\Omega\eta) + C_2Q(\Omega\eta) + C_3R(\Omega\eta) + C_4\hat{S}(\Omega\eta)\right) \\
y_h''(\eta) &= \Omega^2\left(C_1\hat{S}(\Omega\eta) + C_2T(\Omega\eta) + C_3Q(\Omega\eta) + C_4R(\Omega\eta)\right) \\
y_h'''(\eta) &= \Omega^3\left(C_1R(\Omega\eta) + C_2\hat{S}(\Omega\eta) + C_3T(\Omega\eta) + C_4Q(\Omega\eta)\right)
\end{aligned} \tag{f}
$$

To obtain the solution to Eqs. (a) and (b), we first obtain a set of orthogonal eigenfunctions in the following manner. Using Eq. (f) in the boundary conditions given by Eq. (b), it is found that

$$
\begin{aligned}
y(0) &= C_1 = 0 \\
y'(0) &= C_2 = 0 \\
y''(1) &= C_3Q(\Omega) + C_4R(\Omega) = 0 \\
y'''(1) &= C_3T(\Omega) + C_4Q(\Omega) = 0
\end{aligned} \tag{g}
$$

or in matrix form

$$
\begin{pmatrix} Q(\Omega) & R(\Omega) \\ T(\Omega) & Q(\Omega) \end{pmatrix} \begin{Bmatrix} C_3 \\ C_4 \end{Bmatrix} = 0 \tag{h}
$$

To determine the eigenvalue Ω_n, we set the determinant of the coefficients in Eq. (h) to zero and find that Ω_n is the solution to

$$
Q^2(\Omega_n) - R(\Omega_n)T(\Omega_n) = 0 \tag{i}
$$

and the eigenfunction ψ_n is [recall Eqs. (6.107) to (6.109)],

$$
\psi_n(\eta) = S(\Omega_n\eta) - \frac{Q(\Omega_n)}{R(\Omega_n)}T(\Omega_n\eta) \tag{j}
$$

and ψ_n is a solution to

$$
\frac{d^4\psi_n}{d\eta^4} - \Omega_n^4\psi_n = 0 \tag{k}
$$

Therefore, using superposition, the homogeneous solution to Eq. (a) is

$$
y_h = \sum_{n=1}^{\infty} c_n\psi_n(\eta)
$$

Consequently, we assume that the solution to the nonhomogeneous equation is

$$
y = \sum_{n=1}^{\infty} c_n\psi_n(\eta) \tag{l}
$$

Substituting Eq. (l) into Eq. (a) yields

$$\sum_{n=1}^{\infty} c_n \left(\frac{d^4\psi_n}{d\eta^4} - \Omega^4 \psi_n \right) = f(\eta) \tag{m}$$

Using Eq. (k) in Eq. (m) gives

$$\sum_{n=1}^{\infty} c_n \left(\Omega_n^4 - \Omega^4 \right) \psi_n = f(\eta) \tag{n}$$

Since ψ_n is an orthogonal function on the interval $0 \le \eta \le 1$, we multiply Eq. (n) by ψ_m and integrate over the interval to obtain

$$\sum_{n=1}^{\infty} c_n \left(\Omega_n^4 - \Omega^4 \right) \int_0^1 \psi_n(\eta)\psi_m(\eta)d\eta = \int_0^1 f(\eta)\psi_m(\eta)d\eta$$

$$\sum_{n=1}^{\infty} c_n \left(\Omega_n^4 - \Omega^4 \right) N_{4n}\delta_{nm} = \int_0^1 f(\eta)\psi_m(\eta)d\eta$$

$$c_n = \frac{1}{N_{4n}\left(\Omega_n^4 - \Omega^4 \right)} \int_0^1 f(\eta)\psi_n(\eta)d\eta$$

where

$$N_{4n} = \int_0^1 \psi_n^2(\eta)d\eta \tag{o}$$

Hence, the solution is

$$y(\eta) = \sum_{n=1}^{\infty} \frac{\psi_n(\eta)}{N_{4n}\left(\Omega_n^4 - \Omega^4 \right)} \int_0^1 f(\eta)\psi_n(\eta)d\eta$$

provided that $\Omega \neq \Omega_n$.

This result could have also been obtained by using Eqs. (i), (j), (o), and (6.152).

Example 6.3

Consider the equation

$$\frac{d^2 y_m}{dr^2} + \frac{dy_m}{rdr} + \left(k^2 - \frac{m^2}{r^2} \right) y_m = F(r) \tag{a}$$

with the boundary conditions

$$y(a) = 0 \quad \text{and} \quad y(b) = 0 \tag{b}$$

where $0 < a \le r \le b$ and $m = 0, 1, 2, \ldots$

When $F(r) = 0$, this is a Bessel equation, which is a special case of the Sturm–Liouville equation given by Eq. (6.47) with $\kappa^2 = k^2$. The solution to the homogeneous equation is obtained from Table 5.1 as

$$y_{h,m}(r) = AJ_m(kr) + BY_m(kr) \quad r > 0 \tag{c}$$

Substituting the homogeneous solution into the boundary conditions given by Eq. (b), we obtain in matrix form,

$$\begin{pmatrix} J_m(\beta) & Y_m(\beta) \\ J_m(\alpha\beta) & Y_m(\alpha\beta) \end{pmatrix} \begin{Bmatrix} A \\ B \end{Bmatrix} = 0 \tag{d}$$

where $\alpha = b/a > 1$ and $\beta = ka$. To obtain a non-trivial solution, we set the determinant of the coefficients of Eq. (d) to zero and find that $\beta = \beta_{mn}$, $n = 1, 2, 3, \ldots$, where β_{mn} are solutions to

$$J_m(\beta_{mn})Y_m(\alpha\beta_{mn}) - J_m(\alpha\beta_{mn})Y_m(\beta_{mn}) = 0 \tag{e}$$

for a given α and m. The corresponding orthogonal eigenfunction is obtained from Eq. (c) and the first equation of Eq. (d) as [recall Eq. (6.32)]

$$\varphi_{nm}(r) = J_m(\beta_{mn}r / a) - \frac{J_m(\beta_{mn})}{Y_m(\beta_{mn})} Y_m(\beta_{mn}r / a) \tag{f}$$

Since there is an infinite number of solutions, the solution to the homogeneous equation and boundary conditions is obtained by using superposition to obtain

$$y_{h,m}(r) = \sum_{n=1}^{\infty} c_n \varphi_{nm}(r) \tag{g}$$

Then, as indicated previously, we assume a solution to Eq. (a) of the form

$$y_m(r) = \sum_{n=1}^{\infty} c_n \varphi_{nm}(r) \tag{h}$$

Substituting Eq. (h) into the governing equation, Eq. (a), yields

$$\sum_{n=1}^{\infty} c_n \left\{ \frac{d^2\varphi_{nm}}{dr^2} + \frac{d\varphi_{nm}}{rdr} + \left(k^2 - \frac{m^2}{r^2} \right) \varphi_{nm} \right\} = F(r) \tag{i}$$

However, we know that φ_{mn} is a solution to

$$\frac{d^2\varphi_{nm}}{dr^2} + \frac{d\varphi_{nm}}{rdr} - \frac{m^2}{r^2}\varphi_{nm} = -\beta_{mn}^2\varphi_{nm}$$

and, therefore, Eq. (i) becomes

$$\sum_{n=1}^{\infty} c_n \left(k^2 - \beta_{mn}^2 \right) \varphi_{nm}(r) = F(r) \tag{j}$$

Since the weight factor for the Bessel equation is r, we multiply Eq. (j) by $r\varphi_{mj}dr$ and integrate over the region to obtain

$$\sum_{n=1}^{\infty} c_n\left(k^2-\beta_{mn}^2\right)\int_a^b r\varphi_{nm}(r)\varphi_{nj}(r)dr=\int_a^b rF(r)\varphi_{nj}(r)dr$$

$$c_n=\frac{1}{N_n\left(k^2-\beta_{mn}^2\right)}\int_a^b rF(r)\varphi_{mn}(r)dr \tag{k}$$

where

$$N_n=\int_a^b r\varphi_{mn}^2(r)dr$$

Hence, the solution is

$$y_m(r)=\sum_{n=1}^{\infty}\frac{\varphi_{nm}(r)}{N_n\left(k^2-\beta_{mn}^2\right)}\int_a^b rF(r)\varphi_{mn}(r)dr \tag{l}$$

where φ_{mn} is given by Eq. (f) and provided that $k \neq \beta_{mn}$.

6.2 ORTHOGONAL FUNCTIONS FOR COUPLED SYSTEMS: TWO DEPENDENT VARIABLES

There are physical systems that require more than one dependent quantity to describe them. One such system is a beam described by the Timoshenko theory which requires two coupled quantities, one to describe the transverse motion of the beam, and the other the rotation of the beam's cross section. In this case, one can still generate orthogonal functions under certain conditions. We shall discuss what these conditions are.

Let us consider a system whose total energy can be expressed as

$$L=T-U$$

where T is the total kinetic energy and U is the total potential energy. We assume that L is a function of one spatial coordinate x, time t, and the dependent variables u_1 and u_2 and their derivatives up to the second order as follows

$$L=L\left(x,t,u_1,\dot{u}_1,u_{1,x},u_{1,xx},u_2,\dot{u}_2,u_{2,x},u_{2,xx}\right) \tag{6.153}$$

where the over dot indicates the derivative with respect to time, and the subscript ",x" indicates the derivative with respect to x. When the calculus of variations is used on Eq. (6.153), Eq. (6.153) produces the governing equations and the boundary conditions.* In operator notation, we shall assume that the governing equations are of the form

* Magrab EB (2012) *Vibrations of Elastic Systems: With Application to MEMS and NEMS*, Solid Mechanics and Its Applications 184, Springer Science+Business Media B.V., New York, Appendix B and Chapter 5.

$$L_k[u_1,u_2]-r_k(x)\ddot{u}_k=0 \quad k=1,2 \tag{6.154}$$

where the L_j, j = 1, 2 are differential operators in x and the boundary conditions are given by

$$B_C[u_1(x_p,t),u_2(x_p,t)]=0 \quad p=1,2 \tag{6.155}$$

The quantities x_1 and x_2 are the end points of the interval.

To get Eq. (6.154) and (6.155) into the appropriate form, we assume that the system is oscillating harmonically at a frequency ω as

$$u_k(x,t)=U_k(x)e^{j\omega t} \quad k=1,2 \tag{6.156}$$

Then, upon substituting Eq. (6.156) into Eqs. (6.154) and (6.155), we obtain

$$L_k[U_1,U_2]+\omega^2 r_k(x)U_k=0 \quad k=1,2 \tag{6.157}$$

and

$$B_C[U_1(x_p),U_2(x_p)]=0 \quad p=1,2 \tag{6.158}$$

When the form of L given by Eq. (6.153) is a *symmetric quadratic*, then the solution to Eq. (6.157) with the homogeneous boundary conditions given by Eq. (6.158) will yield orthogonal functions.* These orthogonal functions are denoted as

$$U_{p,n}(x) \quad p=1,2 \quad n=1,2,\ldots \tag{6.159}$$

and are the solutions to

$$L_k[U_{1,n},U_{2,n}]+\omega_n^2 r_k(x)U_{k,n}=0 \quad k=1,2 \tag{6.160}$$

and the boundary conditions

$$B_C[U_{1,n}(x_k),U_{2,n}(x_k)]=0 \quad k=1,2 \tag{6.161}$$

The quantity ω_n is the corresponding eigenvalue. Then, under these conditions, the orthogonality condition is

$$\int_{x_1}^{x_2}\left(r_1(x)U_{1,n}(x)U_{1,m}(x)+r_2(x)U_{2,n}(x)U_{2,m}(x)\right)dx=\delta_{nm}N_n \tag{6.162}$$

where

$$N_n=\int_{x_1}^{x_2}\left(r_1(x)U_{1,n}^2(x)+r_2(x)U_{1,n}^2(x)\right)dx \tag{6.163}$$

These results are illustrated in Example 8.20.

---

* Magrab, 2012, *ibid.*

EXERCISES

Section 6.1.3

6.1 Place the following equations in Sturm–Liouville form:

(a) Chebyshev's equation

$$\left(1-x^2\right)\frac{d^2y}{dx^2}-x\frac{dy}{dx}+\alpha^2 y=0$$

(b) Euler's equation

$$x^2\frac{d^2y}{dx^2}+bx\frac{dy}{dx}+cy=0$$

(c) Hermite's equation

$$\frac{d^2y}{dx^2}-2x\frac{dy}{dx}+2\alpha y=0$$

(d) Laguerre's equation

$$x\frac{d^2y}{dx^2}+(1-x)\frac{dy}{dx}+\alpha y=0$$

Section 6.1.2

6.2 Consider the Cauchy–Euler equation

$$x^2\frac{d^2y}{dx^2}+x\frac{dy}{dx}+(\lambda+2)y=0$$

where it is assumed that $\lambda + 2 > 0$. The boundary conditions are $dy(1)/dx = 0$ and $dy(2)/dx = 0$. Determine the eigenvalues and eigenfunctions.

7 Partial Differential Equations

7.1 INTRODUCTION TO SECOND-ORDER PARTIAL DIFFERENTIAL EQUATIONS

7.1.1 Classification of Linear Second-Order Partial Differential Equations

Consider the following general second-order partial differential equation

$$a(x,y)\frac{\partial^2\Phi}{\partial x^2}+2b(x,y)\frac{\partial^2\Phi}{\partial x\partial y}+c(x,y)\frac{\partial^2\Phi}{\partial y^2}+d(x,y)\frac{\partial\Phi}{\partial x}$$
$$+e(x,y)\frac{\partial\Phi}{\partial y}+f(x,y)\Phi=g(x,y) \tag{7.1}$$

where the 2 in $2b(x,y)$ has been introduced for subsequent algebraic convenience.

In order to classify the types of solutions one should expect from Eq. (7.1), we introduce two new coordinate variables $\eta=\eta(x,y)$ and $\xi=\xi(x,y)$ such that the Jacobian of the transformation does not vanish, that is,

$$J=\det\begin{pmatrix}\dfrac{\partial\eta}{\partial x} & \dfrac{\partial\eta}{\partial y}\\ \dfrac{\partial\xi}{\partial x} & \dfrac{\partial\xi}{\partial y}\end{pmatrix}\neq 0 \tag{7.2}$$

Using chain differentiation, we find that

$$\frac{\partial\Phi}{\partial x}=\frac{\partial\Phi}{\partial\xi}\frac{\partial\xi}{\partial x}+\frac{\partial\Phi}{\partial\eta}\frac{\partial\eta}{\partial x}$$
$$\frac{\partial\Phi}{\partial y}=\frac{\partial\Phi}{\partial\xi}\frac{\partial\xi}{\partial y}+\frac{\partial\Phi}{\partial\eta}\frac{\partial\eta}{\partial y} \tag{7.3}$$

and

$$\frac{\partial^2\Phi}{\partial x^2}=\frac{\partial^2\Phi}{\partial\eta^2}\left(\frac{\partial\eta}{\partial x}\right)^2+2\frac{\partial\eta}{\partial x}\frac{\partial\xi}{\partial x}\frac{\partial^2\Phi}{\partial\eta\partial\xi}+\frac{\partial^2\Phi}{\partial\xi^2}\left(\frac{\partial\xi}{\partial x}\right)^2+\frac{\partial^2\eta}{\partial x^2}\frac{\partial\Phi}{\partial\eta}+\frac{\partial^2\xi}{\partial x^2}\frac{\partial\Phi}{\partial\xi}$$
$$\frac{\partial^2\Phi}{\partial x\partial y}=\frac{\partial\eta}{\partial x}\frac{\partial\eta}{\partial y}\frac{\partial^2\Phi}{\partial\eta^2}+\left(\frac{\partial\eta}{\partial x}\frac{\partial\xi}{\partial y}+\frac{\partial\eta}{\partial y}\frac{\partial\xi}{\partial x}\right)\frac{\partial^2\Phi}{\partial\eta\partial\xi}+\frac{\partial\xi}{\partial x}\frac{\partial\xi}{\partial y}\frac{\partial^2\Phi}{\partial\xi^2}+\frac{\partial^2\eta}{\partial x\partial y}\frac{\partial\Phi}{\partial\eta}$$
$$+\frac{\partial^2\xi}{\partial x\partial y}\frac{\partial\Phi}{\partial\xi} \tag{7.4}$$
$$\frac{\partial^2\Phi}{\partial y^2}=\frac{\partial^2\Phi}{\partial\eta^2}\left(\frac{\partial\eta}{\partial y}\right)^2+2\frac{\partial\eta}{\partial y}\frac{\partial\xi}{\partial y}\frac{\partial^2\Phi}{\partial\eta\partial\xi}+\frac{\partial^2\Phi}{\partial\xi^2}\left(\frac{\partial\xi}{\partial y}\right)^2+\frac{\partial^2\eta}{\partial y^2}\frac{\partial\Phi}{\partial\eta}+\frac{\partial^2\xi}{\partial y^2}\frac{\partial\Phi}{\partial\xi}$$

Upon substituting Eqs. (7.3) and (7.4) into Eq. (7.1) and rearranging the results, Eq. (7.1) can be written as

$$A\frac{\partial^2\Phi}{\partial\xi^2}+2B\frac{\partial^2\Phi}{\partial\eta\partial\xi}+C\frac{\partial^2\Phi}{\partial\eta^2}+G(\xi,\eta,\Phi,\Phi_\xi,\Phi_\eta)=0 \tag{7.5}$$

where

$$G(\xi,\eta,\Phi,\Phi_\xi,\Phi_\eta)=D\frac{\partial\Phi}{\partial\eta}+E\frac{\partial\Phi}{\partial\xi}+f(\xi,\eta)\Phi-g(\xi,\eta) \tag{7.6}$$

and

$$\begin{aligned}
A&=a\left(\frac{\partial\xi}{\partial x}\right)^2+2b\frac{\partial\xi}{\partial x}\frac{\partial\xi}{\partial y}+c\left(\frac{\partial\xi}{\partial y}\right)^2\\
B&=a\frac{\partial\eta}{\partial x}\frac{\partial\xi}{\partial x}+b\left(\frac{\partial\eta}{\partial x}\frac{\partial\xi}{\partial y}+\frac{\partial\eta}{\partial y}\frac{\partial\xi}{\partial x}\right)+c\frac{\partial\eta}{\partial y}\frac{\partial\xi}{\partial y}\\
C&=a\left(\frac{\partial\eta}{\partial x}\right)^2+2b\frac{\partial\eta}{\partial x}\frac{\partial\eta}{\partial y}+c\left(\frac{\partial\eta}{\partial y}\right)^2\\
D&=a\frac{\partial^2\eta}{\partial x^2}+2b\frac{\partial^2\eta}{\partial x\partial y}+c\frac{\partial^2\eta}{\partial y^2}+d\frac{\partial\eta}{\partial x}+e\frac{\partial\eta}{\partial y}\\
E&=a\frac{\partial^2\xi}{\partial x^2}+2b\frac{\partial^2\xi}{\partial x\partial y}+c\frac{\partial^2\xi}{\partial y^2}+d\frac{\partial\xi}{\partial x}+e\frac{\partial\xi}{\partial y}
\end{aligned} \tag{7.7}$$

Equation (7.7) is verified with Mathematica procedure M7.1.

It is seen that A and C are of the same form. Therefore, we try to choose a transformation that will make either $A = 0$, or $B = 0$, or $C = 0$ or any two of them equal to zero. Thus, we assume that $\eta(x,y)$ = constant and $\xi(x,y)$ = constant. Then

$$\begin{aligned}
d\eta&=0=\frac{\partial\eta}{\partial x}dx+\frac{\partial\eta}{\partial y}dy\\
\frac{\partial\eta}{\partial x}&=-\frac{dy}{dx}\frac{\partial\eta}{\partial y}
\end{aligned} \tag{7.8}$$

and

$$\begin{aligned}
d\xi&=0=\frac{\partial\xi}{\partial x}dx+\frac{\partial\xi}{\partial y}dy\\
\frac{\partial\xi}{\partial x}&=-\frac{dy}{dx}\frac{\partial\xi}{\partial y}
\end{aligned} \tag{7.9}$$

From Eqs. (7.7) and (7.9), we want to find a solution such that $A = 0$; thus

$$
\begin{aligned}
A &= a\left(\frac{\partial\xi}{\partial x}\right)^2 + 2b\frac{\partial\xi}{\partial x}\frac{\partial\xi}{\partial y} + c\left(\frac{\partial\xi}{\partial y}\right)^2 = 0 \\
&= a\left(-\frac{dy}{dx}\frac{\partial\xi}{\partial y}\right)^2 - 2b\frac{dy}{dx}\left(\frac{\partial\xi}{\partial y}\right)^2 + c\left(\frac{\partial\xi}{\partial y}\right)^2 = 0 \\
&= \left(\frac{\partial\xi}{\partial y}\right)^2\left[a\left(\frac{dy}{dx}\right)^2 - 2b\frac{dy}{dx} + c\right] = 0
\end{aligned}
\tag{7.10}
$$

An identical equation for C is obtained by using Eqs. (7.7) and (7.8). The solution to Eq. (7.10) is

$$
\frac{dy}{dx} = \frac{1}{a(x,y)}\left[b(x,y) \pm \sqrt{\Delta(x,y)}\right] \tag{7.11}
$$

where

$$
\Delta(x,y) = b^2(x,y) - a(x,y)c(x,y) \tag{7.12}
$$

is called the *discriminant.*

The characteristic curves are determined from the integration of Eq. (7.11). Since a, b, and c are, in general, functions of x and y, a general solution cannot be obtained. However, when a, b, and c are constants, Eq. (7.11) can be integrated in this general form to give

$$
y = \frac{x}{a}\left[b \pm \sqrt{\Delta}\right] + C_1
$$

or

$$
C_2 = \frac{x}{a}\left[b \pm \sqrt{\Delta}\right] - y \tag{7.13}
$$

provided that $\Delta > 0$. Since Eq. (7.13) represents two equations, one for η and one for ξ, we let the plus sign be associated with ξ and the minus sign be associated with η. Thus,

$$
\begin{aligned}
\xi &= \frac{x}{a}\left[b + \sqrt{\Delta}\right] - y = \gamma x - y \\
\eta &= \frac{x}{a}\left[b - \sqrt{\Delta}\right] - y = \varepsilon x - y
\end{aligned}
\tag{7.14}
$$

where

$$
\begin{aligned}
\gamma &= \frac{b + \sqrt{\Delta}}{a} \\
\varepsilon &= \frac{b - \sqrt{\Delta}}{a}
\end{aligned}
\tag{7.15}
$$

When the characteristic curves are of the form given by Eq. (7.14), it is easily shown from Eq. (7.7) that

$$\begin{aligned} D &= d\varepsilon - e \\ E &= d\gamma - e \end{aligned} \tag{7.16}$$

We also note the following identity

$$B^2 - AC = \left(b^2 - ac\right)J^2 = \Delta J^2 \tag{7.17}$$

where J is given by Eq. (7.2). Thus, the sign of the discriminant does not change with the coordinate transformation. Also note that the coordinate transformation must be such that $J \neq 0$. Equation (7.17) is verified at the end of Mathematica procedure M7.1.

We now discuss the three possible ranges of values of the discriminant Δ. The solutions to these three types of equations are quite different.

$\Delta > 0$, Hyperbolic Equations: For this case, $A = 0$, $C = 0$, and $B \neq 0$ and when a, b, and c are constants Eq. (7.5) can be written as

$$\frac{\partial^2\Phi}{\partial\eta\partial\xi} + \frac{G}{2B} = 0 \tag{7.18}$$

Equation (7.18) is called the *first canonical form* for hyperbolic equations. Using a second coordinate transformation, Eq. (7.18) can be written as

$$\frac{\partial^2\Phi}{\partial\alpha^2} - \frac{\partial^2\Phi}{\partial\beta^2} + \frac{G}{2B} = 0 \tag{7.19}$$

This is known as the *second canonical form.*

$\Delta = 0$, Parabolic Equations: For this case, $A = 0$, $B = 0$, and $C \neq 0$ and when a, b, and c are constants Eq. (7.5) becomes

$$\frac{\partial^2\Phi}{\partial\eta^2} + \frac{G}{C} = 0$$

$\Delta < 0$, Elliptic Equations: For this case, $A = C \neq 0$, and $B = 0$ and when a, b, and c are constants Eq. (7.5) becomes

$$\frac{\partial^2\Phi}{\partial\xi^2} + \frac{\partial^2\Phi}{\partial\eta^2} + \frac{G}{A} = 0 \tag{7.20}$$

7.1.2 Representative Application Areas

There are several partial differential equations that describe phenomena in many different application areas. These equations are often very similar in form and typically involve the Laplacian operator, which is denoted ∇^2. This operator has different forms depending on the coordinate system selected, which in our case will be the Cartesian, polar cylindrical, and spherical coordinate systems. The Laplacian operators for these three coordinate systems are listed in Table 7.1.

TABLE 7.1
The Laplacian Operator in Three Coordinate Systems

Coordinate System	Laplacian: ∇^2
Cartesian	$\frac{\partial^2}{\partial x^2}+\frac{\partial^2}{\partial y^2}+\frac{\partial^2}{\partial z^2}$
Polar cylindrical	$\frac{\partial^2}{\partial r^2}+\frac{\partial}{r\partial r}+\frac{\partial^2}{r^2\partial \theta^2}+\frac{\partial^2}{\partial z^2}$
Spherical	$\frac{1}{r^2}\frac{\partial}{\partial r}\left(r^2\frac{\partial}{\partial r}\right)+\frac{1}{r^2\sin\varphi}\frac{\partial}{\partial \varphi}\left(\sin\varphi\frac{\partial}{\partial\varphi}\right)+\frac{1}{r^2\sin^2\varphi}\frac{\partial^2}{\partial\theta^2}$ or
	$\frac{\partial^2}{\partial r^2}+\frac{2}{r}\frac{\partial}{\partial r}+\frac{\cot\varphi}{r^2}\frac{\partial}{\partial\varphi}+\frac{1}{r^2}\frac{\partial^2}{\partial\varphi^2}+\frac{1}{r^2\sin^2\varphi}\frac{\partial^2}{\partial\theta^2}$

Many of the partial differential equations involving the Laplacian fall into one of the three types previously mentioned: Parabolic, hyperbolic, or elliptic. In addition, these three types of equations are identified by the following names: (i) the Laplace equation; (ii) the Helmholtz equation; (iii) the diffusion equation; (iv) the wave equation; and (v) the Poisson equation. A sixth equation that we shall consider is the bi-harmonic equation, which involves ∇^4. These six equations are given in Table 7.2 along with representative application areas. In Section 7.2, we shall obtain the general solution to these equations in three coordinate systems and use these general solutions to solve the respective equations subject to various combinations of boundary conditions and initial conditions.

7.2 SEPARATION OF VARIABLES AND THE SOLUTION TO PARTIAL DIFFERENTIAL EQUATIONS OF ENGINEERING AND PHYSICS

7.2.1 Introduction

Separation of variables is a technique for finding solutions to certain linear partial differential equations subject to prescribed boundary conditions and initial conditions. In this regard, we shall be concerned primarily with the Laplace equation, the Helmholtz equation, the diffusion equation, the Poisson equation, and the wave equation, all of which are listed in Table 7.2. It turns out that these equations are separable.

The separation of variables technique assumes that the dependent variable can be expressed as the product of up to four functions, with each function being a function of only one independent variable. In the case of Cartesian coordinates where the dependent variable is $u(x,y,z,t)$, then it would be assumed that $u(x,y,z,t) = X(x)Y(y)Z(z)T(t)$. When the coordinate system is a polar cylindrical coordinate system with the dependent variable $u(r,\theta,z,t)$, then it would be assumed that $u(r,\theta,z,t) = R(r)\Theta(\theta)Z(z)T(t)$. Lastly, when the coordinate system is a spherical coordinate system with the dependent variable $u(r,\varphi,\theta,t)$, then it would be assumed that $u(r,\varphi,\theta,t) = R(r)\Phi(\varphi)\Theta(\theta)T(t)$.

It will be seen that as part of the separation of variable method, separation constants are generated, the signs of which are arbitrary. In our case, the sign of these constants will be chosen so that the resulting separation equations are of the Sturm–Liouville type, which in actuality, is equivalent to selecting the sign of the weighting factor $r(x)$. Therefore, when the boundary

TABLE 7.2
Representative Application Areas of the Most Common Partial Differential Equations Involving the Laplacian

Equation Type	Equation		Application Areas: Φ
Laplace (elliptic)	$\nabla^2\Phi = 0$	(T7.2a)	Stream function – flow around an object Temperature – steady-state temperature Piezometric head – steady flow in an aquifer
Helmholtz (elliptic)	$\nabla^2\Phi + \Omega^2\Phi = G$	(T7.2b)	Harmonic oscillations of the wave equation
Diffusion (parabolic)	$\nabla^2\Phi = \dfrac{\partial\Phi}{\alpha\partial t} + G$	(T7.2c)	Temperature – heat flow in solids Pressure – air damping in a gap Concentration – diffusion between two gases Stream function – object oscillating in a viscous fluid Voltage – telegraph equation Piezometric head – unsteady flow in an aquifer
Wave (hyperbolic)	$\nabla^2\Phi = k^2\dfrac{\partial^2\Phi}{\partial t^2} + G$	(T7.2d)	Potential function – Rayleigh wave propagation Displacement – Love wave propagation, transverse oscillation of membranes, axial oscillations of bars, torsional oscillations of shafts, transverse oscillations of strings Voltage – undamped telegraph equation
Poisson (elliptic)	$\nabla^2\Phi = G$	(T7.2e)	Stress function – torsion of solid shaft ($G = -2$) Gravitational field (G = charge distribution) Piezometric head – steady groundwater flow (G = sink/source) Electric potential field – charge density distribution (G = free charge density)
Biharmonic*	$\nabla^4\Phi = G$	(T7.2f)	Static displacement of rectangular and circular plates Airy stress function for 2D stress with no body forces ($G = 0$)
Biharmonic*	$\nabla^4\Phi + k^2\dfrac{\partial^2\Phi}{\partial t^2} = G$	(T7.2g)	Dynamic displacement of rectangular and circular plates

Note: The equations presented here can be expressed in any of the coordinate systems given in Table 7.1. The quantity G in the table is, in general, a function of one or more of the independent variables. When applicable, each equation is identified as being either parabolic, elliptic, or hyperbolic.

*$\nabla^4 = \nabla^2\nabla^2$ is called the *bi-harmonic operator* and is usually employed as a two spatial dimension operator in either the Cartesian coordinate system or the polar coordinate system.

conditions are homogeneous, one will be able to generate orthogonal functions, which will turn out to be an extremely useful property in implementing the separation of variables method.

In obtaining the solution by the method of separation of variables, we shall be dealing with only well-posed problems, that is, problems in which the prescribed initial conditions and boundary conditions are such that all unknown solution constants can be determined.

The approach that we shall employ is to consider each of the following six partial differential equations separately: The Laplace equation, the homogeneous Helmholtz equation, the homogeneous diffusion equation, the homogeneous wave equation, the Poisson equation, and the bi-harmonic equation. For the first four of these equations the general solutions* will be presented in a table, one for each of these equations. In each table, the solution for the Cartesian coordinate system, the polar or polar cylindrical coordinate system, and the spherical coordinate system will be presented. The tables also contain the intermediate separation equations generated during the separation process and their respective solutions. The details for how one goes about doing this for all the equations and coordinate systems will not be provided. Instead, we shall show how one employs the method for each of the four equations as follows: The Laplace equation in the Cartesian coordinate system, the Helmholtz equation in the polar coordinate system, the diffusion equation in polar cylindrical coordinates, and the wave equation in the Cartesian coordinate system. These examples should provide one with the skills to reproduce all the results that are given in these tables. Lastly, the Poisson equation will be solved in the Cartesian coordinate system and the bi-harmonic equation in the Cartesian and polar coordinate systems.

The general solutions will then be used to obtain explicit solutions to various combinations of coordinate systems, boundary conditions, and initial conditions.

7.2.2 Laplace Equation

The general solutions to the Laplace equation in three coordinate systems are given in Table 7.3 along with some of the intermediate results required to obtain these general solutions. To illustrate how the results in these tables were obtained, we shall present the details of the separation of variable method for the two-dimensional Laplace equation in the Cartesian coordinate system. The solutions for the other coordinates are obtained in a similar manner.

The Laplace equation in the Cartesian coordinate system in terms of the variable $u = u(x,y)$ is obtained from Tables 7.1 and 7.2 as

$$\frac{\partial^2 u}{\partial x^2}+\frac{\partial^2 u}{\partial y^2}=0 \tag{7.21}$$

We assume a solution to Eq. (7.21) of the form

$$u(x,y)=u_0(x,y)+w(x,y) \tag{7.22}$$

where

$$\begin{aligned} u_0(x,y)&=X_0(x)Y_0(y)\\ w(x,y)&=X(x)Y(y) \end{aligned} \tag{7.23}$$

* It is mentioned that there are other forms for the solutions to these partial differential equations than have been presented in these four tables. The reader is referred to the following compendium of such solutions: Polyanin AD (2002) *Handbook of Linear Partial Differential Equations for Engineers and Scientists*. Chapman & Hall/CRC Press, pp. 43, 59, 161, 279, 299, 467, 494, and 503.

TABLE 7.3
Solutions to the Laplace Equation in Three Coordinate Systems

Cartesian Coordinates	**Assumed Separable Form**	**Separable Equations**
$\frac{\partial^2 u}{\partial x^2}+\frac{\partial^2 u}{\partial y^2}=0$	$u(x,y)=X_0(x)Y_0(y)+X(x)Y(y)$	$\frac{d^2X}{Xdx^2}+\frac{d^2Y}{Ydy^2}=0,\quad \frac{d^2X_0}{X_0dx^2}+\frac{d^2Y_0}{Y_0dy^2}=0$
Separated equations	**Solutions to separated equations**	**Solution and special cases**
$\frac{d^2X}{dx^2}+\lambda^2X=0$ (S-L)	$X(x)=A\cos(\lambda x)+B\sin(\lambda x)$	
$\frac{d^2Y}{dx^2}-\lambda^2Y=0$	$Y(y)=Ce^{\lambda y}+De^{-\lambda y}$ or $=C\cosh(\lambda y)+D\sinh(\lambda y)$	$u(x,y)=X_0(x)Y_0(y)+X(x)Y(y)$ (T7.3a)
$\frac{d^2Y_0}{dy^2}=0$	$X_0(x)=a_0+b_0x$	
$\frac{d^2X_0}{dx^2}=0$	$Y_0(y)=c_0+d_0y$	
Polar cylindrical coordinates	**Assumed separable form**	**Separable equations**
$\frac{\partial^2 u}{\partial r^2}+\frac{\partial u}{r\partial r}+\frac{\partial^2 u}{r^2\partial\theta^2}+\frac{\partial^2 u}{\partial z^2}=0$	$u(r,\theta,z)=Z_0(z)\Theta_0(\theta)+R(r)\Theta(\theta)Z(z)$	$\frac{1}{R}\left(\frac{d^2R}{dr^2}+\frac{dR}{rdr}\right)+\frac{d^2\Theta}{r^2\Theta d\theta^2}+\frac{d^2Z}{Zdz^2}=0$ $\frac{d^2Z_0}{Z_0dz^2}+\frac{d^2\Theta_0}{\Theta_0r^2d\theta^2}=0$
Separated equations	**Solutions to separated equations**	**Solution and special cases**
$\frac{d^2Z}{dz^2}-\kappa^2Z=0$	$Z(z)=E\cosh(\kappa z)+F\sinh(\kappa z)$ or $=Ee^{\kappa z}+Fe^{-\kappa z}$	

(Continued)

TABLE 7.3 (CONTINUED)
Solutions to the Laplace Equation in Three Coordinate Systems

$\frac{d^2\Theta}{d\theta^2}+m^2\Theta=0$ (S-L, $m\neq 0$)	$\Theta(\theta,m)=A\cos m\theta+B\sin m\theta$	
$r^2\frac{d^2R}{dr^2}+r\frac{dR}{dr}+\left(\kappa^2r^2-m^2\right)R=0$ (S-L, $\kappa\neq 0$)	$R_m(r,\kappa)=CJ_m(\kappa r)+DY_m(\kappa r)$ $R_m(r,0)=C_1r^m+D_1r^{-m}$ $R_0(r,0)=C_2+D_2\ln r$	$u(r,\theta,z)=Z_0(z)\Theta_0(\theta)$ $+R_m(r,\kappa)\Theta(\theta,m)Z(z)$ (T7.3b) $u(r,\theta)=\Theta_0(\theta)+R_m(r,0)\Theta(\theta,m)$ (T7.3c)
$\frac{d^2Z_0}{dz^2}=0$	$Z_0(z)=a_0+b_0z$	$u(r,z)=Z_0(z)+R_0(r,\kappa)Z(z)$ (T7.3d) $u(r)=R_0(r,0)$ (T7.3e)
$\frac{d^2\Theta_0}{d\theta^2}=0$	$\Theta_0(\theta)=c_0+d_0\theta$	
Spherical coordinates	**Assumed separable form**	**Separable equations**
$\frac{\partial^2u}{\partial r^2}+\frac{2}{r}\frac{\partial u}{\partial r}+\frac{\cot\varphi}{r^2}\frac{\partial u}{\partial\varphi}+\frac{1}{r^2}\frac{\partial^2u}{\partial\varphi^2}+\frac{1}{r^2\sin^2\varphi}\frac{\partial^2u}{\partial\theta^2}=0$	$u(r,\theta,z)=R(r)\Phi(\varphi)\Theta(\theta)$	$\frac{r^2}{R}\left(\frac{d^2R}{dr^2}+\frac{2}{r}\frac{dR}{dr}\right)+\frac{1}{\Phi}\left(\cot\varphi\frac{d\Phi}{d\varphi}+\frac{d^2\Phi}{d\varphi^2}\right)$ $+\frac{1}{\sin^2\varphi}\frac{d^2\Theta}{\Theta d\theta^2}=0$
Separated equations	**Solutions to separated equations**	**Solution and special cases**
$\frac{d^2\Theta}{d\theta^2}+m^2\Theta=0$ (S-L, $m\neq 0$)	$\Theta(\theta)=A\cos m\theta+B\sin m\theta$	
$\frac{d^2R}{dr^2}+\frac{2}{r}\frac{dR}{dr}-\frac{n(n+1)}{r^2}R=0$	$R(r)=Cr^n+Dr^{-n-1}$	$u(r,\varphi,\theta)=R(r)\Phi(\varphi)\Theta(\theta)$ (T7.3f)
$\frac{d^2\Phi}{d\varphi^2}+\cot\varphi\frac{d\Phi}{d\varphi}+\left(n(n+1)-\frac{m^2}{\sin^2\varphi}\right)\Phi=0$ (S-L)	$\Phi(\varphi)=EP_n^m(\cos\varphi)+FQ_n^m(\cos\varphi)$ $\Phi_0(\varphi)=E_0P_n(\cos\varphi)+F_0Q_n(\cos\varphi)$	$u(r,\varphi)=R(r)\Phi_0(\varphi)$ (T7.3g)

Note: The quantities m^2, λ^2, and κ^2 are real constants greater than zero and n is zero or a positive integer. The notation (S–L) indicates that this will be a Sturm–Liouville system when the boundary conditions are those given by Eq. (6.24).

The role of $u_0(x,y)$ will become apparent subsequently, where it will be seen that its primary purpose will be to convert certain nonhomogeneous boundary conditions into homogeneous ones. Substituting Eq. (7.22) into Eq. (7.21) gives

$$\frac{d^2u_0}{dx^2}+\frac{d^2u_0}{dy^2}+\frac{d^2w}{dx^2}+\frac{d^2w}{dy^2}=0 \tag{7.24}$$

or upon using Eq. (7.23)

$$\frac{d^2X_0}{dx^2}Y_0+\frac{d^2Y_0}{dy^2}X_0+\frac{d^2X}{dx^2}Y+\frac{d^2Y}{dy^2}X=0 \tag{7.25}$$

Note the switch from the partial derivative symbol to the regular derivative symbol. This is an important distinction because the integration constants will be constants, not a function of the other independent variable.

A solution to Eq. (7.25) can be obtained if we let $w(x,y)$ be a solution to

$$\frac{d^2X}{dx^2}Y+\frac{d^2Y}{dy^2}X=0 \tag{7.26}$$

and $u_0(x,y)$ be a solution to

$$\frac{d^2X_0}{dx^2}+\frac{d^2Y_0}{dy^2}=0 \tag{7.27}$$

We rewrite Eq. (7.26) as

$$\frac{d^2X}{Xdx^2}=-\frac{d^2Y}{Ydy^2}=-\lambda^2 \quad \lambda^2>0 \tag{7.28}$$

where λ^2 is called the *separation constant* and the minus sign is chosen so that the resulting equation in the variable x will be of the Sturm–Liouville type. Equation (7.28) is a result of the fact that each differential equation in Eq. (7.28) is a function of a different independent variable and the only way that they can be equal is for each to be equal to a quantity that is independent of x and y, that is, a constant. Therefore, Eq. (7.28) yields

$$\frac{d^2X}{dx^2}+\lambda^2X=0 \tag{7.29}$$

and

$$\frac{d^2Y}{dx^2}-\lambda^2Y=0 \tag{7.30}$$

It is seen that Eq. (7.29) is a differential equation of the Sturm–Liouville type when the boundary conditions are those given by Eq. (6.24) with y replaced by X. On the other hand, Eq. (7.30) is *not* a differential equation of the Sturm–Liouville type. Recall Eq. (6.44).

The solution to Eqs. (7.29) is obtained from Table 4.2 as

$$X(x)=A\cos(\lambda x)+B\sin(\lambda x) \tag{7.31}$$

and that to Eq. (7.30) is obtained from Table 4.2 as

$$\begin{aligned} Y(y) &= C\cosh(\lambda y) + D\sinh(\lambda y) \\ &= Ce^{\lambda y} + De^{-\lambda y} \quad \text{(alternate form)} \end{aligned} \tag{7.32}$$

We re-write Eq. (7.27) as

$$\frac{d^2X_0}{X_0dx^2} + \frac{d^2Y_0}{Y_0dy^2} = 0 \tag{7.33}$$

Therefore, the solution to Eq. (7.33) is

$$\frac{d^2X_0}{dx^2} = 0 \tag{7.34}$$

and

$$\frac{d^2Y_0}{dy^2} = 0 \tag{7.35}$$

In this case, we have set the separation constants to zero since the non-zero case has been considered in obtaining the solution to Eq. (7.26). The solutions to Eqs. (7.34) and (7.35), respectively, are

$$X_0(x) = a_0 + b_0x \tag{7.36}$$

and

$$Y_0(y) = c_0 + d_0y \tag{7.37}$$

Thus,

$$\begin{aligned} u_0(x,y) &= \left(a_0 + b_0x\right)\left(c_0 + d_0y\right) \\ w(x,y) &= \left(A\cos(\lambda x) + B\sin(\lambda x)\right)\left(C\cosh(\lambda y) + D\sinh(\lambda y)\right) \end{aligned} \tag{7.38}$$

or, if the alternate form for $Y(y)$ is used,

$$\begin{aligned} u_0(x,y) &= \left(a_0 + b_0x\right)\left(c_0 + d_0y\right) \\ w(x,y) &= \left(A\cos(\lambda x) + B\sin(\lambda x)\right)\left(Ce^{\lambda y} + De^{-\lambda y}\right) \end{aligned} \tag{7.39}$$

Thus, the solution to Eq. (7.21) is obtained by substituting Eqs. (7.38) or Eq. (7.39) into Eq. (7.22).

In the problems that we shall consider in the remainder of this section, the value of λ will be determined by satisfying homogeneous boundary conditions. The remaining constants will be determined from the satisfaction of the remaining boundary conditions.

We shall now solve the Laplace equation in different coordinate systems for different combinations of boundary conditions. The applications considered are summarized in Table 7.4.

TABLE 7.4
Summary of Examples of Applications of the Laplace Equation: Boundary Conditions and Coordinates Systems

Example	Laplace Equation	Boundary Conditions
7.1	$\frac{\partial^2 u}{\partial x^2}+\frac{\partial^2 u}{\partial y^2}=0$	$u(0,y)=0, \quad u(a,y)=0$ $u(x,0)=0, \quad u(x,b)=g(x)$
7.2	$\frac{\partial^2 u}{\partial x^2}+\frac{\partial^2 u}{\partial y^2}=0$	$\frac{\partial u(0,y)}{\partial x}=0, \quad \frac{\partial u(a,y)}{\partial x}=0$ $u(x,0)=0, \quad u(x,b)=g(x)$
7.3	$\frac{\partial^2 u}{\partial x^2}+\frac{\partial^2 u}{\partial y^2}=0$	$u(0,y)=0, \quad u(a,y)=U_o$ $u(x,0)=f(x), \quad u(x,y\to\infty)=\text{finite}$
7.4	$\frac{\partial^2 u}{\partial r^2}+\frac{\partial u}{r\partial r}+\frac{\partial^2 u}{r^2\partial\theta^2}=0$	$u(r,0)=0, \quad u(r,\alpha)=0$ $u(a,\theta)=0, \quad u(b,\theta)=f(\theta)$
7.5	$\frac{\partial^2 u}{\partial r^2}+\frac{\partial u}{r\partial r}+\frac{\partial^2 u}{\partial z^2}=0$	$u(a,z)=0, \quad u(0,z)=\text{finite}$ $u(r,0)=0, \quad u(r,L)=V$
7.6	$\frac{\partial^2 u}{\partial r^2}+\frac{2}{r}\frac{\partial u}{\partial r}+\frac{\cot\varphi}{r^2}\frac{\partial u}{\partial\varphi}+\frac{1}{r^2}\frac{\partial^2 u}{\partial\varphi^2}=0$	$u(a,\varphi)=f(\varphi) \quad 0\le\varphi\le\pi$

Example 7.1: Laplace equation in Cartesian coordinates with Dirichlet boundary conditions

We shall obtain a solution to

$$\frac{\partial^2 u}{\partial x^2}+\frac{\partial^2 u}{\partial y^2}=0 \quad 0<x<a \quad 0<y<b \tag{a}$$

over the region indicated with the following boundary conditions

$$\begin{aligned} &u(0,y)=0\\ &u(a,y)=0\\ &u(x,0)=0\\ &u(x,b)=g(x) \quad \text{where } g(0)=g(a)=0 \end{aligned} \tag{b}$$

The requirements on $g(x)$ are necessary so that there is no discontinuity at the corners.

The solution to Eq. (a) is obtained from Eq. (T7.3a) of Table 7.3 as

$$u(x,y)=X_0(x)Y_0(y)+X(\lambda,x)Y(\lambda,y) \tag{c}$$

where, from Table 7.3,

$$
\begin{aligned}
X_0(x) &= a_0 + b_0 x \\
Y_0(y) &= c_0 + d_0 y \\
X(\lambda, x) &= A\cos(\lambda x) + B\sin(\lambda x) \\
Y(\lambda, y) &= C\cosh(\lambda y) + D\sinh(\lambda y)
\end{aligned} \tag{d}
$$

Using Eqs. (c) and (d), the first boundary condition given in Eq. (b) yields

$$a_0 Y_0(y) + AY(\lambda, y) = 0 \tag{e}$$

It is seen that Eq. (e) is only satisfied for arbitrary y when $a_0 = 0$, and $A = 0$. Thus, Eq. (c) becomes

$$u(x, y) = b_0 x Y_0(y) + B\sin(\lambda x)Y(\lambda, y) \tag{f}$$

Substituting Eq. (f) into the second boundary condition of Eq. (b) yields

$$b_0 a Y_0(y) + B\sin(\lambda a)Y(\lambda, y) = 0 \tag{g}$$

Equation (g) is satisfied for arbitrary y if $b_0 = 0$ and

$$\sin(\lambda a) = 0$$

which is the characteristic equation. The solution to this equation is

$$\lambda a = n\pi$$

or

$$\lambda \to \lambda_n = n\pi / a \quad n = 1, 2, \ldots$$

where λ_n is the eigenvalue. Equation (g) can also be satisfied by setting $B = 0$; however, this results in $u(x,y) = 0$, which is a trivial solution. Thus, since $a_0 = b_0 = 0$, $X_0(x) = 0$ and Eq. (f) simplifies further to

$$u_n(x, y) = \sin(\lambda_n x)Y(\lambda_n, y) \tag{h}$$

where the constant B has been subsumed into the constants of $Y(\lambda_n, y)$. The quantity $\sin(\lambda_n x)$ is the eigenfunction.

From Table 7.3, the function X satisfies the Sturm–Liouville equation

$$\frac{d^2X}{dx^2} + \lambda^2 X = 0$$

From the first two boundary conditions in Eq. (b), it is seen that the boundary conditions are

$$
\begin{aligned}
X(\lambda, 0) &= 0 \\
X(\lambda, a) &= 0
\end{aligned}
$$

The solution to this homogeneous equation with homogeneous boundary conditions is the eigenfunction $\sin(\lambda_n x)$ with the corresponding eigenvalue λ_n. However, based on the material in Section 6.1.4, it was shown that the eigenfunction is an orthogonal eigenfunction on the interval $0 \le x \le a$. From Section 6.1.3, it was found that the weighting factor is one. We shall make use of the orthogonality property subsequently.

Since Eq. (h) satisfies the differential equation and boundary conditions for all n, we use superposition and sum all these solutions to obtain

$$\begin{aligned} u(x,y) &= \sum_{n=1}^{\infty} \sin(\lambda_n x) Y(\lambda_n, y) \\ &= \sum_{n=1}^{\infty} \sin(n\pi x/a)\big(C_n \cosh(n\pi y/a) + D_n \sinh(n\pi y/a)\big) \end{aligned} \tag{i}$$

where we have used Eq. (d).

We now substitute Eq. (i) into the third boundary condition of Eq. (b) and find that

$$\sum_{n=1}^{\infty} C_n \sin(n\pi x/a) = 0 \tag{j}$$

which is satisfied for arbitrary x if $C_n = 0$. Then, Eq. (i) reduces to

$$u(x,y) = \sum_{n=1}^{\infty} D_n \sinh(n\pi y/a)\sin(n\pi x/a) \tag{k}$$

Lastly, we substitute Eq. (k) into the last boundary condition of Eq. (b) and find that

$$\sum_{n=1}^{\infty} D_n \sinh(n\pi b/a)\sin(n\pi x/a) = g(x) \tag{l}$$

Since $\sin(n\pi x/a)$ is an orthogonal function, we can multiply Eq. (l) by $\sin(m\pi x/a)$ and integrate over the interval to obtain

$$\begin{aligned} \sum_{n=1}^{\infty} D_n \sinh(n\pi b/a)\int_0^a \sin(n\pi x/a)\sin(m\pi x/a)dx \\ = \int_0^a g(x)\sin(m\pi x/a)dx \quad m = 1,2,\ldots \\ \frac{a}{2}\sum_{n=1}^{\infty} D_n \sinh(n\pi b/a)\delta_{nm} = \int_0^a g(x)\sin(m\pi x/a)dx \\ D_n = \frac{2}{a\sinh(n\pi b/a)}\int_0^a g(x)\sin(n\pi x/a)dx \end{aligned} \tag{m}$$

where δ_{nm} is the Kronecker delta that we used in Eq. (3.20). Then, using Eq. (m) in Eq. (k), the solution is

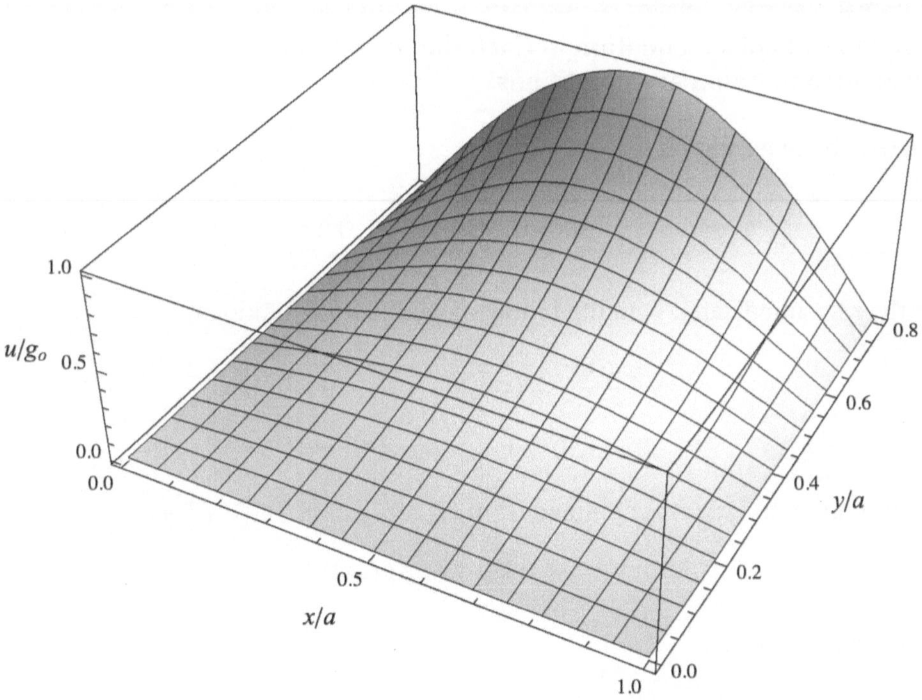

FIGURE 7.1 Numerical evaluation of Eq. (o) of Example 7.1 for $b/a = 0.8$.

$$u(x,y) = \frac{2}{a}\sum_{n=1}^{\infty}\frac{\sin(n\pi x/a)\sinh(n\pi y/a)}{\sinh(n\pi b/a)}\int_0^a g(x)\sin(n\pi x/a)dx \tag{n}$$

We illustrate this result by letting $g(x) = g_o\sin(\pi x/a)$, which happens to be equal to one of the orthogonal solution functions. Then

$$\int_0^a g(x)\sin(n\pi x/a)dx = g_o\int_0^a \sin(\pi x/a)\sin(n\pi x/a)dx = \frac{ag_o}{2}\delta_{1n}$$

and Eq. (n) becomes

$$u(x,y) = g_o\frac{\sin(\pi x/a)\sinh(\pi y/a)}{\sinh(\pi b/a)} \tag{o}$$

Notice that Eq. (o) is a one-term solution. This is because we chose $g(x)$ to be one of the orthogonal functions of the solution $X(x)$; consequently, all the other terms in the solution are orthogonal to this function except $\sin(\pi x/a)$ resulting in all the D_n being equal to zero except D_1.

Using Eqs. (a) and Eq. (b) and the expression for $g(x)$, Eq. (o) is verified with Mathematica procedure M7.2.

A plot of $u(x,y)/g_o$ for $b/a = 0.8$ is shown in Figure 7.1. This figure was obtained with Mathematica procedure M7.3.

Example 7.2: Laplace equation in Cartesian coordinates with Neumann boundary conditions

We shall obtain a solution to

$$\frac{\partial^2 u}{\partial x^2}+\frac{\partial^2 u}{\partial y^2}=0 \quad 0<x<a \quad 0<y<b \tag{a}$$

over the region indicated with the following boundary conditions

$$\begin{aligned}\frac{\partial u(0,y)}{\partial x}&=0, & \frac{\partial u(a,y)}{\partial x}&=0\\ u(x,0)&=0, & u(x,b)&=g(x)\end{aligned} \tag{b}$$

The solution to Eq. (a) is obtained from Eq. (T7.3a) of Table 7.3 as

$$u(x,y)=X_0(x)Y_0(y)+X(\lambda,x)Y(\lambda,y) \tag{c}$$

where, from Table 7.3,

$$\begin{aligned}X_0(x)&=a_0+b_0x\\ Y_0(y)&=c_0+d_0y\\ X(\lambda,x)&=A\cos(\lambda x)+B\sin(\lambda x)\\ Y(\lambda,y)&=C\cosh(\lambda y)+D\sinh(\lambda y)\end{aligned} \tag{d}$$

Noting that

$$\frac{\partial u(x,y)}{\partial x}=b_0Y_0(y)+\left(-A\lambda\sin(\lambda x)+B\lambda\cos(\lambda x)\right)Y(\lambda,y) \tag{e}$$

we find from the first boundary condition of Eq. (b) that

$$b_0Y_0(y)+B\lambda Y(\lambda,y)=0 \tag{f}$$

To satisfy Eq. (f) for arbitrary y it is necessary that $b_0=B=0$. Then, Eq. (c) becomes

$$u(x,y)=Y_0(y)+\cos(\lambda x)Y(\lambda,y) \tag{g}$$

where a_0 and A, respectively, have been incorporated into the remaining constants of Y_0 and Y.

Substituting Eq. (g) into the second boundary condition given by Eq. (b) yields

$$-\lambda\sin(\lambda a)Y(\lambda,y)=0 \tag{h}$$

To avoid a trivial solution, that is, $Y=0$, Eq. (h) is satisfied if

$$\sin(\lambda a)=0$$

which is the characteristic equation. The solution to this equation is

$$\lambda a = n\pi \quad n = 1,2,\ldots$$

or

$$\lambda \to \lambda_n = n\pi / a$$

where λ_n is the eigenvalue. Then, Eq. (g) can be written as

$$u_n(x,y) = Y_0(y) + \cos(\lambda_n x)Y(\lambda_n, y) \quad n = 1,2,\ldots$$

From Table 7.3, the function X satisfies the Sturm–Liouville equation

$$\frac{d^2X}{dx^2} + \lambda^2 X = 0$$

From the first two boundary conditions in Eq. (b), it is seen that the boundary conditions are

$$X'(\lambda,0) = 0$$

$$X'(\lambda,a) = 0$$

The solution to this homogeneous equation and homogeneous boundary conditions is the eigenfunction $\cos(\lambda_n x)$ with the corresponding eigenvalue λ_n. However, based on the material in Section 6.1.4, it was shown that the eigenfunction is an orthogonal eigenfunction on the interval $0 \le x \le a$. From Section 6.1.3, it was found that the weighting factor is one.

Since Eq. (i) satisfies the differential equation and the first two boundary conditions of Eq. (b) for all n, we use superposition to obtain

$$\begin{aligned} u(x,y) &= Y_0(y) + \sum_{n=1}^{\infty} \cos(\lambda_n x)Y(\lambda_n, y) \\ &= \left(c_0 + d_0 y\right) \\ &\quad + \sum_{n=1}^{\infty} \left(E_n \cosh(n\pi y/a) + F_n \sinh(n\pi y/a)\right)\cos(n\pi x/a) \end{aligned} \tag{j}$$

where we have used Eq. (d).

We now substitute Eq. (j) into the third boundary condition of Eq. (b) to find that

$$c_0 + \sum_{n=1}^{\infty} E_n \cos(n\pi x/a) = 0$$

which for arbitrary x the equation is only satisfied when $E_n = 0$ and, consequently, $c_0 = 0$. Therefore, Eq. (j) becomes

$$u(x,y) = d_0 y + \sum_{n=1}^{\infty} F_n \sinh(n\pi y/a)\cos(n\pi x/a) \tag{k}$$

To satisfy the last boundary condition given in Eq. (b), we substitute Eq. (k) into this boundary condition and obtain

$$d_0 b + \sum_{n=1}^{\infty} F_n \cos(n\pi x/a)\sinh(n\pi b/a) = g(x) \tag{l}$$

To determine d_0, we first integrate Eq. (l) over the interval to find that

$$\begin{aligned} d_0 b\int_0^a dx + \sum_{n=1}^{\infty} F_n \sinh(n\pi b/a)\int_0^a \cos(n\pi x/a)dx &= \int_0^a g(x)dx \\ d_0 &= \frac{1}{ab}\int_0^a g(x)dx \end{aligned} \tag{m}$$

where we have used Eq. (3.20), that is,

$$\int_0^a \cos(n\pi x/a)dx = 0$$

This operation is equivalent to determining a_0 in the Fourier series expansion given by Eq. (3.18). To determine F_n, we multiply Eq. (l) by $\cos(m\pi x/a)$ and integrate over the interval to arrive at

$$\begin{aligned} d_0 b\int_0^a \cos(m\pi x/a)dx + \sum_{n=1}^{\infty} F_n \sinh(n\pi b/a)\int_0^a \cos(n\pi x/a)\cos(m\pi x/a)dx \\ = \int_0^a g(x)\cos(m\pi x/a)dx \quad m = 1,2,\ldots \end{aligned}$$

Therefore,

$$F_n = \frac{2}{a\sinh(n\pi b/a)}\int_0^a g(x)\cos(n\pi x/a)dx \tag{n}$$

where we have used Eq. (3.20).

The final solution is obtained by using Eqs. (m) and (n) in Eq. (k) to obtain

$$u(x,y) = \frac{y}{ab}\int_0^a g(x)dx + \frac{2}{a}\sum_{n=1}^{\infty} \frac{\cos(n\pi x/a)\sinh(n\pi y/a)}{\sinh(n\pi b/a)}\int_0^a g(x)\cos(n\pi x/a)dx \tag{o}$$

Using Eqs. (a) and Eq. (b), Eq. (o) is verified with Mathematica procedure M7.4.

To illustrate this result, we let

$$g(x) = 0 \qquad 0 \le x < a/4$$
$$= g_o \qquad a/4 \le x \le 3a/4$$
$$= 0 \qquad 3a/4 < x \le a$$

Then

$$\int_0^a g(x)\cos(n\pi x/a)dx = g_o \int_{a/4}^{3a/4} \cos(n\pi x/a)dx$$

$$= \frac{g_o a}{n\pi}\left[\sin(3n\pi/4) - \sin(n\pi/4)\right]$$

and

$$\int_0^a g(x)dx = g_o \int_{a/4}^{3a/4} dx = \frac{g_o a}{2}$$

Then, Eq. (o) becomes

$$u(x,y) = \left(\frac{y}{a}\right)\frac{g_o}{2b/a} + \frac{2g_o}{\pi}\sum_{n=1}^{\infty}\frac{\cos(n\pi x/a)\sinh(n\pi y/a)}{n\sinh(n\pi b/a)}\left(\sin(3n\pi/4) - \sin(n\pi/4)\right) \qquad \text{(p)}$$

A plot of $u(x,y)/g_o$ for $b/a = 0.8$ is shown in Figure 7.2. This figure was obtained with Mathematica procedure M7.5.

Example 7.3: Laplace equation in Cartesian coordinates with nonhomogeneous boundary conditions

We shall obtain a solution to

$$\frac{\partial^2 u}{\partial x^2} + \frac{\partial^2 u}{\partial y^2} = 0 \quad 0 < x < a \quad 0 < y < \infty \qquad \text{(a)}$$

on a semi-infinite strip indicated with the following boundary conditions

$$u(0,y) = 0$$
$$u(a,y) = U_o$$
$$u(x,0) = f(x) \qquad \text{(b)}$$
$$u(x,\infty) = \text{finite}$$

The solution to Eq. (a) is obtained from Eq. (T7.3a) of Table 7.3, that is,

$$v(x,y) = (a_0 + b_0 x)(c_0 + d_0 y) + (A\cos(\lambda x) + B\sin(\lambda x))(Ce^{\lambda y} + De^{-\lambda y}) \qquad \text{(c)}$$

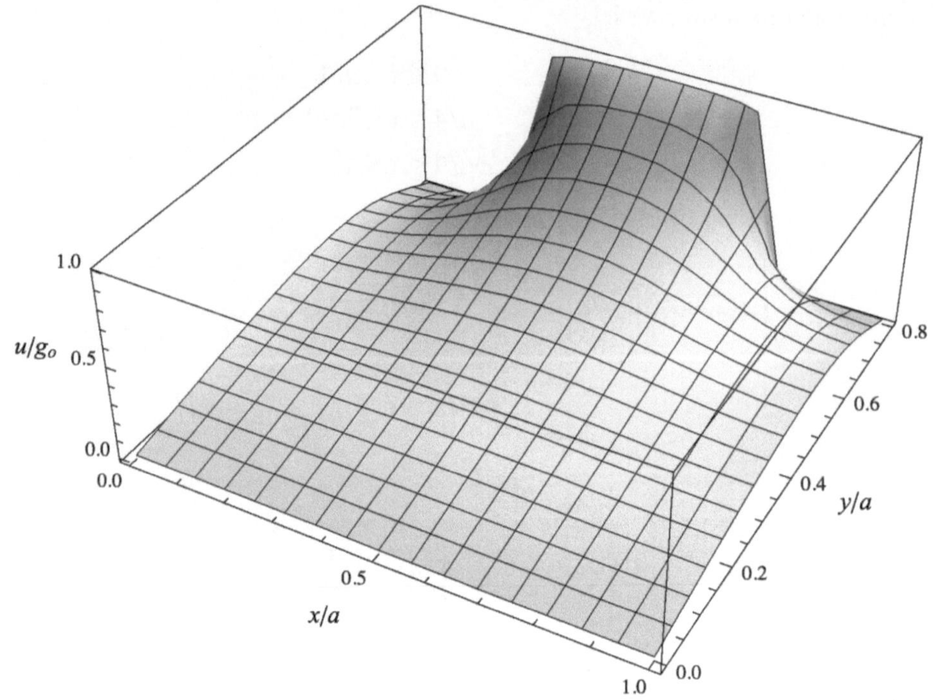

FIGURE 7.2 Numerical evaluation of Eq. (p) of Example 7.2 for $b/a = 0.8$.

where we have chosen the exponential form of the solution in y in anticipation of boundary condition at infinity. We see that for the solution to remain finite as $y \to \infty$, we must set $C = 0$ and $d_0 = 0$ in Eq. (c). Then Eq. (c) becomes

$$v(x,y) = a_0 + b_0 x + \left(A\cos(\lambda x) + B\sin(\lambda x)\right)e^{-\lambda y} \tag{d}$$

and Eq. (d) now satisfies the fourth boundary condition given in Eq. (b). The constants D and c_0 have been incorporated into the remaining constants.

Substituting Eq. (d) into the first boundary condition of Eq. (b) yields

$$a_0 + Ae^{-\lambda y} = 0$$

Therefore, for this boundary condition to be satisfied for arbitrary y, $A = 0$ and, therefore, $a_0 = 0$. Then, Eq. (d) simplifies further to

$$v(x,y) = b_0 x + B\sin(\lambda x)e^{-\lambda y} \tag{e}$$

Substituting Eq. (e) into the second boundary condition of Eq. (b) yields

$$b_0 a + B\sin(\lambda a)e^{-\lambda y} = U_0 \tag{f}$$

A solution to Eq. (f) that does not lead to a trivial solution is to select $b_0 = U_0/a$ and to set

$$\sin(\lambda a) = 0$$

which is the characteristic equation. The solution to this equation is

$$\lambda \to \lambda_n = n\pi / a \quad n = 1,2,\ldots$$

where λ_n is the eigenvalue. With these values, Eq. (e) becomes

$$v_n(x,y) = U_o \frac{x}{a} + B_n \sin(\lambda_n x)e^{-\lambda_n y}$$

Since the solutions are valid for all n, we use superposition to obtain

$$v(x,y) = U_o \frac{x}{a} + \sum_{n=1}^{\infty} B_n \sin(n\pi x/a)e^{-n\pi y/a} \tag{g}$$

It is noted from Table 7.3 that the function X is the solution to a Sturm–Liouville equation and the homogeneous boundary conditions

$$X(0) = 0, \quad X(a) = 0$$

The solution to this homogeneous equation and homogeneous boundary conditions is the eigenfunction $\sin(\lambda_n x)$ with the corresponding eigenvalue λ_n. However, based on the material in Section 6.1.4, it was shown that the eigenfunction is an orthogonal eigenfunction on the interval $0 \le x \le a$. From Section 6.1.3, it was found that the weighting factor is one.

Substituting Eq. (g) into the third boundary condition of Eq. (b) leads to

$$\sum_{n=1}^{\infty} B_n \sin(n\pi x/a) = f(x) - U_o x / a \tag{h}$$

To determine those B_n that satisfy this relation, we multiply Eq. (h) by $\sin(m\pi x/a)$, and integrate over the interval to obtain

$$\sum_{n=1}^{\infty} B_n \int_0^a \sin(n\pi x/a)\sin(m\pi x/a)dx = \int_0^a \left[f(x) - U_o x / a\right]\sin(m\pi x/a)dx \qquad m = 1,2,\ldots$$

$$B_n = \frac{2}{a}\int_0^a \left[f(x) - U_o x/a\right]\sin(n\pi x/a)dx \tag{i}$$

$$= \frac{2}{a}\int_0^a f(x)\sin(n\pi x/a)dx - \frac{2U_o(-1)^n}{n\pi}$$

and we have used Eq. (3.20). Then the final solution is obtained by using Eqs. (g) and (i) to arrive at

$$u(x,y) = U_o x/a + \sum_{n=1}^{\infty}\left[f_n - \frac{2U_o(-1)^n}{n\pi}\right]e^{-n\pi y/a}\sin(n\pi x/a) \tag{j}$$

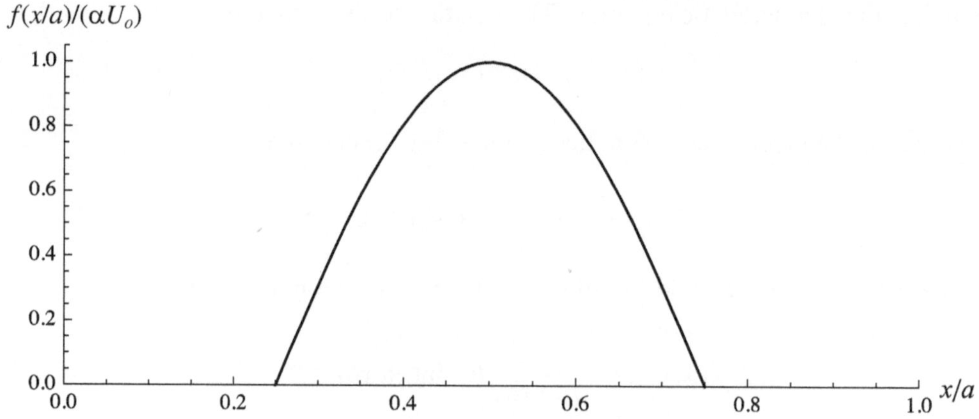

FIGURE 7.3 Boundary condition at $y = 0$ used to obtain Eq. (l) of Example 7.3.

where

$$f_n = \frac{2}{a}\int_0^a f(x)\sin(n\pi x/a)dx \qquad \text{(k)}$$

With a little rearrangement of the Eqs. (a) and (b), Eq. (j) is verified with Mathematica procedure M7.6.

To illustrate these results, we let

$$\begin{aligned} f(x) &= 0 & 0 \le x < a/4 \\ &= \alpha U_o \sin[2\pi(x/a - 1/4)] & a/4 \le x \le 3a/4 \\ &= 0 & 3a/4 < x \le a \end{aligned}$$

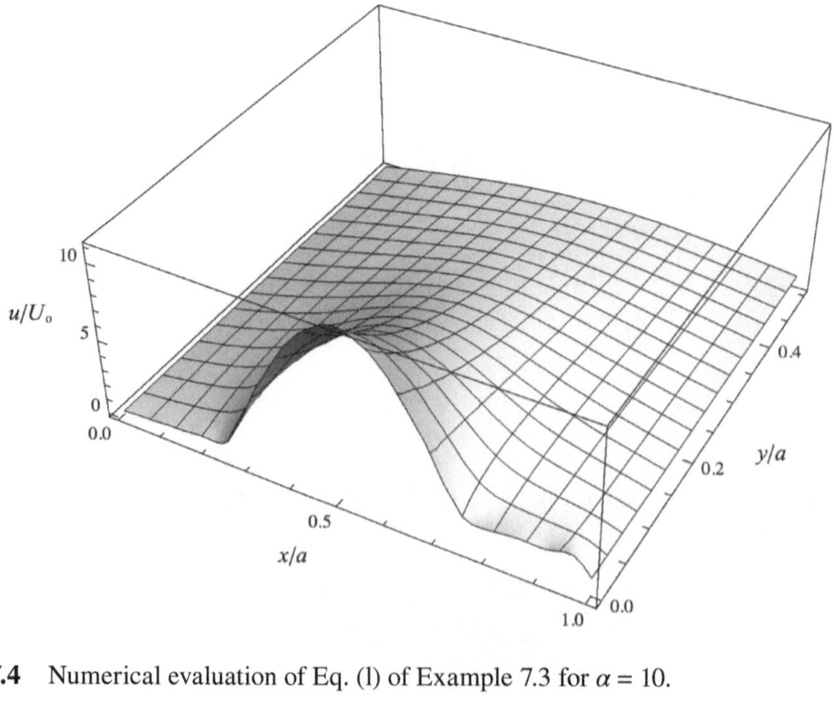

FIGURE 7.4 Numerical evaluation of Eq. (l) of Example 7.3 for $\alpha = 10$.

which is shown in Figure 7.3. Then, from Eq. (k)

$$f_n = \frac{2}{a}\int_0^a f(x)\sin(n\pi x/a)dx = \frac{2\alpha U_o}{a}\int_{a/4}^{3a/4}\sin[2\pi(x/a-1/4)]\sin(n\pi x/a)dx$$

$$= -\frac{4\alpha U_o}{\pi} g_n$$

where

$$g_n = \frac{\sin(n\pi/4)+\sin(3n\pi/4)}{n^2-4} \qquad n \neq 2$$

$$= 0 \qquad n = 2$$

Then, Eq. (j) becomes

$$\frac{u(x,y)}{U_o} = \frac{x}{a} - \frac{2}{\pi}\sum_{n=1}^{\infty}\left(2\alpha g_n + \frac{(-1)^n}{n}\right)e^{-n\pi y/a}\sin(n\pi x/a) \tag{l}$$

A plot of Eq. (l) for $\alpha = 10$ is shown in Figure 7.4. This figure was obtained with Mathematica procedure M7.7.

Example 7.4: Laplace equation in polar coordinates with Dirichlet boundary conditions

We shall obtain a solution to

$$\frac{\partial^2 u}{\partial r^2} + \frac{\partial u}{r\partial r} + \frac{\partial^2 u}{r^2\partial\theta^2} = 0 \quad 0 < a < r < b \quad 0 < \theta < \alpha < 2\pi \tag{a}$$

over the annular sector shown in Figure 7.5 and with the following boundary conditions

$$\begin{aligned} u(r,0) &= 0 \\ u(r,\alpha) &= 0 \\ u(a,\theta) &= 0 \\ u(b,\theta) &= f(\theta) \end{aligned} \tag{b}$$

The solution to Eq. (a) is obtained from Eq. (T7.3c) of Table 7.3 as

$$u(r,\theta) = \Theta_0(\theta) + R_m(r,0)\Theta(\theta,m) \tag{c}$$

where, from Table 7.3,

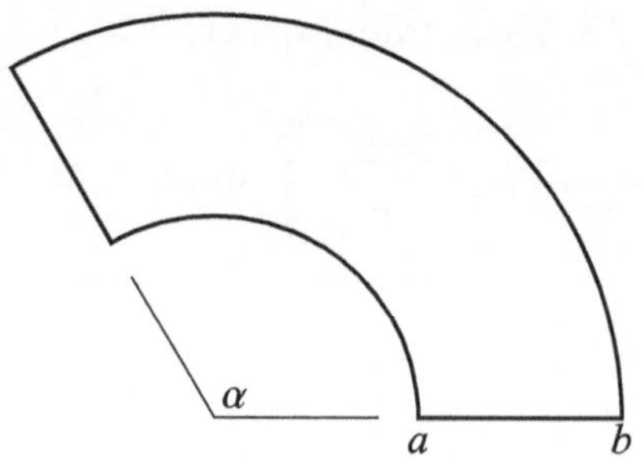

FIGURE 7.5 Annular sector of the region examined in Example 7.4.

$$R_m(r,0) = Cr^m + Dr^{-m}$$

$$\Theta(\theta,m) = A\cos m\theta + B\sin m\theta \tag{d}$$

$$\Theta_0(\theta) = c_0 + d_0\theta$$

Substituting Eq. (c) into the first boundary condition of Eq. (b), we obtain

$$c_0 + R_m(r,0)A = 0$$

which, for arbitrary r, is satisfied when $A = 0$ and, therefore, $c_0 = 0$. Then Eq. (c) becomes

$$u(r,\theta) = d_0\theta + R_m(r,0)\sin m\theta \tag{e}$$

where B has been incorporated into the remaining constants.

Substituting Eq. (e) into the second boundary condition given by Eq. (b), we find

$$d_0\alpha + R_m(r,0)\sin m\alpha = 0$$

To obtain a non-trivial solution for arbitrary r, we set

$$\sin m\alpha = 0$$

which is the characteristic equation and, therefore, $d_0 = 0$. The solution to this equation is

$$m \to m_n = n\pi/\alpha \quad n = 1,2,\ldots$$

where m_n is the eigenvalue. Then, using Eq. (d), Eq. (e) can be written as

$$u_n(r,\theta) = \left(C_n r^{n\pi/\alpha} + D_n r^{-n\pi/\alpha}\right)\sin\left(n\pi\theta/\alpha\right)$$

Since this result is valid for all n, we use superposition and obtain

$$u(r,\theta) = \sum_{n=1}^{\infty}\left(C_n r^{n\pi/\alpha} + D_n r^{-n\pi/\alpha}\right)\sin\left(n\pi\theta/\alpha\right) \tag{f}$$

It is seen from Table 7.3 that Θ satisfies the Sturm–Liouville equation

$$\frac{d^2\Theta}{d\theta^2} + m^2\Theta = 0 \tag{g}$$

From the first two boundary conditions in Eq. (b), it is seen that $\Theta(\theta)$ satisfies the Dirichlet boundary conditions

$$\begin{aligned} \Theta(0) &= 0 \\ \Theta(\alpha) &= 0 \end{aligned} \tag{h}$$

The solution to this homogeneous equation and homogeneous boundary conditions is the eigenfunction $\sin(m_n\theta)$ with the corresponding eigenvalue m_n. However, based on the material in Section 6.1.4, it was shown that the eigenfunction is an orthogonal eigenfunction on the interval $0 \le \theta \le \alpha < 2\pi$. From Section 6.1.3, it was found that the weighting factor is one.

Substituting Eq. (f) into the third boundary condition of Eq. (b) gives

$$u(a,\theta) = \sum_{n=1}^{\infty}\left(C_n a^{n\pi/\alpha} + D_n a^{-n\pi/\alpha}\right)\sin\left(n\pi\theta/\alpha\right) = 0 \tag{i}$$

which means that to satisfy Eq. (i) for arbitrary θ

$$D_n = -C_n a^{2n\pi/\alpha}$$

Using this result in Eq. (f) gives

$$u(r,\theta) = \sum_{n=1}^{\infty} C_n\left(r^{n\pi/\alpha} - a^{2n\pi/\alpha} r^{-n\pi/\alpha}\right)\sin\left(n\pi\theta/\alpha\right) \tag{j}$$

To obtain the final constant, we substitute Eq. (j) into the fourth boundary condition of Eq. (b) and arrive at

$$\sum_{n=1}^{\infty} C_n\left(b^{n\pi/\alpha} - a^{2n\pi/\alpha} b^{-n\pi/\alpha}\right)\sin\left(n\pi\theta/\alpha\right) = f(\theta)$$

We multiply this equation by $\sin(k\pi\theta/\alpha)$ and integrate from 0 to α and obtain

$$C_n = \frac{2}{\alpha\left(b^{n\pi/\alpha} - a^{2n\pi/\alpha} b^{-n\pi/\alpha}\right)}\int_0^{\alpha} f(\theta)\sin\left(n\pi\theta/\alpha\right)d\theta \tag{k}$$

In arriving at Eq. (k), we have used Eq. (3.20). The final expression is obtained by using Eq. (k) in Eq. (j), which can be written as

$$u(\xi,\psi) = \frac{2}{\alpha}\sum_{n=1}^{\infty} \xi^{n\pi/\alpha}\frac{1-(\beta/\xi)^{2n\pi/\alpha}}{1-\beta^{2n\pi/\alpha}}\sin\left(n\pi\psi\right)\int_0^{\alpha} f(\theta)\sin\left(n\pi\theta/\alpha\right)d\theta \tag{l}$$

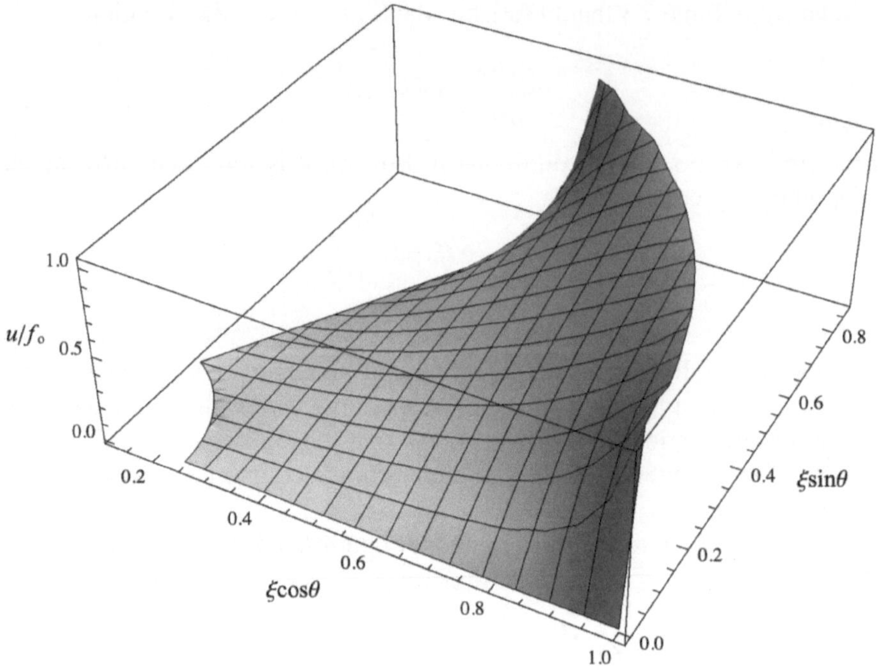

FIGURE 7.6 Numerical evaluation of Eq. (m) of Example 7.4 for $\beta = 0.25$ and $\alpha = \pi/3$.

where $\xi = r/b$, $\beta \le \xi \le 1$, $\beta = a/b < 1$, and $\psi = \theta/\alpha$, $0 \le \psi \le 1$.

There are several special cases of Eq. (l): (i) $\alpha = 2\pi$, we have an annular disk; (ii) $a = 0$ ($\beta = 0$), we have a sector of a solid disk; (iii) $\alpha = 2\pi$ and $a = 0$ ($\beta = 0$), we have a solid disk. These cases are straightforward to obtain from Eq. (l).

To illustrate these results, we let $f(\theta) = f_o$ for $0 \le \theta \le \alpha$; then

$$\int_0^\alpha f(\theta)\sin(n\pi\theta/\alpha)d\theta = f_o\int_0^\alpha \sin(n\pi\theta/\alpha)d\theta = \frac{2\alpha}{n\pi}f_o \quad n = 1,3,5,\ldots$$

and Eq. (l) becomes

$$u(r,\theta) = \frac{4f_o}{\pi}\sum_{n=1,3,5}^{\infty}\frac{\xi^{n\pi/\alpha}}{n}\frac{1-(\beta/\xi)^{2n\pi/\alpha}}{1-\beta^{2n\pi/\alpha}}\sin(n\pi\theta/\alpha) \tag{m}$$

A plot of $u(r,\theta)/f_o$ for $\beta = 0.25$ and $\alpha = \pi/3$ is shown in Figure 7.6. This figure was obtained with Mathematica procedure M7.8.

Example 7.5: Laplace equation in polar cylindrical coordinates with Dirichlet boundary conditions

We shall obtain a solution to

$$\frac{\partial^2 u}{\partial r^2} + \frac{\partial u}{r\partial r} + \frac{\partial^2 u}{\partial z^2} = 0 \quad 0 < r < a \quad 0 < z < L \tag{a}$$

over the region indicated and with the following boundary conditions

$$\begin{aligned} u(a,z) &= 0 \\ u(0,z) &= \text{finite} \\ u(r,0) &= 0 \\ u(r,L) &= V \end{aligned} \tag{b}$$

The solution to Eq. (a) is obtained from Eq. (T7.3d) of Table 7.3 as

$$u(r,z) = Z_0(z) + R_0(\kappa,r)Z(\kappa,z) \tag{c}$$

where, from Table 7.3,

$$\begin{aligned} R_0(\kappa,r) &= CJ_0(\kappa r) + DY_0(\kappa r) \\ Z(\kappa,z) &= E\cosh(\kappa z) + F\sinh(\kappa z) \\ Z_0(z) &= a_0 + b_0 z \end{aligned} \tag{d}$$

From the second boundary condition of equation of Eq. (b), we see that u must remain finite at $r = 0$; therefore, we set $D = 0$. Then Eq. (c) becomes

$$u(r,z) = a_0 + b_0 z + J_0(\kappa r)Z(\kappa,z) \tag{e}$$

and the constant C is incorporated into the constants in $Z(\kappa,z)$. Substituting Eq. (e) into the first boundary condition of Eq. (b) gives

$$a_0 + b_0 z + J_0(\kappa a)Z(\kappa,z) = 0$$

For arbitrary z, this equation is satisfied when $b_0 = 0$,

$$J_0(\kappa a) = 0$$

and, therefore, $a_0 = 0$. The solutions to this equation are those values of λ_n that satisfy

$$J_0(\lambda_n) = 0 \quad n = 1,2,\ldots \tag{f}$$

where λ_n is the eigenvalue. Therefore, $\kappa a = \lambda_n$ are the values that satisfy Eq. (f) and, hence, $\kappa = \lambda_n/a$ and Eq. (e) becomes

$$u_n(r,z) = J_0(\lambda_n r / a)Z(\lambda_n / a, z)$$

Using superposition, we obtain

$$u(r,z) = \sum_{n=1}^{\infty} J_0(\lambda_n r / a)\left(E_n \cosh(\lambda_n z/a) + F_n \sinh(\lambda_n z/a)\right) \tag{g}$$

where we have used Eq. (d).

From Table 7.3 it is seen that $R_0(\kappa,r)$ satisfies the Sturm–Liouville equation

$$r^2\frac{d^2R_0}{dr^2}+r\frac{dR_0}{dr}+\kappa^2r^2R_0=0 \tag{h}$$

From the first boundary condition in Eq. (b), it is seen that $R_0(r)$ satisfies

$$R_0(a)=0 \tag{i}$$

The second boundary condition corresponds to Special Case II of Section 6.1.4, which considers the boundedness requirement. Hence, the solution to this homogeneous equation and homogeneous boundary condition and boundedness requirement is the orthogonal eigenfunction $J_0(\lambda_n r/a)$ on the interval $0 \le x \le a$ with the corresponding eigenvalue λ_n. From Section 6.1.3, it was found that the weighting factor is r.

We now substitute Eq. (g) into the third boundary condition given by Eq. (b) and obtain

$$\sum_{n=1}^{\infty}E_nJ_0(\lambda_n r/a)=0 \tag{j}$$

which, for arbitrary r, is satisfied when $E_n = 0$. Hence, Eq. (g) further reduces to

$$u(r,z)=\sum_{n=1}^{\infty}F_nJ_0(\lambda_n r/a)\sinh(\lambda_n z/a) \tag{k}$$

Upon substituting Eq. (k) into the last boundary condition given in Eq. (b), we arrive at

$$\sum_{n=1}^{\infty}F_nJ_0(\lambda_n r/a)\sinh(\lambda_n L/a)=V \tag{l}$$

We multiply Eq. (k) by $J_0(\lambda_k r/a)rdr$ and integrate over the interval to obtain

$$\sum_{n=1}^{\infty}F_n\sinh(\lambda_n L/a)\int_0^a J_0(\lambda_n r/a)J_0(\lambda_k r/a)rdr=V\int_0^a J_0(\lambda_k r/a)rdr \quad k=1,2,\ldots \tag{m}$$

From Eqs. (6.88), (6.89), and (5.105), we find that

$$\begin{aligned}&\int_0^a J_0(\lambda_n r/a)J_0(\lambda_k r/a)rdr=\delta_{nk}\frac{a^2}{2}J_0'^2(\lambda_k)=\delta_{nk}\frac{a^2}{2}J_1^2(\lambda_k)\\&\int_0^a J_0(\lambda_k r/a)rdr=\frac{a^2}{\lambda_k}J_1(\lambda_k)\end{aligned} \tag{n}$$

where δ_{nk} is the Kronecker delta and we have used Eq. (f). Then, Eq. (m) becomes

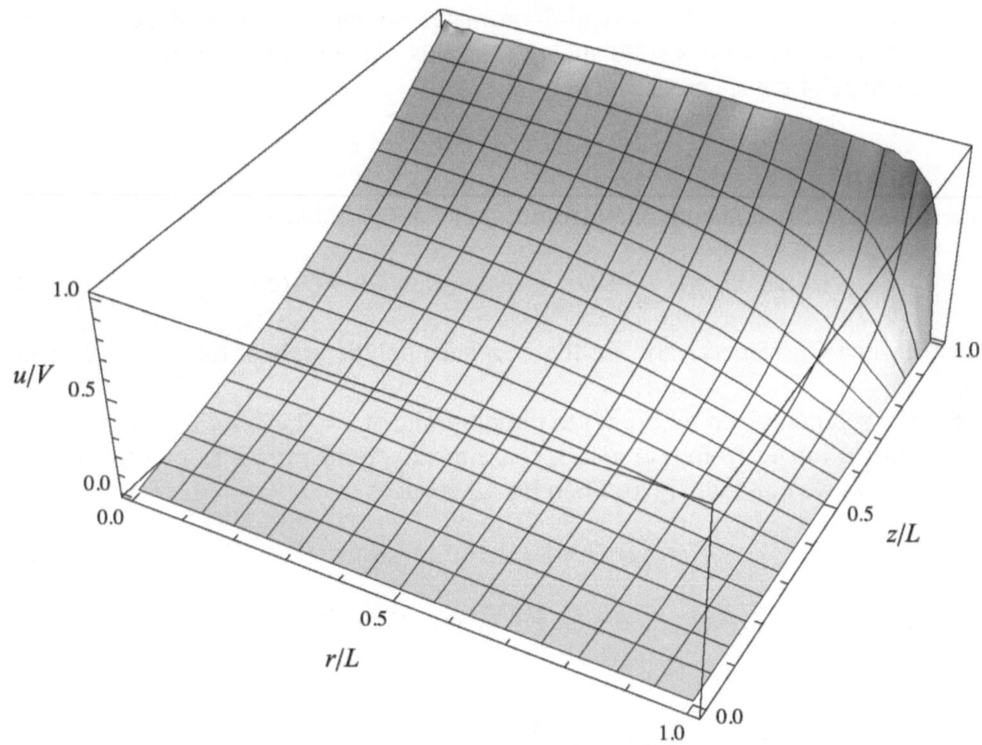

FIGURE 7.7 Numerical evaluation of Eq. (p) of Example 7.5 for $\gamma = 1.1$.

$$\frac{a^2}{2}\sum_{n=1}^{\infty} F_n \sinh(\lambda_n L / a)\delta_{nk} J_1^2(\lambda_k) = \frac{a^2 V}{\lambda_k} J_1(\lambda_k)$$

$$F_n = \frac{2V}{\lambda_n \sinh(\lambda_n L / a) J_1(\lambda_n)} \qquad \text{(o)}$$

Upon using Eq. (o) in Eq. (k), the final result is

$$u(\rho, \xi) = 2V \sum_{n=1}^{\infty} \frac{J_0(\lambda_n \rho)\sinh(\lambda_n \gamma \xi)}{\lambda_n J_1(\lambda_n)\sinh(\lambda_n \gamma)} \qquad \text{(p)}$$

where $\rho = r/a$, $\xi = z/L$, and $\gamma = L/a$.

A plot of $u(r,z)/V$ for $\gamma = 1.1$ is shown in Figure 7.7. This figure was obtained with Mathematica procedure M7.9.

Example 7.6: Laplace equation in spherical coordinates

We shall obtain a solution to

$$\frac{\partial^2 w}{\partial r^2} + \frac{2}{r}\frac{\partial w}{\partial r} + \frac{\cot\varphi}{r^2}\frac{\partial w}{\partial \varphi} + \frac{1}{r^2}\frac{\partial^2 w}{\partial \varphi^2} = 0 \quad 0 < r < a \quad 0 < \varphi < \pi \qquad \text{(a)}$$

which is the Laplacian in spherical coordinates without any dependence on θ. We assume that the sphere is a solid sphere of radius a and that the boundary condition on the surface of the sphere is

$$w(a,\varphi) = f(\varphi) \quad 0 \le \varphi \le \pi \tag{b}$$

The solution to Eq. (a) is obtained from Eq. (T7.3g) of Table 7.3 as

$$w(r,\varphi) = \left(Cr^n + Dr^{-n-1}\right)\left(EP_n(\cos\varphi) + FQ_n(\cos\varphi)\right) \tag{c}$$

where P_n and Q_n, respectively, are the Legendre polynomials of the first and second kind.

For the solution given by Eq. (c) to remain finite at $r = 0$, we set $D = 0$. In Eq. (b), we see that φ ranges from 0 to π and, consequently, $\cos\varphi$ ranges from 1 to -1. Since $Q_n(\pm 1) \to \infty$, we must set $F = 0$. Then Eq. (c) reduces to

$$w_n(r,\varphi) = A_n r^n P_n(\cos\varphi) \tag{d}$$

However, since Eq. (d) is a solution to Eq. (a) for all integer values of n, we use superposition to obtain

$$w(r,\varphi) = \sum_{n=1}^{\infty} A_n r^n P_n(\cos\varphi) \tag{e}$$

Substituting Eq. (e) into the boundary condition given by Eq. (b), we arrive at

$$\sum_{n=1}^{\infty} A_n a^n P_n(\cos\varphi) = f(\varphi) \tag{f}$$

We multiply Eq. (f) by $P_m(\cos\varphi)\sin\varphi$, integrate over the interval, and use Eqs. (6.89) and (5.196) to obtain

$$\begin{aligned} \sum_{n=1}^{\infty} A_n a^n \int_0^{\pi} P_n(\cos\varphi)P_m(\cos\varphi)\sin\varphi d\varphi &= \int_0^{\pi} f(\varphi)P_m(\cos\varphi)\sin\varphi d\varphi \\ A_n &= \frac{2n+1}{2a^n}\int_0^{\pi} f(\varphi)P_n(\cos\varphi)\sin\varphi d\varphi \end{aligned} \tag{g}$$

The final solution is obtained by using Eq. (g) in Eq. (e). Thus,

$$w(r,\varphi) = \frac{1}{2}\sum_{n=1}^{\infty}(2n+1)\left(\frac{r}{a}\right)^n P_n(\cos\varphi)\int_0^{\pi} f(\varphi)P_n(\cos\varphi)\sin\varphi d\varphi \tag{h}$$

If we let

$$\begin{aligned} f(\varphi) &= f_o \qquad 0 \le \varphi \le \pi/2 \\ &= 0 \qquad \pi/2 < \varphi \le \pi \end{aligned}$$

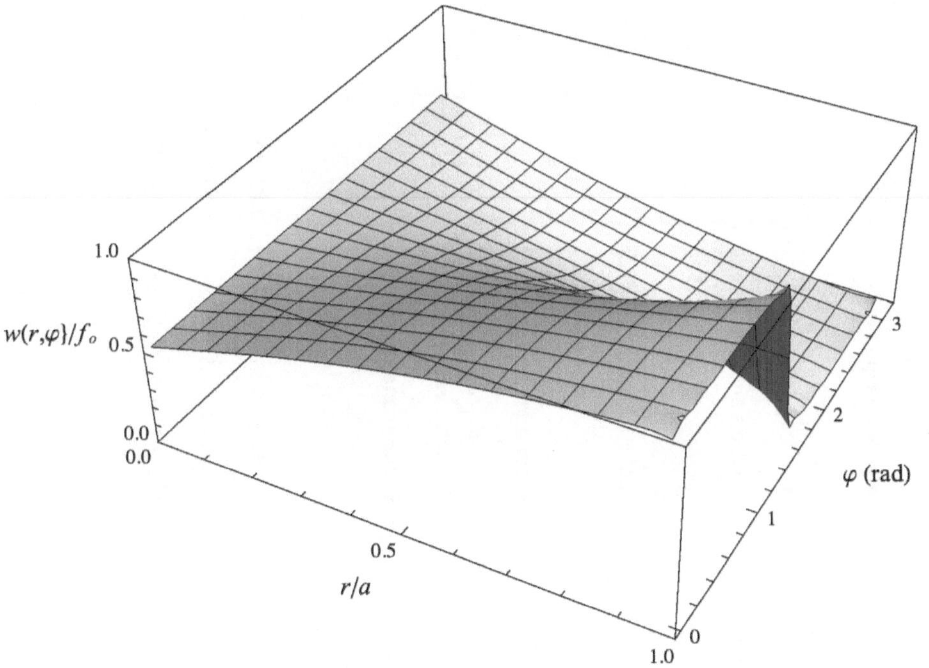

FIGURE 7.8 Numerical evaluation of Eq. (i) of Example 7.6.

then using Mathematica procedure M7.10, we find that

$$\int_0^{\pi} f(\varphi)P_n(\cos\varphi)\sin\varphi d\varphi = f_o\int_0^{\pi/2} P_n(\cos\varphi)\sin\varphi d\varphi$$

$$= \frac{f_o}{2n+1}\left(P_{n-1}(0) - P_{n+1}(0)\right)$$

The solution given by Eq. (h) becomes

$$w(r,\varphi) = \frac{f_o}{2}\sum_{n=1}^{\infty}\left(P_{n-1}(0) - P_{n+1}(0)\right)\left(\frac{r}{a}\right)^n P_n(\cos\varphi) \qquad \text{(i)}$$

A plot of $w(r,\varphi)/f_o$ is shown in Figure 7.8. This figure was obtained with Mathematica procedure M7.11.

7.2.3 Helmholtz Equation

The general solutions to the Helmholtz equation in three coordinate systems are given in Table 7.5 along with some of the intermediate results required to obtain these general solutions. To illustrate how the results in these tables were obtained, we shall present the details of the separation of variable method for the Helmholtz equation in the polar coordinate system. The solutions for the other coordinates are obtained in a similar manner.

TABLE 7.5
Solutions to the Helmholtz Equation in Three Coordinate Systems

Cartesian Coordinates	**Assumed Separable Form**	**Separable Equation**
$\frac{\partial^2 u}{\partial x^2}+\frac{\partial^2 u}{\partial y^2}+\Omega^2 u=0$	$u(x,y)=X(x)Y(y)$	$\frac{d^2Y}{Ydy^2}+\frac{d^2X}{Xdx^2}+\Omega^2=0$
Separated equations	**Solutions to separated equations**	**Solution and special cases**
$\frac{d^2X}{dx^2}+\mu^2 X=0$ (S-L)	$X(x)=A\cos(\mu x)+B\sin(\mu x)$	
$\frac{d^2Y}{dy^2}+\lambda^2 Y=0 \quad \lambda^2=\Omega^2-\mu^2$	$Y(y)=C\cos(\lambda y)+D\sin(\lambda y) \quad \lambda^2>0$ $=C\cosh(\lambda y)+D\sinh(\lambda y) \quad \lambda^2<0$	$u(x,y)=X(x)Y(y)$ (T7.5a)
Polar cylindrical coordinates	**Assumed separable form**	**Separable equation**
$\frac{\partial^2 u}{\partial r^2}+\frac{\partial u}{r\partial r}+\frac{\partial^2 u}{r^2\partial\theta^2}+\frac{\partial^2 u}{\partial z^2}+\Omega^2 u=0$	$u(r,\theta,z)=R(r)\Theta(\theta)Z(z)$	$\frac{1}{R}\left(\frac{d^2R}{dr^2}+\frac{dR}{rdr}\right)+\frac{d^2\Theta}{r^2\Theta d\theta^2}+\frac{d^2Z}{Zdz^2}+\Omega^2=0$
Separated equations	**Solutions to separated equations**	**Solution and special cases**
$\frac{d^2Z}{dz^2}-\kappa^2 Z=0$	$Z(z)=E\cosh(\kappa z)+F\sinh(\kappa z)$ or $=Ee^{\kappa z}+Fe^{-\kappa z}$	
$\frac{d^2\Theta}{d\theta^2}+m^2\Theta=0$ (S-L)	$\Theta(\theta)=A\cos m\theta+B\sin m\theta$	$u(r,\theta,z)=R_m(r,\gamma)\Theta(\theta)Z(z)$ (T7.5b) $u(r,\theta)=R_m(r,\Omega)\Theta(\theta)$ (T7.5c)
$r^2\frac{d^2R}{dr^2}+r\frac{dR}{dr}+\left(\gamma^2 r^2-m^2\right)R=0$ (S-L) $\gamma^2=\kappa^2+\Omega^2$	$R_m(r,\gamma)=CJ_m(\gamma r)+DY_m(\gamma r)$	$u(r,z)=R_0(r,\gamma)Z(z)$ (T7.5d)

(Continued)

TABLE 7.5 (CONTINUED)
Solutions to the Helmholtz Equation in Three Coordinate Systems

Spherical coordinates	Assumed separable form	Separable equation
$\frac{\partial^2 u}{\partial r^2}+\frac{2}{r}\frac{\partial u}{\partial r}+\frac{\cot\varphi}{r^2}\frac{\partial u}{\partial\varphi}+\frac{1}{r^2}\frac{\partial^2 u}{\partial\varphi^2}$ $+\frac{1}{r^2\sin^2\varphi}\frac{\partial^2 u}{\partial\theta^2}+\Omega^2 u=0$	$u(r,\varphi,\theta)=R(r)\Phi(\varphi)\Theta(\theta)$	$\frac{r^2}{R}\left(\frac{d^2R}{dr^2}+\frac{2}{r}\frac{dR}{dr}\right)+\frac{1}{\Phi}\left(\cot\varphi\frac{d\Phi}{d\varphi}+\frac{d^2\Phi}{d\varphi^2}\right)$ $+\frac{1}{\sin^2\varphi}\frac{d^2\Theta}{\Theta d\theta^2}+\Omega^2 r^2=0$
Separated equations	**Solutions to separated equations**	**Solution and special cases**
$\frac{d^2\Theta}{d\theta^2}+m^2\Theta=0 \quad m=\text{integer} \quad \text{(S-L)}$	$\Theta(\theta)=A\cos m\theta+B\sin m\theta$	
$\frac{d^2\Phi}{d\varphi^2}+\cot\varphi\frac{d\Phi}{d\varphi}+\left(n(n+1)-\frac{m^2}{\sin^2\varphi}\right)\Phi=0 \quad \text{(S-L)}$	$\Phi(\varphi,m)=EP_n^m(\cos\varphi)+FQ_n^m(\cos\varphi)$ $\Phi(\varphi,0)=EP_n(\cos\varphi)+FQ_n(\cos\varphi)$	$u(r,\varphi,\theta)=\Phi(\varphi,m)\Theta(\theta)R(r) \quad \text{(T7.5e)}$ $u(r,\varphi)=\Phi(\varphi,0)R(r) \quad \text{(T7.5f)}$
$r^2\frac{d^2R}{dr^2}+2r\frac{dR}{dr}+\left(\Omega^2 r^2-n(n+1)\right)R=0 \quad \text{(S-L)}$	$R(r)=Cj_n(\Omega r)+Dy_n(\Omega r)$	

Note: The quantities m^2, Ω^2, κ^2, and μ^2 are real positive constants and n is zero or a positive integer. The notation (S–L) indicates that this will be a Sturm–Liouville system when the boundary conditions are those given by Eq. (6.24).

The homogeneous Helmholtz equation in the polar coordinate system in terms of the variable $u = u(r,\theta)$ is obtained from Tables 7.1 and 7.2 as

$$\frac{\partial^2 u}{\partial r^2} + \frac{\partial u}{r\partial r} + \frac{\partial^2 u}{r^2\partial\theta^2} + \Omega^2 u = 0 \quad \Omega^2 > 0 \tag{7.40}$$

For this coordinate system, we select

$$u(r,\theta) = R(r)\Theta(\theta) \tag{7.41}$$

Substituting Eq. (7.41) into Eq. (7.40) gives

$$r^2\frac{d^2R}{dr^2}\Theta + r\frac{dR}{dr}\Theta + \frac{d^2\Theta}{d\theta^2}R + \Omega^2 r^2 R\Theta = 0 \tag{7.42}$$

To obtain a solution to Eq. (7.42), we rewrite it as

$$\frac{r^2}{R}\left(\frac{d^2R}{dr^2} + \frac{dR}{rdr}\right) + \frac{d^2\Theta}{\Theta d\theta^2} + \Omega^2 r^2 = 0 \tag{7.43}$$

Therefore, we start with

$$\frac{r^2}{R}\left(\frac{d^2R}{dr^2} + \frac{dR}{rdr}\right) + \Omega^2 r^2 = -\frac{d^2\Theta}{\Theta d\theta^2} = m^2 \quad m^2 > 0 \tag{7.44}$$

Hence, Θ is a solution to

$$\frac{d^2\Theta}{d\theta^2} + m^2\Theta = 0 \quad m^2 > 0 \tag{7.45}$$

It is seen that by selecting a positive separation constant, Eq. (7.45) is a differential equation of the Sturm–Liouville system [recall Eq. (6.41)] when the boundary conditions are those given by Eq. (6.24) with y replaced by Θ. In Eq. (7.45), m is not necessarily an integer. Then using Eq. (7.45) in Eq. (7.44), Eq. (7.44) becomes

$$r^2\frac{d^2R}{dr^2} + r\frac{dR}{dr} + \left(\Omega^2 r^2 - m^2\right)R = 0 \tag{7.46}$$

Equation (7.46) is a Bessel's equation of order m and is a differential equation of the Sturm–Liouville type [recall Eq. (6.45)] when the boundary conditions are those given by Eq. (6.24) with y replaced by R. Another advantage of selecting m^2 positive is that the order of the Bessel equation is real.

The solutions to Eqs. (7.45) and (7.46), respectively, are

$$\Theta(\theta) = A\cos(m\theta) + B\sin(m\theta) \tag{7.47}$$

and

$$R(r) = CJ_m(\Omega r) + DY_m(\Omega r) \tag{7.48}$$

TABLE 7.6
Summary of Examples of Applications of the Helmholtz Equation: Boundary Conditions and Coordinates Systems

Example	Helmholtz Equation	Boundary Conditions
7.7	$\dfrac{\partial^2 u}{\partial x^2}+\dfrac{\partial^2 u}{\partial y^2}+\Omega^2 u=0$	$u(0,y)=u(a,y)=0$ $u(x,0)=u(x,b)=0$
7.8	$\dfrac{\partial^2 u}{\partial r^2}+\dfrac{\partial u}{r\partial r}+\dfrac{\partial^2 u}{r^2\partial\theta^2}+\Omega^2 u=0$	$u(a,\theta)=u(b,\theta)=0$ $u(r,0)=u(r,2\pi)=0$

Equations (7.47) and (7.48) are then used in Eq. (7.41) to yield

$$\begin{aligned} u(r,\theta) &= \left(A\cos(m\theta)+B\sin(m\theta)\right)\left(CJ_m(\Omega r)+DY_m(\Omega r)\right) \\ u(r) &= C_0 J_0(\Omega r)+D_0 Y_0(\Omega r) \end{aligned} \tag{7.49}$$

The second solution in Eq. (7.49) is applicable when Eq. (7.40) is independent of θ.

In the problems that we shall consider in the remainder of this section, the value of Ω will be determined by satisfying homogeneous boundary conditions. Consequently, $\Omega = \Omega_{mn}$ will be the eigenvalues and the solutions at these values will be the eigenfunctions.

We shall now solve the Helmholtz equation in two different coordinate systems. The applications considered are summarized in Table 7.6.

Example 7.7: Helmholtz equation in Cartesian coordinates with Dirichlet boundary conditions

We shall obtain a solution to

$$\frac{\partial^2 u}{\partial x^2}+\frac{\partial^2 u}{\partial y^2}+\Omega^2 u=0 \quad 0<x<a \quad 0<y<b \tag{a}$$

over the region indicated and subject to the boundary conditions

$$\begin{aligned} u(0,y)&=u(a,y)=0 \\ u(x,0)&=u(x,b)=0 \end{aligned} \tag{b}$$

The solution to Eq. (a) is obtained from Eq. (T7.5a) of Table 7.5 as

$$u(x,y)=\left(A\cos(\mu x)+B\sin(\mu x)\right)\left(C\cos(\lambda y)+D\sin(\lambda y)\right) \tag{c}$$

where

$$\lambda^2=\Omega^2-\mu^2 \tag{d}$$

and we have assumed that $\lambda^2 > 0$. Substituting Eq. (c) into the first set of boundary conditions given by Eq. (b) yields

$$A\big(C\cos(\lambda y)+D\sin(\lambda y)\big)=0$$

$$B\sin(\mu a)\big(C\cos(\lambda y)+D\sin(\lambda y)\big)=0$$

Therefore, for arbitrary y, $A = 0$, and to avoid a trivial solution, in the second equation we set

$$\sin(\mu a)=0 \tag{e}$$

which is the characteristic equation. Equation (e) is satisfied when

$$\mu = n\pi/a \quad n=1,2,3,\ldots$$

which is the eigenvalue. Then, Eq. (c) becomes

$$u(x,y)=\big(C\cos(\lambda y)+D\sin(\lambda y)\big)\sin(n\pi x/a) \tag{f}$$

where, based on the discussions in several of the previous examples, $\sin(n\pi x/a)$ is an orthogonal eigenfunction.

Upon substituting Eq. (f) into the second set of boundary conditions given by Eq. (b), we obtain

$$C\sin(n\pi x/a)=0$$

$$\big(C\cos(\lambda b)+D\sin(\lambda b)\big)\sin(n\pi x/a)=0$$

Therefore, for arbitrary x, $C = 0$, and to avoid a trivial solution we set

$$\sin(\lambda b)=0 \tag{g}$$

which is the characteristic equation. Equation (g) is satisfied when

$$\lambda = m\pi/b \quad m=1,2,3,\ldots$$

which is the eigenvalue. Then, the corresponding eigenfunction is $\sin(n\pi y/b)$, which is an orthogonal eigenfunction. From Eq. (d),

$$\Omega^2 \to \Omega_{nm}^2 = \pi^2\left(\frac{n^2}{a^2}+\frac{m^2}{b^2}\right) \quad m,n=1,2,3,\ldots \tag{h}$$

Consequently, a nontrivial solution to Eq. (a) with the boundary conditions given by Eq. (b) only exists when $\Omega = \Omega_{nm}$ and

$$u_{nm}(x,y)=A_{nm}\varphi_{nm}(x,y) \tag{i}$$

where

$$\varphi_{nm}(x,y)=\sin(n\pi x/a)\sin(m\pi y/b) \tag{j}$$

The quantity Ω_{nm} is the eigenvalue and $\varphi_{nm}(x,y)$ is the corresponding eigenfunction. Upon using superposition, the solution is

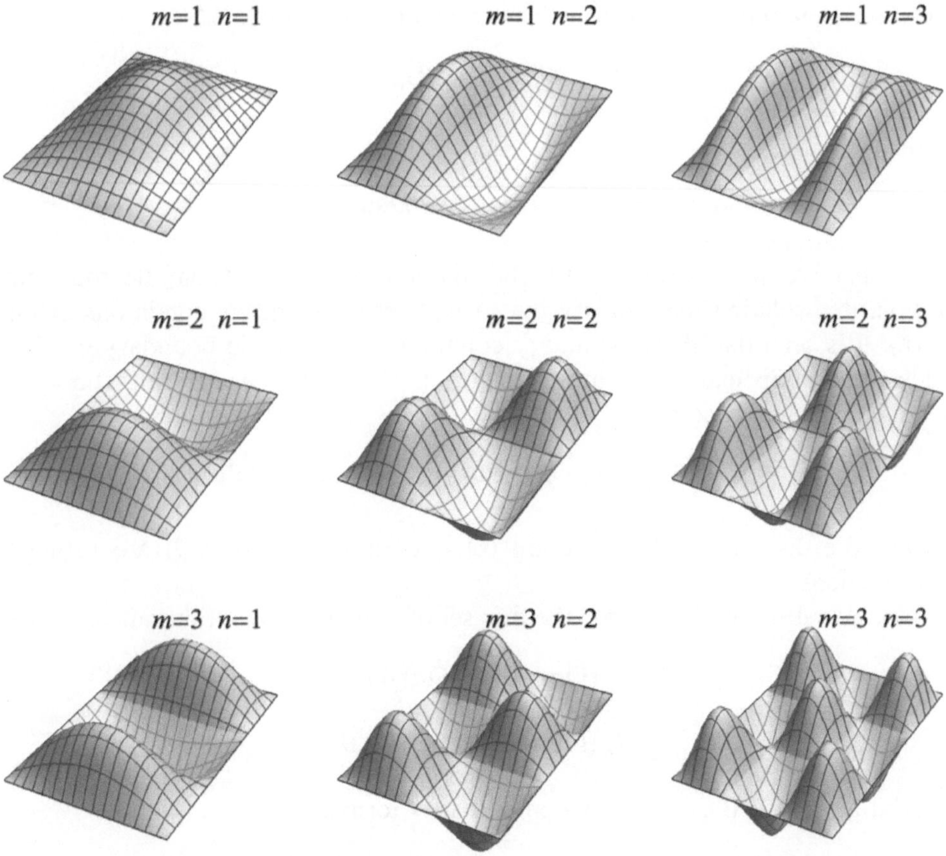

FIGURE 7.9 Eigenfunctions of Example 7.7 for $b/a = 1.5$ and for the values of n and m indicated.

$$u(x,y) = \sum_{m}^{\infty} \sum_{n=1}^{\infty} A_{nm} \sin(n\pi x/a) \sin(m\pi y/b)$$

The eigenfunction given by Eq. (j) is plotted in Figure 7.9 for the case when $b/a = 1.5$ and $m = 1, 2, 3$ and $n = 1, 2, 3$.

Example 7.8: Helmholtz equation in polar coordinates with Dirichlet boundary conditions

We shall obtain a solution to

$$\frac{\partial^2 u}{\partial r^2} + \frac{\partial u}{r \partial r} + \frac{\partial^2 u}{r^2 \partial \theta^2} + \Omega^2 u = 0 \quad a < r < b \quad 0 < \theta < 2\pi \tag{a}$$

over the annular region indicated and subject to the boundary conditions

$$\begin{aligned} u(a,\theta) &= u(b,\theta) = 0 \\ u(r,0) &= u(r,2\pi) = 0 \end{aligned} \tag{b}$$

The solution to Eq. (a) is obtained from Eq. (T7.5c) of Table 7.5 as

$$u(r,\theta) = \left(AJ_\alpha(\Omega r) + BY_\alpha(\Omega r)\right)\Theta(\theta) \tag{c}$$

where

$$\Theta(\theta) = C\cos(\alpha\theta) + D\sin(\alpha\theta) \tag{d}$$

From Table 7.5, the function $\Theta(\theta)$ is the solution to a Sturm–Liouville equation. In addition, this solution must satisfy the second set of boundary conditions given by Eq. (b). It is seen that these boundary conditions are periodic boundary conditions as described in Special Case I in Section 6.1.4. Therefore, we require that $\alpha = m$, an integer, and Eq. (c) becomes

$$u(r,\theta) = \left(AJ_m(\Omega r) + BY_m(\Omega r)\right)\left(C\cos(m\theta) + D\sin(m\theta)\right) \tag{e}$$

Furthermore, as indicated in Section 6.1.4, $\cos(m\theta)$ and $\sin(m\theta)$ are orthogonal eigenfunctions.

Next, we substitute Eq. (e) into the first set of boundary conditions and arrive at

$$\begin{aligned} &\left(AJ_m(\Omega a) + BY_m(\Omega a)\right)\Theta(\theta) = 0 \\ &\left(AJ_m(\Omega b) + BY_m(\Omega b)\right)\Theta(\theta) = 0 \end{aligned} \tag{f}$$

For arbitrary θ, Eq. (f), can be written in matrix form as

$$\begin{pmatrix} J_m(\Omega a) & Y_m(\Omega a) \\ J_m(\Omega b) & Y_m(\Omega b) \end{pmatrix} \begin{Bmatrix} A \\ B \end{Bmatrix} = \begin{Bmatrix} 0 \\ 0 \end{Bmatrix} \tag{g}$$

Equation (g) will have a nontrivial solution when the determinant of the coefficients of A and B equals zero, that is, when λ_{nm} are solutions to

$$\begin{aligned} &J_m(\alpha\lambda_{nm})Y_m(\lambda_{nm}) - J_m(\lambda_{nm})Y_m(\alpha\lambda_{nm}) = 0 \\ &\qquad\qquad m = 0,1,2,\ldots \quad n = 1,2,3\ldots \end{aligned} \tag{h}$$

where $\lambda_{nm} = b\Omega_{nm}$ and $\alpha = a/b$, $0 < \alpha < 1$. Using Eq. (6.33), the corresponding eigenfunction can be written as

$$R_{nm}(r) = J_m(\lambda_{nm}r/b) - \frac{J_m(\lambda_{nm})}{Y_m(\lambda_{nm})}Y_m(\lambda_{nm}r/b)$$

From Table 7.5, it is seen that $R_{nm}(r)$ is the solution to a Sturm–Liouville equation when the boundary conditions are given by Eq. (6.24) with y replaced by R. Therefore, based on the material in Section 6.1.4, the eigenfunction $R_{nm}(r)$ is an orthogonal eigenfunction. From Section 6.1.3, it was found that the corresponding weighting factor is r.

Consequently, a nontrivial solution to Eq. (a) with the boundary conditions given by Eq. (b) only exists when $\Omega_{nm} = \lambda_{nm}/b$ and

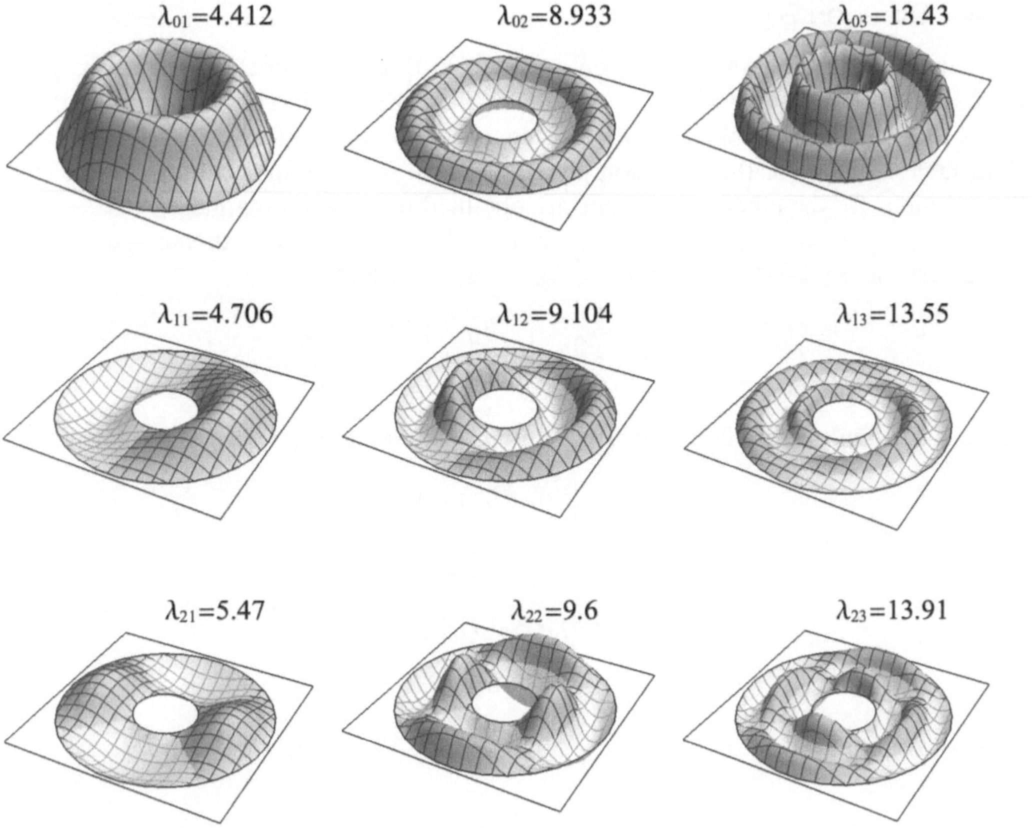

FIGURE 7.10 Eigenvalues λ_{nm} and the corresponding eigenfunctions given by Eq. (i) of Example 7.8 for $\alpha = 0.3$.

$$u_{nm}(r,\theta) = \left(J_m(\lambda_{nm} r/b) - \frac{J_m(\lambda_{nm})}{Y_m(\lambda_{nm})} Y_m(\lambda_{nm} r/b) \right) \cos(m\theta) \quad \text{(i)}$$

and

$$u_{nm}(r,\theta) = \left(J_m(\lambda_{nm} r/b) - \frac{J_m(\lambda_{nm})}{Y_m(\lambda_{nm})} Y_m(\lambda_{nm} r/b) \right) \sin(m\theta) \quad \text{(j)}$$

which are the corresponding eigenfunctions. Using superposition, the complete solution is

$$u(r,\theta) = \sum_{m=0}^{\infty} \sum_{n=1}^{\infty} C_{nm} \left(J_m(\lambda_{nm} r/b) - \frac{J_m(\lambda_{nm})}{Y_m(\lambda_{nm})} Y_m(\lambda_{nm} r/b) \right) \cos(m\theta)$$

$$+ \sum_{m=0}^{\infty} \sum_{n=1}^{\infty} D_{nm} \left(J_m(\lambda_{nm} r/b) - \frac{J_m(\lambda_{nm})}{Y_m(\lambda_{nm})} Y_m(\lambda_{nm} r/b) \right) \sin(m\theta)$$

The eigenfunction given by Eq. (i) and the eigenvalues obtained from Eq. (h) are plotted in Figure 7.10 for $\alpha = 0.3$, $m = 0, 1, 2$ and $n = 1, 2, 3$.

7.2.4 Diffusion Equation

The general solutions to the diffusion equation in three coordinate systems are given in Table 7.7 along with some of the intermediate results required to obtain them. To illustrate how the results in these tables were obtained, we shall present the details of the separation of variable method for the diffusion equation in the polar cylindrical coordinate system. The solutions for the other coordinates are obtained in a similar manner.

The homogeneous diffusion equation in the polar cylindrical coordinate system in terms of the variable $u = u(r,\theta,z,t)$ is obtained from Tables 7.1 and 7.2 as

$$\frac{\partial^2 u}{\partial r^2}+\frac{1}{r}\frac{\partial u}{\partial r}+\frac{1}{r^2}\frac{\partial^2 u}{\partial \theta^2}+\frac{\partial^2 u}{\partial z^2}=\frac{1}{\alpha}\frac{\partial u}{\partial t} \tag{7.50}$$

where α is a real positive constant. We assume a solution to Eq. (7.50) of the form

$$u(r,\theta,z,t)=u_0(\theta,z)+w(r,\theta,z,t) \tag{7.51}$$

where

$$\begin{aligned} u_0(\theta,z)&=\Theta_0(\theta)Z_0(z)\\ w(r,\theta,z,t)&=\Theta(\theta)R(r)Z(z)T(t) \end{aligned} \tag{7.52}$$

Upon substituting Eq. (7.51) into Eq. (7.50) and using Eq. (7.52), we obtain

$$\begin{aligned} &Z_0\frac{d^2\Theta_0}{d\theta^2}+r^2\Theta_0\frac{d^2Z_0}{dz^2}+r^2\frac{d^2R}{dr^2}\Theta ZT+r\frac{dR}{dr}\Theta ZT\\ &\qquad+\frac{d^2\Theta}{d\theta^2}RZT+r^2\frac{d^2Z}{dz^2}R\Theta T=\frac{r^2}{\alpha}\frac{dT}{dt}R\Theta Z \end{aligned} \tag{7.53}$$

A solution to Eq. (7.53) can be obtained if we let $w(r,\theta,z,t)$ be a solution to

$$r^2\frac{d^2R}{dr^2}\Theta ZT+r\frac{dR}{dr}\Theta ZT+\frac{d^2\Theta}{d\theta^2}RZT+r^2\frac{d^2Z}{dz^2}R\Theta T=\frac{r^2}{\alpha}\frac{dT}{dt}R\Theta Z \tag{7.54}$$

and we let $u_0(\theta,z)$ be a solution to

$$\frac{d^2\Theta_0}{r^2\Theta_0 d\theta^2}+\frac{d^2Z_0}{Z_0 dz^2}=0 \tag{7.55}$$

To obtain a solution to Eq. (7.54), we re-write it as

$$\frac{1}{R}\left(\frac{d^2R}{dr^2}+\frac{1}{r}\frac{dR}{dr}\right)+\frac{1}{\Theta r^2}\frac{d^2\Theta}{d\theta^2}+\frac{d^2Z}{Zdz^2}=\frac{1}{\alpha}\frac{dT}{Tdt} \tag{7.56}$$

and start with

$$\frac{1}{R}\left(\frac{d^2R}{dr^2}+\frac{1}{r}\frac{dR}{dr}\right)+\frac{1}{\Theta r^2}\frac{d^2\Theta}{d\theta^2}+\frac{d^2Z}{Zdz^2}=\frac{1}{\alpha T}\frac{dT}{dt}=-\mu^2 \quad \mu^2>0 \tag{7.57}$$

TABLE 7.7
Solutions to the Diffusion Equation in Three Coordinate Systems

Cartesian Coordinates	**Assumed Separable Form**	**Separable Equations**
$\frac{\partial^2 u}{\partial x^2}+\frac{\partial^2 u}{\partial y^2}=\frac{1}{\alpha}\frac{\partial u}{\partial t}$	$u(x,y,t)=X_0(x)Y_0(y)+X(x)Y(y)T(t)$	$\frac{d^2X}{Xdx^2}+\frac{d^2Y}{Ydy^2}-\frac{dT}{\alpha Tdt}=0,\quad \frac{d^2X_0}{X_0dx^2}+\frac{d^2Y_0}{Y_0dy^2}=0$
Separated equations	**Solutions to separated equations**	**Solution and special cases**
$\frac{dT}{dt}+\mu^2\alpha T=0$	$T(t)=Ae^{-\mu^2\alpha t}$	
$\frac{d^2X}{dx^2}+\nu^2X=0$ (S-L)	$X(x,\nu)=B\cos(\nu x)+C\sin(\nu x)$	
$\frac{d^2X_0}{dx^2}=0$	$X_0(x)=a_0+b_0x$	$u(x,y,t)=X_0(x)Y_0(y)+X(x,\nu)Y(y)T(t)$ (T7.7a) $u(x,t)=X_0(x)+X(x,\mu)T(t)$ (T7.7b)
$\frac{\partial^2 Y}{\partial y^2}+\beta^2Y=0\quad \beta^2=\mu^2-\nu^2$	$Y(y)=D\cos(\beta y)+E\sin(\beta y)\quad \beta^2>0$ $Y(y)=D\cosh(\beta y)+E\sinh(\beta y)\quad \beta^2<0$	
$\frac{\partial^2 Y_0}{\partial y^2}=0$	$Y_0(y)=c_0+d_0y$	
Polar cylindrical coordinates	**Assumed separable form**	**Separable equations**
$\frac{\partial^2 u}{\partial r^2}+\frac{1}{r}\frac{\partial u}{\partial r}+\frac{1}{r^2}\frac{\partial^2 u}{\partial \theta^2}+\frac{\partial^2 u}{\partial z^2}=\frac{1}{\alpha}\frac{\partial u}{\partial t}$	$u(r,\theta,z,t)=\Theta_0(\theta)Z_0(z)+R(r)\Theta(\theta)Z(z)T(t)$	$\frac{1}{R}\left(\frac{d^2R}{dr^2}+\frac{1}{r}\frac{dR}{dr}\right)+\frac{1}{r^2}\frac{d^2\Theta}{\Theta d\theta^2}+\frac{d^2Z}{Zdz^2}-\frac{1}{\alpha T}\frac{dT}{dt}=0$ $\frac{1}{r^2}\frac{d^2\Theta_0}{\Theta_0 d\theta^2}+\frac{d^2Z_0}{Z_0dz^2}=0$

(Continued)

TABLE 7.7 (CONTINUED)
Solutions to the Diffusion Equation in Three Coordinate Systems

Separated equations	Solutions to separated equations	Solution and special cases
$\frac{dT}{dt} + \mu^2 \alpha T = 0$	$T(t) = Ae^{-\mu^2 \alpha t}$	
$\frac{d^2\Theta}{d\theta^2} + m^2\Theta = 0$ (S-L)	$\Theta(\theta) = B\cos m\theta + C\sin m\theta$	
$\frac{d^2\Theta_0}{d\theta^2} = 0$	$\Theta_0(\theta) = B_0 + C_0\theta$	$u(r,\theta,z,t) = \Theta_0(\theta)Z_0(z) + R(r,\beta,m)\Theta(\theta)Z(z)T(t)$ (T7.7c)
$\frac{d^2Z}{dz^2} + \eta^2 Z = 0$ (S-L)	$Z(z) = D\cos(\eta z) + E\sin(\eta z)$	$u(r,\theta,t) = \Theta_0(\theta) + R(r,\mu,m)\Theta(\theta)T(t)$ (T7.7d)
		$u(r,z,t) = Z_0(z) + R(r,\beta,0)Z(z)T(t)$ (T7.7e)
$\frac{d^2Z_0}{dz^2} = 0$	$Z_0(z) = D_0 + E_0 z$	$u(r,t) = R(r,\mu,0)T(t)$ (T7.7f)
$r^2\frac{d^2R}{dr^2} + r\frac{dR}{dr} + \left(r^2\beta^2 - m^2\right)R = 0$ (S-L)	$R(r,\beta,m) = FJ_m(\beta r) + GY_m(\beta r)$	
$\beta^2 = \mu^2 - \eta^2$		
Spherical coordinates	**Assumed separable form**	**Separable equations**
$\frac{\partial^2 u}{\partial r^2} + \frac{2}{r}\frac{\partial u}{\partial r} + \frac{\cot\varphi}{r^2}\frac{\partial u}{\partial \varphi} + \frac{1}{r^2}\frac{\partial^2 u}{\partial \varphi^2} - \frac{1}{\alpha}\frac{\partial u}{\partial t} = 0$	$u(r,\varphi,t) = R(r)\Phi(\varphi)T(t)$	$\frac{r^2}{R}\left(\frac{d^2R}{dr^2} + \frac{2}{r}\frac{dR}{dr}\right) + \frac{1}{\Phi}\left(\cot\varphi\frac{d\Phi}{d\varphi} + \frac{d^2\Phi}{d\varphi^2}\right) - \frac{r^2}{\alpha}\frac{dT}{Tdt} = 0$
Separated equations	**Solutions to separated equations**	**Solution and special cases**
$\frac{\partial T}{\partial t} + \mu^2 \alpha T = 0$	$T(t) = Ee^{-\mu^2 \alpha t}$	
$\frac{d^2\Phi}{d\varphi^2} + \cot\varphi\frac{d\Phi}{d\varphi} + n(n+1)\Phi = 0$ (S-L)	$\Phi(\varphi) = CP_n(\cos\varphi) + DQ_n(\cos\varphi)$	$u(r,\varphi,t) = R_n(r,\mu)\Phi(\varphi)T(t)$ (T7.7g)
		$u(r,t) = R_0(r,\mu)T(t)$ (T7.7h)
$r^2\frac{d^2R}{dr^2} + 2r\frac{dR}{dr} + \mu^2 r^2 R - n(n+1)R = 0$ (S-L)	$R_n(r,\mu) = Aj_n(\mu r) + By_n(\mu r)$	

Note: The quantities m^2, η^2, α, v^2, and μ^2 are real positive constants and n is zero or a positive integer. The notation (S–L) indicates that this will be a Sturm–Liouville system when the boundary conditions are those given by Eq. (6.24).

Hence, T is a solution to

$$\frac{\partial T}{\partial t} + \mu^2 \alpha T = 0 \tag{7.58}$$

Then, Eq. (7.57) can be written as

$$\frac{r^2}{R}\left(\frac{\partial^2 R}{\partial r^2} + \frac{\partial R}{r\partial r}\right) + \mu^2 r^2 + r^2 \frac{d^2 Z}{Zdz^2} = -\frac{1}{\Theta}\frac{d^2\Theta}{d\theta^2} = m^2 \quad m^2 > 0 \tag{7.59}$$

Thus, Θ is a solution to

$$\frac{d^2\Theta}{d\theta^2} + m^2\Theta = 0 \tag{7.60}$$

Equation (7.60) is a differential equation of the Sturm–Liouville type [recall Eq. (6.41)] when the boundary conditions are those given by Eq. (6.24) with y replaced by Θ. Then, Eq. (7.59) can be written as

$$\frac{1}{R}\left(\frac{\partial^2 R}{\partial r^2} + \frac{\partial R}{r\partial r}\right) + \mu^2 - \frac{m^2}{r^2} = -\frac{d^2 Z}{Zdz^2} = \eta^2 \quad \eta^2 > 0 \tag{7.61}$$

Thus, Z is a solution to

$$\frac{d^2 Z}{dz^2} + \eta^2 Z = 0 \tag{7.62}$$

Thus, by selecting the separation constant as $+\eta^2$, Eq. (7.62) is a differential equation of the Sturm–Liouville type [recall Eq. (6.41)] when the boundary conditions are those given by Eq. (6.24) with y replaced by Z. Then Eq. (7.59) becomes

$$r^2\frac{\partial^2 R}{\partial r^2} + r\frac{\partial R}{\partial r} + \left(r^2\beta^2 - m^2\right)R = 0 \tag{7.63}$$

where $\beta^2 = \mu^2 - \eta^2$. Equation (7.63) is Bessel's equation of order m and a differential equation of the Sturm–Liouville type (recall Section 6.1.3) when $\beta^2 > 0$ and the boundary conditions are those given by Eq. (6.24) with y replaced by R. When $\beta^2 < 0$, Eq. (7.63) is *not* of the Sturm–Liouville type.

The solutions to Eqs. (7.58), (7.60), (7.62), and (7.63), respectively, are

$$T(t) = Ae^{-\mu^2 \alpha t} \tag{7.64}$$

and

$$\Theta(\theta, m) = B\cos(m\theta) + C\sin(m\theta) \tag{7.65}$$

and

$$Z(z, \eta) = D\cos(\eta z) + E\sin(\eta z) \tag{7.66}$$

and

$$R_m(r,\beta) = FJ_m(\beta r) + GY_m(\beta r) \tag{7.67}$$

It is noted that the quantity $u_0(\theta,z)$ is not a function the independent variables r and t. This is because if the separation constant μ^2 is zero, then the solution is independent of t, which is not a case of interest. It is independent of r because the differential equation in r does not have its own separation constant but is a function of three previously introduced separation constants.

Equation (7.55) is satisfied if Z_0 and Θ_0, respectively, are solutions to

$$\begin{aligned} \frac{d^2\Theta_0}{d\theta^2} &= 0 \\ \frac{d^2 Z_0}{dz^2} &= 0 \end{aligned} \tag{7.68}$$

In this case, we have set the separation constants to zero since the non-zero case has been considered in the solution to Eq. (7.54). Thus, from the straightforward integration of Eq. (7.60) and Eqs. (7.64) to (7.67), Eq. (7.52) becomes

$$\begin{aligned} u_0(\theta,z) &= (a_0 + b_0\theta)(c_0 + d_0 z) \\ w(r,\theta,z,t) &= (FJ_m(\beta r) + GY_m(\beta r))(B\cos(m\theta) + C\sin(m\theta)) \\ &\quad \times (D\cos(\eta z) + E\sin(\eta z))e^{-\mu^2\alpha t} \end{aligned} \tag{7.69}$$

Equation (7.69) has the following special cases. When Eq. (7.50) is independent of z, we have that

$$\begin{aligned} u_0(\theta) &= a_0 + b_0\theta \\ w(r,\theta,t) &= (FJ_m(\mu r) + GY_m(\mu r))(B\cos(m\theta) + C\sin(m\theta))e^{-\mu^2\alpha t} \end{aligned} \tag{7.70}$$

When Eq. (7.50) is independent of θ, we have that

$$\begin{aligned} u_0(z) &= (c_0 + d_0 z) \\ w(r,z,t) &= (FJ_0(\beta r) + GY_0(\beta r))(D\cos(\eta z) + E\sin(\eta z))e^{-\mu^2\alpha t} \end{aligned} \tag{7.71}$$

Lastly, when Eq. (7.50) is independent of z and θ, we have that

$$\begin{aligned} u_0(\theta,z) &= 0 \\ w(r,t) &= (FJ_0(\mu r) + GY_0(\mu r))e^{-\mu^2\alpha t} \end{aligned} \tag{7.72}$$

Thus, the solutions to Eq. (7.50) and its special cases are

$$\begin{aligned} u(r,\theta,z,t) &= u_0(\theta,z) + w(r,\theta,z,t) \\ u(r,\theta,t) &= u_0(\theta) + w(r,\theta,t) \\ u(r,z,t) &= u_0(z) + w(r,z,t) \\ u(r,t) &= w(r,t) \end{aligned} \tag{7.73}$$

TABLE 7.8
Summary of Examples of Applications of the Diffusion Equation: Boundary Conditions, Initial Conditions, and Coordinates Systems

Example	Diffusion Equation	Boundary Conditions	Initial Condition
7.9	$\frac{\partial^2 u}{\partial x^2} = \frac{1}{\alpha}\frac{\partial u}{\partial t}$	$u(0,t) = 0, \quad u(L,t) = 0$	$u(x,0) = f(x)$
7.10	$\frac{\partial^2 u}{\partial x^2} = \frac{1}{\alpha}\frac{\partial u}{\partial t}$	$\frac{\partial u(0,t)}{\partial x} = 0, \quad u(L,t) = U_o$	$u(x,0) = f(x)$
7.11	$\frac{\partial^2 u}{\partial r^2} + \frac{\partial u}{r\partial r} = \frac{1}{\alpha}\frac{\partial u}{\partial t}$	$\frac{\partial u(a,t)}{\partial r} + hu(a,t) = 0$ $u(0,t) = \text{finite}$	$u(r,0) = f(r)$
7.12	$\frac{\partial^2 u}{\partial x^2} = \frac{1}{\alpha}\frac{\partial u}{\partial t} - \frac{1}{\beta}Q(x,t)$	$u(0,t) = a, \quad u(L,t) = b$	$u(x,0) = f(x)$
7.13	$\frac{\partial^2 u}{\partial x^2} + \frac{\partial^2 u}{\partial y^2} = \frac{1}{\alpha}\frac{\partial u}{\partial t}$	$u(0,y,t) = 0, \quad u(a,y,t) = 0$ $u(x,b,t) = 0, \quad u(x,b,t) = 0$	$u(x,y,0) = f(x,y)$
7.14	$\frac{\partial}{r^2\partial r}\left(r^2\frac{\partial u}{\partial r}\right) = \frac{1}{\alpha}\frac{\partial u}{\partial t}$	$u(0,t) = \text{finite}$ $u(a,t) = 0$	$u(r,0) = f(r)$

We shall now solve the diffusion equation in different coordinate systems for different combinations of boundary conditions and initial conditions. The applications considered are summarized in Table 7.8.

Example 7.9: Diffusion equation in Cartesian coordinates

We shall obtain a solution to

$$\frac{\partial^2 u}{\partial x^2} = \frac{1}{\alpha}\frac{\partial u}{\partial t} \quad 0 < x < L \quad t > 0 \tag{a}$$

over the interval indicated and subject to the initial condition

$$u(x,0) = f(x) \tag{b}$$

and the boundary conditions

$$u(0,t) = 0$$
$$u(L,t) = 0 \tag{c}$$

The solution to Eq. (a) is obtained from Eq. (T7.7b) of Table 7.7 with $v = \lambda$ as

$$u(x,t) = a_o + b_o x + \left(A\cos(\lambda x) + B\sin(\lambda x)\right)e^{-\lambda^2 \alpha t} \tag{d}$$

Substituting Eq. (d) into the first boundary condition of Eq. (c) results in

$$a_o + Ae^{-\lambda^2 \alpha t} = 0$$

Therefore, for arbitrary t, $A = 0$ and, hence, $a_o = 0$ and Eq. (d) becomes

$$u(x,t) = b_o x + B\sin(\lambda x)e^{-\lambda^2 \alpha t} \tag{e}$$

Substituting Eq. (e) into the second boundary condition of Eq. (c) gives

$$b_o L + B\sin(\lambda L)e^{-\lambda^2 \alpha t} = 0$$

In order not to have a solution that is independent of time, we keep B arbitrary and set

$$\sin(\lambda L) = 0 \tag{f}$$

and, therefore, $b_o = 0$. Equation (f) is the characteristic equation. The solution to this equation is

$$\lambda \to \lambda_n = n\pi / L \quad n = 1,2,\ldots$$

where λ_n is the eigenvalue. Then, Eq. (e) can be written as

$$u_n(x,t) = B_n \sin(\lambda_n x)e^{-\lambda_n^2 \alpha t}$$

Since this solution is valid for all n, we use superposition to obtain

$$u(x,t) = \sum_{n=1}^{\infty} B_n \sin(n\pi x/L)e^{-n^2\pi^2 \alpha t/L^2} \tag{g}$$

As was shown in several of the preceding examples, for these boundary conditions $\sin(\lambda_n x)$ is an orthogonal function with a weighting factor of one.

Lastly, we substitute Eq. (g) into the initial condition given by Eq. (b) to arrive at

$$\sum_{n=1}^{\infty} B_n \sin(n\pi x/L) = f(x)$$

Multiplying this equation by $\sin(m\pi x/L)$ and integrating over the interval results in

$$\begin{aligned}
&\sum_{n=1}^{\infty} B_n \int_0^L \sin(n\pi x/L)\sin(m\pi x/L)dx = \int_0^L f(x)\sin(m\pi x/L)dx \quad m = 1,2,\ldots \\
&B_n = \frac{2}{L}\int_0^L f(x)\sin(n\pi x/L)dx
\end{aligned} \tag{h}$$

where we have used Eq. (3.20). Substituting Eq. (h) into Eq. (g) yields

$$u(x,t) = \frac{2}{L}\sum_{n=1}^{\infty} \sin(n\pi x/L) e^{-n^2\pi^2\alpha t/L^2} \int_0^L f(x)\sin(n\pi x/L)dx \qquad \text{(i)}$$

Equation (i) is verified with Mathematica procedure M7.12.

Example 7.10: Diffusion equation in Cartesian coordinates with nonhomogeneous boundary conditions

We shall obtain a solution to

$$\frac{\partial^2 u}{\partial x^2} = \frac{1}{\alpha}\frac{\partial u}{\partial t} \qquad 0 < x < L \quad t > 0 \qquad \text{(a)}$$

over the interval indicated and subject to the initial condition

$$u(x,0) = f(x) \qquad \text{(b)}$$

and the boundary conditions

$$\frac{\partial u(0,t)}{\partial x} = 0$$
$$u(L,t) = U_o \qquad \text{(c)}$$

The solution to Eq. (a) is obtained from Eq. (T7.7b) of Table 7.7 with $v = \lambda$, that is,

$$u(x,t) = a_o + b_o x + \left(A\cos(\lambda x) + B\sin(\lambda x)\right)e^{-\lambda^2\alpha t} \qquad \text{(d)}$$

Differentiating Eq. (d), we obtain

$$\frac{\partial u}{\partial x} = b_o + \lambda\left(-A\sin(\lambda x) + B\cos(\lambda x)\right)e^{-\lambda^2\alpha t} \qquad \text{(e)}$$

Using Eq. (e) in the first boundary condition of Eq. (c) gives

$$b_o + \lambda B e^{-\lambda^2\alpha t} = 0$$

To satisfy this equation for arbitrary t, we set $B = 0$ and, therefore, $b_o = 0$. Then Eq. (d) simplifies to

$$u(x,t) = a_o + A\cos(\lambda x)e^{-\lambda^2\alpha t} \qquad \text{(f)}$$

Substituting this result into the second boundary condition of Eq. (c) yields

$$a_o + A\cos(\lambda L)e^{-\lambda^2\alpha t} = U_o \qquad \text{(g)}$$

Equation (g) can be satisfied for arbitrary t if

$$\cos(\lambda L) = 0$$

and, therefore, $a_o = U_o$. This equation is the characteristic equation the solution to which is

$$\lambda \to \lambda_n = \frac{n\pi}{2L} \quad n = 1,3,5,\ldots$$

where λ_n is the eigenvalue and $\cos(\lambda_n x)$ is the eigenfunction. Then, Eq. (f) becomes

$$u(x,t) = U_o + A_n \cos(\lambda_n x) e^{-\lambda_n^2 \alpha t} \quad n = 1,3,5,\ldots \tag{h}$$

Since this solution is valid for all odd n, we use superposition to obtain

$$u(x,t) = U_o + \sum_{n=1,3,5}^{\infty} A_n \cos(\lambda_n x) e^{-\lambda_n^2 \alpha t} \tag{i}$$

It is noted that with $a_o = U_o$, $\cos(\lambda_n x)$ satisfies a pair of homogeneous boundary conditions. From Table 7.7, it is seen that $\cos(\lambda_n x)$ is a solution to a Sturm–Liouville equation. Therefore, from the material in Section 6.1.4, the eigenfunction $\cos(\lambda_n x)$ is an orthogonal eigenfunction.

Lastly, we substitute Eq. (i) into the initial condition given by Eq. (b) and arrive at

$$\sum_{n=1,3,5}^{\infty} A_n \cos(\lambda_n x) = f(x) - U_o \tag{j}$$

Multiplying Eq. (j) by $\cos(\lambda_m x)$ and integrating over the interval, we obtain

$$\sum_{n=1,3,5}^{\infty} A_n \int_0^L \cos(\lambda_n x)\cos(\lambda_m x)dx = \int_0^L \left(f(x) - U_o\right)\cos(\lambda_m x)dx \quad m = 1,3,5\ldots$$

$$A_n = \frac{2}{L}\int_0^L \left(f(x) - U_o\right)\cos(\lambda_n x)dx \tag{k}$$

where we have used Eq. (3.20). Substituting Eq. (k) into Eq. (i) yields the solution

$$u(x,t) = U_o + \frac{2}{L}\sum_{n=1,3,5}^{\infty}\left\langle \int_0^L \left(f(x) - U_o\right)\cos(\lambda_n x)dx \right\rangle \cos(\lambda_n x) e^{-\lambda_n^2 \alpha t} \tag{l}$$

Equation (l) is verified with Mathematica procedure M7.13.

Example 7.11: Diffusion equation in cylindrical coordinates with Robin boundary conditions

We shall obtain a solution to

$$\frac{\partial^2 u}{\partial r^2} + \frac{\partial u}{r\partial r} = \frac{1}{\alpha}\frac{\partial u}{\partial t} \quad 0 < r < a \quad t > 0 \tag{a}$$

on the interval indicated with the boundary conditions

$$\frac{\partial u(a,t)}{\partial r} + hu(a,t) = 0$$
$$u(0,t) = \text{finite} \tag{b}$$

where h is a constant, and the initial condition

$$u(r,0) = f(r) \tag{c}$$

The solution to Eq. (a) is obtained from Eq. (T7.7f) of Table 7.7 with $\mu = \lambda$, that is,

$$u(r,t) = \left(CJ_0(\lambda r) + DY_0(\lambda r)\right)e^{-\lambda^2 \alpha t} \tag{d}$$

For Eq. (d) to remain finite at $r = 0$, we set $D = 0$. Then Eq. (d) simplifies to

$$u(r,t) = CJ_0(\lambda r)e^{-\lambda^2 \alpha t} \tag{e}$$

We now substitute Eq. (e) into the first boundary condition of Eq. (b) and arrive at

$$C\left(\frac{\partial}{\partial r}J_0(\lambda a) + hJ_0(\lambda a)\right)e^{-\lambda^2 \alpha t} = 0$$

To avoid a trivial solution, that is $C = 0$, we set

$$-\kappa J_1(\kappa) + haJ_0(\kappa) = 0 \tag{f}$$

In Eq. (f), $\kappa = \lambda a$ and we have used Eq. (5.100) with $m = 0$. Equation (f) is the characteristic equation. If κ_n is a root of Eq. (f), that is, κ_n satisfies

$$-\kappa_n J_1(\kappa_n) + ahJ_0(\kappa_n) = 0 \tag{g}$$

for specified values of a and h, then Eq. (e) becomes

$$u_n(r,t) = C_n J_0(\kappa_n r / a)e^{-\kappa_n^2 \alpha t / a^2}$$

Since this solution is valid for all n, we use superposition to obtain

$$u(r,t) = \sum_{n=1}^{\infty} C_n J_0(\kappa_n r / a)e^{-\kappa_n^2 \alpha t / a^2} \tag{h}$$

As explained in Example 7.5, $J_0(\kappa_m r/a)$ is an orthogonal eigenfunction with a weighting factor r.

Upon substituting Eq. (g) into the initial condition given by Eq. (c), we arrive at

$$\sum_{n=1}^{\infty} C_n J_0(\kappa_n r / a) = f(r) \tag{i}$$

Multiplying Eq. (h) by $J_0(\kappa_m r/a)rdr$ and integrating over the interval, we obtain

$$\sum_{n=1}^{\infty} C_n \int_0^a J_0(\kappa_n r / a) J_0(\kappa_m r / a) r dr = \int_0^a f(r) J_0(\kappa_m r / a) r dr \quad m = 1, 2, \ldots \tag{j}$$

Noting from Eqs. (6.88) and (6.89) that

$$\int_0^a J_0(\kappa_n r / a) J_0(\kappa_m r / a) r dr = \frac{a^2}{2}\left[J_0^2(\kappa_n) + J_1^2(\kappa_n)\right]\delta_{nm}$$

$$= \frac{a^2}{2}\left[J_0^2(\kappa_n) + \frac{h^2 a^2}{\kappa_n^2} J_0^2(\kappa_n)\right]\delta_{nm}$$

$$= \frac{a^2}{2\kappa_n^2}\left[\kappa_n^2 + h^2 a^2\right] J_0^2(\kappa_n)\delta_{nm}$$

where we have used Eq. (g), Eq. (j) yields

$$C_n = \frac{2\kappa_n^2}{a^2\left[\kappa_n^2 + h^2 a^2\right] J_0^2(\kappa_n)} \int_0^a f(r) J_0(\kappa_m r / a) r dr \tag{k}$$

Upon substituting Eq. (k) into Eq. (h), we arrive at

$$u(r,t) = \frac{2}{a^2} \sum_{n=1}^{\infty} \frac{\kappa_n^2 J_0(\kappa_n r / a) e^{-\kappa_n^2 \alpha t / a^2}}{\left[\kappa_n^2 + h^2 a^2\right] J_0^2(\kappa_n)} \int_0^a f(r) J_0(\kappa_n r / a) r dr \tag{l}$$

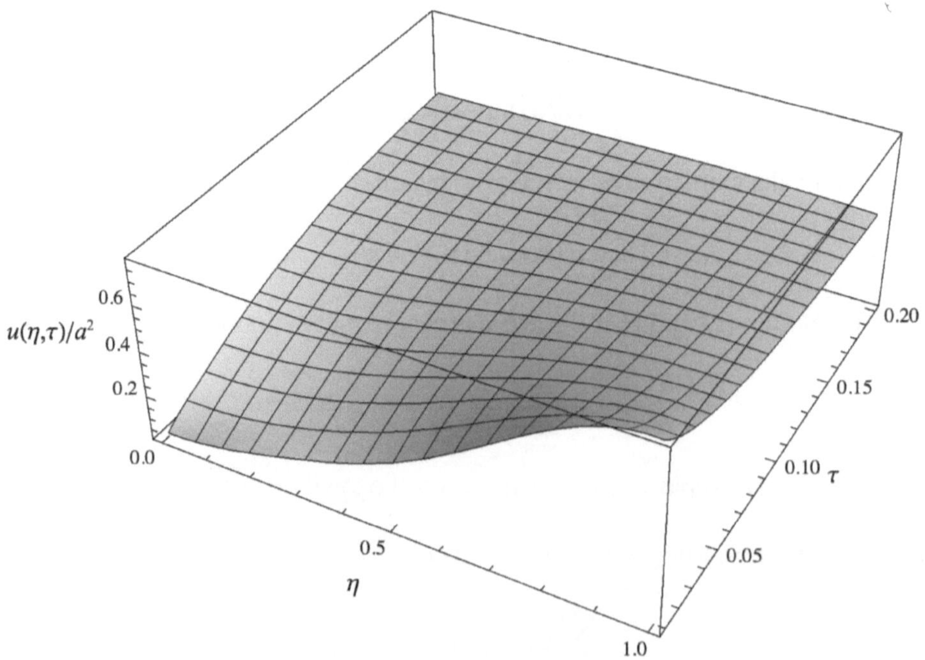

FIGURE 7.11 Numerical evaluation of Eq. (m) of Example 7.11 for $ha = 0.2$.

To illustrate the results, we assume that $f(r) = r^2$. Then

$$\int_0^a f(r)J_0(\kappa_m r/a)rdr = \int_0^a J_0(\kappa_m r/a)r^3dr = \frac{a^4}{\kappa_m^2}\left(2J_2(\kappa_m) - \kappa_m J_3(\kappa_m)\right)$$

and Eq. (l) becomes

$$u(r,t) \to u(\eta,\tau) = 2a^2\sum_{n=1}^{\infty}\frac{\left(2J_2(\kappa_m) - \kappa_m J_3(\kappa_m)\right)J_0(\kappa_n\eta)e^{-\kappa_n^2\tau}}{\left[\kappa_n^2 + h^2a^2\right]J_0^2(\kappa_n)} \qquad \text{(m)}$$

where $\eta = r/a$ and $\tau = \alpha t/a^2$. A plot of Eq. (m) for $ha = 0.2$ is shown in Figure 7.11. This figure was obtained with Mathematica procedure M7.14.

Example 7.12: Nonhomogeneous diffusion equation

We shall obtain a solution to

$$\frac{\partial^2 u}{\partial x^2} = \frac{1}{\alpha}\frac{\partial u}{\partial t} - \frac{1}{\beta}Q(x,t) \quad 0 < x < L \quad t > 0 \qquad \text{(a)}$$

with the initial condition

$$u(x,0) = f(x) \qquad \text{(b)}$$

and the boundary conditions

$$u(0,t) = a$$
$$u(L,t) = b \qquad \text{(c)}$$

The solution to Eq. (a) when $Q(x,t) = 0$ (the homogenous solution) is obtained from Eq. (T7.7b) of Table 7.7 with $\nu = \mu = \lambda$, that is,

$$u(x,t) = a_o + b_o x + \left(A\cos(\lambda x) + B\sin(\lambda x)\right)T(t) \qquad \text{(d)}$$

In the approach taken here, $T(t)$ is considered an unknown function to be determined. The reason for this approach will become apparent as we progress though the solution. Substituting Eq. (d) into the first boundary condition of Eq. (c) gives

$$a_o + AT(t) = a$$

For this relation to be valid for any value of t, $A = 0$ and, therefore, $a_o = a$. Thus, Eq. (d) becomes

$$u(x,t) = a + b_o x + \sin(\lambda x)T(t) \qquad \text{(e)}$$

Substituting Eq. (e) into the second boundary condition of Eq. (c) results in

$$a + b_o L + \sin(\lambda L)T(t) = b$$

For this relation to be valid for any value of t, we set

$$\sin(\lambda L) = 0 \rightarrow \lambda = \lambda_n = \frac{n\pi}{L}$$

and, therefore, $b_o = (b - a)/L$. The equation above is the characteristic equation whose solutions are the eigenvalues λ_n. Thus, Eq. (e) becomes

$$u_n(x,t) = a + (b-a)\frac{x}{L} + \sin(n\pi x/L)T_n(t) \quad n = 1,2,\ldots$$

Then, using superposition, the solution becomes

$$u(x,t) = a + (b-a)\frac{x}{L} + \sum_{n=1}^{\infty} T_n(t)\sin(n\pi x/L) \tag{f}$$

It is noted that with $a_o = a$ and $b_o = (b - a)/L$, $\sin(n\pi x/L)$ satisfies a pair of homogeneous boundary conditions. From Table 7.7, it is seen that $\sin(n\pi x/L)$ is a solution to a Sturm–Liouville equation. Therefore, from the material in Section 6.1.4, the eigenfunction $\sin(n\pi x/L)$ is an orthogonal eigenfunction. From Section 6.1.3, it was found that the weighting factor is one.

We now substitute Eq. (f) into Eq. (a) and obtain

$$\begin{aligned}
&-\sum_{n=1}^{\infty}\left(\frac{n^2\pi^2}{L^2}\right)\sin(n\pi x/L)T_n(t) = \frac{1}{\alpha}\sum_{n=1}^{\infty}\sin(n\pi x/L)\frac{dT_n}{dt} - \frac{1}{\beta}Q(x,t) \\
&\sum_{n=1}^{\infty}\left(\frac{dT_n}{dt} + \frac{n^2\pi^2\alpha}{L^2}T_n\right)\sin(n\pi x/L) = \frac{\alpha}{\beta}Q(x,t)
\end{aligned} \tag{g}$$

Multiplying Eq. (g) by $\sin(m\pi x/L)$ and integrating over the interval, we arrive at

$$\begin{aligned}
&\sum_{n=1}^{\infty}\left(\frac{dT_n}{dt} + \frac{n^2\pi^2\alpha}{L^2}T_n\right)\int_0^L \sin(n\pi x/L)\sin(m\pi x/L)dx = \frac{\alpha}{\beta}\int_0^L Q(x,t)\sin(m\pi x/L)dx \\
&\frac{dT_n}{dt} + \frac{n^2\pi^2\alpha}{L^2}T_n = q_n(t)
\end{aligned} \tag{h}$$

where we have used Eq. (3.20) and

$$q_n(t) = \frac{2\alpha}{\beta L}\int_0^L Q(x,t)\sin(m\pi x/L)dx \tag{i}$$

The solution to Eq. (h) is obtained from Eq. (4.25) as

$$T_n = \left(G_n + p_n(t)\right)e^{-n^2\pi^2\alpha t/L^2} \tag{j}$$

where G_n is determined from the initial condition and

$$p_n(t) = \int_0^t q_n(\tau) e^{n^2\pi^2\alpha\tau/L^2} d\tau \tag{k}$$

Thus, Eq. (f) becomes

$$u(x,t) = a + (b-a)\frac{x}{L} + \sum_{n=1}^{\infty} \left(G_n + p_n(t)\right) e^{-n^2\pi^2\alpha t/L^2} \sin(n\pi x/L) \tag{l}$$

To obtain G_n, we substitute Eq. (l) into Eq. (b) and obtain

$$a + (b-a)\frac{x}{L} + \sum_{n=1}^{\infty} G_n \sin(n\pi x/L) = f(x) \tag{m}$$

where, from Eq. (k), we have used the fact that $p_n(0) = 0$. Multiplying Eq. (m) by $\sin(m\pi x/L)$ and integrating over the interval, we arrive at

$$\sum_{n=1}^{\infty} G_n \int_0^L \sin(n\pi x/L)\sin(m\pi x/L)dx = \int_0^L \left(f(x) - a - (b-a)\frac{x}{L} \right) \sin(m\pi x/L)dx \tag{n}$$

Thus,

$$G_n = \frac{2}{L}\int_0^L \left(f(x) - a - (b-a)\frac{x}{L} \right) \sin(m\pi x/L)dx \tag{o}$$

and we have used Eq. (3.20).

To illustrate this result, we shall assume that $a = 0$, $b = 1$, and

$$Q(x,t) = e^{-t}\sin(3\pi x/L)$$

The initial condition $f(x)$ shall remain unspecified. Then, from Eq. (i)

$$q_n(t) = \frac{2\alpha}{\beta L} e^{-t} \int_0^L \sin(3\pi x/L)\sin(m\pi x/L)dx = \frac{\alpha}{\beta} e^{-t}\delta_{3n}$$

Substituting this result into Eq. (k) yields

$$p_n(t) = \frac{\alpha}{\beta}\delta_{3n} \int_0^t e^{-t} e^{n^2\pi^2\alpha t/L^2} dt = \frac{\alpha}{\beta}\delta_{3n} \left(\frac{1 - e^{-t} e^{n^2\pi^2\alpha t/L^2}}{1 - n^2\pi^2\alpha / L^2} \right)$$

In addition, Eq. (o) simplifies to

$$G_n = \frac{2}{L}\int_0^L \left(f(x) - \frac{x}{L} \right) \sin(m\pi x/L)dx$$

Then, Eq. (l) gives

$$u(x,t) = \frac{x}{L} + \sum_{n=1}^{\infty} \left(G_n + \frac{\alpha}{\beta} \delta_{3n} \left(\frac{1 - e^{-t} e^{n^2\pi^2\alpha t/L^2}}{1 - n^2\pi^2\alpha/L^2} \right) \right) e^{-n^2\pi^2\alpha t/L^2} \sin(n\pi x/L)$$

$$= \frac{x}{L} + \frac{\alpha}{\beta} \left(\frac{e^{-9\pi^2\alpha t/L^2} - e^{-t}}{1 - 9\pi^2\alpha/L^2} \right) \sin(3\pi x/L) + \sum_{n=1}^{\infty} G_n e^{-n^2\pi^2\alpha t/L^2} \sin(n\pi x/L)$$

Example 7.13: Diffusion equation in a rectangular region

We shall obtain a solution to

$$\frac{\partial^2 u}{\partial x^2} + \frac{\partial^2 u}{\partial y^2} = \frac{1}{\alpha} \frac{\partial u}{\partial t} \qquad 0 < x < a \quad 0 < y < b \quad t > 0 \tag{a}$$

over the region indicated with the initial condition

$$u(x, y, 0) = f(x, y) \tag{b}$$

and the boundary conditions

$$\begin{aligned} u(0, y, t) &= 0, \quad & u(a, y, t) &= 0 \\ u(x, b, t) &= 0, \quad & u(x, b, t) &= 0 \end{aligned} \tag{c}$$

The solution to Eq. (a) is obtained from Eq. (T7.7a) of Table 7.7 as

$$u(x, y, t) = X_0(x)Y_0(y) + X(x)Y(y)T(t) \tag{d}$$

where, from Table 7.7,

$$\begin{aligned} T(t) &= Ae^{-\mu^2\alpha t} \\ X_0(x) &= a_0 + b_0 x \\ Y_0(y) &= c_0 + d_0 y \\ X(x) &= B\cos(\nu x) + C\sin(\nu x) \\ Y(y) &= D\cos(\beta y) + E\sin(\beta y) \end{aligned} \tag{e}$$

and

$$\beta^2 = \mu^2 - \nu^2 \tag{f}$$

Upon substituting Eq. (d) into the first set of boundary conditions given in Eq. (c), we obtain with the use of Eq. (e)

$$a_0 Y_0(y) + BY(y)T(t) = 0$$

$$(a_0 + b_0 a)Y_0(y) + C\sin(\nu a)Y(y)T(t) = 0$$

We see that for arbitrary y and t and in order to have a nontrivial solution, we set $a_0 = b_0 = B = 0$ and

$$\sin(\nu a) = 0$$

which is the characteristic equation. Therefore, $\nu = n\pi/a$, n =1, 2, …, and Eq. (d) reduces to

$$u(x,y,t) = \sin(n\pi x / a)Y(y)T(t) \tag{g}$$

where the constant has been incorporated into $Y(y)$. As shown in previous examples, for these boundary conditions $\sin(n\pi x/a)$ is an orthogonal eigenfunction.

Upon substituting Eq. (g) into the second set of boundary conditions given by Eq. (c), we arrive at

$$DT(t)\sin(n\pi x / a) = 0$$

$$\big(D\cos(\beta b) + E\sin(\beta b)\big)T(t)\sin(n\pi x / a) = 0$$

Therefore, we see that for arbitrary x and t, $D = 0$, and in order not to have a trivial solution, we set

$$\sin(\beta b) = 0$$

which is the characteristic equation. Therefore, $\beta = m\pi/b$, m =1, 2, …, and Eq. (g) becomes

$$u(x,y,t) = \sin(n\pi x/a)\sin(m\pi y/b)T(t) \tag{h}$$

where the constant E has been incorporated into $T(t)$. Again, for these boundary conditions $\sin(n\pi x/b)$ is an orthogonal eigenfunction.

From Eq. (f), we find that

$$\mu^2 \to \mu_{nm}^2 = \pi^2\left(\frac{n^2}{a^2} + \frac{m^2}{b^2}\right) \tag{i}$$

Hence,

$$T(t) = A_{nm}e^{-\mu_{nm}^2 \alpha t}$$

and Eq. (h) can be written as

$$u_{nm}(x,y,t) = A_{nm}\sin(n\pi x/a)\sin(m\pi y/b)e^{-\mu_{nm}^2 \alpha t} \tag{j}$$

Since Eq. (j) is valid for all combinations of n and m, we use superposition to obtain

$$u(x,y,t) = \sum_{n=1}^{\infty}\sum_{m=1}^{\infty} A_{nm}\sin(n\pi x/a)\sin(m\pi y/b)e^{-\mu_{nm}^2 \alpha t} \tag{k}$$

We now substitute Eq. (k) into the initial condition given by Eq. (b) and obtain

$$\sum_{n=1}^{\infty}\sum_{m=1}^{\infty} A_{nm}\sin(n\pi x/a)\sin(m\pi y/b) = f(x,y) \tag{l}$$

Multiplying Eq. (l) by $\sin(p\pi x/a)\sin(s\pi x/b)dxdy$, where p and s are integers, and integrating over the region yields

$$\sum_{n=1}^{\infty}\sum_{m=1}^{\infty} A_{nm}\int_0^a \sin(n\pi x/a)\sin(p\pi x/a)dx\int_0^b \sin(m\pi y/b)\sin(s\pi y/b)dy$$

$$= \int_0^a\int_0^b f(x,y)\sin(p\pi x/a)\sin(s\pi y/b)dxdy$$

Upon using Eq. (3.20), this result reduces to

$$A_{nm} = \frac{4}{ab}\int_0^a\int_0^b f(x,y)\sin(n\pi x/a)\sin(m\pi y/b)dxdy \tag{m}$$

Thus, substitution of Eq. (m) into Eq. (k) is the solution to Eqs. (a) to (c).

Example 7.14: Diffusion equation in spherical coordinates

We shall obtain a solution to

$$\frac{\partial}{r^2\partial r}\left(r^2\frac{\partial u}{\partial r}\right) = \frac{1}{\alpha}\frac{\partial u}{\partial t} \qquad 0 < r < a \quad t > 0 \tag{a}$$

over the interval indicted and subject to the initial condition

$$u(r,0) = f(r) \tag{b}$$

and the boundary conditions

$$\begin{aligned} u(0,t) &= \text{finite} \\ u(a,t) &= 0 \end{aligned} \tag{c}$$

The solution to Eq. (a) is obtained from Eq. (T7.7h) of Table 7.7 with $\mu = \lambda$, that is,

$$u(r,t) = \left(Aj_0(\lambda r) + By_0(\lambda r)\right)e^{-\lambda^2\alpha t}$$

or, upon using Eqs. (5.118) and (5.120)

$$u(r,t) = \left(A\frac{\sin(\lambda r)}{\lambda r} + B\frac{\cos(\lambda r)}{\lambda r}\right)e^{-\lambda^2\alpha t} \tag{d}$$

To satisfy the first boundary condition of Eq. (c), we must set $B = 0$. Then, Eq. (d) becomes

$$u(r,t) = A\frac{\sin(\lambda r)}{\lambda r}e^{-\lambda^2 \alpha t} \tag{e}$$

Upon substituting Eq. (e) into the second boundary condition of Eq. (c), we find that

$$A\frac{\sin(\lambda a)}{\lambda a}e^{-\lambda^2 \alpha t} = 0$$

In order not to have a trivial solution, we choose $\gamma = \lambda a$, where γ satisfies

$$\sin(\gamma) = 0$$

Therefore, $\gamma = \gamma_n = n\pi$ and $\lambda = n\pi/a$. Then, Eq. (e) can be written as

$$u_n(r,t) = c_n \frac{\sin(n\pi r / a)}{r}e^{-n^2\pi^2 \alpha t/a^2} \tag{f}$$

According to Special Case II and the discussion of Eq. (6.82), $\sin(n\pi r/a)/r$ is an orthogonal eigenfunction with, as shown in Eq. (6.51), a weighting factor of r^2. Since Eq. (f) is valid for all n, we use superposition to arrive at

$$u(r,t) = \sum_{n=1}^{\infty} c_n \frac{\sin(n\pi r/a)}{r}e^{-n^2\pi^2 \alpha t/a^2} \tag{g}$$

To determine c_n, we substitute Eq. (g) into the initial condition given by Eq. (b) and obtain

$$\sum_{n=1}^{\infty} c_n \frac{\sin(n\pi r/a)}{r} = f(r) \tag{h}$$

We multiply Eq. (h) by $(\sin(m\pi r/a)/r)r^2 dr = \sin(m\pi r/a) r dr$ and integrate over the interval to obtain

$$\begin{aligned} \sum_{n=1}^{\infty} c_n \int_0^a \sin(n\pi r/a)\sin(m\pi r/a)dr &= \int_0^a f(r)\sin(m\pi r/a) r dr \\ c_n &= \frac{2}{a}\int_0^a f(r)\sin(n\pi r/a) r dr \end{aligned} \tag{i}$$

Then, Eq. (g) becomes

$$u(r,t) = \frac{2}{a}\sum_{n=1}^{\infty} \frac{\sin(n\pi r/a)}{r}e^{-n^2\pi^2 \alpha t/a^2} \int_0^a f(r)\sin(n\pi r/a) r dr \tag{j}$$

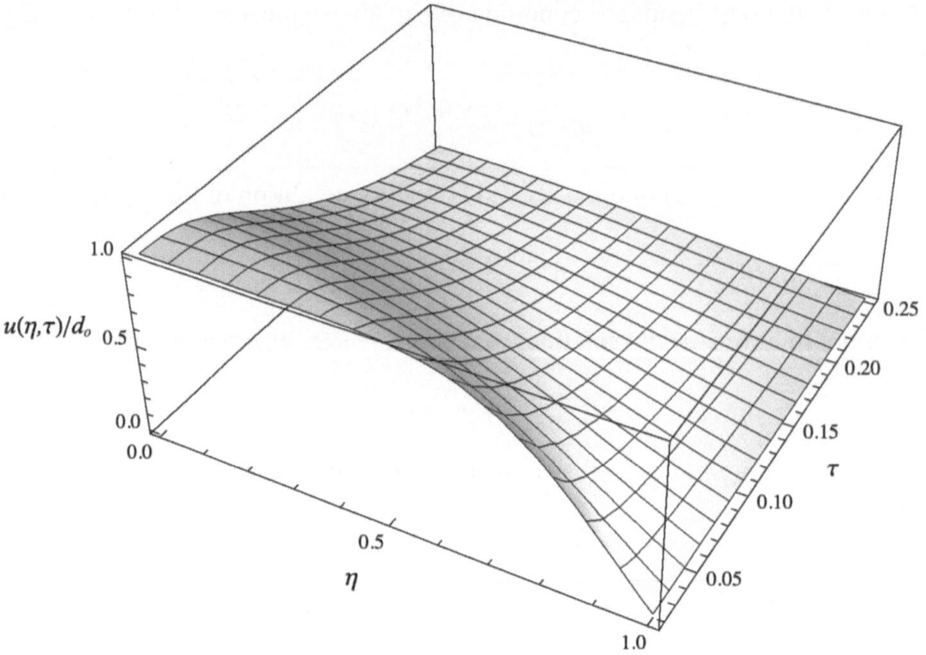

FIGURE 7.12 Numerical evaluation of Eq. (k) of Example 7.14.

If $f(r) = d_o$, then Eq. (j) becomes

$$\frac{u(r,t)}{d_o} = \frac{2}{\pi}\sum_{n=1}^{\infty}(-1)^{n+1}\frac{\sin(n\pi\eta)}{n\eta}e^{-n^2\pi^2\tau} \tag{k}$$

where $\eta = r/a$, $\tau = \alpha t/a^2$, and we have used Mathematica procedure M7.15 to perform the integration in Eq. (j). Using Mathematica procedure M7.16, Eq. (k) is plotted in Figure 7.12.

7.2.5 Wave Equation

The general solutions to the wave equation in three coordinate systems are given in Table 7.9 along with some of the intermediate results required to obtain them. To illustrate how the results in these tables were obtained, we shall present the details of the separation of variable method for the wave equation in the Cartesian coordinate system. The solutions for the other coordinates are obtained in a similar manner.

The homogeneous wave equation in the Cartesian coordinate system in terms of the variable $u = u(x,y)$ is obtained from Tables 7.1 and 7.2 as

$$\frac{\partial^2 u}{\partial x^2} + \frac{\partial^2 u}{\partial y^2} - k^2\frac{\partial^2 u}{\partial t^2} = 0 \quad k^2 > 0 \tag{7.74}$$

We assume a solution of the form

$$u(x,y,t) = X(x)Y(y)T(t) \tag{7.75}$$

TABLE 7.9
Solutions to the Wave Equation in Three Coordinate Systems

Cartesian Coordinates	**Assumed Separable Form**	**Separable Equation**
$\frac{\partial^2 u}{\partial x^2}+\frac{\partial^2 u}{\partial y^2}-k^2\frac{\partial^2 u}{\partial t^2}=0$	$u(x,y)=X(x)Y(y)T(t)$	$\frac{d^2X}{Xdx^2}+\frac{d^2Y}{Ydy^2}-k^2\frac{d^2T}{Tdt^2}=0$
Separated equations	**Solutions to separated equations**	**Solution and special cases**
$\frac{d^2T}{dt^2}+\mu^2T=0$ (S-L)	$T(t,\mu)=E\cos(\mu t)+F\sin(\mu t)$	
$\frac{d^2X}{dy^2}+\nu^2X=0$ (S-L)	$X(x,\nu)=B\cos(\nu x)+C\sin(\nu x)$	$u(x,y,t)=X(x,\nu)Y(y,\beta)T(t,\mu)$ (T7.9a) $u(x,t)=X(x,\mu)T(t,\mu)$ (T7.9b)
$\frac{d^2Y}{dx^2}+\beta^2Y=0 \quad \beta^2=k^2\mu^2-\nu^2$	$Y(y,\beta)=D\cos(\beta y)+E\sin(\beta y) \quad \beta^2>0$ $Y(y,\beta)=D\cosh(\beta y)+E\sinh(\beta y) \quad \beta^2<0$	
Polar coordinates	**Assumed separable form**	**Separable equation**
$\frac{\partial^2 u}{\partial r^2}+\frac{\partial u}{r\partial r}+\frac{\partial^2 u}{r^2\partial\theta^2}-k^2\frac{\partial^2 u}{\partial t^2}=0$	$u(r,\theta,z)=R(r)\Theta(\theta)T(t)$	$\frac{1}{R}\left(\frac{d^2R}{dr^2}+\frac{dR}{rdr}\right)+\frac{d^2\Theta}{r^2\Theta d\theta^2}-k^2\frac{\partial^2T}{T\partial t^2}=0$
Separated equations	**Solutions to separated equations**	**Solution and special cases**
$\frac{\partial^2T}{\partial t^2}+\mu^2T=0$ (S-L)	$T(t)=E\cos(\mu t)+F\sin(\mu t)$	
$\frac{d^2\Theta}{d\theta^2}+m^2\Theta=0$ (S-L)	$\Theta(\theta)=A\cos m\theta+B\sin m\theta$	$u(r,\theta,t)=R_m(r)\Theta(\theta)T(t)$ (T7.9c) $u(r,t)=R_0(r)T(t)$ (T7.9d)
$r^2\frac{d^2R}{dr^2}+r\frac{dR}{dr}+\left(\mu^2k^2r^2-m^2\right)R=0$ (S-L)	$R_m(r)=CJ_m(k\mu r)+DY_m(k\mu r)$	

(Continued)

TABLE 7.9 (CONTINUED)

Solutions to the Wave Equation in Three Coordinate Systems

Spherical coordinates	Assumed separable form	Separable equations
$\frac{\partial^2 u}{\partial r^2}+\frac{2}{r}\frac{\partial u}{\partial r}+\frac{\cot\varphi}{r^2}\frac{\partial u}{\partial\varphi}+\frac{1}{r^2}\frac{\partial^2 u}{\partial\varphi^2}-k^2\frac{\partial^2 u}{\partial t^2}=0$	$u(r,\varphi,t)=R(r)\Phi(\varphi)T(t)$	$\frac{r^2}{R}\left(\frac{d^2R}{dr^2}+\frac{2}{r}\frac{dR}{dr}\right)+\frac{1}{\Phi}\left(\cot\varphi\frac{d\Phi}{d\varphi}+\frac{d^2\Phi}{d\varphi^2}\right)-k^2r^2\frac{d^2T}{Tdt^2}=0$
Separated equations	**Solutions to separated equations**	**Solution and special cases**
$\frac{\partial^2 T}{\partial t^2}+\frac{\mu^2}{k^2}T=0$ (S-L)	$T(t)=E\cos(\mu t)+F\sin(\mu t)$	
$\frac{d^2\Phi}{d\varphi^2}+\cot\varphi\frac{d\Phi}{d\varphi}+n(n+1)\Phi=0$ (S-L)	$\Phi(\varphi)=CP_n(\cos\varphi)+DQ_n(\cos\varphi)$	$u(r,\varphi,t)=R(r)\Phi(\varphi)T(t)$ (T7.9e)
$r^2\frac{d^2R}{dr^2}+2r\frac{dR}{dr}+\mu^2k^2r^2R-n(n+1)R=0$ (S-L)	$R(r)=Aj_n(\mu kr)+By_n(\mu kr)$	

Note: The quantities m^2, ν^2, k^2, and μ^2 are real positive constants and n is zero or a positive integer. The notation (S–L) indicates that this will be a Sturm–Liouville system when the boundary conditions are those given by Eq. (6.24).

Upon substituting Eq. (7.75) into Eq. (7.74), we obtain

$$\frac{d^2X}{dx^2}YT+\frac{d^2Y}{dy^2}XT-k^2\frac{d^2T}{dt^2}XY=0 \tag{7.76}$$

To obtain a solution to Eq. (7.76), Eq. (7.76) is rewritten as

$$\frac{1}{k^2}\left(\frac{d^2X}{Xdx^2}+\frac{d^2Y}{Ydy^2}\right)=\frac{d^2T}{Tdt^2}=-\mu^2 \quad \mu^2>0 \tag{7.77}$$

Hence, T is a solution to

$$\frac{d^2T}{dt^2}+\mu^2T=0 \tag{7.78}$$

Equation (7.78) is a differential equation of the Sturm–Liouville type [recall Eq. (6.41)] when the boundary conditions are those given by Eq. (6.24) with y replaced by T. Then, Eq. (7.77) can be written as

$$\frac{d^2X}{Xdx^2}+k^2\mu^2=-\frac{d^2Y}{Ydy^2}=\nu^2 \quad \nu^2>0 \tag{7.79}$$

Thus, Y is a solution to

$$\frac{d^2Y}{dy^2}+\nu^2Y=0 \tag{7.80}$$

Thus, by choosing the separation as $\nu^2 > 0$, Eq. (7.80) is a differential equation of the Sturm–Liouville system [recall Eq. (6.41)] when the boundary conditions are those given by Eq. (6.24) with y replaced by Y. Then Eq. (7.79) becomes

$$\frac{d^2X}{dx^2}+\lambda^2X=0 \tag{7.81}$$

where

$$\lambda^2=k^2\mu^2-\nu^2 \tag{7.82}$$

Equation (7.81) is a differential equation of the Sturm–Liouville type [recall Eq. (6.41)] when $\lambda^2 > 0$ and the boundary conditions are those given by Eq. (6.24) with y replaced by X. When $\lambda^2 < 0$, Eq. (7.81) is *not* of the Sturm–Liouville type.

The solutions to Eqs. (7.78), (7.80), and (7.82), respectively, are

$$T(t)=E\cos(\mu t)+F\sin(\mu t) \tag{7.83}$$

and

$$Y(y)=C\cos(\nu y)+D\sin(\nu y) \tag{7.84}$$

and

$$X(x)=A\cos(\lambda x)+B\sin(\lambda x) \tag{7.85}$$

TABLE 7.10
Summary of Examples of Applications of the Wave Equation: Boundary Conditions, Initial Conditions, and Coordinates Systems

Example	Wave Equation	Boundary Conditions	Initial Condition
7.15	$c_o^2 \dfrac{\partial^2 u}{\partial x^2} - \dfrac{\partial^2 u}{\partial t^2} = 0$		$u(x,0) = f(x)$ $\dfrac{\partial u(x,0)}{\partial t} = g(x)$
7.16	$c_o^2 \dfrac{\partial^2 u}{\partial x^2} - \dfrac{\partial^2 u}{\partial t^2} = 0$	$u(0,t) = 0, \quad u(L,t) = 0$	$u(x,0) = f(x)$ $\dfrac{\partial u(x,0)}{\partial t} = g(x)$
7.17	$\dfrac{\partial^2 u}{\partial x^2} + \dfrac{\partial^2 u}{\partial y^2} - \dfrac{\partial^2 u}{c^2 \partial t^2} = 0$	$u(0,y,t) = 0, \quad u(a,y,t) = 0$ $u(x,0,t) = 0, \quad u(x,b,t) = 0$	$u(x,y,0) = f(x,y)$ $\dfrac{\partial u(x,y,0)}{\partial t} = g(x,y)$
7.18	$\dfrac{\partial^2 u}{\partial r^2} + \dfrac{\partial u}{r \partial r} + \dfrac{\partial^2 u}{r^2 \partial \theta^2}$ $- \dfrac{1}{c^2} \dfrac{\partial^2 u}{\partial t^2} = 0$	$u(0,\theta,t) = \text{finite}$ $u(a,\theta,t) = 0$	$u(r,\theta,0) = f(r,\theta)$ $\dfrac{\partial u(r,\theta,0)}{\partial t} = g(r,\theta)$
7.19	$\dfrac{\partial^2 u}{\partial \tau^2} - \dfrac{\partial^2 u}{\partial \eta^2} = \alpha^2 F(\eta,\tau)$	$u(0,\tau) = 0, \quad u(1,\tau) = 0$	$u(\eta,0) = 0$ $\dfrac{\partial u(\eta,0)}{\partial \tau} = 0$

Thus, the solution to Eq. (7.74) is

$$\begin{aligned} u(x,y,t) &= X(x)Y(y)T(t) \\ u(x,t) &= X(x)T(t) \quad \lambda = k\mu \end{aligned} \tag{7.86}$$

The second equation of Eq. (7.86) is the case when Eq. (7.74) is independent of y.

We shall now solve the wave equation in different coordinate systems for different combinations of boundary conditions and initial conditions. The applications considered are summarized in Table 7.10.

Example 7.15: D'Alembert's solution of the wave equation

We shall obtain a solution to

$$c_o^2 \frac{\partial^2 u}{\partial x^2} - \frac{\partial^2 u}{\partial t^2} = 0 \tag{a}$$

subject to the initial conditions

$$\begin{aligned} u(x,0) &= f(x) \\ \frac{\partial u(x,0)}{\partial t} &= g(x) \end{aligned} \tag{b}$$

Comparing Eq. (a) to the form given in Table 7.9, we see that $k = 1/c_o$. Notice that no boundary conditions have been specified.

Before solving, we compare Eq. (a) with Eq. (7.1) and find that $a = c_o^2$, $c = -1$, and $b = d = e = f = g = 0$. Then the discriminant given by Eq. (7.32) is

$$\Delta = b^2 - ac = c_o^2 \rightarrow \sqrt{\Delta} = c_o$$

Since $\Delta > 0$, Eq. (a) is a hyperbolic equation. It is noted that Eq. (a) is in the second canonical form for a hyperbolic equation [recall Eq. (7.19)]. We shall use its first canonical form given by Eq. (7.18). To this end, we use Eq. (7.14) to obtain the characteristic curves, which with $x = x$ and $y = t$ gives

$$\begin{aligned} \xi &= \frac{x}{a}\left[b + \sqrt{\Delta}\right] - t = \frac{x}{c_o^2} c_o - t = \frac{x}{c_o} - t \quad \rightarrow \xi = x - c_o t \\ \eta &= \frac{x}{a}\left[b - \sqrt{\Delta}\right] - t = -\frac{x}{c_o^2} c_o - t = -\frac{x}{c_o} - t \quad \rightarrow \eta = x + c_o t \end{aligned} \tag{c}$$

The first canonical form is given by Eq. (7.18) as

$$\frac{\partial^2 u}{\partial \eta \partial \xi} = 0 \tag{d}$$

The solution to Eq. (d) is

$$u(\xi, \eta) = F(\eta) + G(\xi) \tag{e}$$

or

$$u(x,t) = F(x + c_o t) + G(x - c_o t) \tag{f}$$

To determine F and G, we use the initial conditions given by Eq. (b). Hence, from Eq. (f), the first initial condition gives

$$u(x,0) = F(x) + G(x) = f(x) \tag{g}$$

From Eq. (e)

$$\begin{aligned} \frac{\partial u(\eta,\xi)}{\partial t} &= \frac{\partial F(\eta)}{\partial t} + \frac{\partial G(\xi)}{\partial t} = \frac{dF(\eta)}{d\eta}\frac{d\eta}{dt} + \frac{dG(\xi)}{d\xi}\frac{d\xi}{dt} \\ &= c_o \frac{dF(\eta)}{d\eta} - c_o \frac{dG(\xi)}{d\xi} \end{aligned}$$

Then, from the second initial condition,

$$\begin{aligned} \frac{\partial u(x,0)}{\partial t} &= \left[c_o \frac{dF(x + c_o t)}{d(x + c_o t)} - c_o \frac{dG(x - c_o t)}{d(x - c_o t)} \right]_{t=0} \\ &= c_o \frac{dF(x)}{dx} - c_o \frac{dG(x)}{dx} = c_o \frac{d}{dx}\big(F(x) - G(x)\big) = g(x) \end{aligned} \tag{h}$$

We integrate Eq. (h) with respect to x from zero to any point x and obtain

$$F(x)-G(x)=\frac{1}{c_o}\int_0^x g(\alpha)d\alpha + C \qquad \text{(i)}$$

We now use Eqs. (g) and (i) to solve for F and G, respectively, by adding and subtracting them, and find that

$$\begin{aligned} F(x)&=\frac{f(x)}{2}+\frac{1}{2c_o}\int_0^x g(\alpha)d\alpha+\frac{C}{2} \\ G(x)&=\frac{f(x)}{2}-\frac{1}{2c_o}\int_0^x g(\alpha)d\alpha-\frac{C}{2} \end{aligned} \qquad \text{(j)}$$

The final solution is obtained by using Eq. (j) in Eq. (f) to arrive at

$$\begin{aligned} u(x,t)&=F(x+c_ot)+G(x-c_ot) \\ &=\frac{f(x+c_ot)}{2}+\frac{1}{2c_o}\int_0^{x+c_ot} g(\alpha)d\alpha+\frac{C}{2} \\ &+\frac{f(x-c_ot)}{2}-\frac{1}{2c_o}\int_0^{x-c_ot} g(\alpha)d\alpha-\frac{C}{2} \\ &=\frac{1}{2}\left(f(x+c_ot)+f(x-c_ot)\right) \\ &+\frac{1}{2c_o}\int_0^{x+c_ot} g(\alpha)d\alpha-\frac{1}{2c_o}\int_0^{x-c_ot} g(\alpha)d\alpha \\ &=\frac{1}{2}\left(f(x+c_ot)+f(x-c_ot)\right)+\frac{1}{2c_o}\int_{x-c_ot}^{x+c_ot} g(\alpha)d\alpha \end{aligned} \qquad \text{(k)}$$

Equation (k) is known as *D'Alembert's formula* for the solution to the one-dimensional wave equation. Equation (k) is verified with Mathematica procedure M7.17.

Example 7.16: Wave equation in Cartesian coordinates: One spatial dimension

We shall obtain a solution to

$$c^2\frac{\partial^2 u}{\partial x^2}-\frac{\partial^2 u}{\partial t^2}=0 \quad 0<x<L \quad t>0 \qquad \text{(a)}$$

over the interval indicated and subject to the initial conditions

$$u(x,0) = f(x)$$
$$\frac{\partial u(x,0)}{\partial t} = g(x) \tag{b}$$

and the boundary conditions

$$u(0,t) = 0$$
$$u(L,t) = 0 \tag{c}$$

The solution to Eq. (a) is obtained from Eq. (T7.9b) of Table 7.9 as

$$u(x,t) = X(x,\mu)T(t,\mu) \tag{d}$$

where

$$X(\mu,x) = A\cos(\mu x / c) + B\sin(\mu x / c)$$
$$T(\mu,t) = E\cos(\mu t) + F\sin(\mu t) \tag{e}$$

and we have used the fact that $k = 1/c$. Substituting Eq. (d) into the first boundary condition of Eq. (c), we obtain

$$AT(\mu,t) = 0$$

Therefore, for arbitrary t, $A = 0$. Then Eq. (d) simplifies to

$$u(x,t) = \sin(\mu x / c)T(t,\mu,) \tag{f}$$

where the constant B has been incorporated into the remaining constants. Substituting Eq. (f) into the second boundary condition of Eq. (c) gives

$$\sin(\mu L / c)T(t,\mu) = 0$$

Therefore, so as not to have a trivial solution, we set

$$\sin(\mu L / c) = 0$$

which is the characteristic equation. The solution to this equation is

$$\mu \to \mu_n = n\pi c / L$$

where μ_n is the eigenvalue. Thus, Eq. (f) becomes

$$u_n(x,t) = \sin(n\pi x / L)T(t,\mu_n)$$

Since this result is valid for all values of n, we use superposition to obtain

$$u(x,t) = \sum_{n=1}^{\infty}\left(E_n \cos(n\pi ct/L) + F_n \sin(n\pi ct/L)\right)\sin(n\pi x/L) \tag{g}$$

where we have used Eq. (e).

From Table 7.9, the function X satisfies the Sturm–Liouville equation

$$\frac{d^2X}{dx^2} + \mu^2 X = 0$$

From the boundary conditions in Eq. (c), it is seen that the boundary conditions are

$$X(\mu,0) = 0$$
$$X(\mu,L) = 0$$

The solution to this homogeneous equation and homogeneous boundary conditions is the eigenfunction $\sin(n\pi x/L)$ with the corresponding eigenvalue μ_n. However, based on the material in Section 6.1.4, it was shown that the eigenfunction is an orthogonal eigenfunction on the interval $0 \le x \le L$. From Section 6.1.3, it was found that the weighting factor is one.

To determine E_n and F_n, we first substitute Eq. (g) into the initial conditions given by Eq. (b), to obtain

$$\begin{aligned}\sum_{n=1}^{\infty} E_n \sin(n\pi x/L) &= f(x) \\ \sum_{n=1}^{\infty} (n\pi c/L) F_n \sin(n\pi x/L) &= g(x)\end{aligned} \tag{h}$$

Each of these equations is multiplied by $\sin(m\pi x/L)$ and integrated over the interval to arrive at

$$\begin{aligned}\sum_{n=1}^{\infty} E_n \int_0^L \sin(n\pi x/L)\sin(m\pi x/L)dx &= \int_0^L f(x)\sin(m\pi x/L)dx \quad m = 1,2,\ldots \\ E_n &= \frac{2}{L}\int_0^L f(x)\sin(n\pi x/L)dx\end{aligned} \tag{i}$$

and

$$\begin{aligned}\sum_{n=1}^{\infty} (n\pi c/L) F_n \int_0^L \sin(n\pi x/L)\sin(m\pi x/L)dx &= \int_0^L g(x)\sin(m\pi x/L)dx \quad m = 1,2,\ldots \\ F_n &= \frac{2}{n\pi c}\int_0^L g(x)\sin(n\pi x/L)dx\end{aligned} \tag{j}$$

where we have used Eq. (3.20).

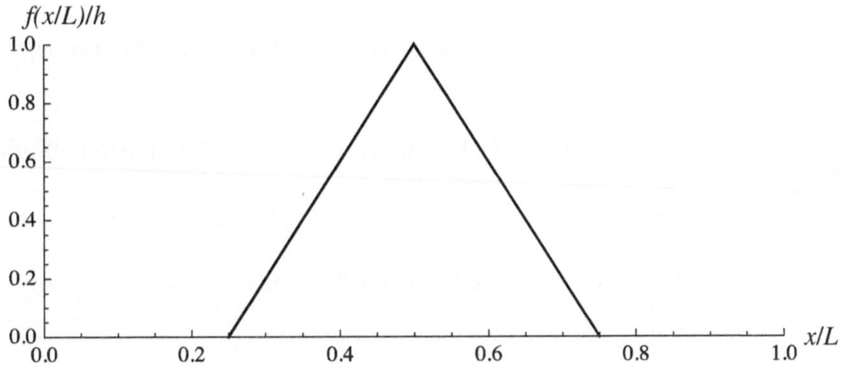

FIGURE 7.13 Initial condition for Example 7.16.

Hence, upon substituting Eqs. (i) and (j) into Eq. (g), the solution is

$$u(x,t)=\frac{2}{L}\sum_{n=1}^{\infty}\cos(n\pi ct/L)\sin(n\pi x/L)\int_0^L f(x)\sin(n\pi x/L)dx$$
$$+\frac{2}{n\pi c}\sum_{n=1}^{\infty}\sin(n\pi ct/L)\sin(n\pi x/L)\int_0^L g(x)\sin(n\pi x/L)dx \tag{k}$$

To illustrate these results, we assume that $g(x) = 0$ and that

$$f(x)=\begin{cases} 0 & 0\le x\le L/4 \\ 4h(x/L-1/4) & L/4\le x\le L/2 \\ 4h(3/4-x/L) & L/2\le x\le 3L/4 \\ 0 & 3L/4\le x\le L \end{cases}$$

which is shown in Figure 7.13. Then, from Eq. (j), $F_n = 0$ and from Eq. (i)

$$E_n=\frac{8h}{L}\int_{L/4}^{L/2}(x/L-1/4)\sin(n\pi x/L)dx$$
$$+\frac{8h}{L}\int_{L/2}^{3L/4}(3/4-x/L)\sin(n\pi x/L)dx \tag{l}$$
$$=\frac{8h}{n^2\pi^2}\left(2\sin(n\pi/2)-\sin(3n\pi/4)-\sin(n\pi/4)\right)$$
$$=\frac{64h}{n^2\pi^2}\left(\cos(n\pi/8)+\cos(3n\pi/8)\right)\sin^3(n\pi/8)$$

where we have used Mathematica procedure M7.18 to obtain this result. Substituting Eq. (l) into Eq. (g) gives

$$u(\eta,\tau)=\frac{64h}{\pi^2}\sum_{n=1,3,5}^{\infty}\frac{1}{n^2}\big(\cos(n\pi/8)+\cos(3n\pi/8)\big)\sin^3(n\pi/8)\cos(n\pi\tau)\sin(n\pi\eta) \qquad \text{(m)}$$

where $\eta = x/L$, $0 \le \eta \le 1$, and $\tau = ct/L$. Equation (l) is verified with Mathematica procedure M7.19.

Another way that Eq. (m) can be written is by using the trigonometric identity

$$2\cos[(a-b)/2]\sin[(a+b)/2]=\sin a+\sin b$$

If we let

$$\frac{1}{2}(a+b)=n\pi x/L$$

$$\frac{1}{2}(a-b)=n\pi ct/L$$

then

$$a=\frac{n\pi}{L}(x+ct)$$

$$b=\frac{n\pi}{L}(x-ct)$$

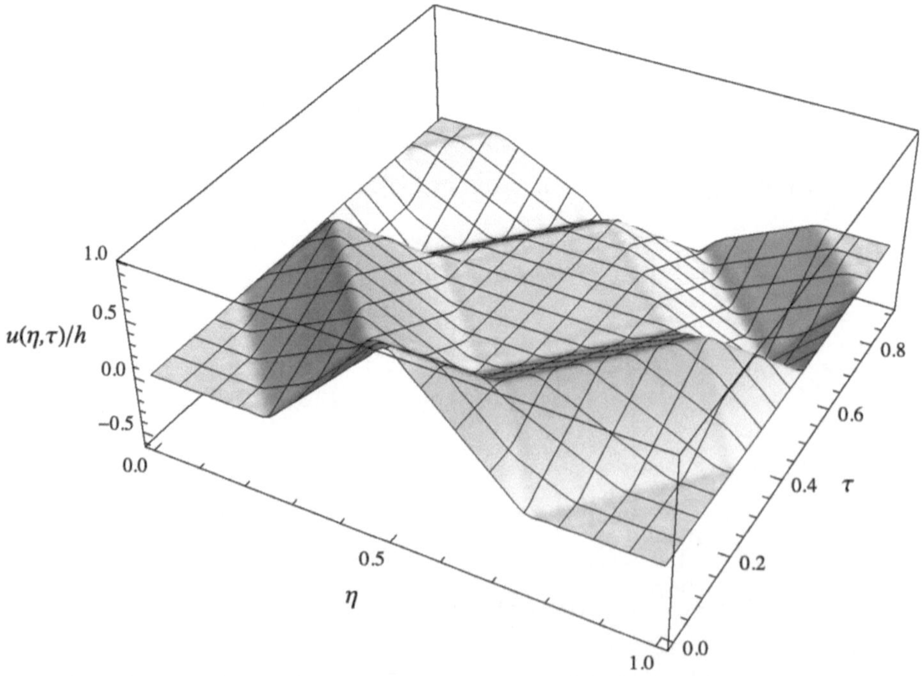

FIGURE 7.14 Numerical evaluation of Eq. (m) of Example 7.16.

and Eq. (m) can be written as

$$u(x,t) = \frac{16h}{\pi^2} \sum_{n=1,3,5}^{\infty} \frac{(-1)^{(n-1)/2}}{n^2} \sin^2(n\pi/8) \left[\sin\left(\frac{n\pi}{L}(x+ct)\right) + \sin\left(\frac{n\pi}{L}(x-ct)\right) \right]$$

which has the features of D'Alembert's formula on a term by term basis.

A plot of $u(x,t)/h$ given by Eq. (m) is shown in Figure 7.14. This figure was obtained with Mathematica procedure M7.20.

Example 7.17: Wave equation in Cartesian coordinates: Two spatial dimensions

We shall obtain a solution to

$$\frac{\partial^2 u}{\partial x^2} + \frac{\partial^2 u}{\partial y^2} - \frac{\partial^2 u}{c^2 \partial t^2} = 0 \quad 0 < x < a \quad 0 < y < b \tag{a}$$

over the region indicated when subject to the initial conditions

$$\begin{aligned} u(x,y,0) &= f(x,y) \\ \frac{\partial u(x,y,0)}{\partial t} &= g(x,y) \end{aligned} \tag{b}$$

and the boundary conditions

$$\begin{aligned} u(0,y,t) &= 0 \\ u(a,y,t) &= 0 \\ u(x,0,t) &= 0 \\ u(x,b,t) &= 0 \end{aligned} \tag{c}$$

The solution to Eq. (a) is obtained from Eq. (T7.9a) of Table 7.9, with $k = 1/c$, that is,

$$u(x,y,t) = X(x,\lambda)Y(y,\nu)T(t,\mu) \tag{d}$$

where, from Table 7.9,

$$\begin{aligned} X(x,\lambda) &= A\cos(\lambda x) + B\sin(\lambda x) \\ Y(t,\nu) &= C\cos(\nu y) + D\sin(\nu y) \\ T(t,\mu) &= E\cos(\mu t) + F\sin(\mu t) \end{aligned} \tag{e}$$

and

$$\lambda = \sqrt{\mu^2 / c^2 - \nu^2} \tag{f}$$

From prior examples, we see that substitution of Eq. (d) into the first two boundary conditions of Eq. (c) gives that $A = 0$ and $\nu \to \nu_n = n\pi/a$, where ν_n is the eigenvalue. Then, using superposition, Eq. (d) becomes

$$u(x,y,t) = \sum_{n=1}^{\infty} \sin(n\pi x/a) Y(y,\lambda) T(t,\mu) \qquad \text{(g)}$$

Similarly, when Eq. (g) is substituted into the last two boundary conditions given by Eq. (c) it is found that $C = 0$ and $\lambda \to \lambda_m = m\pi/b$, where λ_m is the eigenvalue. Then Eq. (g) reduces further to

$$u(x,y,t) = \sum_{m=1}^{\infty}\sum_{n=1}^{\infty} \sin(n\pi x/a)\sin(m\pi y/b)\left(E_{nm}\cos(\mu_{nm}t) + F_{nm}\sin(\mu_{nm}t)\right) \qquad \text{(h)}$$

where we have used Eq. (e) and, from Eq. (f),

$$\mu_{nm}^2 = c^2\pi^2\left(\frac{n^2}{a^2} + \frac{m^2}{b^2}\right) \qquad \text{(i)}$$

It has been shown in the previous examples that with these homogeneous boundary conditions $\sin(m\pi y/b)$ and $\sin(n\pi x/a)$ are orthogonal functions and their respective weighting factor is one.

Substituting Eq. (h) into the initial conditions given by Eq. (b) yields

$$\begin{aligned} &\sum_{m=1}^{\infty}\sum_{n=1}^{\infty} E_{nm}\sin(n\pi x/a)\sin(m\pi y/b) = f(x,y) \\ &\sum_{m=1}^{\infty}\sum_{n=1}^{\infty} \mu_{nm}F_{nm}\sin(n\pi x/a)\sin(m\pi y/b) = g(x,y) \end{aligned} \qquad \text{(j)}$$

We multiply each equation in Eq. (j) by $\sin(p\pi x/a)\sin(s\pi y/b)$, where p and s are integers, and integrate over the region to arrive at

$$\begin{aligned} &\sum_{m=1}^{\infty}\sum_{n=1}^{\infty} E_{nm}\int_0^a \sin(n\pi x/a)\sin(p\pi x/a)dx\int_0^b \sin(m\pi y/b)\sin(s\pi y/b)dy \\ &\qquad = \int_0^b\int_0^a f(x,y)\sin(p\pi x/a)\sin(s\pi y/b)dxdy \\ &\sum_{m=1}^{\infty}\sum_{n=1}^{\infty} E_{nm}\left(\frac{a}{2}\delta_{np}\right)\left(\frac{b}{2}\delta_{ms}\right) = \int_0^b\int_0^a f(x,y)\sin(p\pi x/a)\sin(s\pi y/b)dxdy \\ &E_{nm} = \frac{4}{ab}\int_0^b\int_0^a f(x,y)\sin(n\pi x/a)\sin(m\pi y/b)dxdy \end{aligned} \qquad \text{(k)}$$

In a similar manner, we obtain

$$F_{nm} = \frac{4}{ab\mu_{nm}} \int_0^b \int_0^a g(x,y)\sin(n\pi x/a)\sin(m\pi y/b)\,dxdy \tag{l}$$

In arriving at these results, we have used Eq. (3.20). The final solution is obtained by substituting Eqs. (k) and (l) into Eq. (h).

Equations (h), (k), and (l) are verified with Mathematica procedure M7.21.

To illustrate the results, we set $g(x,y) = 0$ and $f(x,y) = f_o(x/a)(x/a - 1)(y/b)(y/b - 1)$. Then $F_{nm} = 0$ and

$$\begin{aligned} E_{nm} &= \frac{4f_o}{ab} \int_0^a \frac{x}{a}\left(\frac{x}{a} - 1\right)\sin(n\pi x/a)\,dx \int_0^b \frac{y}{b}\left(\frac{y}{b} - 1\right)\sin(n\pi y/b)\,dy \\ &= \frac{64}{n^3 m^3 \pi^6} f_o \qquad n,m = 1,3,5,\ldots \end{aligned}$$

Thus, Eq. (h) can be written as

$$\frac{u(x,y,t)}{f_o} = \frac{64}{\pi^6} \sum_{m=1,3,5}^{\infty} \sum_{n=1,3,5}^{\infty} \frac{1}{n^3 m^3} \sin(n\pi\eta)\sin(m\pi\xi/\alpha)\cos(\hat{\mu}_{nm}\tau) \tag{m}$$

where $\tau = ct/a$, $\alpha = b/a$, $\eta = x/a$, $\xi = y/b$, and

$$\hat{\mu}_{nm} = \pi^2\left(n^2 + \frac{m^2}{\alpha^2}\right)$$

Equation (m) is plotted in Figure 7.15 for $b/a = 1.5$ and $\tau = 0.25$, 0.425, 0.45, and 1.25. This figure was obtained with Mathematica procedure M7.22.

Example 7.18: Wave equation in polar coordinates

We shall obtain a solution to

$$\frac{\partial^2 u}{\partial r^2} + \frac{\partial u}{r\partial r} + \frac{\partial^2 u}{r^2\partial\theta^2} - \frac{1}{c^2}\frac{\partial^2 u}{\partial t^2} = 0 \quad 0 < r < a \quad 0 < \theta < 2\pi \tag{a}$$

over the region indicated when subject to the initial conditions

$$\begin{aligned} u(r,\theta,0) &= f(r,\theta) \\ \frac{\partial u(r,\theta,0)}{\partial t} &= g(r,\theta) \end{aligned} \tag{b}$$

and the boundary conditions

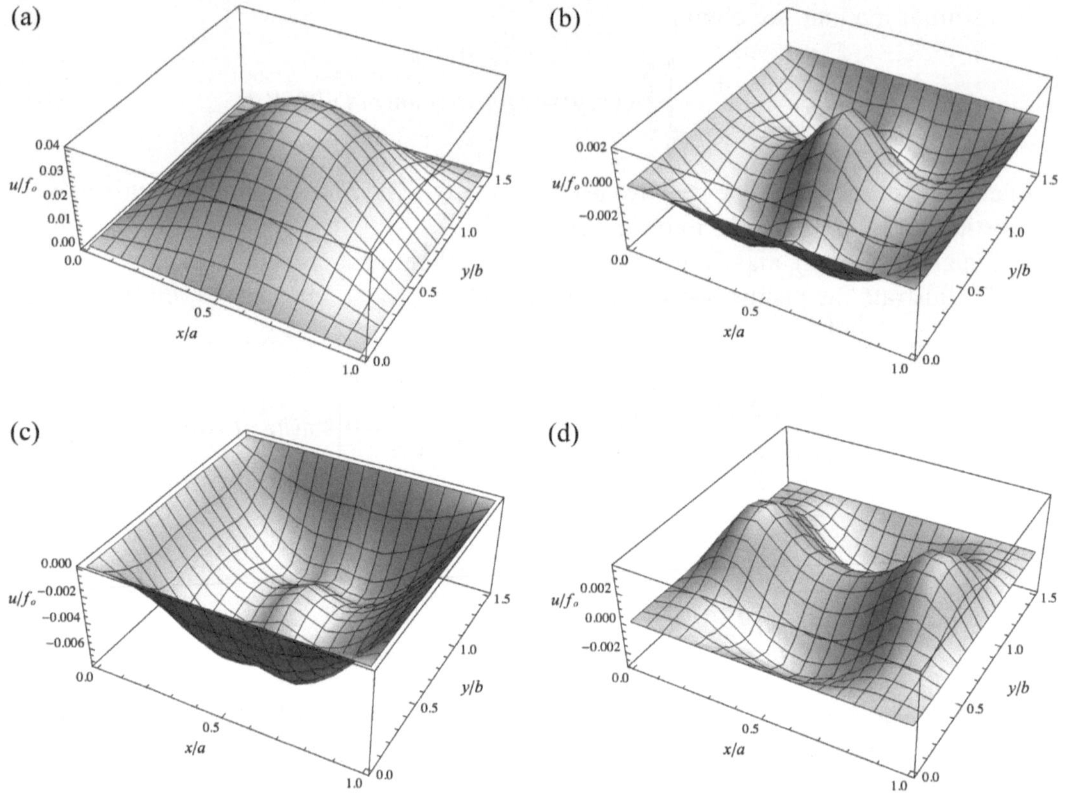

FIGURE 7.15 Evaluation of Eq. (m) of Example 7.17 for $b/a = 1.5$ and (a) $\tau = 0.25$; (b) $\tau = 0.425$; (c) $\tau = 0.45$; and (d) $\tau = 1.25$.

$$\begin{aligned} u(0,\theta,t) &= \text{finite} \\ u(a,\theta,t) &= 0 \\ u(r,\theta,t) &= u(r,\theta+2\pi,t) \end{aligned} \qquad \text{(c)}$$

The solution to Eq. (a) is obtained from Eq. (T7.9c) of Table 7.9 with $k = 1/c$, that is,

$$u(r,\theta,t) = R_m(r)\Theta_m(\theta)T(t) \qquad \text{(d)}$$

where, from Table 7.9,

$$\begin{aligned} R_m(r) &= CJ_m(\mu r/c) + DY_m(\mu r/c) \\ \Theta_m(\theta) &= A\cos m\theta + B\sin m\theta \\ T(t) &= E\cos(\mu t) + F\sin(\mu t) \end{aligned} \qquad \text{(e)}$$

and recalling the discussion in Example 7.8, m in Eq. (e) is an integer. In addition, $\cos(m\theta)$ and $\sin(m\theta)$ are orthogonal eigenfunctions.

To satisfy the first boundary condition in Eq. (c), we set $D = 0$. With these modifications, Eq. (d) can be written as

$$u_m(r,\theta,t) == J_m(\mu r/c)\Theta_m(\theta)T(t) \quad m=0,1,2,\ldots \tag{f}$$

where the constant C is incorporated into the remaining constants. Then, using superposition, the solution for all values of m is

$$u(r,\theta,t) = \sum_{m=0}^{\infty} J_m(\mu r/c)\Theta_m(\theta)T(t) \tag{g}$$

Substituting Eq. (g) into the second boundary condition given in Eq. (c), it is found that

$$\sum_{m=0}^{\infty} J_m(\mu a/c)\Theta_m(\theta)T(t) = 0$$

Therefore, for arbitrary θ and t,

$$J_m(\mu a / c) = 0 \quad m=0,1,2,\ldots$$

which is the characteristic equation. The solutions to this equation are denoted λ_{mn}, where λ_{mn} are the eigenvalues that satisfy

$$J_m(\lambda_{mn}) = 0 \quad m=0,1,2,\ldots \quad n=1,2,\ldots \tag{h}$$

Thus, since

$$\lambda_{mn} = \mu a/c$$

we find that

$$\mu = c\lambda_{mn}/a \tag{i}$$

As discussed in Example 7.5, $J_m(\lambda_{mn}r/a)$ is an orthogonal eigenfunction and $\mu \to \mu_{mn} = c\lambda_{mn}/a$ is the corresponding eigenvalue. The weighting factor is r.

Then, using Eq. (e) and superposition, Eq. (g) can be written as

$$\begin{aligned} u(r,\theta,t) &= \sum_{n=1}^{\infty}\sum_{m=0}^{\infty} J_m(\lambda_{mn}r/a)\cos m\theta\left(E_{cmn}\cos(c\lambda_{mn}t/a) + F_{cmn}\sin(c\lambda_{mn}t/a)\right) \\ &+ \sum_{n=1}^{\infty}\sum_{m=1}^{\infty} J_m(\lambda_{mn}r/a)\sin m\theta\left(E_{smn}\cos(c\lambda_{mn}t/a) + F_{smn}\sin(c\lambda_{mn}t/a)\right) \end{aligned} \tag{j}$$

Equation (j) is the solution to Eq. (a) and the boundary conditions given in Eq. (c). The four constants are determined from the initial conditions given by Eq. (b). Substituting Eq. (j) into the first equation of Eq. (b), we obtain

$$\sum_{n=1}^{\infty}\sum_{m=0}^{\infty} E_{cmn}J_m(\lambda_{mn}r/a)\cos m\theta + \sum_{n=1}^{\infty}\sum_{m=1}^{\infty} E_{smn}J_m(\lambda_{mn}r/a)\sin m\theta = f(r,\theta) \tag{k}$$

We multiply Eq. (k) by $J_p(\lambda_{ps}r/a)\cos(p\theta)rdrd\theta$, where p and s are integers, and integrate over the region to obtain

$$\begin{aligned}&\sum_{n=1}^{\infty}\sum_{m=0}^{\infty}E_{cmn}\int_0^a J_m(\lambda_{mn}r/a)J_p(\lambda_{ps}r/a)rdr\int_0^{2\pi}\cos m\theta\cos p\theta d\theta\\&+\sum_{n=1}^{\infty}\sum_{m=1}^{\infty}E_{smn}\int_0^a J_m(\lambda_{mn}r/a)J_p(\lambda_{ps}r/a)rdr\int_0^{2\pi}\sin m\theta\cos p\theta d\theta\\&=\int_0^{2\pi}\int_0^a f(r,\theta)J_m(\lambda_{mn}r/a)\cos m\theta rdrd\theta\quad p=1,2,\ldots\end{aligned} \tag{l}$$

$$E_{cmn}=\frac{2}{\pi(1+\delta_{0m})a^2J_{m+1}^2(\lambda_{mn})}\int_0^{2\pi}\int_0^a f(r,\theta)J_m(\lambda_{mn}r/a)\cos m\theta rdrd\theta$$

where we have used Eq. (3.20) for the trigonometric functions and for the Bessel functions we have used Eqs. (6.88) and (6.89), that is,

$$\int_0^a J_m(\lambda_{mn}r/a)J_m(\lambda_{ms}r/a)rdr=\delta_{ns}\frac{a^2}{2}J_{m+1}^2(\lambda_{mn})$$

In arriving at this expression, we have used Eq. (h). In a similar manner, we determine that

$$E_{smn}=\frac{2}{\pi a^2J_{m+1}^2(\lambda_{mn})}\int_0^{2\pi}\int_0^a f(r,\theta)J_m(\lambda_{mn}r/a)\sin m\theta rdrd\theta \tag{m}$$

To determine F_{cmn} and F_{smn}, we substitute Eq. (j) into the second equation of Eq. (b) and obtain

$$\begin{aligned}&\sum_{n=1}^{\infty}\sum_{m=1}^{\infty}\frac{c\lambda_{mn}}{a}F_{cmn}J_m(\lambda_{mn}r/a)\cos m\theta\\&+\sum_{n=1}^{\infty}\sum_{m=1}^{\infty}\frac{c\lambda_{mn}}{a}F_{smn}J_m(\lambda_{mn}r/a)\sin m\theta=g(r,\theta)\end{aligned} \tag{n}$$

Proceeding in a manner similar to that used to obtain Eqs. (l) and (m), we arrive at

$$\begin{aligned}F_{cmn}&=\frac{2}{\pi(1+\delta_{0m})ac\lambda_{mn}J_{m+1}^2(\lambda_{mn})}\int_0^{2\pi}\int_0^a g(r,\theta)J_m(\lambda_{mn}r/a)\cos m\theta rdrd\theta\\F_{smn}&=\frac{2}{\pi ac\lambda_{mn}J_{m+1}^2(\lambda_{mn})}\int_0^{2\pi}\int_0^a g(r,\theta)J_m(\lambda_{mn}r/a)\sin m\theta rdrd\theta\end{aligned} \tag{o}$$

The final solution is obtained by substituting Eqs. (l), (m), and (o) into Eq. (j).

To obtain the case when u is independent of θ, that is, when $f(r,\theta) \to f(r)$, and $g(r,\theta) \to g(r)$, we set $m = 0$ in the previous results and find that $\lambda_{mn} \to \lambda_n$, where λ_n is the solution to

$$J_0(\lambda_n) = 0 \tag{p}$$

In this case, Eq. (j) reduces to

$$u(r,t) = \sum_{n=1}^{\infty} J_0(\lambda_n r/a)\left(E_n \cos(c\lambda_n t/a) + F_n \sin(c\lambda_n t/a)\right) \tag{q}$$

where

$$E_n = \frac{1}{a^2 J_1^2(\lambda_n)} \int_0^a f(r) J_0(\lambda_n r/a) r dr$$

$$F_n = \frac{2}{ac\lambda_n J_1^2(\lambda_n)} \int_0^a g(r) J_0(\lambda_n r/a) r dr$$

Equation (q) is verified with Mathematica procedure M7.23.

Example 7.19: Nonhomogeneous wave equation

We shall determine the solution to the following nonhomogeneous wave equation, which in non-dimensional form is given by

$$\frac{\partial^2 w}{\partial \tau^2} - \frac{\partial^2 w}{\partial \eta^2} = \alpha^2 F(\eta,\tau) \tag{a}$$

The boundary conditions are

$$\begin{aligned} w(0,\tau) &= 0 \\ w(1,\tau) &= 0 \end{aligned} \tag{b}$$

and the initial conditions are zero, that is,

$$w(\eta,0) = 0, \quad \frac{dw(\eta,0)}{d\tau} = 0$$

We start with the solution to Eq. (a) when $F = 0$, which is obtained from Eq. (T7.9b) of Table 7.9 with $k = 1$. In the present notation, we have

$$w(\eta,\tau) = E(\mu,\eta)\varphi(\tau) \tag{c}$$

where

$$E(\mu,\eta) = A\cos(\mu\eta) + B\sin(\mu\eta) \tag{d}$$

and $\varphi(\tau)$ is to be determined. Upon substituting Eq. (c) into the first boundary condition of Eq. (b), it is found that

$$A\varphi(\tau)=0$$

Therefore, for arbitrary τ, we must set A = 0. Then, using Eq. (d) in Eq. (c), Eq. (c) becomes

$$w(\eta,\tau)=B\sin(\mu\eta)\varphi(\tau) \tag{e}$$

Upon substituting Eq. (e) into the second boundary condition of Eq. (b), it is found that

$$B\sin(\mu)\varphi(\tau)=0$$

In order not to have a trivial solution, we set

$$\sin\mu=0$$

which is the characteristic equation. The solution to this equation is

$$\mu\rightarrow\mu_n=n\pi \quad n=1,2,\ldots$$

where μ_n is the eigenvalue. Then, Eq. (e) becomes

$$w_n(\eta,\tau)=\sin(n\pi\eta)\varphi_n(\tau) \tag{f}$$

where the constant B has been incorporated in the unknown function $\varphi_n(\tau)$. Since Eq. (f) is a valid solution for all n, we use superposition to obtain

$$u(\eta,\tau)=\sum_{n=1}^{\infty}\varphi_n(\tau)\sin n\pi\eta \tag{g}$$

As shown in previous examples, $\sin(n\pi\eta)$ is an orthogonal eigenfunction.
We now substitute Eq. (g) into Eq. (a) and obtain

$$\sum_{n=1}^{\infty}\left(\frac{\partial^2\varphi_n}{\partial\tau^2}+n^2\pi^2\varphi_n\right)\sin n\pi\eta=\alpha^2F(\eta,\tau) \tag{h}$$

Multiplying Eq. (h) by $\sin(m\pi\eta)$ and integrating over the interval, we obtain

$$\frac{\partial^2\varphi_n}{\partial\tau^2}+n^2\pi^2\varphi_n=g_n(\tau) \tag{i}$$

where

$$g_n(\tau)=2\alpha^2\int_0^1 F(\eta,\tau)\sin n\pi\eta\, d\eta \tag{j}$$

The solution to Eq. (i) is given in Table 4.3 with ζ = 0 and which, in the present notation, is

$$\varphi_n(\tau) = A_1 \cos n\pi\tau + B_1 \sin n\pi\tau + \frac{1}{n\pi}\int_0^{\tau} g_n(\xi)\sin(n\pi[\tau-\xi])d\xi \tag{k}$$

For zero initial conditions, it is straightforward to show that $A_1 = B_1 = 0$. Then, Eq. (k) becomes

$$\varphi_n(\tau) = \frac{1}{n\pi}\int_0^{\tau} g_n(\xi)\sin(n\pi[\tau-\xi])d\xi \tag{l}$$

Substituting Eq. (l) into Eq. (g) gives

$$w(\eta,\tau) = \sum_{n=1}^{\infty}\frac{\sin n\pi\eta}{n\pi}\int_0^{\tau} g_n(\xi)\sin(n\pi[\tau-\xi])d\xi \tag{m}$$

and substituting Eq. (j) into Eq. (m) gives the final solution as

$$w(\eta,\tau) = \frac{2\alpha^2}{\pi}\sum_{n=1}^{\infty}\frac{\sin n\pi\eta}{n}\int_0^{\tau}\left(\int_0^{1} F(\eta,\xi)\sin n\pi\eta d\eta\right)\sin(n\pi[\tau-\xi])d\xi \tag{n}$$

To illustrate these results, we shall examine two nonhomogeneous functions. For the first function, we assume that

$$F(\eta,\tau) = \delta(\eta-\eta_o)\delta(\tau) \tag{o}$$

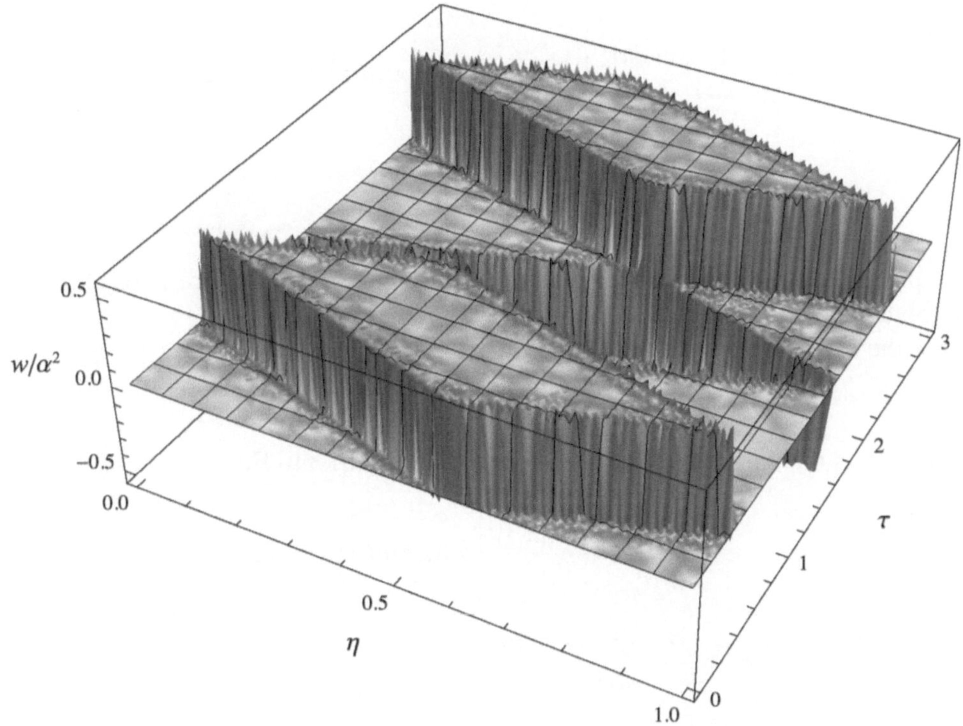

FIGURE 7.16 Numerical evaluation of Eq. (p) of Example 7.19, which is for an impulse at $\eta = 0.6$.

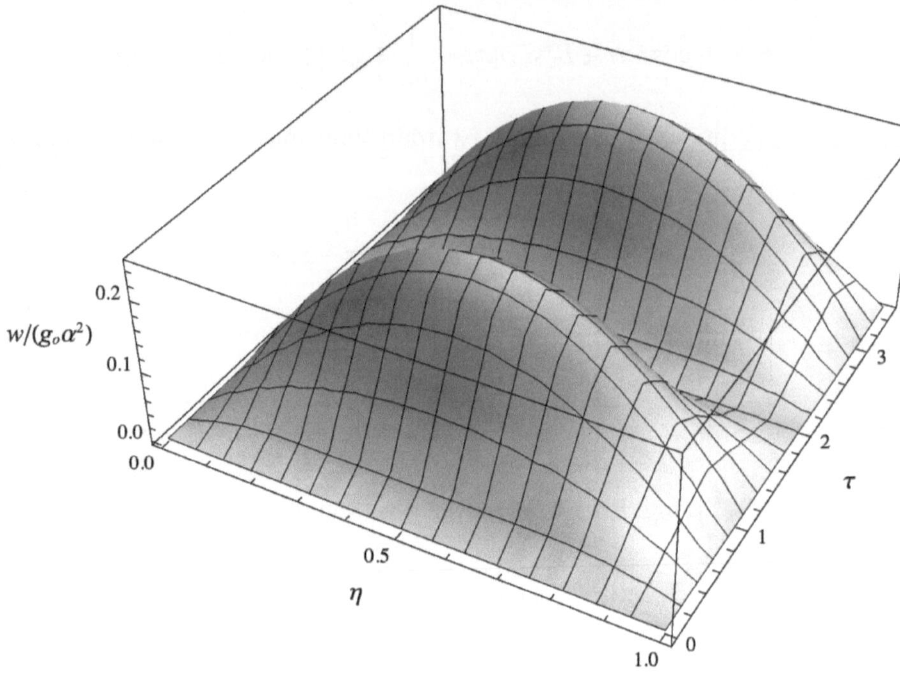

FIGURE 7.17 Numerical evaluation of Eq. (r) of Example 7.19, which is for a uniformly distributed step function.

Substituting Eq. (o) into Eq. (n), we arrive at

$$
\begin{aligned}
w(\eta,\tau) &= \frac{2\alpha^2}{\pi}\sum_{n=1}^{\infty}\frac{\sin n\pi\eta}{n}\int_0^1 \delta(\eta-\eta_o)\sin n\pi\eta\, d\eta \int_0^{\tau}\delta(\xi)\sin(n\pi[\tau-\xi])d\xi \\
&= \frac{2\alpha^2}{\pi}\sum_{n=1}^{\infty}\frac{\sin n\pi\eta \sin n\pi\eta_o}{n}\sin(n\pi\tau)
\end{aligned} \tag{p}
$$

A plot of $w(\eta,\tau)/\alpha^2$ given by Eq. (p) for $\eta_o = 0.6$ is shown in Figure 7.16. This figure was obtained with Mathematica procedure M7.24.

For the second function, we assume that

$$
F(\eta,\tau) = g_o u(\tau) \tag{q}
$$

where $u(\tau)$ is the unit step function. Substituting Eq. (q) into Eq. (n) yields

$$
\begin{aligned}
w(\eta,\tau) &= \frac{2g_o\alpha^2}{\pi}\sum_{n=1}^{\infty}\frac{\sin(n\pi\eta)}{n}\int_0^1 \sin n\pi\eta\, d\eta \int_0^{\tau}\sin(n\pi[\tau-\xi])d\xi \\
&= \frac{4g_o\alpha^2}{\pi^3}\sum_{n=1,3,5}^{\infty}\frac{\sin(n\pi\eta)}{n^3}\left(1-\cos(n\pi\tau)\right)
\end{aligned} \tag{r}
$$

A plot of $w(\eta,\tau)/(\alpha^2 g_o)$ given by Eq. (r) is shown in Figure 7.17. This figure was obtained with Mathematica procedure M7.25.

7.2.6 Poisson Equation

We shall illustrate by an example one way of obtaining a solution to Poisson's equation.

Example 7.20: Poisson's equation in Cartesian coordinates

Consider the following equation, which is a nonhomogeneous version of the Laplace equation in Cartesian coordinates

$$\nabla^2 u = \frac{\partial^2 u}{\partial x^2} + \frac{\partial^2 u}{\partial y^2} = f(x,y) \quad 0 \le x \le a \quad 0 \le y \le b \tag{a}$$

We assume that the boundary conditions are

$$\begin{aligned} u(0,y) &= 0 \\ u(a,y) &= 0 \\ u(x,0) &= 0 \\ u(x,b) &= 0 \end{aligned} \tag{b}$$

To solve Eq. (a), we first solve a similar problem

$$\nabla^2 u = \frac{\partial^2 u}{\partial x^2} + \frac{\partial^2 u}{\partial y^2} = -\lambda u \quad 0 < x < a \quad 0 < y < b \tag{c}$$

which is the Helmholtz equation. The quantity λ is to be determined. The boundary conditions for Eq. (c) are those given by Eq. (b). The solution to Eq. (c) for the boundary conditions given by Eq. (i), which are the same as those given by Eq.(b) of Example 7.7, is

$$u_{nm}(x,y) = A_{nm} \sin(n\pi x/a)\sin(m\pi y/b) \tag{d}$$

where, from Eq. (h) of that example,

$$\lambda \to \lambda_{nm} = \pi^2 \left(\frac{n^2}{a^2} + \frac{m^2}{b^2} \right) \tag{e}$$

Thus, substituting Eqs. (d) and (e) into Eq. (c), we find that

$$\nabla^2 u_{nm}(x,y) = -\lambda_{nm} u_{nm}(x,y) \tag{f}$$

Using superposition and collecting all the solutions to Eq. (f), we have that

$$\begin{aligned} \nabla^2 \sum_{m=1}^{\infty}\sum_{n=1}^{\infty} u_{nm}(x,y) &= -\sum_{m=1}^{\infty}\sum_{n=1}^{\infty} \lambda_{nm} u_{nm}(x,y) \\ \nabla^2 u(x,y) &= -\sum_{m=1}^{\infty}\sum_{n=1}^{\infty} \lambda_{nm} u_{nm}(x,y) \end{aligned} \tag{g}$$

where

$$u(x,y)=\sum_{m=1}^{\infty}\sum_{n=1}^{\infty}u_{nm}(x,y)=\sum_{m=1}^{\infty}\sum_{n=1}^{\infty}A_{nm}\sin(n\pi x/a)\sin(m\pi y/b) \qquad \text{(h)}$$

Since Eq. (h) satisfies the boundary conditions given by Eq. (b), we can insert Eq. (h) into Eq. (a) to arrive at

$$\nabla^2\sum_{m=1}^{\infty}\sum_{n=1}^{\infty}u_{nm}(x,y)=f(x,y) \qquad \text{(i)}$$

From Eqs. (d) and (f), Eq. (i) becomes

$$\sum_{m=1}^{\infty}\sum_{n=1}^{\infty}\lambda_{nm}A_{nm}\sin(n\pi x/a)\sin(m\pi y/b)=-f(x,y) \qquad \text{(j)}$$

To determine A_{nm}, we multiply Eq. (j) by $\sin(k\pi x/a)\sin(l\pi y/b)$, integrate over the region, and use Eq. (3.20) to arrive at

$$A_{nm}=-\frac{4F_{nm}}{ab\lambda_{nm}} \qquad \text{(k)}$$

where

$$F_{nm}=\int_0^b\int_0^a f(x,y)\sin(k\pi x/a)\sin(l\pi y/b)dxdy \qquad \text{(l)}$$

Therefore, substituting Eqs. (k) and (l) into Eq. (h) yields

$$u(x,y)=-\frac{4}{ab}\sum_{m=1}^{\infty}\sum_{n=1}^{\infty}\left(\frac{n^2\pi^2}{a^2}+\frac{m^2\pi^2}{b^2}\right)^{-1}\sin(n\pi x/a)\sin(m\pi y/b)$$

$$\times\int_0^b\int_0^a f(x,y)\sin(k\pi x/a)\sin(l\pi y/b)dxdy$$

7.2.7 Bi-Harmonic Equation

We shall determine under what conditions one can obtain a separable solution to the bi-harmonic equation in Cartesian and polar coordinates, and for each coordinate system we shall give an example.

Bi-Harmonic Equation in Cartesian Coordinates: Referring to Tables 7.1 and 7.2, we consider the bi-harmonic equation in Cartesian coordinates with $G = k = 0$; thus,

$$\frac{\partial^4 w}{\partial x^4}+2\frac{\partial^4 w}{\partial x^2\partial y^2}+\frac{\partial^4 w}{\partial y^4}=0 \quad 0<x<a \quad 0<y<b \qquad (7.87)$$

In general, Eq. (7.87) is not separable except under certain circumstances. To show what these circumstances are, we let

$$w(x,y) = X(x)Y(y) \tag{7.88}$$

and find that Eq. (7.87) can be written as

$$\frac{d^4X}{Xdx^4} + 2\frac{d^2X}{Xdx^2}\frac{d^2Y}{Ydy^2} + \frac{d^4Y}{Ydy^4} = 0 \tag{7.89}$$

In order for this equation to be separable either

$$\frac{\partial^2 X}{X\partial x^2} = -\alpha^2 \tag{7.90}$$

or

$$\frac{\partial^2 Y}{Y\partial y^2} = -\beta^2 \tag{7.91}$$

or both Eqs. (7.90) and (7.91) have to be satisfied. The solutions to Eqs. (7.90) and (7.91), respectively, are

$$X(x) \to X_s(x) = A\cos\alpha x + B\sin\alpha x \tag{7.92}$$

and

$$Y(y) \to Y_s(y) = C\cos\beta y + D\sin\beta y \tag{7.93}$$

Thus, Eq. (7.87) can only be separated by one of the following functions:

$$w(x,y) = Y(y)X_s(x) \text{ or } w(x,y) = Y_s(y)X(x) \text{ or } w(x,y) = Y_s(y)X_s(x) \tag{7.94}$$

It should be noted that the equation in X is a fourth-order equation and that the equation in Y is also of the fourth order. Therefore, to have a well-posed boundary-value problem, one must specify eight boundary conditions, two on each of the four boundaries. But from Eqs. (7.92) and (7.93), we only have two solution constants. So, for the separable solutions to be of any use, one must have boundary conditions that can be satisfied by Eq. (7.94). One candidate set of boundary conditions that can overcome this limitation is either

$$w(0,y) = \frac{d^2w(0,y)}{dx^2} = 0 \quad \text{and} \quad w(a,y) = \frac{d^2w(a,y)}{dx^2} = 0 \tag{7.95}$$

or

$$w(x,0) = \frac{d^2w(x,0)}{dy^2} = 0 \quad \text{and} \quad w(x,b) = \frac{d^2w(x,b)}{dy^2} = 0 \tag{7.96}$$

or both of them. Substituting Eq. (7.94) into Eq. (7.95), we find that

$$Y(y)X_s(0) = -\alpha^2 Y(y)X_s(0) = 0 \quad \text{and} \quad Y(y)X_s(a) = -\alpha^2 Y(y)X_s(a) = 0$$

Thus, satisfying $X_s(0) = X_s(a) = 0$ will satisfy the remaining two boundary conditions. Similarly, for the boundary conditions given by Eq. (7.96).

Assuming that the separable solution is given by the first equation of Eq. (7.94), we substitute this solution into Eq. (7.87) and arrive at

$$\left(\frac{d^4Y}{dy^4} - 2\alpha^2\frac{d^2Y}{dy^2} + \alpha^4 Y\right)X_s = 0$$

Therefore, for a nontrivial solution, we must obtain a solution to

$$\frac{d^4Y}{dy^4} - 2\alpha^2\frac{d^2Y}{dy^2} + \alpha^4 Y = 0 \tag{7.97}$$

The solution to Eq. (7.97) is given in Table 4.2, which in the present notation is

$$Y(y) = C_1\cosh(\alpha y) + C_2\sinh(\alpha y) + y\left(C_3\cosh(\alpha y) + C_4\sinh(\alpha y)\right) \tag{7.98}$$

and the separable solution is

$$\begin{aligned} w(x,y) &= X_s(x)Y(y) \\ &= \left(A\cos\alpha x + B\sin\alpha x\right) \\ &\quad \times\left(C_1\cosh(\alpha y) + C_2\sinh(\alpha y) + y\left(C_3\cosh(\alpha y) + C_4\sinh(\alpha y)\right)\right) \end{aligned} \tag{7.99}$$

In a similar manner, using the second solution of Eq. (7.94), we arrive at

$$X(x) = D_1\cosh(\beta x) + D_2\sinh(\beta x) + x\left(D_3\cosh(\beta x) + D_4\sinh(\beta x)\right) \tag{7.100}$$

and the separable solution is

$$\begin{aligned} w(x,y) &= Y_s(y)X(x) \\ &= \left(C\cos\beta y + D\sin\beta y\right) \\ &\quad \times\left(E_1\cosh(\beta x) + E_2\sinh(\beta x) + x\left(E_3\cosh(\beta x) + E_4\sinh(\beta x)\right)\right) \end{aligned} \tag{7.101}$$

We shall use these results in Example 7.21.

We now consider the bi-harmonic equation in Cartesian coordinates with only $G = 0$; thus,

$$\frac{\partial^4 w}{\partial x^4} + 2\frac{\partial^4 w}{\partial x^2\partial y^2} + \frac{\partial^4 w}{\partial y^4} + k^2\frac{\partial^2 w}{\partial t^2} = 0 \quad 0 < x < a \quad 0 < y < b \tag{7.102}$$

In general, Eq. (7.102) is not separable except under certain circumstances. To show what these circumstances are, we let

$$w(x,y,t) = X(x)Y(y)T(t) \tag{7.103}$$

and find that Eq. (7.87) can be written as

$$\frac{d^4X}{Xdx^4}+2\frac{d^2X}{Xdx^2}\frac{d^2Y}{Ydy^2}+\frac{d^4Y}{Ydy^4}+k^2\frac{d^2T}{Tdt^2}=0 \tag{7.104}$$

Thus, in Eq. (7.104), we set

$$k^2\frac{d^2T}{Tdt^2}=-\gamma^2 \tag{7.105}$$

and Eq. (7.104) becomes

$$\frac{d^4X}{Xdx^4}+2\frac{d^2X}{Xdx^2}\frac{d^2Y}{Ydy^2}+\frac{d^4Y}{Ydy^4}-\gamma^2=0 \tag{7.106}$$

Therefore, Eq. (7.106) is spatially separable only for the same conditions for which Eq. (7.89) is separable.

Bi-Harmonic Equation in Polar Coordinates: Referring to Tables 7.1 and 7.2, the bi-harmonic equation in polar coordinates with $G = k = 0$ can be written as

$$\nabla^4_{r\theta}w=0 \tag{7.107}$$

where

$$\nabla^2_{r\theta}=\nabla^2_r+\frac{1}{r^2}\frac{\partial^2}{\partial\theta^2} \tag{7.108}$$

and

$$\nabla^2_r=\frac{\partial^2}{\partial r^2}+\frac{1}{r}\frac{\partial}{\partial r} \tag{7.109}$$

Expanding Eq. (7.107), we have that

$$\begin{aligned}\nabla^4_{r\theta}w=\nabla^2_{r\theta}\nabla^2_{r\theta}w&=\left(\nabla^2_r+\frac{1}{r^2}\frac{\partial^2}{\partial\theta^2}\right)\left(\nabla^2_r w+\frac{1}{r^2}\frac{\partial^2 w}{\partial\theta^2}\right)\\&=\nabla^4_r w+\frac{\partial^2}{\partial\theta^2}\left(\frac{1}{r^2}\nabla^2_r w+\nabla^2_r\left(\frac{w}{r^2}\right)\right)+\frac{1}{r^4}\frac{\partial^4 w}{\partial\theta^4}\end{aligned} \tag{7.110}$$

We assume a separable solution to Eq. (7.110) of the form

$$w=R(r)\Theta(\theta) \tag{7.111}$$

Upon substituting Eq. (7.111) into Eq. (7.110), we obtain

$$\nabla^4_{r\theta}w=\Theta\nabla^4_r R+\frac{d^2\Theta}{d\theta^2}\left(\frac{1}{r^2}\nabla^2_r R+\nabla^2_r\left(\frac{R}{r^2}\right)\right)+\frac{1}{r^4}\frac{d^4\Theta}{d\theta^4}R=0 \tag{7.112}$$

If we let

$$\frac{d^2\Theta}{d\theta^2}=-m^2\Theta \tag{7.113}$$

where m is an integer (see the discussion in Example 7.8), then Eq. (7.112) can be written as

$$\begin{gathered}
\nabla_r^4 R - m^2\left(\frac{1}{r^2}\nabla_r^2 R + \nabla_r^2\left(\frac{R}{r^2}\right)\right) + \frac{m^4}{r^4}R = 0 \\
\left(\nabla_r^2 - \frac{m^2}{r^2}\right)\left(\nabla_r^2 R - \frac{m^2}{r^2}R\right) = 0 \\
\left(\frac{d^2}{dr^2} + \frac{1}{r}\frac{d}{dr} - \frac{m^2}{r^2}\right)\left(\frac{d^2 R}{dr^2} + \frac{1}{r}\frac{dR}{dr} - \frac{m^2}{r^2}R\right) = 0
\end{gathered} \tag{7.114}$$

where we have cancelled the common factor Θ. Thus, for a region for which $0 \le \theta \le 2\pi$, Eq. (7.107) is separable with no restrictions on the boundary conditions. The solution to Eq. (7.114), which is a function of m, is given in Table 4.2.

The solution to Eq. (7.113) is

$$\Theta = A\cos m\theta + B\sin m\theta \tag{7.115}$$

and the solution to Eq. (7.114) is

$$w = R_m(r)\left(A\cos m\theta + B\sin m\theta\right) \tag{7.116}$$

where, from Table 4.2,

$$R_0 = \alpha_{10} + \alpha_{20}\ln r + \alpha_{30}r^2 + \alpha_{40}r^2\ln r$$

$$R_1 = \alpha_{11}r + \alpha_{21}r\ln r + \alpha_{31}r^{-1} + \alpha_{41}r^3$$

$$R_m = \alpha_{1m}r^m + \alpha_{2m}r^{-m} + \alpha_{3m}r^{2-m} + \alpha_{4m}r^{2+m} \qquad m > 1$$

We now consider the bi-harmonic equation in polar coordinates with only $G = 0$; thus, Eq. (7.107) can be written as

$$\nabla_{r\theta}^4 w + k^2\frac{d^2 w}{dt^2} = 0 \tag{7.117}$$

We assume a separable solution to Eq. (7.119) of the form

$$w = R(r)\Theta(\theta)T(t) \tag{7.118}$$

Upon substituting Eq. (7.120) into Eq. (7.119), we obtain

$$\Theta T\nabla_r^4 R + \frac{d^2\Theta}{d\theta^2}\left(\frac{1}{r^2}\nabla_r^2 R + \nabla_r^2\left(\frac{R}{r^2}\right)\right)T + \frac{1}{r^4}\frac{d^4\Theta}{d\theta^4}RT + k^2 R\Theta\frac{d^2 T}{dt^2} = 0 \tag{7.119}$$

If we let

$$\begin{gathered}
\frac{d^2\Theta}{d\theta^2} = -m^2\Theta \\
\frac{d^2 T}{dt^2} = -\frac{\kappa^4}{k^2}T
\end{gathered} \tag{7.120}$$

then, Eq. (7.119) becomes

$$\nabla_r^4 R - m^2\left(\frac{1}{r^2}\nabla_r^2 R + \nabla_r^2\left(\frac{R}{r^2}\right)\right) + \left(\frac{m^4}{r^4} - \kappa^4\right)R = 0 \tag{7.121}$$

where we have cancelled the common factors Θ and T.

If we define

$$\nabla_{r,m}^2 = \frac{d^2}{dr^2} + \frac{1}{r}\frac{d}{dr} - \frac{m^2}{r^2} \tag{7.122}$$

then Eq. (7.121) can be written as

$$\left(\nabla_{r,m}^2 - \kappa^2\right)\left(\nabla_{r,m}^2 + \kappa^2\right)R = 0 \tag{7.123}$$

We note from Eq. (7.123) that if we let $R = R_1$ be the solution to

$$\nabla_{r,m}^2 R_1 + \kappa^2 R_1 = 0 \tag{7.124}$$

and we let $R = R_2$ be the solution to

$$\nabla_{r,m}^2 R_2 - \kappa^2 R_2 = 0 \tag{7.125}$$

then, the solution to Eq. (7.123) is $R = R_1 + R_2$.

From Table 5.1, the solution to Eq. (7.124) is

$$R_1(r) = C_1 J_m(\kappa r) + C_2 Y_m(\kappa r) \tag{7.126}$$

and the solution to Eq. (7.125) is

$$R_2(r) = C_3 I_m(\kappa r) + C_4 K_m(\kappa r) \tag{7.127}$$

Hence,

$$R(r) = C_1 J_m(\kappa r) + C_2 Y_m(\kappa r) + C_3 I_m(\kappa r) + C_4 K_m(\kappa r) \tag{7.128}$$

The solutions to Eq. (7.120) are

$$\begin{aligned}\Theta &= A\cos m\theta + B\sin m\theta \\ T &= C\cos\left(\kappa^2 t / k\right) + D\sin\left(\kappa^2 t / k\right)\end{aligned} \tag{7.129}$$

Therefore, from Eqs. (7.128) and (7.129), the solution to Eq. (7.117) is

$$\begin{aligned}w(r,\theta,t) &= \left(C_1 J_m(\kappa r) + C_2 Y_m(\kappa r) + C_3 I_m(\kappa r) + C_4 K_m(\kappa r)\right) \\ &\quad \times\left(A\cos m\theta + B\sin m\theta\right)\left(C\cos\left(\kappa^2 t / k\right) + D\sin\left(\kappa^2 t / k\right)\right)\end{aligned} \tag{7.130}$$

Example 7.21: Approximate solution to a uniformly loaded rectangular plate with clamped edges

To show the degree of difficulty that can arise when a partial differential is not separable, we shall consider the case of a rectangular plate that is clamped along its four edges and subjected to a uniform load p_o over its entire surface. An approximate solution will be obtained by use of what is sometimes called the assumed solution method. The governing equation is given by

$$\frac{\partial^4 w}{\partial x^4} + 2\frac{\partial^4 w}{\partial x^2 \partial y^2} + \frac{\partial^4 w}{\partial y^4} = \frac{p_o}{D} \quad 0 < x < a \quad 0 < y < b \tag{a}$$

where $w(x,y)$ is the displacement of the plate and D is a constant. For a plate whose coordinate system is placed at its center as shown in Figure 7.18, the boundary conditions are given as

$$\begin{aligned} w(\pm a, y) &= 0 \\ w(x, \pm b) &= 0 \end{aligned} \tag{b}$$

and

$$\begin{aligned} \frac{\partial w(\pm a, y)}{\partial x} &= 0 \\ \frac{\partial w(x, \pm b)}{\partial y} &= 0 \end{aligned} \tag{c}$$

which represent four pairs of boundary conditions, one pair at each edge. We see that these boundary conditions are not those that permit a separable solution as indicated by Eqs. (7.95) and (7.96).

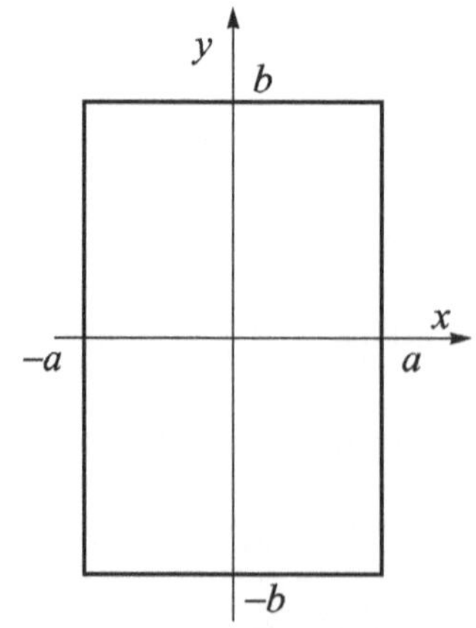

FIGURE 7.18 Coordinate system for a rectangular plate.

One method* that will yield an approximate solution is now described. We assume that the particular solution to Eq. (a) is given by

$$w_p(x,y) = \frac{p_o}{8D}(a^2 - x^2)(b^2 - y^2) \tag{d}$$

which was chosen because it also satisfies the boundary conditions given by Eq. (b).

Then, the total solution is assumed to be

$$w(x,y) = \frac{p_o}{8D}\left\{(a^2 - x^2)(b^2 - y^2) + \Psi\right\} \tag{e}$$

where Ψ is to be determined such that it is a solution to Eq. (a) and the boundary conditions given by Eqs. (b) and (c). Based on many of the results in this chapter, in particular Eqs. (7.99) and (7.101), we choose Ψ in the form

$$\Psi = \sum_{n=1,3,5}^{\infty} \left\{A_n Y_n(y)\cos\alpha_n x + B_n X_n(x)\cos\beta_n y\right\} \tag{f}$$

where

$$\alpha_n = \frac{n\pi}{2a}, \quad \beta_n = \frac{n\pi}{2b} \tag{g}$$

It is seen that at $x = \pm a$, $\cos(\pm a\alpha_n) = 0$ and that at $y = \pm b$, $\cos(\pm b\beta_r) = 0$ for $n = 1, 3, 5, \ldots$ Thus, we have the following tasks: Find $Y_n(y)$ and $X_n(x)$ such that they satisfy the following conditions: Eq. (a) with $p_o = 0$, $Y_n(\pm b) = 0$, and $X_n(\pm a) = 0$.

To determine a suitable function for $Y_n(y)$, we assume that

$$w(x,y) = Y_n(y)\cos\alpha_n x \tag{h}$$

and substitute it into Eq. (a) with $p_o = 0$ and obtain

$$\frac{d^4 Y_n}{dy^4} - 2\alpha_n^2 \frac{d^2 Y_n}{dy^2} + \alpha_n^4 Y_n = 0 \tag{i}$$

From Eq. (7.98), we have that the solution to Eq. (i) with $\alpha = \alpha_n$ is

$$Y_n(y) = C_{n1}\sinh\alpha_n y + C_{n2}\cosh\alpha_n y + y\left(C_{n3}\sinh\alpha_n y + C_{n4}\cosh\alpha_n y\right) \tag{j}$$

In a similar manner, using

$$w(x,y) = X_n(x)\cos\beta_n y \tag{k}$$

we have, upon using Eq. (7.100) with $\beta = \beta_n$,

$$X_n(x) = D_{n1}\sinh\beta_n x + D_{n2}\cosh\beta_n x + x\left(D_{n3}\sinh\beta_n x + D_{n4}\cosh\beta_n x\right) \tag{l}$$

---

* Introduced by Hencky for his PhD dissertation in 1913 at Darmstadt and discussed in A. E. H. Love, *A Treatise of the Mathematical Theory of Elasticity*, 4th ed. 1927, American Edition 1947, pp. 493–495.

Upon substituting Eq. (j) into the second boundary condition of Eq. (b), it is found that

$$\begin{aligned} w(x,b) &= Y_n(b)\cos\alpha_n x = \{C_{n1}\sinh\alpha_n b + C_{n2}\cosh\alpha_n b + \\ &\quad b(C_{n3}\sinh\alpha_n b + C_{n4}\cosh\alpha_n b)\}\cos\alpha_n x = 0 \\ w(x,-b) &= Y_n(-b)\cos\alpha_n x = \{-C_{n1}\sinh\alpha_n b + C_{n2}\cosh\alpha_n b - \\ &\quad b(-C_{n3}\sinh\alpha_n b + C_{n4}\cosh\alpha_n b)\}\cos\alpha_n x = 0 \end{aligned} \tag{m}$$

First adding these two equations and then subtracting these two equations, respectively, we arrive at

$$2C_{n2}\cosh\alpha_n b + 2bC_{n3}\sinh\alpha_n b = 0$$

or

$$C_{n2} = -\frac{b\sinh\alpha_n b}{\cosh\alpha_n b}C_{n3} \tag{n}$$

and

$$2C_{n1}\sinh\alpha_n b + 2bC_{n4}\cosh\alpha_n b = 0$$

or

$$C_{n4} = -\frac{\sinh\alpha_n b}{b\cosh\alpha_n b}C_{n1} \tag{o}$$

Using Eq. (n) in Eq. (j), we obtain

$$\begin{aligned} Y_n(y) &= C_{n1}\left(\sinh\alpha_n y - \frac{y\sinh\alpha_n b}{b\cosh\alpha_n b}\cosh\alpha_n y\right) \\ &\quad + C_{n3}\left(y\sinh\alpha_n y - \frac{b\sinh\alpha_n b}{\cosh\alpha_n b}\cosh\alpha_n y\right) \end{aligned} \tag{p}$$

It is noted that in Eq. (p), the terms multiplying C_{n1} and C_{n3} are zero when $y = \pm b$. Therefore, without loss of generality, we shall choose (from the C_{n3} term)

$$Y_n(y) = \bar{Y}_n(y) = b\sinh\alpha_n b\cosh\alpha_n y - y\cosh\alpha_n b\sinh\alpha_n y \tag{q}$$

As expected, it is seen that in Eq. (q), $Y_n(\pm b) = 0$. A similar set of operations gives

$$X_n(x) = \bar{X}_n(x) = a\sinh\beta_n a\cosh\beta_n x - x\cosh\beta_n a\sinh\beta_n x \tag{r}$$

As expected, it is seen that in Eq. (r), $X_n(\pm a) = 0$.

Taking the derivatives of Eqs. (q) and (r), respectively, we arrive at

$$\frac{\partial\bar{Y}_n(y)}{\partial y} = b\alpha_n\sinh(b\alpha_n)\sinh(\alpha_n y) - \cosh(\alpha_n b)\big(\sinh(\alpha_n y) + \alpha_n y\cosh(\alpha_n y)\big) \tag{s}$$

and

$$\frac{\partial \bar{X}_n(x)}{\partial x} = a\beta_n \sinh(a\beta_n)\sinh(\beta_n x) - \cosh(a\beta_n)\big(\sinh(\beta_n x) + \beta_n x\cosh(\beta_n x)\big) \quad \text{(t)}$$

We shall have use of these relations subsequently.

Then, the solution up to this point becomes

$$w(x,y) = \frac{p_o}{8D}\left\{(a^2 - x^2)(b^2 - y^2) + \sum_{n=1,3,5}^{\infty}\{A_n\bar{Y}_n(y)\cos\alpha_n x + B_n\bar{X}_n(x)\cos\beta_n y\}\right\} \quad \text{(u)}$$

which satisfies the differential equation given by Eq. (a) and the boundary conditions given by Eq. (b). Still to be satisfied are the boundary conditions given by Eq. (c). Substituting Eq. (u) into Eq. (c), we obtain

$$\begin{aligned}
&\sum_{n=1,3,5}^{\infty}\left\{-A_n\alpha_n\bar{Y}_n(y)\sin\alpha_n a + B_n\frac{\partial \bar{X}_n(a)}{\partial x}\cos\beta_n y\right\} = 2a(b^2 - y^2) \\
&\sum_{n=1,3,5}^{\infty}\left\{A_n\frac{\partial \bar{Y}_n(b)}{\partial y}\cos\alpha_n x - B_n\beta_n\bar{X}_n(x)\sin\beta_n b\right\} = 2b(a^2 - x^2)
\end{aligned} \quad \text{(v)}$$

since, as seen below,

$$\frac{\partial X_n(-a)}{\partial x} = -\frac{\partial X_n(a)}{\partial x}$$

$$\frac{\partial Y_n(-b)}{\partial y} = -\frac{\partial Y_n(b)}{\partial y}$$

From Eqs. (s) and (t), we note that

$$\begin{aligned}
\frac{\partial \bar{Y}_n(b)}{\partial y} &= b\alpha_n \sinh(b\alpha_n)\sinh(\alpha_n b) - \cosh(\alpha_n b)\big(\sinh(\alpha_n b) + \alpha_n b\cosh(\alpha_n b)\big) \\
&= -\left(\alpha_n b + \frac{1}{2}\sinh(2\alpha_n b)\right) = -\frac{\partial \bar{Y}_n(-b)}{\partial y}
\end{aligned} \quad \text{(w)}$$

and

$$\begin{aligned}
\frac{\partial \bar{X}_n(a)}{\partial x} &= a\beta_n \sinh(a\beta_n)\sinh(\beta_n a) - \cosh(a\beta_n)(\sinh(\beta_n a) + \beta_n a\cosh(\beta_n a)) \\
&= -\left(\beta_n a + \frac{1}{2}\sinh(2\beta_n a)\right) = -\frac{\partial \bar{X}_n(-a)}{\partial x}
\end{aligned} \quad \text{(x)}$$

Then Eq. (v) can be written as

TABLE 7.11
Definitions of Coefficients Appearing in Eqs. (z) and (bb) of Example 7.21: $b_a = b/a$

$$a_{mn} = \alpha_n \sin(\alpha_n a)\int_{-b}^{b} \bar{Y}_n(y)\cos(\beta_m y)dy = \frac{b^3}{a^2}\bar{a}_{mn} \qquad \bar{a}_{mn} = -\frac{8mn^2\sin(n\pi/2)\sin(m\pi/2)\cosh^2(\pi b_a n/2)}{\pi\left(m^2+n^2b_a^2\right)^2}$$

$$b_m = \left(\beta_n a + \frac{1}{2}\sinh(2\beta_n a)\right)\delta_{nm}\int_{-b}^{b}\cos^2(\beta_m y)dy = b\bar{b}_m \qquad \bar{b}_m = -\frac{1}{2}\left(\pi m/b_a + \sinh(\pi m/b_a)\right)$$

$$e_m = 2a\int_{-b}^{b}\left(y^2-b^2\right)\cos(\beta_m y)dy = ab^3\bar{e}_m \qquad \bar{e}_m = \frac{64\sin(m\pi/2)}{\pi^3 m^3}$$

$$c_m = \left(\alpha_n b + \frac{1}{2}\sinh(2\alpha_n b)\right)\delta_{nm}\int_{-a}^{a}\cos^2(\alpha_m x)dx = a\bar{c}_m \qquad \bar{c}_m = -\frac{1}{2}\left(\pi m b_a + \sinh(\pi m b_a)\right)$$

$$d_{mn} = \beta_n \sin(\beta_n b)\int_{-a}^{a}\bar{X}_n(x)\cos(\alpha_m x)dx = \frac{b^2}{a}\bar{d}_{mn} \qquad \bar{d}_{mn} = -\frac{8mn^2\sin(n\pi/2)\sin(m\pi/2)\cosh^2(\pi n/(2b_a))}{\pi\left(n^2+m^2b_a^2\right)^2}$$

$$f_m = 2b\int_{-a}^{a}\left(x^2-a^2\right)\cos(\alpha_m x)dx = a^3b\bar{f}_m \qquad \bar{f}_m = \frac{64\sin(m\pi/2)}{\pi^3 m^3}$$

$$\sum_{n=1,3,5}^{\infty}\left\{A_n\alpha_n\bar{Y}_n(y)\sin\alpha_n a + B_n\left(\beta_n a + \frac{1}{2}\sinh(2\beta_n a)\right)\cos\beta_n y\right\} = 2a(y^2-b^2)$$
$$\sum_{n=1,3,5}^{\infty}\left\{A_n\left(\alpha_n b + \frac{1}{2}\sinh(2\alpha_n b)\right)\cos\alpha_n x + B_n\beta_n\bar{X}_n(x)\sin\beta_n b\right\} = 2b(x^2-a^2) \qquad \text{(y)}$$

To determine the unknown constants A_n and B_n, we multiply the first equation of Eq. (y) by $\cos(\beta_m y)dy$ and integrate from $-b$ to b and we multiply the second equation of Eq. (y) by $\cos(\alpha_m x)dx$ and integrate from $-a$ to a. Performing these operations gives

$$\sum_{n=1,3,5}^{\infty} a_{mn}A_n + b_m B_m = e_m$$

$$c_m A_m + \sum_{n=1,3,5}^{\infty} d_{mn}B_n = f_m \quad m = 1,3,5,\ldots \tag{z}$$

where the terms appearing in Eq. (z) are given in Table 7.11. These constants were determined with Mathematica procedure M7.26.

Equation (z) can be expressed in matrix form as

$$\begin{pmatrix} a_{11} & a_{13} & a_{15} & \cdots & b_1 & 0 & 0 & \cdots \\ a_{31} & a_{33} & a_{35} & \cdots & 0 & b_3 & 0 & \cdots \\ a_{51} & a_{53} & a_{55} & \cdots & 0 & 0 & b_5 & \cdots \\ & & & \vdots & & & & \\ c_1 & 0 & 0 & \cdots & d_{11} & d_{13} & d_{15} & \cdots \\ 0 & c_3 & 0 & \cdots & d_{31} & d_{33} & d_{35} & \cdots \\ 0 & 0 & c_5 & \cdots & d_{51} & d_{53} & d_{55} & \cdots \\ & & & \vdots & & & & \end{pmatrix} \begin{Bmatrix} A_1 \\ A_3 \\ A_3 \\ \vdots \\ B_1 \\ B_3 \\ B_5 \\ \vdots \end{Bmatrix} = \begin{Bmatrix} e_1 \\ e_3 \\ e_5 \\ \vdots \\ f_1 \\ f_3 \\ f_5 \\ \vdots \end{Bmatrix} \tag{aa}$$

It is mentioned that if Eq. (a) were separable, then Eq. (aa) would have been a diagonal matrix.

Before proceeding with a numerical evaluation of Eq. (z), we shall convert Eq. (z) to non-dimensional form by introducing the definitions in the second column of Table 7.11 and the following definitions: $b_a = b/a$ and

$$\hat{A}_n = \frac{bA_m}{a^4}$$

$$\hat{B}_m = \frac{B_m}{a^3}$$

Then, Eq. (z) can be written as

$$\sum_{n=1,3,5}^{\infty} \bar{a}_{mn}\hat{A}_n + \left(\bar{b}_m / b_a\right)\hat{B}_m = b_a \bar{e}_m$$

$$\left(\bar{c}_m / b_a^3\right)\hat{A}_m + \sum_{n=1,3,5}^{\infty} \bar{d}_{mn}\hat{B}_n = \bar{f}_m / b_a \quad m = 1,3,5,\ldots \tag{bb}$$

and Eq. (u) can be written as

$$w(x,y) = \frac{p_o a^2}{8D}\left\{ b_a^2 \left(1 - \frac{x^2}{a^2}\right)\left(1 - \frac{y^2}{b^2}\right) + \sum_{n=1,3,5}^{\infty} \left\{ \hat{A}_n \hat{Y}_n(y)\cos\alpha_n x + \hat{B}_n \hat{X}_n(x)\cos\beta_n y \right\} \right\} \tag{cc}$$

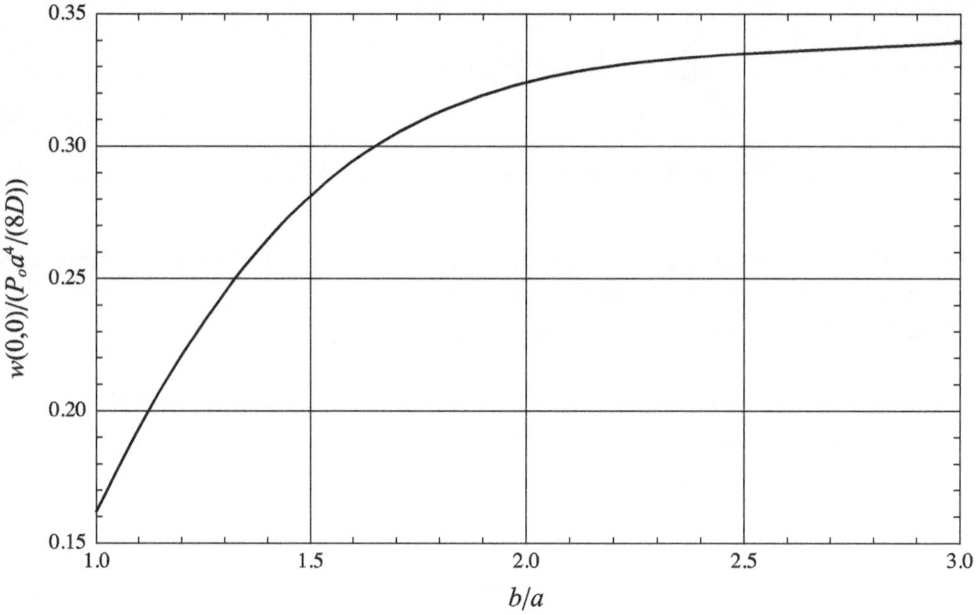

FIGURE 7.19 Maximum displacement of a rectangular plate clamped on all four edges as a function of its aspect ratio. The maximum displacement occurs at the center of the plate.

where

$$\hat{X}_n(x) = \overline{X}_n(x)/a$$

$$\hat{Y}_n(y) = \overline{Y}_n(y)/a$$

Equation (cc) is evaluated with Mathematica procedure M7.27 at the plate's center as a function of *b*/*a* and the results are shown in Figure 7.19. For this loading, the maximum displacement occurs at the plate's center.

Example 7.22: Solution to the bi-harmonic equation in polar coordinates: Natural frequencies and modes shapes of a clamped solid circular plate

The governing equation for a solid circular plate of radius a (m), Young's modulus E (N/m^2), Poisson's ratio ν, density ρ (kg/m^3), and thickness h (m) that is undergoing symmetric (angular-independent) harmonic oscillations at a frequency ω can be written as*

$$\left(\nabla_\xi^2 - \Omega^2\right)\left(\nabla_\xi^2 + \Omega^2\right)w(\xi) = 0 \qquad \text{(a)}$$

where $w(\xi)$ is the transverse displacement,

$$\nabla_\xi^2 = \frac{d^2}{d\xi^2} + \frac{1}{\xi}\frac{d}{d\xi} \qquad \text{(b)}$$

* E. B. Magrab, *Vibrations of Elastic Systems: with Application to MEMS and NEMS*, Solid Mechanics and Its Applications 184, Springer Science+Business Media B.V., New York (February 2012) Section 4.2.

and

$$\xi = \frac{r}{a}, \quad \Omega^2 = \omega t_p, \quad t_p^2 = \frac{\rho h a^4}{D} \text{ s}^2 \quad D = \frac{Eh^3}{12(1-\nu^2)} \text{ Nm}$$

The boundary conditions are that the plate is clamped along its outer perimeter. These boundary conditions are expressed as

$$\begin{aligned} &w(0) \text{ and } w'(0) \text{ are finite} \\ &w(1) = w'(1) = 0 \end{aligned} \tag{c}$$

where the prime (′) denotes the derivative with respect ξ.

The solution to Eq. (a) is given by Eq. (7.128) with $m = 0$; thus

$$R(\xi) = C_1 J_0(\Omega\xi) + C_2 Y_0(\Omega\xi) + C_3 I_0(\Omega\xi) + C_4 K_0(\Omega\xi) \tag{d}$$

Based on the boundedness conditions given by the first two equations of Eq. (c), we must set $C_2 = C_4 = 0$. Then, Eq. (d) reduces to

$$R(\xi) = C_1 J_0(\Omega\xi) + C_3 I_0(\Omega\xi) \tag{e}$$

Upon substituting Eq. (e) into the second pair of boundary conditions given by Eq. (c), we find that

$$\begin{pmatrix} J_0(\Omega) & I_0(\Omega) \\ -J_1(\Omega) & I_1(\Omega) \end{pmatrix} \begin{Bmatrix} C_1 \\ C_3 \end{Bmatrix} = 0 \tag{f}$$

where we have used Eqs. (5.100) and (5.132). The natural frequencies (eigenvalues) $\Omega = \Omega_n$ are determined by setting the determinant of the coefficients in Eq. (f) to zero, which results in

$$J_0(\Omega_n) I_1(\Omega_n) + I_0(\Omega_n) J_1(\Omega_n) = 0 \tag{g}$$

The corresponding mode shapes (eigenfunctions) $R_n(r)$ are obtained from Eq. (e) and the first equation of Eq. (f) as

$$R_n(\xi) = J_0(\Omega_n\xi) - \frac{J_0(\Omega_n)}{I_0(\Omega_n)} I_0(\Omega_n\xi) \quad 0 \le \xi \le 1 \tag{h}$$

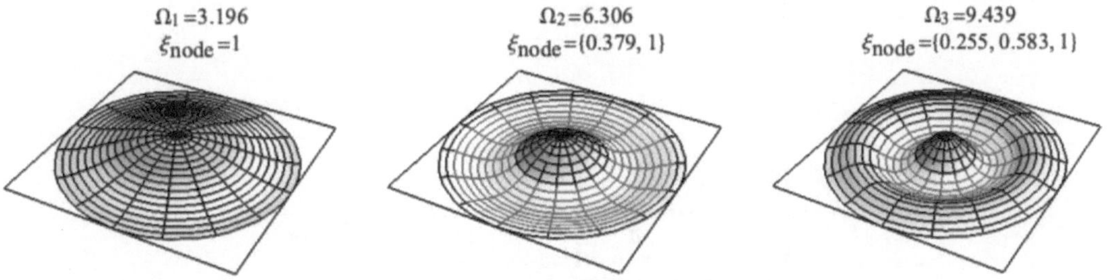

FIGURE 7.20 Lowest three natural frequencies, locations of nodal circles, and modes shapes of a solid circular plate whose motion is independent of its angular coordinate.

The values of $\xi = \xi_{node}$ for which $R_n(\xi_{node}) = 0$ are called nodal radii and are the radii of the nodal circles where the displacements are zero. The lowest three natural frequencies and their corresponding mode shapes are shown in Figure 7.20. This figure was obtained with Mathematica procedure M7.28.

7.3 PLACING PARTIAL DIFFERENTIAL EQUATIONS INTO NON-DIMENSIONAL FORM

In this section, we shall illustrate with several examples how to convert partial differential equations to non-dimensional form. Additional examples can be found in Example 8.19 and Sections 9.2.1, 9.4.1, 9.5.1, 9.6.1, and 9.7.2. It is mentioned that in most of the examples that follow only the governing differential equations are made non-dimensional; one still has to make the boundary conditions and initial conditions non-dimensional.

Example 7.23: Commonality of wave equation for bars, shafts, and strings

In this example, we shall show that when the wave equation for the longitudinal motion of a bar, the torsional motion of a circular shaft, and the transverse motion of a string are placed in non-dimensional form, the resulting wave equation will be

$$\frac{\partial^2 \bar{u}}{\partial \eta^2} - \frac{\partial^2 \bar{u}}{\partial \tau^2} = g(\eta,\tau)$$

The governing equations for these three systems are given in Figure 7.21. The definition of the symbols appearing in the equations in Figure 7.21 are given in Table 7.12.

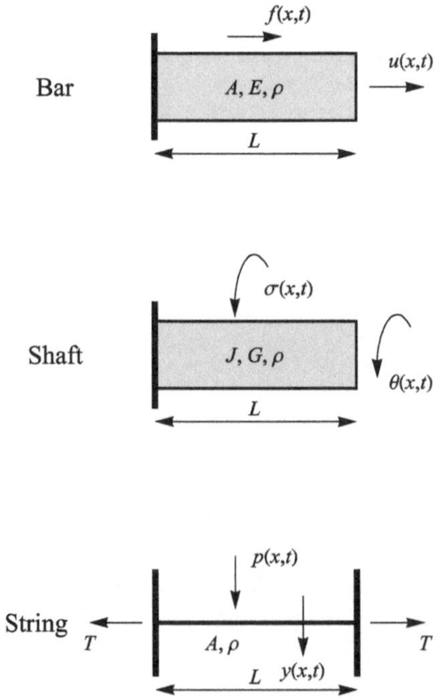

$$AE\frac{\partial^2 u}{\partial x^2} - A\rho\frac{\partial^2 u}{\partial t^2} = f(x,t)$$

$$JG\frac{\partial^2 \theta}{\partial x^2} - J\rho\frac{\partial^2 \theta}{\partial t^2} = \sigma(x,t)$$

$$T\frac{\partial^2 y}{\partial x^2} - \rho A\frac{\partial^2 y}{\partial t^2} = p(x,t)$$

FIGURE 7.21 Nomenclature for three distinctly different systems whose motion is described by the one-dimensional wave equation.

The parameters that are used to make these equations non-dimensional are shown in Table 7.13. We shall show how the results in Table 7.13 were obtained for the case of the longitudinal motions of a bar. It is noted that this equivalence is for the governing equation only, for the boundary conditions for each of these equations are quite different in both form and physical significance.

The governing equation describing the longitudinal motion of a bar is

$$AE\frac{\partial^2 u}{\partial x^2} - A\rho\frac{\partial^2 u}{\partial t^2} = f(x,t) \quad \text{N/m} \tag{a}$$

Making the substitution $\eta = x/L$, Eq. (a) can be written as

$$\frac{AE}{L^2}\frac{\partial^2 u}{\partial \eta^2} - A\rho\frac{\partial^2 u}{\partial t^2} = f(x,t) \quad \text{N/m}$$

$$\frac{\partial^2 u}{\partial \eta^2} - \frac{\rho L^2}{E}\frac{\partial^2 u}{\partial t^2} = \frac{L^2}{AE}f(\eta,t) \quad \text{m} \tag{b}$$

$$\frac{\partial^2 \bar{u}}{\partial \eta^2} - \frac{\rho L^2}{E}\frac{\partial^2 \bar{u}}{\partial t^2} = \frac{L}{AE}f(\eta,t)$$

where $\bar{u} = u/L$. Using the definitions in Table 7.13, namely,

$$t_o^2 = \frac{\rho L^2}{E} \rightarrow \frac{\text{kg}}{\text{m}^3}\frac{\text{m}^2}{\text{N/m}^2} = \text{s}^2$$

$$g(\eta,\tau) = \frac{L}{AE}f(\eta,\tau)$$

TABLE 7.12
Definitions of the Symbols in Figure 7.21

Symbol	Units	Description
x	m	Position
t	s	Time
T	N	Tension in string
L	m	Length
A	m^2	Area
J	m^4	Area moment of inertia
E	N/m^2	Tensile modulus
G	N/m^2	Shear modulus
ρ	kg/m^3	Density
$u(x,t)$	m	Axial displacement of bar
$y(x,t)$	m	Transverse displacement of string
$\theta(x,t)$	rad	Rotation of shaft
$f(x,t)$	N/m	Externally applied axial force
$\sigma(x,t)$	N rad	Externally applied moment
$p(x,t)$	N/m	Externally applied transverse force

TABLE 7.13
Definitions of the Non-Dimensional Quantities for the Systems in Figure 7.21

System	$\tau = t/t_o$	$g(\eta,\tau)$	$\bar{u}(\eta,\tau)$
Bar	$t_o = \sqrt{\rho L^2/E}$ s	$f(\eta,\tau)L/(AE)$	$u(\eta,\tau)/L$
Shaft	$t_o = \sqrt{\rho L^2/G}$ s	$\sigma(\eta,\tau)L^2/(JG)$ rad	-
String	$t_o = \sqrt{\rho A L^2/T}$ s	$p(\eta,\tau)L/T$	$y(\eta,\tau)/L$

Note: For all systems $\eta = x/L$.

and noting that

$$\frac{\rho L^2}{E}\frac{\partial^2 \bar{u}}{\partial t^2} = t_o^2 \frac{\partial^2 \bar{u}}{\partial t^2} = \frac{\partial^2 \bar{u}}{\partial \tau^2}$$

where $\tau = t/t_o$, Eq. (b) can be written as

$$\frac{\partial^2 \bar{u}}{\partial \eta^2} - \frac{\partial^2 \bar{u}}{\partial \tau^2} = g(\eta,\tau)$$

Example 7.24: Equation for the forced vibrations of beams

Consider a thin elastic beam that has a length L (m), cross-sectional area moment of inertia I (m^4), Young's modulus E (N/m^2), cross sectional area A (m^2), and density ρ (kg/m^3). If the transverse displacement of the beam is $y = y(x)$ (m), the distance along the beam is x (m), and t (s) is time, then the governing equation describing the motion of the beam is

$$EI\frac{\partial^4 y}{\partial x^4} + \rho A\frac{\partial^2 y}{\partial t^2} = F_o f(x,t) \text{ N/m} \tag{a}$$

where F_o (N/m) is the magnitude of an externally applied force per unit length and $f(x,t)$ is a non-dimensional shape factor in time and space. The quantity $F_o f(x,t)$ is an externally applied quantity and assumed to be known.

To make Eq. (a) non-dimensional, we start by setting $\eta = x/L$. Then Eq. (a) becomes

$$\begin{aligned} &\frac{EI}{L^4}\frac{\partial^4 y}{\partial \eta^4} + \rho A\frac{\partial^2 y}{\partial t^2} = F_o f(\eta,t) \\ &\frac{\partial^4 y}{\partial \eta^4} + t_o^2\frac{\partial^2 y}{\partial t^2} = d_o f(\eta,t) \end{aligned} \tag{b}$$

where

$$t_o^2 = \frac{\rho A L^4}{EI}, \quad d_o = \frac{F_o L^4}{EI} \tag{c}$$

Examining the units of t_o and d_o, we find that

$$t_o^2 = \frac{\rho A L^4}{EI} \to \frac{\text{kg}}{\text{m}^3}\frac{\text{m}^2 \times \text{m}^4}{\text{m}^4\text{N/m}^2} = \frac{\text{kg m}}{\text{N}} = \text{s}^2$$

$$d_o = \frac{F_o L^4}{EI} \to \frac{\text{N}}{\text{m}}\frac{\text{m}^4}{\text{m}^4\text{N/m}^2} = \text{m} \tag{d}$$

If we set

$$w = \frac{y}{d_o} \quad \text{and} \quad \tau = \frac{t}{t_o}$$

then our final result is obtained from Eq. (b) as

$$\frac{\partial^4 w}{\partial \eta^4} + \frac{\partial^2 w}{\partial \tau^2} = f(\eta,\tau) \tag{e}$$

Notice that we have gone from five independent parameters appearing in Eq. (a) to none in Eq. (e).

Example 7.25: Equation for transient heat conduction

Consider the one-dimensional heat conduction equation

$$\frac{\partial u}{\partial t} = \alpha \frac{\partial^2 u}{\partial x^2} \quad \text{K/s} \tag{a}$$

where

$$\alpha = \frac{k}{c_p \rho} \tag{b}$$

It is assumed that the boundary conditions are

$$\frac{\partial u(0,t)}{\partial x} = 0, \quad u(a,t) = T_2 \tag{c}$$

and the initial condition is

$$u(x,0) = T_1 \tag{d}$$

In Eqs. (a) through (d), k is the thermal conductivity [W/(Km)], c_p is the specific heat at constant pressure [J/(kgK)], u is the change in temperature (K), T_1 and T_2 are constant temperatures (K), ρ is the density (kg/m^3), t is time (s), and x is length (m) that varies from $x = 0$ to $x = a$. [Note: W = watts, K = kelvin, J = joule = Ws.]

We start by noting that

$$\alpha = \frac{k}{c_p \rho} = \frac{\text{W}}{\text{Km}} \frac{\text{kgK}}{\text{J}} \frac{\text{m}^3}{\text{kg}} = \frac{\text{Wm}^2}{\text{Ws}} = \frac{\text{m}^2}{\text{s}}$$

Therefore, we introduce the length a by setting $\eta = x/a$. Then Eq. (a) becomes

$$\frac{\partial u}{\partial t} = \frac{\alpha}{a^2} \frac{\partial^2 u}{\partial \eta^2} \quad \text{K/s} \tag{e}$$

We define

$$t_o = \frac{a^2}{\alpha} \rightarrow \frac{\text{m}^2}{\text{m}^2/\text{s}} = \text{s}$$

and use this definition in Eq. (e) to obtain

$$\frac{\partial u}{\partial t} = \frac{1}{t_o} \frac{\partial^2 u}{\partial \eta^2}$$

Defining $\tau = t/t_o$, we obtain

$$\frac{\partial u}{\partial \tau} = \frac{\partial^2 u}{\partial \eta^2} \quad \text{K} \tag{f}$$

To deal with the boundary conditions, we introduce a non-dimensional temperature

$$\bar{u}(\eta,\tau) = \frac{u(\eta,\tau) - T_1}{T_2 - T_1} \tag{g}$$

Substituting Eq. (g) into Eq. (f) yields

$$\frac{\partial \bar{u}}{\partial \tau} = \frac{\partial^2 \bar{u}}{\partial \eta^2} \tag{h}$$

Using Eq. (g) in the boundary conditions given by Eq. (c), we find that

$$\begin{aligned} &\frac{\partial \bar{u}(0,\tau)}{\partial \eta} = 0 \\ &u(a,t) = T_2 \rightarrow \bar{u}(1,\tau) = 1 \end{aligned} \tag{i}$$

and for the initial condition given by Eq. (d), we obtain

$$u(x,0) = T_1 \rightarrow \bar{u}(\eta,0) = 0 \tag{j}$$

It is mentioned that depending on the boundary conditions and initial conditions the form of Eq. (g) may be different. However, Eq. (h) remains the same.

Example 7.26: Equation for the transverse vibrations of a string subject to a concentrated travelling force

Consider the following equation governing the transverse displacement u (m) of a string of length L (m) and density ρ (kg/m) that is fixed at each end and on which a concentrated load of magnitude P (N) is travelling across the string at a velocity V (m/s)

$$\frac{\partial^2 u}{\partial t^2} = c^2 \frac{\partial^2 u}{\partial x^2} + \frac{P}{\rho}\delta(Vt - x) \ \ \text{m/s}^2 \tag{a}$$

The quantity c (m/s) is the speed at which disturbances in the string travel and the quantities P and V are externally applied quantities and are assumed to be known.

To convert Eq. (a) to non-dimensional form we first let $\eta = x/L$. Then, Eq. (a) becomes

$$\frac{\partial^2 u}{\partial t^2} = \frac{c^2}{L^2}\frac{\partial^2 u}{\partial \eta^2} + \frac{P}{\rho}\delta(Vt - L\eta) \ \ \text{m/s}^2 \tag{b}$$

We note the following

$$t_o = \frac{L}{c} \rightarrow \frac{\text{m}}{\text{m/s}} = \text{s}$$

Therefore, we define $\tau = t/t_o$ and Eq. (b) becomes

$$\frac{\partial^2 u}{\partial \tau^2} = \frac{\partial^2 u}{\partial \eta^2} + \frac{Pt_o^2}{\rho}\delta(Vt_o\tau - L\eta) \tag{c}$$

Referring to Eq. (B.4) of Appendix B, we rewrite the argument of the delta function as

$$\delta(Vt_o\tau - L\eta) = \delta(Vt_o[\tau - \alpha\eta]) = \frac{1}{|Vt_o|}\delta(\tau - \alpha\eta)$$

where $\alpha = c/V$. Then, Eq. (c) is written as

$$\frac{\partial^2 u}{\partial \tau^2} = \frac{\partial^2 u}{\partial \eta^2} + \hat{P}\delta(\tau - \alpha\eta) \ \text{m} \tag{d}$$

where

$$\hat{P} = \frac{Pt_o}{V\rho} \rightarrow \frac{\text{Ns}}{(\text{m/s})(\text{kg/m})} = \text{m}$$

We can proceed one more step and define

$$w = \frac{u}{\hat{P}}$$

Then, Eq. (e) becomes

$$\frac{\partial^2 w}{\partial \tau^2} = \frac{\partial^2 w}{\partial \eta^2} + \delta(\tau - \alpha\eta) \tag{e}$$

which has one parameter: α.

Example 7.27: Forced vibrations of a Timoshenko beam

Consider the following coupled set of partial differential equations that describe the motion of a beam described by Timoshenko's theory

$$\begin{aligned} &\kappa GA\left(\frac{\partial^2 w}{\partial x^2} - \frac{\partial \psi}{\partial x}\right) + P\frac{\partial^2 w}{\partial x^2} - \rho A\frac{\partial^2 w}{\partial t^2} = -F_a(x,t) \quad 0 \le x \le L \quad t > 0 \\ &\kappa GA\left(\frac{\partial w}{\partial x} - \psi\right) + EI\frac{\partial^2 \psi}{\partial x^2} - \rho I\frac{\partial^2 \psi}{\partial t^2} = -M_a(x,t) \end{aligned} \tag{a}$$

where the quantities appearing in Eq. (a) are defined in Table 7.14. The quantities P, F_a, and M_a are externally applied quantities and are assumed to be known.

There are several ways to choose the various non-dimensional quantities and as will be seen in the approach used here the process is somewhat of an art. Referring to Table 7.14, we start by letting

$$\eta = \frac{x}{L}, \quad y = \frac{w}{L}, \quad \tau = \frac{t}{t_o}$$

TABLE 7.14
Engineering Units for the Symbols in Eq. (a) of Example 7.27

Symbol	Units	Description
x	m	Position
L	m	Beam length
w	m	Transverse displacement
ψ	rad	Rotation of cross section due to bending
t	s	Time
ν		Poisson's ratio
κ		Shear correction factor
A	m^2	Area
I	m^4	Area moment of inertia
E	N/m^2	Tensile modulus
$G = E/[2(1+\nu)]$	N/m^2	Shear modulus
ρ	kg/m^3	Density
$F_a(x,t)$	N/m	Externally applied force
$M_a(x,t)$	N	Externally applied moment
P	N	Externally applied axial force

where t_o (s) is to be determined. Then Eq. (a) becomes

$$\frac{\kappa GA}{L}\left(\frac{\partial^2 y}{\partial\eta^2}-\frac{\partial\psi}{\partial\eta}\right)+\frac{P}{L}\frac{\partial^2 y}{\partial\eta^2}-\frac{\rho AL}{t_o^2}\frac{\partial^2 y}{\partial\tau^2}=-F_a(\eta,\tau)\ \text{N/m}$$

$$\kappa GA\left(\frac{\partial y}{\partial\eta}-\psi\right)+\frac{EI}{L^2}\frac{\partial^2\psi}{\partial\eta^2}-\frac{\rho I}{t_o^2}\frac{\partial^2\psi}{\partial\tau^2}=-M_a(\eta,\tau)\ \text{N} \tag{b}$$

Next, we divide the respective equations in Eq. (b) by the constants appearing in their leading terms to obtain

$$\left(\frac{\partial^2 y}{\partial\eta^2}-\frac{\partial\psi}{\partial\eta}\right)+\frac{P}{\kappa GA}\frac{\partial^2 y}{\partial\eta^2}-\frac{\rho AL^2}{t_o^2\kappa GA}\frac{\partial^2 y}{\partial\tau^2}=-\frac{F_a(\eta,\tau)L}{\kappa GA}$$

$$\left(\frac{\partial y}{\partial\eta}-\psi\right)+\frac{EI}{\kappa GAL^2}\frac{\partial^2\psi}{\partial\eta^2}-\frac{\rho I}{t_o^2\kappa GA}\frac{\partial^2\psi}{\partial\tau^2}=-\frac{M_a(\eta,\tau)}{\kappa GA} \tag{c}$$

Using the definition of G given in Table 7.14 in Eq. (c) yields

$$\left(\frac{\partial^2 y}{\partial\eta^2}-\frac{\partial\psi}{\partial\eta}\right)+\frac{2P(1+\nu)}{\kappa EA}\frac{\partial^2 y}{\partial\eta^2}-\frac{2\rho L^2(1+\nu)}{t_o^2\kappa E}\frac{\partial^2 y}{\partial\tau^2}=-\frac{2F_a(\eta,\tau)L(1+\nu)}{\kappa EA}$$

$$\left(\frac{\partial y}{\partial\eta}-\psi\right)+\frac{2EI(1+\nu)}{\kappa EAL^2}\frac{\partial^2\psi}{\partial\eta^2}-\frac{2\rho I(1+\nu)}{t_o^2\kappa EA}\frac{\partial^2\psi}{\partial\tau^2}=-\frac{2M_a(\eta,\tau)(1+\nu)}{\kappa EA} \tag{d}$$

To collect some common terms, the following non-dimensional quantity is introduced into Eq. (d)

$$\gamma=\frac{2(1+\nu)}{\kappa}$$

and Eq. (d) can be written as

$$\left(\frac{\partial^2 y}{\partial\eta^2}-\frac{\partial\psi}{\partial\eta}\right)+\frac{P}{EA}\gamma\frac{\partial^2 y}{\partial\eta^2}-\frac{\rho L^2}{t_o^2 E}\gamma\frac{\partial^2 y}{\partial\tau^2}=-\frac{\gamma F_a(\eta,\tau)L}{EA}$$

$$\left(\frac{\partial y}{\partial\eta}-\psi\right)+\frac{I}{AL^2}\gamma\frac{\partial^2\psi}{\partial\eta^2}-\frac{\rho I}{t_o^2 EA}\gamma\frac{\partial^2\psi}{\partial\tau^2}=-\frac{\gamma M_a(\eta,\tau)}{EA} \tag{e}$$

Examination of Eq. (e) indicates that another non-dimensional quantity to introduce is

$$R_o^2=\frac{I}{AL^2}\rightarrow\frac{\text{m}^4}{\text{m}^2\text{m}^2}=1$$

Then Eq. (e) becomes

$$\left(\frac{\partial^2 y}{\partial \eta^2}-\frac{\partial \psi}{\partial \eta}\right)+\gamma R_o^2 \frac{PL^2}{EI}\frac{\partial^2 y}{\partial \eta^2}-\frac{\rho A L^4}{t_o^2 EI}\gamma R_o^2 \frac{\partial^2 y}{\partial \tau^2}=-\gamma R_o^2 \frac{F_a(\eta,\tau)L^3}{EI}$$

$$\left(\frac{\partial y}{\partial \eta}-\psi\right)+\gamma R_o^2 \frac{\partial^2 \psi}{\partial \eta^2}-\frac{\rho L^2}{t_o^2 E}\gamma R_o^2 \frac{\partial^2 \psi}{\partial \tau^2}=-\gamma R_o^2 \frac{L^2 M_a(\eta,\tau)}{EI} \qquad \text{(f)}$$

In examining the units of the following collection of parameters appearing in Eq. (f), it is found that the following quantities are non-dimensional

$$S=\frac{PL^2}{EI}\rightarrow\frac{\text{Nm}^2}{\text{N/m}^2\cdot\text{m}^4}=1$$

$$f_a(\eta,\tau)=\frac{F_a(\eta,\tau)L^3}{EI}\rightarrow\frac{\text{N/m}\cdot\text{m}^3}{\text{N/m}^2\cdot\text{m}^4}=1$$

$$m_a(\eta,\tau)=\frac{L^2 M_a(\eta,\tau)}{EI}\rightarrow\frac{\text{m}^2\text{N}}{\text{N/m}^2\cdot\text{m}^4}=1$$

Introducing these definitions into Eq. (f) yields

$$\left(\frac{\partial^2 y}{\partial \eta^2}-\frac{\partial \psi}{\partial \eta}\right)+\gamma R_o^2 S\frac{\partial^2 y}{\partial \eta^2}-\frac{\rho A L^4}{t_o^2 EI}\gamma R_o^2 \frac{\partial^2 y}{\partial \tau^2}=-\gamma R_o^2 f_a(\eta,\tau)$$

$$\left(\frac{\partial y}{\partial \eta}-\psi\right)+\gamma R_o^2 \frac{\partial^2 \psi}{\partial \eta^2}-\frac{\rho L^2}{t_o^2 E}\gamma R_o^2 \frac{\partial^2 \psi}{\partial \tau^2}=-\gamma R_o^2 m_a(\eta,\tau) \qquad \text{(g)}$$

We now have to determine the definition of the previously introduced parameter t_o. For the top and bottom equations of Eq. (g), respectively, we note that t_o could be either of the following definitions

$$t_o^2=\frac{\rho A L^4}{EI}\rightarrow\frac{\text{kg/m}^3\cdot\text{m}^2\cdot\text{m}^4}{\text{N/m}^2\cdot\text{m}^4}=\frac{\text{kg}\cdot\text{m}}{\text{N}}=\frac{\text{s}^2\text{m}}{\text{m}}=\text{s}^2$$

$$t_o^2=\frac{\rho L^2}{E}\rightarrow\frac{\text{kg/m}^3\cdot\text{m}^2}{\text{N/m}^2}=\frac{\text{kg}\cdot\text{m}}{\text{N/m}^2}=\frac{\text{s}^2\text{m}}{\text{m}}=\text{s}^2$$

However, if we multiply and divide the term containing t_o in the second equation by R_o^2, then we can use the top equation as the definition of t_o and Eq. (g) becomes

$$\left(\frac{\partial^2 y}{\partial \eta^2}-\frac{\partial \psi}{\partial \eta}\right)+\gamma R_o^2 S\frac{\partial^2 y}{\partial \eta^2}-\gamma R_o^2 \frac{\partial^2 y}{\partial \tau^2}=-\gamma R_o^2 f_a(\eta,\tau)$$

$$\left(\frac{\partial y}{\partial \eta}-\psi\right)+\gamma R_o^2 \frac{\partial^2 \psi}{\partial \eta^2}-\gamma R_o^4 \frac{\partial^2 \psi}{\partial \tau^2}=-\gamma R_o^2 m_a(\eta,\tau) \qquad \text{(h)}$$

which is the desired result. Notice that we now have only two independent non-dimensional parameters: γ and R_o.

7.4 PARTIAL DIFFERENTIAL EQUATIONS AND IRREGULARLY SHAPED REGIONS: NUMERICAL SOLUTIONS USING MATHEMATICA'S FINITE ELEMENT CAPABILITY

In Section 7.2, we have examined numerous examples wherein the shape of the region being analyzed made it possible for the separation of variables technique to yield analytic solutions. Many applications of practical interest are concerned with regions whose shape does not permit one to use this technique. In these cases, one must resort to numerical means. One of the most effective and frequently used methods is the finite element method. In this section, we shall employ Mathematica to illustrate the results of this method by using a few illustrative examples. The three examples are:

(1) Steady-state temperature in a rectangular region containing a heated circular hole, which is governed by the Laplace equation
(2) Torsion of a closed rectangular region with a rectangular cutout, which is governed by the Poisson equation
(3) Response of an elliptic membrane with circular cutouts to an impulse force, which is governed by the wave equation

Example 7.28

We shall determine the steady-state temperature distribution in a rectangular region represented by the non-dimensional coordinates $0 \le \eta \le \eta_e$ and $-\beta \le \xi \le \beta$. Within this region there is a circular hole of non-dimensional radius r_h that has its center at (η_h,ξ_h). The governing equation is

$$\frac{\partial^2 u}{\partial \eta^2} + \frac{\partial^2 u}{\partial \xi^2} = 0 \quad 0 < \eta < \eta_e \quad -\beta < \xi < \beta$$

where $u = u(\eta,\xi)$ is the non-dimensional temperature. The boundary conditions on the perimeter are assumed to be

$$u(\eta,\beta) = 0$$

$$u(\eta,-\beta) = 2\sin\left(\pi\eta / \eta_e\right)$$

and

$$u(0,\xi) = 2\sin\left(\pi(\xi+\beta)/(2\beta)\right)$$

$$u(\eta_e,\xi) = 0$$

The boundary condition on the perimeter of the hole is given by

$$u(\eta,\xi) = u_h \quad (\eta-\eta_h)^2 + (\xi-\xi_h)^2 = r_h^2$$

where u_h is the prescribed constant temperature. These are all Dirichlet boundary conditions. It is assumed that $\eta_e = 1.5$, $\beta = 1$, $r_h = 0.25$, $u_h = 1.8$, $\eta_h = 0.5$, and $\xi_h = 0.35$. The solution is obtained with Mathematica procedure M7.29, and the results are shown in Figure 7.22.

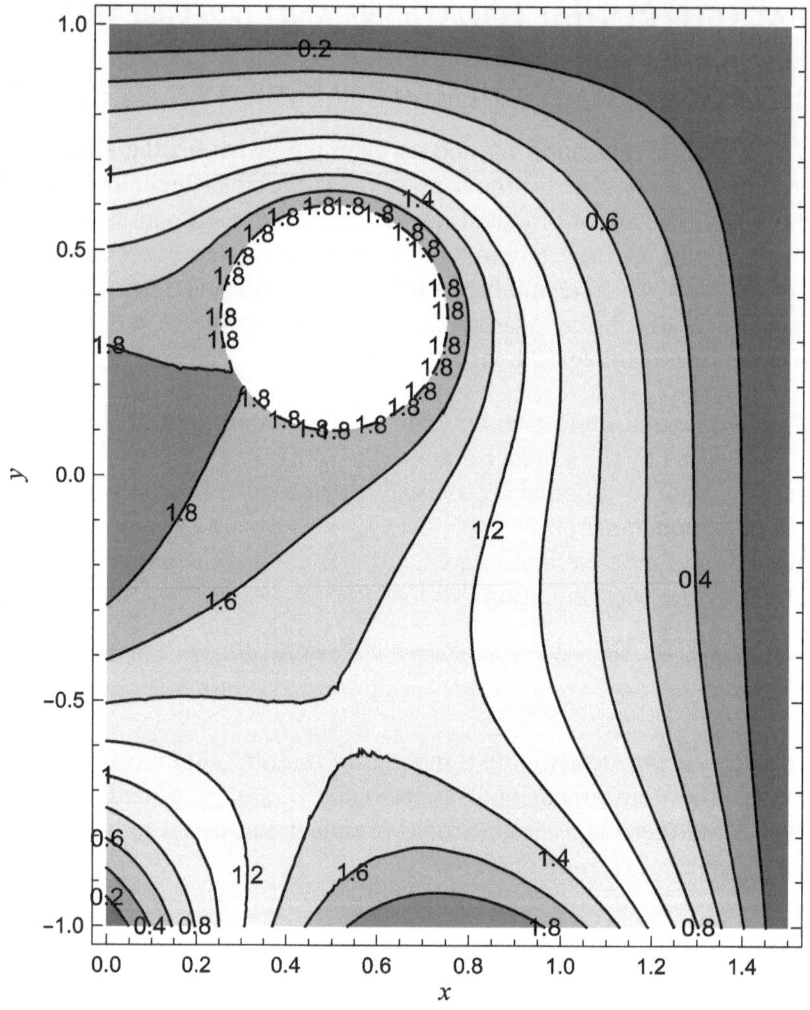

FIGURE 7.22 Temperature distribution in the plate of Example 7.28.

Example 7.29

If the coordinates of an arbitrary cross section of a structural element that is subjected to a twist θ per unit length are the (x,y) coordinates, then the governing equation from which the shear stresses can be determined is

$$\nabla^2\Phi = \frac{\partial^2\Phi}{\partial x^2} + \frac{\partial^2\Phi}{\partial y^2} = -2$$

which is valid over the cross section denoted as a region R. The quantity $\Phi = \phi/(G\theta)$ (m^2), where ϕ (N/m) is the stress function, G (N/m^2) is the shear modulus, and θ (rad/m) is the axial twist (rotation) of the cross section per unit length. The scaled shear stresses are given by

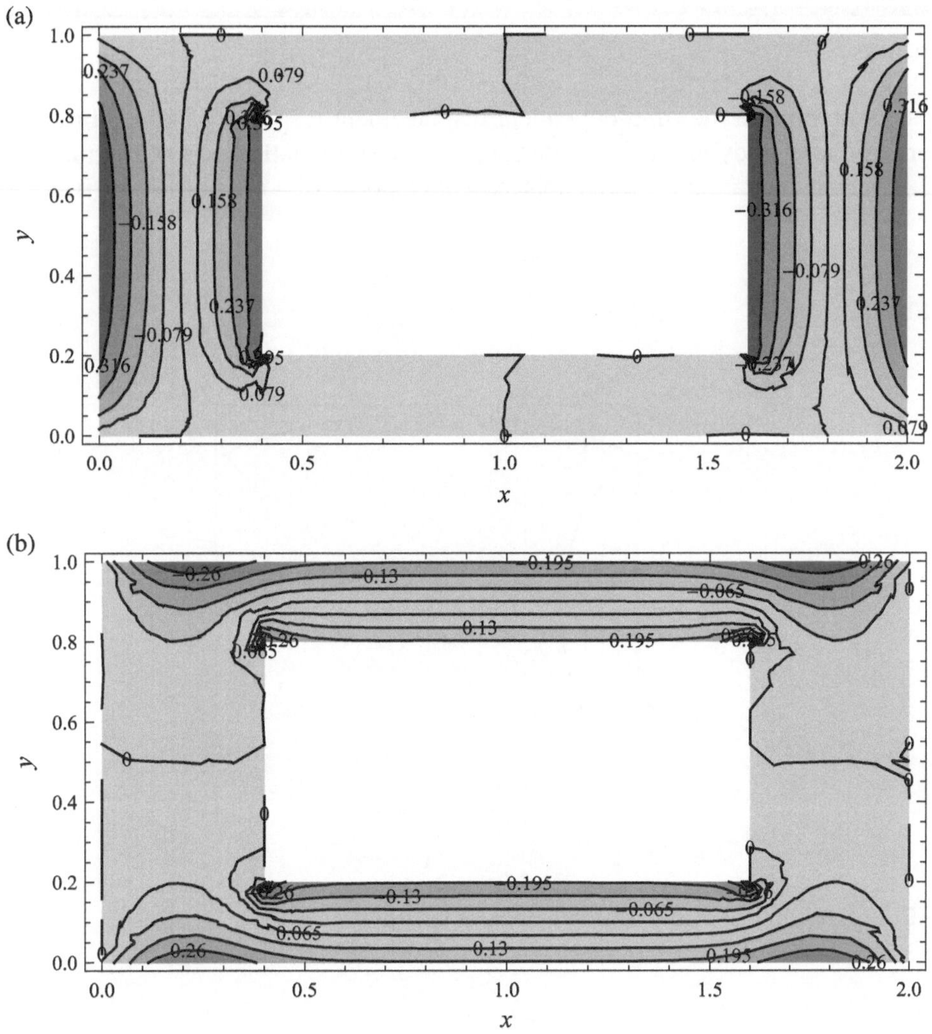

FIGURE 7.23 Shear stresses due to torsion in a rectangular region with a rectangular cutout: (a) $\bar{\tau}_{yz}$ and (b) $\bar{\tau}_{xz}$.

$$\bar{\tau}_{yz} = -\frac{\partial \Phi}{\partial x}\ \mathrm{m}$$

$$\bar{\tau}_{xz} = \frac{\partial \Phi}{\partial y}\ \mathrm{m}$$

where

$$\bar{\tau}_{yz} = \frac{\tau_{yz}}{G\theta}, \quad \bar{\tau}_{xz} = \frac{\tau_{xz}}{G\theta}$$

The boundary conditions are that $\Phi = 0$ on the entire boundary; that is, the Dirichlet boundary conditions have to be satisfied on the boundary. Using Mathematica procedure M7.30, the contours of the shear stresses $\bar{\tau}_{yz}$ and $\bar{\tau}_{xz}$ are shown in Figure 7.23.

Example 7.30

We shall determine the time-dependent displacement response of a membrane to an externally applied dynamic force concentrated over a small region of the membrane. The governing equation for the displacement response of a membrane is given by

$$\frac{\partial^2 w}{\partial \eta^2} + \frac{\partial^2 w}{\partial \xi^2} = \frac{\partial^2 w}{\partial \tau^2} - f(\tau, \eta, \xi)$$

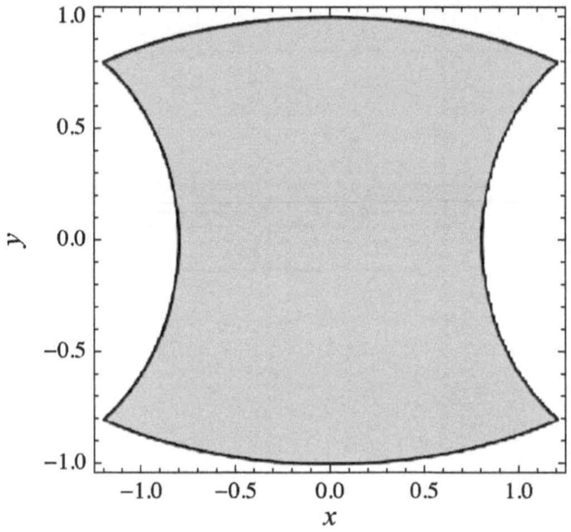

FIGURE 7.24 Elliptical membrane with circular cutouts.

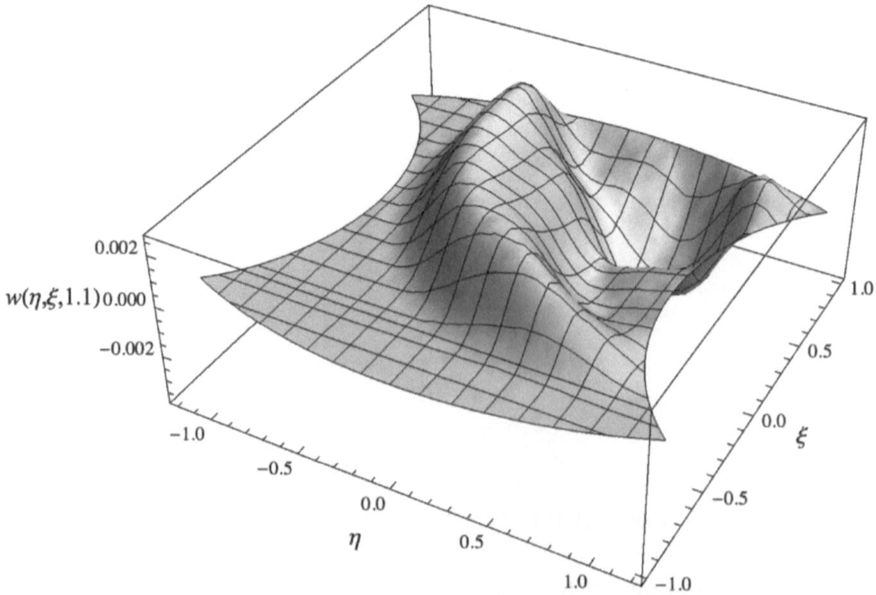

FIGURE 7.25 Response of an elliptical membrane with circular cutouts at $\tau = 1.1$ to an impulse force applied at (0.5,0.5).

where w(τ,η,ξ) is the displacement, $f(\tau,\eta,\xi)$ is force applied to the membrane, η and ξ are the coordinates, and τ is time. These quantities are non-dimensional. The concentrated impulse force is approximated by

$$f(\tau,\eta,\xi)=e^{-50\left(t^2+(\eta-\eta_f)^2+(\xi-\xi_f)^2\right)}$$

where (η_f,ξ_f) are the coordinates of the concentrated force.

The shape of the membrane is created by the intersection of an ellipse with two circles of equal diameter as shown in Figure 7.24. A snapshot of the displacement response at τ = 1.1 when the force is applied at η_f = 0.5, and ξ_f = 0.5 is shown in Figure 7.25. This result is obtained with Mathematica procedure M7.31.

MATHEMATICA PROCEDURES

Note: In order to make the Mathematica procedures that follow a little more readable, the `Clear` statement and many of the graphics-enhancing statements have been omitted.

```
(*M7.1*)
u=eta[x,y];
v=xi[x,y];
px=D[p[u,v],x];
py=D[p[u,v],y];
pxx=D[p[u,v],x,x];
pyy=D[p[u,v],y,y];
pxy=D[p[u,v],y,x];
pde=a pxx+2 b pxy+c pyy+d px+e py;
coe=Coefficient[pde,{D[p[x,y],y,y]/.
{x->eta[x,y],y->xi[x,y]},
D[p[x,y],x,y]/.{x->eta[x,y],y->xi[x,y]},
D[p[x,y],x,x]/.{x->eta[x,y],y->xi[x,y]},
D[p[x,y],x]/.{x->eta[x,y],y->xi[x,y]},
D[p[x,y],y]/.{x->eta[x,y],y->xi[x,y]}}];
Aa=coe[[1]]
Bb=Collect[Expand[coe[[2]]/2],b]
Cc=coe[[3]]
Dd=coe[[4]]
Ee=coe[[5]]
Simplify[Bb^2-Aa Cc]
```

```
(*M7.2*)
eqn=D[u[x,y],x,x]+D[u[x,y],y,y];
FullSimplify[ComplexExpand[DSolveValue[
   {eqn==0,u[x,0]==0,u[x,b]==go Sin[Pi x/a],u[0,y]==0,u[a,y]==0},
   u[x,y],{x,y}]]]
```

```
(*M7.3*)
ba=0.8;
Plot3D[Sin[Pi x] Sinh[Pi y]/Sinh[Pi ba],{x,0,1},{y,0,ba}]
```

```
(*M7.4*)
eqn=D[u[x,y],x,x]+D[u[x,y],y,y];
DSolveValue[{eqn==0,u[x,0]==0,u[x,b]==g[x],
   Derivative[1,0][u][0,y]==0,Derivative[1,0][u][a,y]==0},
   u[x,y],{x,y}]
(*Integrations for a specific g(x)*)
Integrate[Cos[n Pi x/a],{x,a/4,3 a/4}]
Integrate[1,{x,a/4,3 a/4}]

(*M7.5*)
fcn[xa_,ya_,nn_,ba_]:=ya/2/ba+2/Pi Total[Table[Cos[k Pi xa ]*
   Sinh[k Pi ya]/k/Sinh[k Pi ba] (Sin[3 k Pi/4]-Sin[k Pi/4]),
   {k,1,nn}]]
ba=0.8;
Plot3D[fcn[x,y,55,ba],{x,0,1},{y,0,ba}]

(*M7.6*)
sol=DSolveValue[{Laplacian[v[x,y],{x,y}]==0,v[x,0]==f[x]-Uo x/a,
   v[x,b]==0,v[0,y]==0,v[a,y]==0},v[x,y],{x,y}]
```

Note: When, in the output of M7.6, the limit as $b \to \infty$ is taken on the following expression

$$\operatorname{csch}\left(n\pi b/a\right)\sinh\left(n\pi(b-y)/a\right) = \frac{\sinh\left(n\pi(b-y)/a\right)}{\sinh\left(n\pi b/a\right)}$$

it will be found that

$$\lim_{b\to\infty}\left[\frac{\sinh\left(n\pi(b-y)/a\right)}{\sinh\left(n\pi b/a\right)}\right] = e^{-n\pi y/a}$$

This brings the output into a form that mirrors Eq. (j).

```
(*M7.7*)
fn[n_]:=If[n==2,0,(Sin[3 k Pi/4]+Sin[k Pi/4])/(n^2-4)]
fcn[xa_,ya_,nn_,alp_]:=xa+2/ Pi Total[Table[Sin[k Pi xa ]*
   Exp[-k Pi ya] (-2 alp fn[k]-(-1)^k/k),{k,1,nn}]]
Plot3D[fcn[x,y,21,10],{x,0,1},{y,0,0.5},PlotRange->All]

(*M7.8*)
fcn[x_,th_,bet_,alp_,nn_]:=4/Pi Total[Table[Sin[n Pi th/alp]*
   x^(n Pi/alp)/n (1-(bet /x)^(2 n Pi/alp))/(1-bet^(2 n Pi/alp)),
   {n,1,nn,2}]]
bet=0.25; alp= Pi/3;
coord=Table[Table[{x Cos[th],x Sin[th],fcn[x,th,bet,alp,51]},
   {x,Range[bet,1,(1-bet)/16]}],{th,Range[0,alp,alp/20]}];
ListPlot3D[Flatten[coord,1],RegionFunction->(Norm[{#1,#2}]>bet&)]

(*M7.9*)
fcn[r_,z_,gam_,nn_,lam_]:=2 Total[Table[
   BesselJ[0,lam[[n]] r] Sinh[lam[[n]] z gam]/lam[[n]]/
   BesselJ[1,lam[[n]]]/Sinh[lam[[n]] gam],{n,1,nn}]]
```

```
nn=45;    gam=1.1;
lam=N[BesselJZero[0,Range[1,nn]]];
Plot3D[fcn[r,z,gam,nn,lam],{r,0,1},{z,0,1}]

(*M7.10*)
Simplify[Integrate[LegendreP[n,Cos[phi]] Sin[phi],{phi,0,Pi/2}],
   Assumptions->n∈Integers]

(*M7.11*)
w[r_,ph_]:=0.5 Total[Table[(LegendreP[n-1,0]-LegendreP[n+1,0])*
   r^n LegendreP[n,Cos[ph]],{n,0,31}]]
Plot3D[w[rr,pp],{rr,0,0.97},{pp,0,Pi}]

(*M7.12*)
sol=DSolveValue[{D[u[x,t],t]==a^2 D[u[x,t],{x,2}],u[x, 0] ==f[x],
   u[0,t]==0,u[L,t]==0 },u[x,t],{x,t}]

(*M7.13*)
sol=DSolveValue[{ D[u[x,t],t]==a^2 D[u[x,t],{x,2}],u[x,0]==f[x],
   Derivative[1,0][u][0,t]==0,u[L,t]==Uo },u[x,t],{x,t}]

(*M7.14*)
deter[kap_,ha_]:=-kap BesselJ[1,kap]+ha BesselJ[0,kap]
gues={1,4,7,10,16,19,23,26,29};    ha=0.2;    len=Length[gues];
rtes=Table[x/.FindRoot[deter[x,ha],{x,gues[[n]]}],{n,1,len}];
uu[x_,t_]:=2 Total[Table[
   (2 BesselJ[2,rtes[[n]]]-rtes[[n]] BesselJ[3,rtes[[n]]])*
   BesselJ[0,x rtes[[n]]] Exp[-t rtes[[n]]^2]/
   ((rtes[[n]]^2+ha^2) BesselJ[0,rtes[[n]]]^2),{n,1,len}]]
Plot3D[uu[x,t],{x,0,1},{t,0.01,0.2}]

(*M7.15*)
Simplify[Integrate[Sin[n Pi r/a] r,{r,0,a}],
   Assumptions->n∈Integers]

(*M7.16*)
uv[x_,t_]:=2/Pi Total[Table[(-1)^(n+1) Sin[n Pi x]/(n x)*
   Exp[-n^2 Pi^2 t],{n,1,5}]]
Plot3D[uv[x,t],{x,0,1},{t,0.02,0.25}]

(*M7.17*)
eqn=c^2 D[w[x,t],x,x]-D[w[x,t],t,t];
FullSimplify[DSolveValue[{eqn==0,w[x,0]==f[x],
   Derivative[0,1][w][x,0]==g[x]},w[x,t],{x,t}],Assumptions->c>0]

(*M7.18*)
Simplify[8 Integrate[(x/L-1/4) Sin[n Pi x/L],{x,L/4,L/2}]/L+
   8 Integrate[(3/4-x/L) Sin[n Pi x/L],{x,L/2,3 L/4}]/L,
   Assumptions->n∈Integers]
```

```
(*M7.19*)
eqn=D[w[x,t],x,x]-D[w[x,t],t,t];
f=Piecewise[{{0,0<=x<=1/4},{4 (x-1/4),1/4<x<=1/2},
   {4 (3/4-x),1/2<x<=3/4},{0,3/4<x<=1}}];
DSolveValue[{eqn==0,w[x,0]==f,Derivative[0,1][w][x,0]==0,
   w[0,t]==0,w[1,t]==0},w[x,t],{x,t}]

(*M7.20*)
fcn[x_,t_,nn_]:=32/Pi^2 Total[Table[(-1)^((n-1)/2)/n^2*
   Sin[n Pi/8]^2 Cos[n Pi t] Sin[n Pi x],{n,1,nn,2}]]
Plot3D[fcn[x,t,35],{x,0,1},{t,0,0.6}]

(*M7.21*)
weqn=D[u[x,y,t],{t,2}]==c^2 Laplacian[u[x,y,t],{x,y}];
ic={u[x,y,0]==f[x,y],Derivative[0,0,1][u][x,y,0]==g[x,y]};
bc={u[x,0,t]==0,u[0,y,t]==0,u[a,y,t]==0,u[x,b,t]==0};
DSolveValue[{weqn,ic,bc},u[x,y,t],{x,y,t}]

(*M7.22*)
uxy[x_,y_,t_,alp_]:=(nn=11;
   64/Pi^6 Total[Total[Table[Table[Sin[n Pi x] Sin[m Pi*
     y/alp]*Cos[t Pi Sqrt[n^2+m^2/alp^2]]/(n^3 m^3),
   {n,1,nn,2}],{m,1,nn,2}]]])
pllt[t_,alp_]:=Plot3D[uxy[x,y,t,alp],{x,0,1},{y,0,alp}]
f11=pllt[0.25,1.5]
f12=pllt[0.425,1.5]
f21=pllt[0.45,1.5]
f22=pllt[1.25,1.5]

(*M7.23*)
peqn=D[u[r,t],{t,2}]==c^2 ( D[u[r,t],{r,2}]+D[u[r,t],r]/r);
pbc=u[a,t]==0;
pic={u[r,0]==f[r],Derivative[0,1][u][r,0]==g[r]};
FullSimplify[DSolveValue[{peqn,pic, pbc},u[r,t],{r,t}],
   Assumptions->a>0]//TraditionalForm

(*M7.24*)
etao=0.6;
pdel=2/Pi Total[Table[Sin[n Pi eta] Sin[n Pi etao]*
   Sin[n Pi t]/n,{n,1,80}]];
Plot3D[pdel,{eta,0,1},{t,0,3}]

(*M7.25*)
etao=0.6;
pdel=4/Pi^3 Total[Table[Sin[n Pi eta] (1- Cos[n Pi t])/n^3,
   {n,1,81,2}]];
Plot3D[pdel,{eta,0,1},{t,0,3.5}]

(*M7.26*)
alphan=(2 n-1) Pi/(2 a);    betam=(2 m-1) Pi/(2 b);
alpham=(2 m-1) Pi/(2 a);    betan=(2 n-1) Pi/(2 b);
Ybar=b Sinh[alphan b] Cosh[alphan y]-
```

```
    y Cosh[alphan b] Sinh[alphan y];
Xbar=a Sinh[betan a] Cosh[betan x]-
    x Cosh[betan a] Sinh[betan x];
anm=Simplify[alphan Sin[alphan a]Integrate[Ybar Cos[betam y],
    {y,-b,b}],Assumptions->{n,m}∈Integers]
bm=Simplify[KroneckerDelta[m,n] (betan a+Sinh[2 betan a]/2)*
    Integrate[Cos[betam y]^2,{y,-b,b}],
Assumptions->{n,m}∈Integers]
em=Simplify[2 a Integrate[(y^2-b^2) Cos[betam y],{y,-b,b}],
    Assumptions->{n,m}∈Integers]
cm=Simplify[(alphan b+Sinh[2 alphan b]/2)*
    Integrate[Cos[alpham x]^2,{x,-a,a}],
    Assumptions->{n,m}∈Integers]
dnm=Simplify[betan Sin[betan b] Integrate[Xbar Cos[alpham x],
    {x,-a,a}],Assumptions->{n,m}∈Integers]
em=Simplify[2 b Integrate[(x^2-a^2) Cos[alpham x],{x,-a,a}],
    Assumptions->{n,m}∈Integers]

(*M7.27*)
amn[m_,n_,ba_]:=-8 m n^2 Sin[n Pi/2] Sin[m Pi/2]*
    Cosh[Pi ba n/2]^2/(Pi (m^2+n^2 ba^2)^2)
bm[m_,ba_]:=-(m Pi/ba+Sinh[Pi m/ba])/2
em[m_,ba_]:=64 Sin[m Pi/2]/(Pi m)^3
cm[m_,ba_]:=-(m Pi ba+Sinh[Pi m ba])/2
dmn[m_,n_,ba_]:=-8 m n^2 Sin[n Pi/2] Sin[m Pi/2]*
    Cosh[Pi n/2/ba]^2/(Pi (n^2+m^2 ba^2)^2)
fm[m_,ba_]:=64 (Sin[m Pi/2])/(Pi m)^3
Xnn[n_,ba_]:=Sinh[(2 n-1) Pi/ba/2]
Ynn[n_,ba_]:=Sinh[(2 n-1) Pi ba/2]
w00[bba_]:=(nN=3;num=(nN+1)/2;
    co11=Table[Table[amn[n,m,bba],{m,1,nN,2}],{n,1,nN,2}];
    co12=DiagonalMatrix[Table[bm[n,bba]/bba,{n,1,nN,2}]];
    co21=DiagonalMatrix[Table[cm[n,bba]/bba^3,{n,1,nN,2}]];
    co22=Table[Table[dmn[n,m,bba],{m,1,nN,2}],{n,1,nN,2}];
    topp=Table[Flatten[{co11[[mm,1;;num]],co12[[mm,1;;num]]}],
      {mm,1,num}];
    bott=Table[Flatten[{co21[[mm,1;;num]],co22[[mm,1;;num]]}],
      {mm,1,num}];
rhs=Flatten[{Table[em[m,bba] bba,{m,1,nN,2}],
      Table[fm[m,bba]/bba,{m,1,nN,2}]}];
ary=Flatten[{topp,bott},1];
AB=LinearSolve[ary,rhs];
(bba^2+Total[Table[AB[[m]] Ynn[m,bba],{m,1,num}]]+
    Total[Table[AB[[m]] Xnn[m-num,bba],{m,num+1,2 num}]]))
Plot[w00[x],{x,1,3}]

(*M7.28*)
nf[gues_]:=x/.FindRoot[BesselJ[0,x] BesselI[1,x]+
    BesselI[0,x] BesselJ[1,x]==0,{x,gues}]
mode[nf_,x_]:=BesselJ[0,nf x]-BesselJ[0,nf]/BesselI[0,nf]*
    BesselI[0,nf x]
node[nf_,gues_]:=x/.FindRoot[mode[nf,x]==0,{x,gues}]
```

```
lab[n_]:=Column[{Row[{ΩToString[n],"=",NumberForm[nfreq[[n]],4]}],
   Row[{"ξnode=",NumberForm[nod[[n]],3]}]},Center]
plt[n_]:=Show[RevolutionPlot3D[mode[nfreq[[n]],x],{x,0,1},
   Boxed->False,Axes->False,ViewPoint->Above,PlotLabel->lab[n]],
   Graphics3D[{Opacity[0.6],
     InfinitePlane[{{1,1,0},{-1,1,0},{1,-1,0}}]}]]
nfreq={nf[Pi],nf[2 Pi],nf[3 Pi]};
nod={1,{node[nfreq[[2]],0.4],1},
   {node[nfreq[[3]],0.3],node[nfreq[[3]],0.6],1}};
GraphicsRow[{plt[1],plt[2],plt[3]}]

(*M7.29*)
xe=1.5;    bet=1.;    rh=0.25;    xh=0.5;    yh=0.35;
flow1=ImplicitRegion[0<=x<=xe&&-bet<=y<=bet,{x,y}];
flow2=ImplicitRegion[(x-xh)^2+(y-yh)^2<=rh^2,{x,y}];
flow=RegionDifference[flow1,flow2];
dirc={DirichletCondition[temp[x,y]==0,x==xe],
   DirichletCondition[temp[x,y]==2 Sin[0.5 Pi (y+bet)/bet],x==0],
   DirichletCondition[temp[x,y]==0,y==bet],
   DirichletCondition[temp[x,y]==2 Sin[Pi x/ xe],y==-bet],
   DirichletCondition[temp[x,y]==1.8,(x-xh)^2+(y-yh)^2==rh^2]};
flo=NDSolveValue[{Laplacian[temp[x,y],{x,y}]==0,dirc},temp,
   {x,y}∈flow];
ContourPlot[flo[x,y],{x,y}∈flow,AspectRatio->2/1.5]

(*M7.30*)
Needs["NDSolve`FEM`"]
a=2; b=1; dx=0.4;
out=ImplicitRegion[0<=x<=a&&0<=y<=b,{x,y}] ;
in=ImplicitRegion[+dx<=x<=a-dx&&dx/2<=y<=b-dx/2,{x,y}];
reg=RegionDifference[out,in];
oper=Laplacian[z[x,y],{x,y}]+2;
reg1=ToBoundaryMesh[reg];
regm=ToElementMesh[reg1];
dircond=DirichletCondition[z[x,y]==0,True];
sol=NDSolveValue[{oper==0,dircond},z,{x,y}∈regm];
GraphicsRow[{ContourPlot[D[-sol[w,y],w]/.w->x,{x,y}∈regm],
   ContourPlot[D[sol[x,w],w]/.w->y,{x,y}∈regm]}]

(*M7.31*)
reg1=ImplicitRegion[(y)^2+(x/2)^2<=1,{x,y}];
reg2=ImplicitRegion[y^2+(x-1.8)^2<=1,{x,y}];
reg3=ImplicitRegion[y^2+(x+1.8)^2<=1,{x,y}];
reg=RegionDifference[RegionDifference[reg1,reg2],reg3];
xf=0.5; yf=0.5;
f=Exp[-50. ((x-xf)^2+(y-yf)^2+t^2)];
oper=Laplacian[w[t,x,y],{x,y}]-D[w[t,x,y],t,t]+f;
diri=DirichletCondition[w[t,x,y]==0,True];
solw=NDSolveValue[{oper==0,w[0,x,y]==0,
```

```
    Derivative[1,0,0][w][0,x,y]==0,diri},w,{t,0,2 Pi},
    {x,y}∈ reg];
Plot3D[solw[1.1,x,y],{x,y}∈reg]
```

EXERCISES

SECTION 7.2.2

7.1 Solve the Laplace equation

$$\frac{\partial^2\Psi}{\partial r^2}+\frac{\partial\Psi}{r\partial r}+\frac{\partial^2\Psi}{r^2\partial\theta^2}=0$$

subject to the boundary conditions on the semicircular annulus that are given by

$$\Psi(r,0)=\Psi(r,\pi)=0$$

$$\Psi(a,\theta)=\sin\theta$$

$$\Psi(b,\theta)=0 \quad b>a$$

7.2 Consider the rectangular region $0 \le x \le a$ and $0 \le y \le b$ that is governed by the Laplace equation

$$\frac{\partial^2 u}{\partial x^2}+\frac{\partial^2 u}{\partial y^2}=0$$

with the following boundary conditions

$$\frac{\partial u(0,y)}{\partial x}=0, \qquad \frac{\partial u(a,y)}{\partial x}=0$$

$$\frac{\partial u(x,0)}{\partial y}=0, \qquad u(x,b)=\alpha+\beta x$$

Determine an expression for u that satisfies this equation and the boundary conditions.

7.3 Referring to Figure 7.5, obtain a solution to

$$\frac{\partial^2 u}{\partial r^2}+\frac{\partial u}{r\partial r}+\frac{\partial^2 u}{r^2\partial\theta^2}=0 \quad 0<a<r<b \quad 0<\theta<\alpha<2\pi \tag{a}$$

over the annular sector indicated and shown in Figure 7.5 with the following boundary conditions

$$\begin{aligned} u(r,0)&=0\\ u(r,\alpha)&=u_o\\ u(a,\theta)&=0\\ u(b,\theta)&=0 \end{aligned} \tag{b}$$

7.4 Obtain a solution to the following system

$$\frac{1}{r}\frac{\partial}{\partial r}\left(r\frac{\partial u}{\partial r}\right)+\frac{\partial^2 u}{\partial z^2}=0 \quad 0<r<a \quad 0<z<b \tag{a}$$

with the boundary conditions

$$\begin{aligned} &u(0,z)=\text{finite}, && \frac{du(a,z)}{dr}=0 \\ &u(r,0)=f(r), && u(r,b)=g(r) \end{aligned} \tag{b}$$

Section 7.2.3

7.5 Consider the equation

$$\frac{\partial^2 u}{\partial x^2}+\frac{\partial^2 u}{\partial y^2}+\Omega^2 u=f_o\sin(\alpha\pi x/a)\sin(\beta\pi x/b) \quad 0<x<a \quad 0<y<b$$

subject to the boundary conditions

$$u(0,y)=u(a,y)=0$$

$$u(x,0)=u(x,b)=0$$

Using the results of Example 7.7, determine the solution when: (i) α and β are not integers; (ii) $\alpha = k$, an integer and $\beta = l$, an integer.

Section 7.2.4

7.6 Consider the transient symmetrical distribution of non-dimensional temperature Θ in a solid sphere of non-dimensional radius 1 that is subject to a constant non-dimensional heat rate g. The governing equation is

$$\frac{\partial\Theta}{\partial\tau}=\frac{1}{\eta^2}\frac{\partial}{\partial\eta}\left(\eta^2\frac{\partial\Theta}{\partial\eta}\right)+g$$

The boundary conditions are $\Theta(1,\tau) = 1$ and $\Theta(0,\tau)$ is finite and the initial condition is $\Theta(\eta,0) = 0$. Determine the temperature distribution in the sphere.

7.7 Consider the transient symmetrical distribution of non-dimensional temperature Θ in a solid infinitely long cylindrical rod of non-dimensional radius 1 that is subject to a constant non-dimensional heat rate H. The governing equation is

$$\frac{\partial\Theta}{\partial\tau}=\frac{a^2}{\eta}\frac{\partial}{\partial\eta}\left(\eta\frac{\partial\Theta}{\partial\eta}\right)+H \quad 0\le\eta<1 \quad \tau>0$$

where a is a known constant. The steady-state temperature at the center of the rod is θ_c and is an unknown quantity at this point. The boundary condition at $\eta = 1$ is

$$\Theta'(1,\tau)=-h\Theta(1,\tau)$$

where the prime indicates the derivative with respect to η and h is a constant. The boundary condition at $\eta = 0$ is that $\Theta(0,\tau)$ is finite. The initial condition is $\Theta(\eta,0) = 0$. Determine the temperature distribution in the rod. Hint: Assume that $\Theta(\eta,\tau) = \Theta_s(\eta) + \Theta_d(\eta,\tau)$, where $\Theta_s(\eta)$ is the solution to the nonhomogeneous equation containing H. Then, $\Theta_s(0) = \theta_c$ and θ_c is subsequently determined from

$$\Theta_s'(1,\tau) = -h\Theta_s(1,\tau)$$

7.8 Obtain a solution to the following non-dimensional system

$$\frac{\partial \Theta}{\partial \tau} = \frac{1}{\eta}\frac{\partial}{\partial \eta}\left(\eta\frac{\partial \Theta}{\partial \eta}\right) \quad 0 < \alpha \le \eta < 1 \quad \tau > 0$$

with the boundary conditions

$$\Theta'(\alpha,\tau) = -1$$

$$\Theta'(1,\tau) + k\Theta(1,\tau) = 0$$

and the initial condition

$$\Theta(\eta,0) = \gamma$$

The quantities k and γ are constants and the prime (′) denotes the derivative with respect to η.

7.9 Obtain a solution to the following non-dimensional system

$$\frac{\partial \Theta}{\partial \tau} = \frac{1}{\eta}\frac{\partial}{\partial \eta}\left(\eta\frac{\partial \Theta}{\partial \eta}\right) + \frac{\partial^2 \Theta}{\partial z^2} \quad 0 \le \eta \le \alpha \quad 0 \le z < 1 \quad \tau > 0$$

with the boundary conditions

$$\Theta(0,z,\tau) = \text{finite}, \qquad \frac{d\Theta(\alpha,z,\tau)}{d\eta} + k\Theta(\alpha,z,\tau) = 0$$

$$\Theta(\eta,0,\tau) = 1, \qquad \frac{d\Theta(\eta,1,\tau)}{dz} = 0$$

and the initial condition

$$\Theta(\eta,z,0) = f(\eta,z)$$

7.10 Obtain a solution to the following non-dimensional system

$$\frac{\partial u}{\partial \tau} = \frac{1}{\eta}\frac{\partial}{\partial \eta}\left(\eta\frac{\partial u}{\partial \eta}\right) + q_o \quad 0 \le \eta \le 1 \quad \tau > 0$$

with the boundary conditions

$$u(0,\tau) = \text{finite}$$

$$u(1,\tau) = 0$$

and the initial condition

$$u(\eta,0)=0$$

SECTION 7.2.5

7.11 Given the following partial differential equation

$$\frac{\partial^2 u}{\partial t^2}+2\zeta\frac{\partial u}{\partial t}=\frac{\partial^2 u}{\partial x^2}\quad 0\le\zeta<1$$

with the boundary conditions

$$u(0,1)=u(1,t)=0$$

and the initial conditions

$$u(x,0)=0$$

$$\dot{u}(x,0)=x\sin 2\pi x$$

where the over dot indicates the derivative with respect to t. Obtain the solution for u.

7.12 Find the solution to the non-dimensional equation

$$\frac{\partial^2 u}{\partial \tau^2}=\frac{\partial^2 u}{\partial \eta^2}-h_o u\quad 0<\eta<1\quad \tau>0$$

for the boundary conditions $u(0,\tau) = u(1,\tau) = 0$ and for the initial conditions

$$u(\eta,0)=\begin{cases}2u_o\eta & 0\le\eta\le 1/2\\ 2u_o(-\eta+1) & 1/2\le\eta\le 1\end{cases}$$

$$\dot{u}(\eta,0)=0$$

7.13 Consider the following circular region that is governed by the wave equation and subjected to an external excitation $f(\eta,\tau)$. The governing equation is

$$\frac{\partial^2 u}{\partial \eta^2}+\frac{\partial u}{\eta\partial \eta}-\frac{\partial^2 u}{\partial \tau^2}=f(\eta,\tau)\quad 0\le\eta<1\quad 0<\tau$$

subject to the initial conditions

$$u(r,0)=0\quad\text{and}\quad\frac{\partial u(r,0)}{\partial \tau}=0$$

and the boundary conditions

$$u(0,\tau)=\text{finite}$$

$$u(1,\tau)=0$$

Obtain an expression for u that satisfies this equation, boundary conditions, and initial conditions.

Section 7.2.7

7.14 (a) Find the general solution to

$$\nabla^2\left(\nabla^2\psi - \alpha\frac{\partial\psi}{\partial t}\right) = 0$$

where

$$\nabla^2 = \frac{\partial^2}{\partial r^2} + \frac{1}{r}\frac{\partial}{\partial r} + \frac{1}{r^2}\frac{\partial^2}{\partial\theta^2}$$

when

$$\psi(r,\theta,t) = \Psi(r)\sin\theta e^{j\omega t}$$

Hint: Notice that the equation can be written two ways and that both forms must be satisfied.

$$\nabla^2\left(\nabla^2\psi - \alpha\frac{\partial\psi}{\partial t}\right) = \left(\nabla^2 - \alpha\frac{\partial}{\partial t}\right)\nabla^2\psi = 0$$

(b) Select that portion of the solution in (a) that remains finite as $r \to \infty$.

7.15 Consider the following equation describing a square plate subject to a non-dimensional uniform load P

$$\nabla^4 u = P \quad 0 < \eta < 1 \quad 0 < \xi < 1$$

where

$$\nabla^2 = \frac{\partial^2}{\partial\eta^2} + \frac{\partial^2}{\partial\xi^2}$$

On the edges $\eta = 0$ and $\eta = 1$, the boundary conditions are

$$u(0,\xi) = u(1,\xi) = 0$$

$$\frac{\partial^2}{\partial\eta^2}u(0,\xi) = \frac{\partial^2}{\partial\eta^2}u(1,\xi) = 0$$

and those on the edges $\xi = 0$ and $\xi = 1$ are

$$u(\eta,0) = u(\eta,1) = 0$$

$$\frac{\partial}{\partial\xi}u(\eta,0) = \frac{\partial}{\partial\xi}u(\eta,1) = 0$$

Determine the solution to this system.

8 Laplace Transforms

8.1 LAPLACE TRANSFORM

8.1.1 Definition

Consider a class of functions represented by $g(t)$ that is zero for $t < 0$. The Laplace transform of $g(t)$ is defined as

$$L[g(t)] = G(s) = \int_0^{\infty} g(t)e^{-st}\,dt \tag{8.1}$$

where Re[s] is a positive constant.

An expression for the inverse Laplace transform can be obtained from the inverse Fourier transform as follows. If we let $F(\omega)$ denote the Fourier transform of $f(t)$, then from Eq. (3.34)

$$\begin{aligned} F(\omega) &= \int_{-\infty}^{\infty} f(\xi)e^{-j\omega\xi}\,d\xi \\ f(t) &= \frac{1}{2\pi}\int_{-\infty}^{\infty} F(\omega)e^{j\omega t}\,d\omega \end{aligned} \tag{8.2}$$

Therefore,

$$f(t) = \frac{1}{2\pi}\int_{-\infty}^{\infty} e^{j\omega t}\left[\int_{-\infty}^{\infty} f(\xi)e^{-j\omega\xi}\,d\xi\right]d\omega \tag{8.3}$$

Consider the function

$$\begin{aligned} f(t) &= g(t)e^{-\gamma t} \qquad \gamma > 0,\ t > 0 \\ &= 0 \qquad\qquad\quad t \le 0 \end{aligned}$$

When this function is substituted into Eq. (8.3), we obtain

$$\begin{aligned} g(t)e^{-\gamma t} &= \frac{1}{2\pi}\int_{-\infty}^{\infty} e^{j\omega t}\left[\int_0^{\infty} g(\xi)e^{-\gamma\xi}e^{-j\omega\xi}\,d\xi\right]d\omega \\ g(t) &= \frac{1}{2\pi}\int_{-\infty}^{\infty} e^{(\gamma+j\omega)t}\left[\int_0^{\infty} g(\xi)e^{-(\gamma+j\omega)\xi}\,d\xi\right]d\omega \end{aligned} \tag{8.4}$$

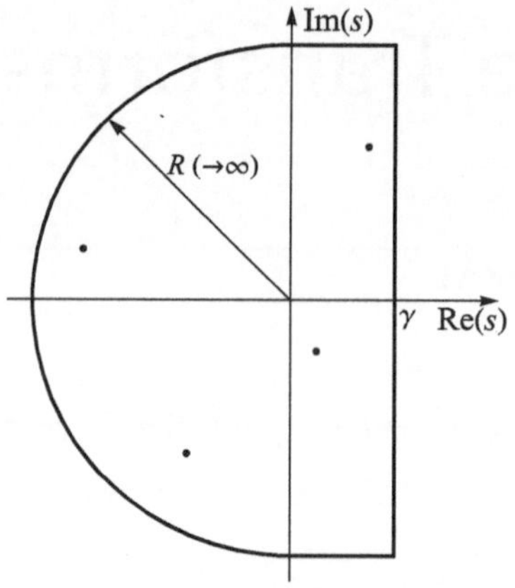

FIGURE 8.1 Contour of Laplace transform inversion integral.

Let $s = \gamma + j\omega$; therefore, $ds = jd\omega$ and Eq. (8.4) becomes

$$g(t) = L^{-1}[G(s)] = \frac{1}{2\pi j}\int_{\gamma - j\infty}^{\gamma + j\infty} e^{st}\left[\int_0^{\infty} g(\xi)e^{-s\xi}\,d\xi\right]ds = \frac{1}{2\pi j}\int_{\gamma - j\infty}^{\gamma + j\infty} e^{st}G(s)ds \tag{8.5}$$

which is the inversion formula for the Laplace transform. In this case, we choose γ to be to the right of the real part of s as shown in Figure 8.1.

8.1.2 Derivation of Laplace Transform Pairs

In this section, we shall obtain several Laplace transform pairs by using the definition given by Eq. (8.1). The results are summarized in Table 8.2 of Appendix 8.1.

Derivatives of functions: The Laplace transform of the nth derivative of $g(t) = f(t)$ is obtained by using integration by parts as follows

$$\int_0^{\infty} \frac{d^n f(t)}{dt^n} e^{-st}dt = e^{-st}\frac{d^{n-1}f(t)}{dt^{n-1}}\bigg|_0^{\infty} + s\int_0^{\infty}\frac{d^{n-1}f(t)}{dt^{n-1}}e^{-st}dt$$

$$= -f^{(n-1)}(0) + s\int_0^{\infty}\frac{d^{n-1}f(t)}{dt^{n-1}}e^{-st}dt$$

$$= -f^{(n-1)}(0) + s\left[-e^{-st}\frac{d^{n-2}f(t)}{dt^{n-2}}\bigg|_0^{\infty} + s\int_0^{\infty}\frac{d^{n-2}f(t)}{dt^{n-2}}e^{-st}dt\right]$$

$$= -f^{(n-1)}(0) - sf^{(n-2)}(0) + s^2\int_0^{\infty}\frac{d^{n-2}f(t)}{dt^{n-2}}e^{-st}dt$$

$$= s^n F(s) - \sum_{k=1}^{n} s^{n-k}f^{k-1}(0) \quad n \geq 1$$

Therefore, we have the following Laplace transform pair

$$\frac{d^n f(t)}{dt^n} \Leftrightarrow s^n F(s) - \sum_{k=1}^{n} s^{n-k} f^{k-1}(0) \quad n \geq 1 \tag{8.6}$$

It is also noted from Eq. (8.1) that

$$\frac{dF(s)}{ds} = -\int_0^\infty tf(t)e^{-st}dt$$

and

$$\frac{d^n F(s)}{ds^n} = (-1)^n \int_0^\infty t^n f(t)e^{-st}dt$$

Therefore,

$$t^n f(t) \Leftrightarrow (-1)^n \frac{d^n F(s)}{ds^n} \tag{8.7}$$

We state without proof the following useful Laplace transform pair

$$\frac{f(t)}{t} \Leftrightarrow \int_s^\infty F(s)ds \tag{8.8}$$

Unit step function: $g(t) = u(t - a)$. In this case, we use Eq. (8.1) to arrive at

$$L[u(t-a)] = \int_0^\infty u(t-a)e^{-st}dt = \int_a^\infty e^{-st}dt = \left.\frac{e^{-st}}{-s}\right|_a^\infty = \frac{e^{-as}}{s}$$

Therefore, we have the following Laplace transform pair

$$u(t-a) \Leftrightarrow \frac{e^{-as}}{s} \tag{8.9}$$

When $a = 0$, it is seen that the Laplace transform of 1 is $1/s$.

This result is verified with Mathematica procedure M8.1.

$g(t) = t^k$. In this case, we use Eq. (8.1) and let $\xi = st$ to arrive at

$$L[t^k] = \int_0^\infty t^k e^{-st}dt = \int_0^\infty \left(\frac{\xi}{s}\right)^k e^{-\xi} \frac{d\xi}{s} = \frac{1}{s^{k+1}} \int_0^\infty \xi^k e^{-\xi} d\xi = \frac{\Gamma(k+1)}{s^{k+1}}$$

where $\Gamma(x)$ is the gamma function [see Eq. (C.1) of Appendix C]. Therefore, we have the following Laplace transform pair

$$t^k \Leftrightarrow \frac{\Gamma(k+1)}{s^{k+1}} \tag{8.10}$$

When k is a positive integer, $\Gamma(k+1)=k!$ and Eq. (8.10) becomes

$$t^k \Leftrightarrow \frac{k!}{s^{k+1}} \tag{8.11}$$

Equation (8.10) is verified with Mathematica procedure M8.2.

$g(t) = \sin(at)$. We use Eq. (8.6) as follows. Since

$$L\left[\frac{d^2 g(t)}{dt^2}\right] = s^2 L[g(t)] - sg(0) - g^{(1)}(0)$$

then, for $g(t) = \sin(at)$,

$$-a^2 L[\sin at] = s^2 L[\sin at] - s\sin(0) - a\cos(0)$$

$$\left(s^2 + a^2\right) L[\sin at] = a$$

Therefore, we have the following Laplace transform pair

$$\sin at \Leftrightarrow \frac{a}{s^2 + a^2} \tag{8.12}$$

In a similar manner, we find that

$$\begin{aligned} \cos at &\Leftrightarrow \frac{s}{s^2 + a^2} \\ \sinh at &\Leftrightarrow \frac{a}{s^2 - a^2} \\ \cosh at &\Leftrightarrow \frac{s}{s^2 - a^2} \end{aligned} \tag{8.13}$$

Equations (8.12) and (8.13) are verified with Mathematica procedure M8.3.

$g(t) = e^{at}$. Employing Eq. (8.1), it is found that

$$L[e^{at}] = \int_0^\infty e^{at} e^{-st}\,dt = \int_0^\infty e^{(a-s)t}\,dt = \left.\frac{e^{-(s-a)t}}{a-s}\right|_0^\infty = \frac{1}{s-a} \quad \text{Re}[s] > \text{Re}[a] \tag{8.14}$$

Therefore, we have the following Laplace transform pair

$$e^{at} \Leftrightarrow \frac{1}{s-a} \quad \text{Re}[s] > \text{Re}[a] \tag{8.15}$$

The requirement that $\text{Re}[s] > \text{Re}[a]$ is needed to ensure that in Eq. (8.14) the exponential goes to zero as $t \to \infty$.

Equation (8.15) is verified with Mathematica procedure M8.4.

$g(t) = f(t)e^{bt}$. Employing Eq. (8.1), it is found that

$$L[f(t)e^{bt}] = \int_0^\infty f(t)e^{bt}e^{-st}dt = \int_0^\infty f(t)e^{-(s-b)t}dt = F(s-b)$$

Therefore, we have the following Laplace transform pair

$$f(t)e^{bt} \Leftrightarrow F(s-b) \tag{8.16}$$

Using Eqs. (8.12) and (8.16), we obtain the additional Laplace transform pairs

$$\begin{aligned} e^{bt}\sin at &\Leftrightarrow \frac{a}{(s-b)^2+a^2} \\ e^{bt}\cos at &\Leftrightarrow \frac{s-b}{(s-b)^2+a^2} \end{aligned} \tag{8.17}$$

and using Eqs. (8.11) and (8.16), we obtain the Laplace transform pair

$$e^{bt}t^m \Leftrightarrow \frac{m!}{(s-b)^{m+1}} \qquad m = 0,1,2,\ldots \tag{8.18}$$

Furthermore, we can extend these results as follows

$$L[f(t)\sin at] = L\left[\frac{f(t)}{2j}\left(e^{jat} - e^{-jat}\right)\right] = \frac{1}{2j}\left[F(s-ja) - F(s+ja)\right]$$

In a similar manner, we can find the Laplace transform pairs for $f(t)\cos(at)$, $f(t)\cosh(at)$, and $f(t)\sinh(at)$.

Equations (8.17) and (8.18) are verified with Mathematica procedure M8.5.

8.1.3 Partial Fractions

Consider a ratio of two polynomials $N(s)/D(s)$ with no common factors and the roots of $D(s)$ are distinct; that is, there are no repeating roots, and the degree of the polynomial in the denominator is greater than the degree of the polynomial in the numerator. The first step in the method of partial fractions is to factor D into the product of lower-order polynomials such that

$$F(s) = \frac{N(s)}{D(s)} = \sum_{k=1}^{m} \frac{A_k}{s-a_k} \tag{8.19}$$

where A_k can be determined from

$$A_k = \lim_{s\to a_k}\left[(s-a_k)\frac{N(s)}{D(s)}\right] = \frac{N(a_k)}{D'(a_k)} \tag{8.20}$$

and the prime denotes the derivative with respect to s. Recall Eq. (2.69).

For roots that are repeated m times, that is, when there is a term in $D(s)$ that appears as $(s - a_o)^m$, $F(s)$ is replaced by

$$F(s) = \sum_{k=1}^{N-m} \frac{A_k}{s - a_k} + \sum_{k=1}^{m} \frac{B_k}{\left(s - a_o\right)^k} \tag{8.21}$$

where B_k can be determined from

$$B_p = \frac{1}{(m-p)!} \frac{d^{(m-p)}}{ds^{(m-p)}} \left(\left(s - a_o\right)^m \frac{N(s)}{D(s)} \right)_{s \to a_o} \qquad p = 1, 2, \ldots, m \tag{8.22}$$

We now use Eqs. (8.15) and (8.18) to find that the inverse Laplace transform of Eq. (8.21) is

$$\frac{N(s)}{D(s)} = \sum_{k=1}^{N-m} \frac{A_k}{s - a_k} + \sum_{k=1}^{m} \frac{B_k}{\left(s - a_o\right)^k} \Leftrightarrow \sum_{k=1}^{N-m} A_k e^{a_k t} + \sum_{k=1}^{m} \frac{B_k}{(k-1)!} e^{a_o t} t^{k-1} \tag{8.23}$$

We shall now illustrate these results with two examples.

Example 8.1

Consider the following polynomial, which is a function of the Laplace transform parameter s

$$F(s) = \frac{s}{\left(s^2 + 1\right)\left(s + 2\right)^2} \tag{a}$$

It is seen from Eq. (a) that there are two distinct roots $a_{1,2} = \pm j$ and two repeating roots $a_o = -2$. We note that

$$D'(s) = 4 + 2s\left(2s^2 + 6s + 5\right)$$

Then, for the non-repeating roots, we use Eq. (8.20) to arrive at

$$A_k = \frac{N(a_k)}{D'(a_k)} = \frac{a_k}{4 + 2a_k\left(2a_k^2 + 6a_k + 5\right)} \qquad k = 1, 2$$

which yields

$$A_1 = \frac{j}{4 + 2j\left(-2 + 6j + 5\right)} = \frac{1}{6 + 8j} = \frac{6 - 8j}{100}$$

$$A_2 = \frac{-j}{4 - 2j\left(-2 - 6j + 5\right)} = \frac{1}{6 - 8j} = \frac{6 + 8j}{100}$$

For the repeating roots, we use Eq. (8.22) with $m = 2$,

$$B_1 = \frac{d}{ds}\left(\frac{s}{s^2+1}\right)_{s\to a_o} = \frac{1-a_o^2}{\left(a_o^2+1\right)^2} = -\frac{3}{25}$$

$$B_2 = \left(\frac{s}{s^2+1}\right)_{s\to a_o} = \frac{a_o}{a_o^2+1} = -\frac{2}{5}$$

Then, from Eq. (8.23),

$$\begin{aligned} f(t) &= \frac{6-8j}{100}e^{jt} + \frac{6+8j}{100}e^{-jt} - \frac{3}{25}e^{-2t} - \frac{2}{5}te^{-2t} \\ &= \frac{3}{25}\cos t + \frac{4}{25}\sin t - \frac{3}{25}e^{-2t} - \frac{2}{5}te^{-2t} \end{aligned} \tag{b}$$

Equation (b) is verified with Mathematica M8.6

Example 8.2

Consider the following polynomial, which is a function of the Laplace transform parameter s

$$F(s) = \frac{s^3+(\varepsilon^2-\delta^2)s}{\left(s^2-\delta^2\right)\left(s^2+\varepsilon^2\right)} = \frac{s\left(s^2+\varepsilon^2-\delta^2\right)}{(s-\delta)(s+\delta)(s+j\varepsilon)(s-j\varepsilon)} \tag{a}$$

Since there are no repeating roots, we use Eq. (8.20). First, we note that

$$D'(s) = 2s\left(2s^2+\varepsilon^2-\delta^2\right)$$

Then, Eq. (8.20) gives

$$A_k = \frac{N(a_k)}{D'(a_k)} = \frac{a_k^2+\varepsilon^2-\delta^2}{2\left(2a_k^2+\varepsilon^2-\delta^2\right)}$$

where $a_{1,2} = \pm\delta$ and $a_{3,4} = \pm j\varepsilon$. Hence

$$A_1 = \frac{\delta^2+\varepsilon^2-\delta^2}{2\left(2\delta^2+\varepsilon^2-\delta^2\right)} = \frac{\varepsilon^2}{2\left(\varepsilon^2+\delta^2\right)} = A_2$$

$$A_3 = \frac{-\varepsilon^2+\varepsilon^2-\delta^2}{2\left(-2\varepsilon^2+\varepsilon^2-\delta^2\right)} = \frac{\delta^2}{2\left(\varepsilon^2+\delta^2\right)} = A_4$$

and

$$\begin{aligned} F(s) &= \frac{A_1}{s-\delta} + \frac{A_2}{s+\delta} + \frac{A_3}{s-j\varepsilon} + \frac{A_4}{s+j\varepsilon} \\ &= A_1\left(\frac{1}{s-\delta} + \frac{1}{s+\delta}\right) + A_3\left(\frac{1}{s-j\varepsilon} + \frac{1}{s+j\varepsilon}\right) \\ &= A_1\left(\frac{2s}{s^2-\delta^2}\right) + A_3\left(\frac{2s}{s^2+\varepsilon^2}\right) \\ &= \frac{1}{\varepsilon^2+\delta^2}\left[\frac{s\varepsilon^2}{\left(s^2-\delta^2\right)} + \frac{s\delta^2}{\left(s^2+\varepsilon^2\right)}\right] \end{aligned} \tag{b}$$

Using Eqs. (8.12) and (8.13), the inverse Laplace transform of Eq. (b) is

$$\begin{aligned} L^{-1}\left[\frac{s^3+(\varepsilon^2-\delta^2)s}{\left(s^2-\delta^2\right)\left(s^2+\varepsilon^2\right)}\right] &= \frac{1}{\varepsilon^2+\delta^2} L^{-1}\left[\frac{s\varepsilon^2}{\left(s^2-\delta^2\right)} + \frac{s\delta^2}{\left(s^2+\varepsilon^2\right)}\right] \\ &= \frac{1}{\varepsilon^2+\delta^2}\left[\delta^2\cos\varepsilon t + \varepsilon^2\cosh\delta t\right] \end{aligned} \tag{c}$$

Equation (c) is verified with Mathematica procedure M8.7.

8.1.4 Convolution Integral

Consider two functions $f(t)$ and $h(t)$ that are zero for $t < 0$ and whose Laplace transforms, respectively, are $F(s)$ and $H(s)$. We would like to determine the inverse Laplace transform of the product $F(s)H(s)$. From Eq. (8.5) with $G(s) = F(s)H(s)$, we find that

$$\begin{aligned} L^{-1}[F(s)H(s)] &= \frac{1}{2\pi j}\int_{\gamma-j\infty}^{\gamma+j\infty} e^{st}F(s)H(s)ds \\ &= \frac{1}{2\pi j}\int_{\gamma-j\infty}^{\gamma+j\infty} e^{st}F(s)\left[\int_0^\infty h(\xi)e^{-s\xi}d\xi\right]ds \\ &= \int_0^\infty h(\xi)\left[\frac{1}{2\pi j}\int_{\gamma-j\infty}^{\gamma+j\infty} e^{s(t-\xi)}F(s)ds\right]d\xi \\ &= \int_0^\infty h(\xi)f(t-\xi)d\xi \end{aligned} \tag{8.24}$$

However, by definition $f(t-\xi)=0$ when $t<\xi$; therefore, Eq. (8.24) can be written as

$$L^{-1}[F(s)H(s)]=\int_0^t h(\xi)f(t-\xi)d\xi \tag{8.25}$$

Then we have the following Laplace transform pair

$$\int_0^t h(\xi)f(t-\xi)d\xi=\int_0^t f(\eta)h(t-\eta)d\eta \Leftrightarrow F(s)H(s) \tag{8.26}$$

where, to obtain the second integral form, we used the change of variables $\eta=t-\xi$. Notice that when $F(s)=H(s)$, Eq. (8.26) gives

$$\int_0^t f(\xi)f(t-\xi)d\xi \Leftrightarrow F(s)^2 \tag{8.27}$$

We now use this result to determine the Laplace transform of the integral

$$\int_0^t f(\xi)d\xi$$

If we consider the convolution integral with $h(t)=1$, then from Eq. (8.26)

$$\int_0^t f(\xi)d\xi \Leftrightarrow \frac{F(s)}{s}$$

since, from Eq. (8.9) or Eq. (8.11), $L[1]=1/s$.

The use of convolution is illustrated with the following examples.

Example 8.3

We shall determine the inverse of

$$\frac{1}{s^2}\frac{1}{s-a} \tag{a}$$

We know from Eqs. (8.11) and (8.15), respectively, that

$$L^{-1}\left[\frac{1}{s^2}\right]=t$$
$$L^{-1}\left[\frac{1}{s-a}\right]=e^{at} \tag{b}$$

Therefore, from Eq. (8.25) with $f(t) = t$ and $g(t) = e^{at}$, we have

$$L^{-1}\left[\frac{1}{s^2}\frac{1}{s-a}\right] = \int_0^t \xi e^{a(t-\xi)} d\xi = e^{at}\int_0^t \xi e^{-a\xi} d\xi = e^{a\xi}\left[-\frac{e^{-a\xi}(1+a\xi)}{a^2}\right]_0^t$$
$$= \frac{1}{a^2}\left[e^{at} - (1+at)\right] \tag{c}$$

This result could have been obtained by using partial fractions as follows. From Eq. (8.21)

$$F(s) = \frac{N(s)}{D(s)} = \frac{1}{s^2(s-a)} = \frac{B_1}{s} + \frac{B_2}{s^2} + \frac{A_1}{s-a} \tag{d}$$

From Eqs. (8.20) and (8.22),

$$A_1 = \lim_{s\to a}\left[(s-a)\frac{1}{s^2(s-a)}\right] = \frac{1}{a^2}$$

$$B_1 = \frac{d}{ds}\left(\frac{s^2}{s^2(s-a)}\right)_{s\to 0} = \left(\frac{-1}{(s-a)^2}\right)_{s\to 0} = -\frac{1}{a^2} \tag{e}$$

$$B_2 = \left(\frac{s^2}{s^2(s-a)}\right)_{s\to 0} = \left(\frac{1}{(s-a)}\right)_{s\to 0} = -\frac{1}{a}$$

Then, using Eq. (8.23) with $N = 3$, $m = 2$, $a_1 = a$, and $a_o = 0$, we obtain

$$L^{-1}\left[\frac{1}{s^2(s-a)}\right] = \frac{1}{a^2}\left[-1 - at + e^{at}\right]$$

This result is verified with Mathematica procedure M8.8.

Example 8.4

We shall determine the inverse transform of

$$\frac{1}{(s^2+k^2)^2}$$

We note from Eq. (8.12) that

$$L[\sin kt] = \frac{k}{s^2+k^2}$$

Therefore, we can use Eq. (8.27) to arrive at

$$L^{-1}\left[\left(s^2+k^2\right)^{-2}\right]=\frac{1}{k^2}\int_0^t \sin[k\xi]\sin[k(t-\xi)]d\xi$$

$$=\frac{1}{2k^3}\left[\sin kt - kt\cos kt\right]$$

This result is verified with Mathematica procedure M8.9.

Example 8.5

Consider a series RL_oC circuit with a voltage source. From Example 4.25, the governing equation for this circuit is

$$Ri(t)+L_o\frac{di(t)}{dt}+\frac{1}{C}\int_0^t i(\tau)d\tau = v(t) \text{ V} \tag{a}$$

where $i(t)$ is the current and $v(t)$ is the voltage from the source. Taking the Laplace transform of Eq. (a) yields

$$R\hat{i}(s)+L_o\left[s\hat{i}(s)-i(0)\right]+\frac{\hat{i}(s)}{Cs}=\hat{v}(s) \text{ Vs}$$

where $i(0)$ is the initial current, $\hat{i}(s)$ is the Laplace transform of $i(\tau)$ and $\hat{v}(s)$ is the Laplace transform of $v(\tau)$. Solving for $\hat{i}(s)$, we obtain

$$\hat{i}(s)=\frac{s\left(\hat{v}(s)+L_o i(0)\right)}{L_o s^2+Rs+1/C} \text{ C} \tag{b}$$

This result is verified with Mathematica procedure M8.10.

To find the inverse Laplace transform of Eq. (b), we convert it to a form that is compatible with the form appearing as Laplace transform pair 37 of Table 8.2 in Appendix 8.1. Thus, we write Eq. (b) as

$$\hat{i}(s)=\frac{si(0)}{s^2+2\zeta\omega_n s+\omega_n^2}+\frac{s\hat{v}(s)}{L_o\left(s^2+2\zeta\omega_n s+\omega_n^2\right)} \tag{c}$$

where

$$2\zeta=\frac{1}{CL_o}, \quad \omega_n=R\sqrt{\frac{C}{L_o}} \text{ s}^{-1}$$

From Laplace transform pair 37 of Table 8.2 in Appendix 8.1, we have that

$$\frac{s}{s^2+2\zeta\omega_n s+\omega_n^2}\Leftrightarrow-\frac{\omega_n}{\omega_d}e^{-\zeta\omega_n t}\sin(\omega_d t-\phi) \tag{d}$$

where

$$\phi = \cos^{-1}\zeta \quad 0 \le \zeta < 1$$

$$\omega_d = \omega_n\sqrt{1-\zeta^2}$$

Then, using Eq. (d) and Eq. (8.26), we obtain for the inverse transform of Eq. (c)

$$i(t) = -\frac{1}{\sqrt{1-\zeta^2}}\left\{ i(0)h(t) + \frac{1}{L_o}\int_0^t h(t-\xi)v(\xi)d\xi \right\} \text{ A}$$

where

$$h(t) = e^{-\zeta\omega_n t}\sin(\omega_d t - \phi)$$

8.1.5 Translation and Scaling

Consider the function

$$f(t) = g(t-a)u(t-a) \tag{8.28}$$

where $u(t)$ is the unit step function. The Laplace transform of this function is

$$F(s) = \int_0^\infty g(t-a)u(t-a)e^{-st}dt = \int_a^\infty g(t-a)e^{-st}dt$$

$$= \int_0^\infty g(\xi)e^{-s(\xi+a)}d\xi = e^{-sa}\int_0^\infty g(\xi)e^{-s\xi}d\xi = e^{-sa}G(s)$$

where we have set $\xi = t - a$. Therefore, we have the following Laplace transform pair

$$g(t-a)u(t-a) \Leftrightarrow e^{-sa}G(s) \tag{8.29}$$

We now consider the Laplace transform of $f(at)$; thus,

$$L[f(at)] = \int_0^\infty f(at)e^{-st}dt = \frac{1}{a}\int_0^\infty f(\xi)e^{-(s/a)\xi}d\xi = \frac{1}{a}F(s/a)$$

where we have set $\xi = at$. Therefore, we have the following Laplace transform pair

$$f(at) \Leftrightarrow \frac{1}{a}F(s/a) \tag{8.30}$$

If we let $c = 1/a$, then

$$f(t/c) \Leftrightarrow cF(cs) \tag{8.31}$$

We now use Eqs. (8.16) and (8.30) to find that

$$\frac{1}{a}f(t/a)e^{bt/a} \Leftrightarrow F(as-b) \tag{8.32}$$

Example 8.6

We shall determine the inverse transform of

$$\frac{\sinh(as)}{s^2\cosh(bs)} \tag{a}$$

We write Eq. (a) as

$$\frac{\sinh(as)}{s^2\cosh(bs)} = \frac{e^{as}-e^{-as}}{s^2\left(e^{bs}+e^{-bs}\right)} = \frac{e^{-s(b-a)}-e^{-s(b+a)}}{s^2\left(1+e^{-2bs}\right)}$$

and use Eq. (A.1) of Appendix A to obtain

$$\begin{aligned}\frac{\sinh(as)}{s^2\cosh(bs)} &= \frac{e^{-s(b-a)}-e^{-s(b+a)}}{s^2}\sum_{n=0}^{\infty}(-1)^n e^{-2nbs} \\ &= \sum_{n=0}^{\infty}\frac{(-1)^n}{s^2}e^{-\alpha_n s} - \sum_{n=0}^{\infty}\frac{(-1)^n}{s^2}e^{-\beta_n s}\end{aligned}$$

where

$$\alpha_n = (2n+1)b - a$$
$$\beta_n = (2n+1)b + a$$

To obtain the inverse Laplace transform, we use Eqs. (8.29) and (8.11) to arrive at

$$\begin{aligned}L^{-1}\left[\frac{\sinh(as)}{s^2\cosh(bs)}\right] &= \sum_{n=0}^{\infty}(-1)^n\left(t-\alpha_n\right)u\left(t-\alpha_n\right) \\ &\quad - \sum_{n=0}^{\infty}(-1)^n\left(t-\beta_n\right)u\left(t-\beta_n\right)\end{aligned} \tag{b}$$

Equation (b) for $a = b = 1$ is plotted in Figure 8.2, which was obtained with Mathematica procedure M8.11.

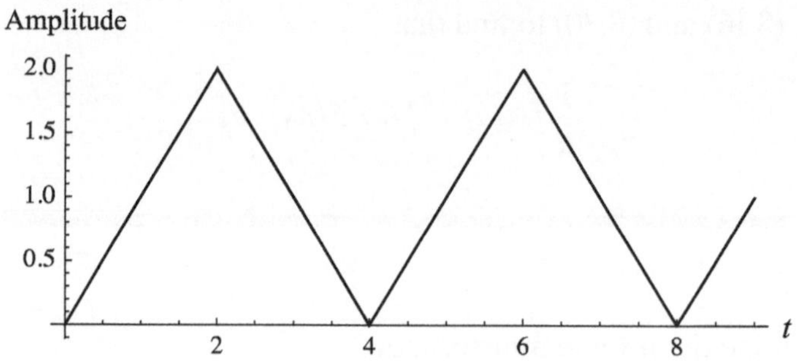

FIGURE 8.2 Plot of the inverse Laplace transform given by Eq. (b) of Example 8.6 for $a = b = 1$.

8.1.6 Periodic Functions

Consider the periodic function defined in Section 3.1 as

$$f(t+nT) = f(t) \quad 0 \le t \le T \quad n = 1,2,3\ldots \tag{8.33}$$

which can be expressed as

$$f(t) = \sum_{n=0}^{\infty} f(t-nT)\left[u(t-nT) - u(t-(n+1)T)\right] \quad t > 0 \tag{8.34}$$

where $u(t)$ is the unit step function. Then the Laplace transform of Eq. (8.34) is

$$F(s) = \sum_{n=0}^{\infty} \int_{nT}^{(n+1)T} f(t-nT)e^{-st}\,dt \tag{8.35}$$

If we let $\tau = t - nT$, then Eq. (8.35) becomes

$$\begin{aligned} F(s) &= \sum_{n=0}^{\infty} \int_0^T f(\tau)e^{-s(\tau+nT)}\,d\tau = \sum_{n=0}^{\infty} e^{-snT} \int_0^T f(\tau)e^{-s\tau}\,d\tau \\ &= \int_0^T f(\tau)e^{-s\tau}\,d\tau \sum_{n=0}^{\infty} e^{-snT} \end{aligned} \tag{8.36}$$

However, it is known that

$$\sum_{n=0}^{\infty} x^n = \frac{1}{1-x} \quad x < 1$$

Thus, with $x = e^{-sT}$, Eq. (8.36) becomes

$$F(s) = \left(1 - e^{-sT}\right)^{-1} \int_0^T f(\tau)e^{-s\tau}\,d\tau \tag{8.37}$$

Equation (8.37) is illustrated with the following example.

Example 8.7

Consider a periodic square wave, which is given by

$$\begin{aligned} f(t) &= 1 \qquad 0 \le t \le T/2 \\ f(t) &= -1 \qquad T/2 \le t \le T \end{aligned} \tag{a}$$

Substituting Eq. (a) into the integral portion of Eq. (8.37), we obtain

$$\int_0^T f(\tau)e^{-s\tau}d\tau = \int_0^{T/2} e^{-s\tau}d\tau - \int_{T/2}^{T} e^{-s\tau}d\tau = -\frac{e^{-s\tau}}{s}\bigg|_0^{T/2} + \frac{e^{-s\tau}}{s}\bigg|_{T/2}^{T}$$

$$= \frac{1}{s}\left(1 - 2e^{-sT/2} + e^{-sT}\right) = \frac{1}{s}\left(1 - e^{-sT/2}\right)^2$$

and Eq. (8.37) yields

$$F(s) = \frac{1}{s}\frac{\left(1-e^{-sT/2}\right)^2}{\left(1-e^{-sT}\right)} = \frac{1}{s}\frac{\left(1-e^{-sT/2}\right)^2}{\left(1-e^{-sT/2}\right)\left(1+e^{-sT/2}\right)}$$

$$= \frac{1}{s}\frac{\left(1-e^{-sT/2}\right)}{\left(1+e^{-sT/2}\right)} = \frac{1}{s}\frac{e^{sT/4}\left(1-e^{-sT/2}\right)}{e^{sT/4}\left(1+e^{-sT/2}\right)} = \frac{1}{s}\frac{\left(e^{sT/4}-e^{-sT/4}\right)}{\left(e^{sT/4}+e^{-sT/4}\right)}$$

$$= \frac{1}{s}\frac{\sinh(sT/4)}{\cosh(sT/4)} = \frac{1}{s}\tanh(sT/4)$$

Therefore, we have the following Laplace transform pair

$$\left.\begin{aligned} f(t) &= 1 \qquad nT \le t \le (n+1)T/2 \\ f(t) &= -1 \qquad (n+1)T/2 \le t \le (n+1)T \end{aligned}\right\} \Leftrightarrow \frac{1}{s}\tanh(sT/4) \tag{b}$$

8.1.7 Inversion Integral Revisited

The inversion integral is given by Eq. (8.4) as

$$g(t) = \frac{1}{2\pi j}\int_{\gamma-j\infty}^{\gamma+j\infty} e^{st}G(s)ds \tag{8.38}$$

where $G(s)$ is the Laplace transform of $g(t)$. Bromwich* showed that one could replace Eq. (8.38) with the contour integral

$$g(t) = \frac{1}{2\pi j}\int_{\gamma-j\infty}^{\gamma+j\infty} e^{zt}G(z)dz \tag{8.39}$$

* Bromwich TJ I'A (1916) Normal Coordinates in Dynamical Systems. *Proceedings of the London Mathematical Society* Ser 2, 15: 401–448.

where z is a complex variable. The integration of Eq. (8.39) is performed by using residues. In this case, we always choose γ to be to the right of the real part of any poles of $G(z)$.

We now illustrate the use of Eq. (8.39) with the following examples. In these examples, we shall use the results of Cauchy's integral formulas given, respectively, by Eqs. (2.67) to (2.69) as

$$\int_C \frac{f(z)}{(z-z_o)^n}dz \to \int_C g(z)dz = \frac{2\pi j}{(n-1)!}\lim_{z\to z_o}\left[\frac{d^{n-1}}{dz^{n-1}}\left\{(z-z_o)^n g(z)\right\}\right]$$

$$\int_C \frac{f(z)}{z-z_o}dz \to \int_C g(z)dz = 2\pi j \lim_{z\to z_o}\left[(z-z_o)g(z)\right]$$

$$\int_C \frac{p(z)}{q(z)}dz \to 2\pi j\frac{p(z_o)}{q'(z_o)} \quad p(z_o)\neq 0 \quad z_o \text{ is a simple pole}$$

Example 8.8

Let

$$G(s) = \frac{1}{s\sinh as} \tag{a}$$

Substituting Eq. (a) into Eq. (8.39), we have that

$$g(t) = \frac{1}{2\pi j}\int_{\gamma-j\infty}^{\gamma+j\infty}\frac{e^{zt}}{z\sinh az}dz \tag{b}$$

We have shown in Eq. (2.37) that the zeros of $\sinh(az)$ are $z_n = \pm jn\pi/a$, $n = 0, 1, \ldots,$ which are simple poles. From Eq. (A.5) of Appendix A, we have that

$$z\sinh z \approx z^2\left(1+\frac{z^2}{3!}+\frac{z^4}{5!}+\cdots\right)$$

Therefore, in Eq. (b), we have a second-order pole at $z = 0$. Then, using the Cauchy integral formulas given above, the evaluation of Eq. (b) is

$$g(t) = \frac{1}{2\pi j}\int_{\gamma-j\infty}^{\gamma+j\infty}\frac{e^{zt}}{z\sinh az}dz$$

$$= \frac{2\pi j}{2\pi j}\left\{\lim_{z\to 0}\frac{d}{dz}\left(\frac{z^2e^{zt}}{z\sinh az}\right)+\sum_{k=1}^{\infty}e^{z_kt}\left(z_k\frac{d}{dz}\left[\sinh az\right]_{z=z_k}\right)^{-1}\right\} \tag{c}$$

$$= \lim_{z\to 0}\left(\frac{1}{\sinh az}+\frac{zt}{\sinh az}-\frac{az\cosh(az)}{\sinh^2 az}\right)e^{zt}+\sum_{k=1}^{\infty}\frac{e^{z_kt}}{az_k\cosh az_k}$$

Again using Eq. (A.5) of Appendix A, we see that for $z << 1$, $\sinh(z) \sim z$ and $\cosh(z) \sim 1$. Then

$$\lim_{z\to 0}\left(\frac{1}{\sinh az}+\frac{zt}{\sinh az}-\frac{az\cosh(az)}{\sinh^2 az}\right)e^{zt}=\lim_{z\to 0}\left(\frac{1}{az}+\frac{zt}{az}-\frac{az}{(az)^2}\right)e^{zt}=\frac{t}{a}$$

Then Eq. (c) becomes

$$g(t)=\frac{t}{a}+\sum_{k=1}^{\infty}\frac{e^{jk\pi t/a}}{(jk\pi)\cosh(jk\pi)}+\sum_{k=1}^{\infty}\frac{e^{-jk\pi t/a}}{(-jk\pi)\cosh(-jk\pi)}$$

$$=\frac{t}{a}+\sum_{k=1}^{\infty}\frac{e^{jk\pi t/a}}{(jk\pi)\cos(k\pi)}-\sum_{k=1}^{\infty}\frac{e^{-jk\pi t/a}}{(jk\pi)\cos(k\pi)}$$

$$=\frac{t}{a}+\sum_{k=1}^{\infty}\frac{e^{jk\pi t/a}}{(jk\pi)(-1)^k}-\sum_{k=1}^{\infty}\frac{e^{-jk\pi t/a}}{(jk\pi)(-1)^k}=\frac{t}{a}+\sum_{k=1}^{\infty}\frac{(-1)^k}{jk\pi}\left[e^{jk\pi t/a}-e^{-jk\pi t/a}\right]$$

$$=\frac{t}{a}+\frac{2}{\pi}\sum_{k=1}^{\infty}\frac{(-1)^k}{k}\sin(k\pi t/a)$$

and we have used the fact that $\cosh(\pm jk\pi)=\cos(k\pi)=(-1)^k$. Recall Eq. (2.38).

This result is verified with Mathematica procedure M8.12.

Example 8.9

We revisit Example 8.6 and again determine the inverse Laplace transform of

$$\frac{\sinh(as)}{s^2\cosh(bs)} \tag{a}$$

This time, however, we shall use Eq. (8.39). Thus,

$$g(t)=\frac{1}{2\pi j}\int_{\gamma-j\infty}^{\gamma+j\infty}\frac{e^{zt}\sinh(az)}{z^2\cosh(bz)}dz \tag{b}$$

We see from Eq. (b) that we have a second order pole at $z = 0$ and, from Eq. (2.38), a multiplicity of simple poles at $\cosh(bz) = 0$, that is, when $z = \pm j\pi(2n-1)/(2b)$. Then, Eq. (b) gives

$$g(t)=\frac{1}{2\pi j}\int_{\gamma-j\infty}^{\gamma+j\infty}\frac{e^{zt}\sinh(az)}{z^2\cosh(bz)}dz$$

$$=\lim_{z\to 0}\frac{d}{dz}\left(\frac{\sinh(az)e^{zt}}{\cosh(bz)}\right)+\sum_{k=1}^{\infty}e^{z_k t}\sinh(az_k)\left(z_k^2\frac{d}{dz}\left[\cosh(bz)\right]_{z=z_k}\right)^{-1}$$

$$+\sum_{k=1}^{\infty}e^{-z_k t}\sinh(az_k)\left(z_k^2\frac{d}{dz}\left[\cosh(bz)\right]_{z=-z_k}\right)^{-1} \tag{c}$$

$$=a+\sum_{k=1}^{\infty}\frac{e^{z_k t}\sinh(az_k)}{z_k^2 b\sinh(bz_k)}-\sum_{k=1}^{\infty}\frac{e^{-z_k t}\sinh(az_k)}{z_k^2 b\sinh(bz_k)}=a+\sum_{k=1}^{\infty}\frac{\sinh(az_k)\left(e^{z_k t}-e^{-z_k t}\right)}{z_k^2 b\sinh(bz_k)}$$

$$=a-\frac{8b}{\pi^2}\sum_{k=1}^{\infty}\frac{\sin\left(a\pi(2n-1)/2\right)\cos\left(\pi t(2n-1)/(2b)\right)}{(2n-1)^2\sin\left(\pi(2n-1)/2\right)}$$

$$=a-\frac{8b}{\pi^2}\sum_{k=1}^{\infty}\frac{(-1)^{n-1}}{(2n-1)^2}\sin\left(a\pi(2n-1)/2\right)\cos\left(\pi t(2n-1)/(2b)\right)$$

where we have used Eq. (2.37) and with the help of Mathematica the fact that

$$\lim_{z\to 0}\frac{d}{dz}\left(\frac{\sinh(az)e^{zt}}{\cosh(bz)}\right)=\lim_{z\to 0}\left\{e^{zt}\operatorname{sech}(bz)\left(a\cosh(az)+\sinh(az)\left(t-b\tanh(bz)\right)\right)\right\}$$

$$=\lim_{z\to 0}\left\{a+2atz+\frac{1}{2}z^2\left(a^3-3ab^2+3at^2\right)+\cdots\right\}=a$$

A plot of the final expression of Eq. (c) with $a = b = 1$ results in Figure 8.2. This result is verified with Mathematica procedure M8.13.

Example 8.10

Let

$$G(s)=\frac{a\cosh\eta\sqrt{s}}{s^{3/2}\sinh b\sqrt{s}} \tag{a}$$

We note from Eq. (A.5) of Appendix A that

$$\cosh\eta\sqrt{s}=1+\frac{s\eta^2}{2!}+\frac{s^2\eta^4}{4!}+\cdots$$

$$s^{3/2}\sinh b\sqrt{s}=bs^2\left(1+\frac{sb^2}{3!}+\frac{s^2b^4}{5!}+\cdots\right) \tag{b}$$

and, therefore, the pole at $s = 0$ is a second-order pole. Also, the poles of $\sinh b\sqrt{s}$ are $\sqrt{s_n}=\pm jn\pi/b$ or $s_n=-n^2\pi^2/b^2$.

Substituting Eq. (a) into Eq. (8.39) gives

$$g(t) = \frac{1}{2\pi j} \int_{\gamma - j\infty}^{\gamma + j\infty} \frac{ae^{zt} \cosh \eta\sqrt{z}}{z^{3/2} \sinh b\sqrt{z}} dz \tag{c}$$

The residues of Eq. (c) yield

$$g(t) = a\frac{2\pi j}{2\pi j}\left\{ \begin{aligned} &\lim_{z\to 0} \frac{d}{dz}\left[\frac{z^2 e^{zt} \cosh \eta\sqrt{z}}{z^{3/2} \sinh b\sqrt{z}}\right] \\ &+ \sum_{n=1}^{\infty} \lim_{z\to z_n}\left(\frac{e^{zt}\cosh \eta\sqrt{z}}{z^{3/2}}\left(\frac{d}{dz}\sinh b\sqrt{z}\right)^{-1}\right) \end{aligned}\right\} \tag{d}$$

$$= a\left\{ \lim_{z\to 0} \frac{d}{dz}\left[\frac{z^2 e^{zt} \cosh \eta\sqrt{z}}{z^{3/2} \sinh b\sqrt{z}}\right] + \sum_{k=1}^{\infty} \lim_{z\to z_n}\left(\frac{2\sqrt{z}e^{zt}\cosh \eta\sqrt{z}}{z^{3/2} b\cosh b\sqrt{z}}\right)\right\}$$

However,

$$\lim_{z\to 0} \frac{d}{dz}\left[\frac{z^2 e^{zt} \cosh \eta\sqrt{z}}{z^{3/2} \sinh b\sqrt{z}}\right] = \lim_{z\to 0}\left[e^{tz}\operatorname{csch} b\sqrt{z}\left(\cosh\eta\sqrt{z}\left(1+2tz - b\sqrt{z}\coth b\sqrt{z}\right) \right.\right.$$
$$\left.\left. + \eta\sqrt{z}\sinh\eta\sqrt{z}\right)/2\sqrt{z}\right] \tag{e}$$

From Eq. (A.5) of Appendix A, we find that for $z << 1$ Eq. (e) becomes

$$\lim_{z\to 0} \frac{d}{dz}\left[\frac{z^2 e^{zt} \cosh \eta\sqrt{z}}{z^{3/2} \sinh b\sqrt{z}}\right] = \lim_{z\to 0}\left[\frac{e^{tz}}{2bz}\left(1+2tz - b\sqrt{z}\left\{\frac{1}{b\sqrt{z}} + \frac{b\sqrt{z}}{3}\right\} + \eta^2 z\right)\right]$$
$$= \frac{1}{b}\left(t + \frac{1}{2}\eta^2 - \frac{b^2}{6}\right) \tag{f}$$

We have used Mathematica procedure M8.14 to perform the operations in Eqs. (e) and (f). Furthermore,

$$\sum_{k=1}^{\infty} \lim_{z\to z_n}\left(\frac{2\sqrt{z}e^{zt}\cosh \eta\sqrt{z}}{z^{3/2} b\cosh b\sqrt{z}}\right)$$
$$= -\frac{2ab}{\pi^2}\sum_{k=1}^{\infty}\frac{e^{-k^2\pi^2 t/b^2}}{k^2}\left[\frac{\cosh(jk\pi\eta/b)}{\cosh(jk\pi)} + \frac{\cosh(-jk\pi\eta/b)}{\cosh(-jk\pi)}\right] \tag{g}$$
$$= \frac{4ab}{\pi^2}\sum_{k=1}^{\infty}\frac{(-1)^{k+1}}{k^2}e^{-k^2\pi^2 t/b^2}\cos(k\pi\eta/b)$$

where we have used the fact that $\cosh(\pm jk\pi) = \cos(k\pi) = (-1)^k$. Using Eqs. (f) and (g) in Eq. (d) yields

$$g(t) = \frac{a}{b}\left\{t + \frac{1}{2}\eta^2 - \frac{b^2}{6} + \frac{4b^2}{\pi^2}\sum_{k=1}^{\infty}\frac{(-1)^{k+1}}{k^2}e^{-k^2\pi^2 t/b^2}\cos(k\pi\eta/b)\right\}$$

Example 8.11

Let

$$G(s) = \frac{2I_1(\lambda\sqrt{s})}{\lambda s^{3/2}I_0(\lambda\sqrt{s})} \tag{a}$$

where $I_m(x)$ is the modified Bessel function of the first kind of order m. Placing Eq. (a) in Eq. (8.39), we arrive at

$$g(t) = \frac{1}{2\pi j}\int_{\gamma-j\infty}^{\gamma+j\infty}\frac{2I_1(\lambda\sqrt{z})e^{zt}}{\lambda z^{3/2}I_0(\lambda\sqrt{z})}dz \tag{b}$$

We note from Eq. (5.124) that

$$I_0(z) = 1 + \frac{z^2}{4} + \frac{z^4}{64} + \cdots$$

$$I_1(z) = \frac{z}{2} + \frac{z^3}{16} + \frac{z^5}{384} + \cdots = \frac{z}{2}\left[1 + \frac{z^2}{8} + \frac{z^4}{192} + \right]$$

Therefore,

$$\begin{aligned}\frac{2I_1(\lambda\sqrt{z})e^{zt}}{\lambda z^{3/2}I_0(\lambda\sqrt{z})} &= \frac{2}{\lambda z^{3/2}}\frac{\left(\lambda z^{1/2}/2\right)\left(1+\lambda^2 z/8\cdots\right)}{1+\lambda^2 z/4\cdots}(1+zt\ldots)\\ &= \frac{\left(1+\lambda^2 z/8\cdots\right)}{z\left(1+\lambda^2 z/4\cdots\right)}(1+zt\ldots) \sim \frac{1}{z} \quad z \ll 1\end{aligned} \tag{c}$$

and we see that we have a simple pole at $z = 0$.

We also note that we have simple poles when

$$I_0(\lambda\sqrt{z}) = 0 \tag{d}$$

Using Eq. (5.123), that is,

$$I_m(\pm x) = (j)^{-m}J_m(\pm jx)$$

we find that

$$I_0(x) = J_0(jx)$$

or

$$I_0(-jy) = J_0(y)$$

where $y = jx$ ($x = -jy$). Therefore, Eq. (d) has a solution when

$$J_0(\alpha_n) = 0 \quad n = 1, 2, \ldots$$

Since $y = \alpha_n$ and $x = \lambda\sqrt{z}$, it is seen that $j\lambda\sqrt{z_k} = \alpha_k$ and, therefore, $z_k = -\alpha_k^2 / \lambda^2$. Then, Eq. (b) becomes

$$\begin{aligned} g(t) &= \frac{1}{2\pi j} \int_{\gamma - j\infty}^{\gamma + j\infty} \frac{2I_1(\lambda\sqrt{z})e^{zt}}{\lambda z^{3/2} I_0(\lambda\sqrt{z})} dz \\ &= \lim_{z \to 0} \left[\frac{2zI_1(\lambda\sqrt{z})e^{zt}}{\lambda z^{3/2} I_0(\lambda\sqrt{z})} \right] \\ &\quad + \frac{2}{\lambda} \sum_{k=1}^{\infty} \lim_{z \to z_k} \left[\frac{I_1(\lambda\sqrt{z})e^{zt}}{z^{3/2}} \left(\frac{d}{dz} I_0(\lambda\sqrt{z}) \right)^{-1} \right] \end{aligned} \tag{e}$$

But, from Eq. (c), we found that for $z << 1$

$$\frac{2I_1(\lambda\sqrt{z})e^{zt}}{\lambda z^{3/2} I_0(\lambda\sqrt{z})} \approx \frac{1}{z}$$

Then, Eq. (e) becomes

$$\begin{aligned} g(t) &= \lim_{z \to 0} \frac{z}{z} + \frac{2}{\lambda} \sum_{k=1}^{\infty} \lim_{z \to z_k} \left[\frac{I_1(\lambda\sqrt{z})e^{zt}}{z^{3/2}} \left(\frac{\lambda}{2z^{1/2}} I_1(\lambda\sqrt{z}) \right)^{-1} \right] \\ &= 1 + \frac{4}{\lambda^2} \sum_{k=1}^{\infty} \frac{e^{-\alpha_k^2 t/\lambda^2}}{-\alpha_k^2 / \lambda^2} = 1 - 4 \sum_{k=1}^{\infty} \frac{e^{-\alpha_k^2 t/\lambda^2}}{\alpha_k^2} \end{aligned} \tag{f}$$

In obtaining Eq. (f), we used the following relation from Table 5.1

$$\frac{d}{dz}\left[I_0(\lambda\sqrt{z}) \right] = \frac{\lambda}{2\sqrt{z}} I_1\left(\lambda\sqrt{z}\right)$$

8.2 APPLICATIONS OF THE LAPLACE TRANSFORM TO ORDINARY AND PARTIAL DIFFERENTIAL EQUATIONS

In this section, we shall use the Laplace transform method to obtain solutions to the following boundary value problems.*

Example 8.12	Displacement of a beam with an in-span spring to a uniform static load
Example 8.13	Response of a two degree-of-freedom system to external forces and initial conditions
Example 8.14	One-dimensional heat conduction in rectangular coordinates with nonhomogeneous boundary conditions and an initial condition
Example 8.15	Diffusion in polar coordinates with time-dependent boundary conditions
Example 8.16	Response of a semi-infinite beam to a time-dependent boundary condition
Example 8.17	Traveling point load on a semi-infinite string
Example 8.18	Longitudinal vibrations of a bar with an attachment at a free end
Example 8.19	Response of a beam with an in-span mass to a dynamic load
Example 8.20	Response of a Timoshenko beam to a dynamic load

The ordinary and partial differential equations, boundary conditions, and initial conditions that appear in each of these examples are summarized in Table 8.1.

Example 8.12: Displacement of a beam with an in-span spring to a uniform static load

Consider a thin elastic beam that has a length L (m), cross-sectional area moment of inertia I (m^4), and Young's modulus E (N/m^2). The beam is restrained by an in-span elastic spring with spring constant k_s (N/m) that is located at L_s (m), $0 < L_s < L$, and has a uniform load of magnitude F_o (N/m) applied over the length of the beam. If the transverse displacement of the beam is $w = w(x)$ (m) and the distance along the beam is x (m), then the governing equation describing the displacement of the beam is

$$EI\frac{d^4w}{dx^4} + k_s\delta(x - L_s)w = F_o \quad \text{N/m} \tag{a}$$

Using the procedure similar to that given in Example 7.26, Eq. (a) is placed in non-dimensional form as

$$\frac{d^4y}{d\eta^4} + K_s\delta(\eta - \eta_s)y = f_o \quad 0 < \eta < 1 \tag{b}$$

* For a very large compendium of solved and partially solved boundary value problems using the Laplace transform and a very large list of journal citations, see Duffy DG (2004) *Transform Methods for Solving Partial Differential Equations*. Chapman & Hall/CRC Press, Chapter 2.

TABLE 8.1
Summary of Examples Using the Laplace Transform as Part of the Solution Method

Example	Governing Ordinary/Partial Differential Equation(s)	Boundary Conditions	Initial Conditions	Variable[a]
8.12	$\frac{d^4y}{d\eta^4}+K_s\delta(\eta-\eta_s)y=f_o$	$y(0)=y'(0)=0$ $y(1)=y'(1)=0$	N/A	η
8.13	$\frac{d^4y}{d\eta^4}+K_s\delta(\eta-\eta_s)y=f_o$	N/A	Unspecified	τ
8.14	$\frac{d\Theta}{d\tau}=\frac{\partial^2\Theta}{\partial\eta^2}$	$\frac{\partial\Theta(0,\tau)}{\partial\eta}=a$ $\Theta(\infty,\tau)=\text{finite}$	$\Theta(\eta,0)=1$	τ
8.15	$\frac{\partial^2\Psi}{\partial r^2}+\frac{\partial\Psi}{r\partial r}+\frac{\partial^2\Psi}{r^2\partial\theta^2}=\frac{1}{\alpha^2}\frac{\partial\Psi}{\partial t}$	$\Psi(a,\theta,t)=\cos(\theta)u(t)$ $\Psi(0,\theta,t)=\text{finite}$	$\Psi(r,\theta,0)=0$	t
8.16	$\frac{d^4w}{d\eta^4}+\frac{d^2w}{d\tau^2}=0$	$\frac{d^2w(0,\tau)}{d\eta^2}=0$ $w(0,\tau)=W_ou(\tau)$ $w(\infty,\tau)=\text{finite}$	$w(\eta,0)=0$ $\dot{w}(\eta,0)=0$	τ
8.17	$\frac{\partial^2y}{\partial\tau^2}=\frac{\partial^2y}{\partial\eta^2}+\hat{P}\delta(\tau-\alpha\eta)$	$y(0,\tau)=0$ $y(\infty,\tau)=\text{finite}$	$y(\eta,0)=0$ $\dot{y}(\eta,0)=0$	τ
8.18	$\frac{\partial^2y}{\partial\eta^2}=\frac{\partial^2y}{\partial\tau^2}$	$y(0,\tau)=0$ $m\frac{\partial^2y(1,\tau)}{\partial\tau^2}=-\frac{\partial y(1,\tau)}{\partial\eta}$	$y(\eta,0)=0$ $\dot{y}(\eta,0)=v$	τ
8.19	$\frac{\partial^4y}{\partial\eta^4}+\left(1+m_o\delta(\eta-\eta_o)\right)\frac{\partial^2y}{\partial\tau^2}=f(\eta,\tau)$	$y(0,\tau)=y'(0,\tau)=0$ $y(1,\tau)=y'(1,\tau)=0$	$y(\eta,0)=0$ $\dot{y}(\eta,0)=0$	η
8.20	$\frac{\partial}{\partial\eta}\left(\frac{\partial y}{\partial\eta}-\psi\right)-\gamma_{bs}R_o^2\frac{\partial^2y}{\partial\tau^2}=-\gamma_{bs}R_o^2\hat{f}_a(\eta,\tau)$ $\frac{\partial y}{\partial\eta}-\psi+\gamma_{bs}R_o^2\frac{\partial^2\psi}{\partial\eta^2}-\gamma_{bs}R_o^4\frac{\partial^2\psi}{\partial\tau^2}=0$	$y(0,\tau)=\psi(0,\tau)=0$ $y(1,\tau)=\psi(1,\tau)=0$	$y(\eta,0)=0$ $\dot{y}(\eta,0)=0$ $\psi(\eta,0)=0$ $\dot{\psi}(\eta,0)=0$	η

Note: The over dot indicates the derivative with respect to τ.
[a] Laplace transform taken with respect to this variable.

where $y = w/L$ and

$$\eta = x/L, \qquad \eta_s = L_s/L$$
$$K_s = \frac{k_s L^3}{EI}, \qquad f_o = \frac{F_o L^3}{EI} \tag{c}$$

The boundary conditions will be specified subsequently.

To obtain the solution to Eq. (b), we take the Laplace transform with respect to η and use Laplace transform pairs 5, 10, and 13 of Table 8.2 in Appendix 8.1 to obtain

$$s^4\bar{Y}(s) = s^3 y(0) + s^2 y'(0) + s y''(0) + y'''(0) - K_s y(\eta_s) e^{-s\eta_s} + \frac{f_o}{s}$$

or

$$\bar{Y}(s) = \frac{1}{s} y(0) + \frac{1}{s^2} y'(0) + \frac{1}{s^3} y''(0) + \frac{1}{s^4} y'''(0) - K_s y(\eta_s) \frac{e^{-s\eta_s}}{s^4} + \frac{f_o}{s^5} \tag{d}$$

where s is the non-dimensional Laplace transform parameter, $\bar{Y}(s)$ is the Laplace transform of $y(\eta)$, and the prime (′) denotes the derivative with respect to η.

Using Laplace transform pairs 6 and 16 of Table 8.2 of Appendix 8.1, the inverse Laplace transform of Eq. (d) is

$$\begin{aligned} y(\eta) &= y(0) + \eta y'(0) + \frac{\eta^2}{2} y''(0) + \frac{\eta^3}{6} y'''(0) \\ &\quad - K_s y(\eta_s) \frac{(\eta - \eta_s)^3}{6} u(\eta - \eta_s) + \frac{\eta^4}{24} f_o \end{aligned} \tag{e}$$

where $u(\eta)$ is the unit step function.

We shall assume that the beam is clamped at both ends; thus, the boundary conditions are

$$\begin{aligned} y(0) &= y'(0) = 0 \\ y(1) &= y'(1) = 0 \end{aligned} \tag{f}$$

Then, using the first set of boundary conditions of Eq. (f), Eq. (e) simplifies to

$$y(\eta) = \frac{\eta^2}{2} y''(0) + \frac{\eta^3}{6} y'''(0) - K_s y(\eta_s) \frac{(\eta - \eta_s)^3}{6} u(\eta - \eta_s) + \frac{\eta^4}{24} f_o \tag{g}$$

Upon substituting Eq. (g) into the second set of boundary conditions given by Eq. (f), we find that

$$\begin{aligned} \frac{1}{2} y''(0) + \frac{1}{6} y'''(0) &= K_s y(\eta_s) \frac{1}{6} (1 - \eta_s)^3 + \frac{1}{24} f_o \\ y''(0) + \frac{1}{2} y'''(0) &= 3K_s y(\eta_s) \frac{1}{6} (1 - \eta_s)^2 + \frac{1}{6} f_o \end{aligned} \tag{h}$$

Solving Eq. (h) for $y''(0)$ and $y'''(0)$, we obtain

$$\begin{aligned} y''(0) &= -K_s y(\eta_s) a_s + \frac{1}{12} f_o \\ y'''(0) &= K_s y(\eta_s) b_s - \frac{1}{2} f_o \end{aligned} \tag{i}$$

where

$$\begin{aligned} a_s &= \eta_s (1-\eta_s)^2 \\ b_s &= (1-\eta_s)^2 (1+2\eta_s) \end{aligned} \tag{j}$$

Then Eq. (g) can be written as

$$y(\eta) = K_s y(\eta_s)\left(-a_s\frac{\eta^2}{2} + \frac{\eta^3}{6} b_s - \frac{(\eta-\eta_s)^3}{6} u(\eta-\eta_s)\right) + \frac{f_o}{24}\eta^2(\eta-1)^2 \tag{k}$$

To determine $y(\eta_s)$, we note that Eq. (k) is valid at $\eta = \eta_s$. Therefore, from Eq. (k), we find that

$$y(\eta_s) = \frac{f_o\eta_s^2(\eta_s-1)^2}{24}\left[1 - K_s\left(-a_s\frac{\eta_s^2}{2} + \frac{\eta_s^3}{6} b_s\right)\right]^{-1} \tag{l}$$

Then substitution of Eq. (l) into Eq. (k) is the solution to Eq. (a) subject to the boundary conditions given by Eq. (f).

It is noted that when $K_s = 0$, Eq. (k) reduces to

$$y(\eta) = \frac{f_o}{24}\eta^2(\eta-1)^2 \tag{m}$$

and when $K_s \to \infty$, Eq. (k) reduces to

$$y(\eta) = \frac{f_o\eta_s^2(\eta_s-1)^2}{24}\left(-a_s\frac{\eta_s^2}{2} + \frac{\eta_s^3}{6} b_s\right)^{-1}\left(a_s\frac{\eta^2}{2} - \frac{\eta^3}{6} b_s + \frac{(\eta-\eta_s)^3}{6} u(\eta-\eta_s)\right) + \frac{f_o}{24}\eta^2(\eta-1)^2 \tag{n}$$

which is the case of a beam with a rigid in-span support. Equation (k) is shown in Figure 8.3 for $\eta_s = 0.35$ and various values of K_s. This figure was obtained with Mathematica procedure M8.15.

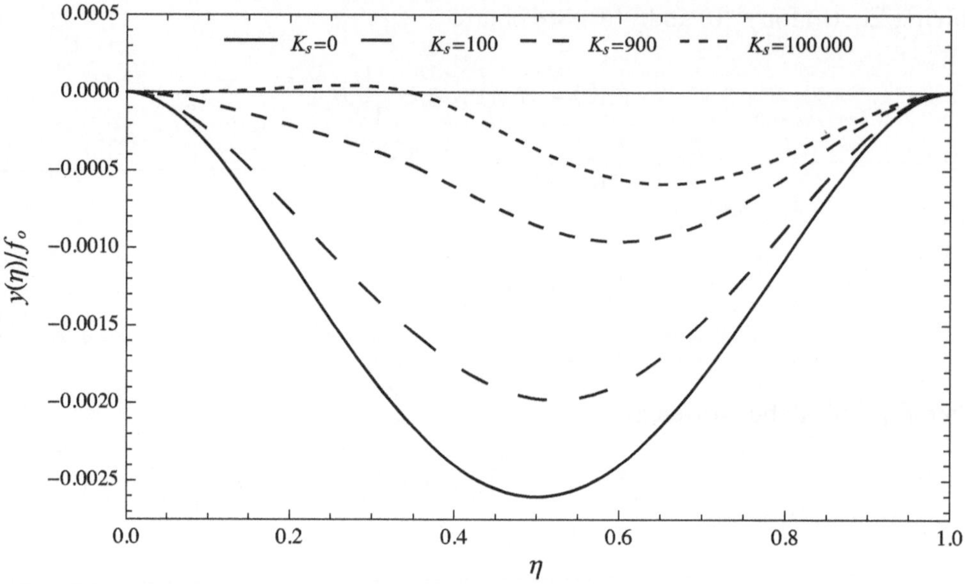

FIGURE 8.3 Displacement response of a beam clamped at both ends with an in-span elastic spring at $\eta_s = 0.35$ and for various values of K_s when the beam is subjected to a uniform load.

Example 8.13: Response of a two degree-of-freedom system to external forces and initial conditions

Consider the following pair of coupled ordinary differential equations that describe the forced motion of a two degree-of-freedom system given by Eq. (d) of Example 4.27

$$\frac{d^2x_1}{d\tau^2}+\left(2\zeta_1+2\zeta_2 m_r\omega_r\right)\frac{dx_1}{d\tau}+\left(1+m_r\omega_r^2\right)x_1$$

$$-2\zeta_2 m_r\omega_r\frac{dx_2}{d\tau}-m_r\omega_r^2 x_2=\frac{f_1(\tau)}{k_1}\quad \text{m} \tag{a}$$

$$\frac{d^2x_2}{d\tau^2}+2\zeta_2\omega_r\frac{dx_2}{d\tau}+\omega_r^2 x_2-2\zeta_2\omega_r\frac{dx_1}{d\tau}-\omega_r^2 x_1=\frac{f_2(\tau)}{k_1 m_r}\quad \text{m}$$

where $f_k(\tau)$, $k = 1, 2$, are known functions. Upon taking the Laplace transform of these equations with respect to τ, we obtain

$$\begin{aligned} A(s)X_1(s)-B(s)X_2(s)&=K_1(s)\\ -C(s)X_1(s)+E(s)X_2(s)&=K_2(s)\end{aligned} \tag{b}$$

where

$$A(s) = s^2 + 2(\zeta_1 + \zeta_2 m_r \omega_r)s + 1 + m_r \omega_r^2$$

$$B(s) = 2\zeta_2 m_r \omega_r s + m_r \omega_r^2$$

$$C(s) = 2\zeta_2 \omega_r s + \omega_r^2$$

$$E(s) = s^2 + 2\zeta_2 \omega_r s + \omega_r^2$$

and

$$K_1(s) = \frac{F_1(s)}{k_1} + \dot{x}_1(0) + \left[s + 2\zeta_1 + 2\zeta_2 m_r \omega_r\right] x_1(0) - 2\zeta_2 m_r \omega_r x_2(0)$$

$$K_2(s) = \frac{F_2(s)}{k_1 m_r} + \dot{x}_2(0) + \left[s + 2\zeta_2 \omega_r\right] x_2(0) - 2\zeta_2 \omega_r x_1(0)$$

In this notation, $\dot{x}_k(0)$ and $x_k(0)$, $k = 1,2$, are the initial conditions, and $F_j(s)$, $j = 1,2$, is the Laplace transform of $f_j(\tau)$.

Solving Eq. (b) for $X_1(s)$ and $X_2(s)$ yields

$$X_1(s) = \frac{K_1(s)E(s)}{D_1(s)} + \frac{K_2(s)B(s)}{D_1(s)}$$
$$X_2(s) = \frac{K_1(s)C(s)}{D_1(s)} + \frac{K_2(s)A(s)}{D_1(s)} \qquad \text{(c)}$$

where

$$D_1(s) = s^4 + \left[2\zeta_1 + 2\zeta_2 \omega_r m_r + 2\zeta_2 \omega_r\right] s^3 + \left[1 + m_r \omega_r^2 + \omega_r^2 + 4\zeta_1 \zeta_2 \omega_r\right] s^2 + \left[2\zeta_2 \omega_r + 2\zeta_1 \omega_r^2\right] s + \omega_r^2$$

Thus, once $f_k(\tau)$, $k = 1, 2$, and the initial conditions are specified and the Laplace transforms of $f_k(\tau)$ are obtained, the inverse Laplace transform of $X_k(s)$ gives $x_k(\tau)$. Notice that we have one solution for all combinations of initial conditions and forcing functions. When the initial conditions are zero,

$$K_1(s) = \frac{F_1(s)}{k_1} \quad K_2(s) = \frac{F_2(s)}{k_1 m_r} \qquad \text{(d)}$$

A general solution for the inverse Laplace transform of Eq. (c) cannot be obtained because D_1 is a quartic. However, if we select numerical values for m_r, ω_r, ζ_1, and ζ_2, then the inverse Laplace transform can be obtained for that set of values. We shall assume that an impulse force of magnitude F_o (N) is applied to mass m_2, no force is applied to m_1, and all the initial conditions are zero. In this case, $F_1(s) = 0$ and from

Laplace transform pair 13 of Table 8.2 in Appendix 8.1 $F_2(s) = F_o$. Therefore, since all the initial conditions are zero, we can use Eq. (d), which becomes

$$K_1(s) = 0 \quad K_2(s) = \frac{F_o}{k_1 m_r}$$

and Eq. (c) simplifies to

$$X_1(s) = \frac{F_o B(s)}{k_1 m_r D_1(s)}$$
$$X_2(s) = \frac{F_o A(s)}{k_1 m_r D_1(s)} \tag{e}$$

The inverse Laplace transform of Eq. (e) is obtained with Mathematica procedure M8.16 for $m_r = 0.3$ $\omega_r = 0.85$, $\zeta_1 = 0.1$, and $\zeta_2 = 0.1$. The results are shown in Figure 8.4.

We now study the scenario where the initial conditions are zero and $f_k(\tau) = F_k e^{j\Omega}$, where $\Omega = \omega/\omega_{n1}$ is a non-dimensional frequency coefficient. In this case, we can assume a solution to the differential equations of the form $x_k(\tau) = X_k e^{j\Omega\tau}$, $k = 1,2$. Upon substituting this assumed solution into Eq. (a), we obtain

$$A_1(j\Omega)X_1 - B_1(j\Omega)X_2 = \frac{F_1}{k_1}$$
$$-C_1(j\Omega)X_1 + E_1(j\Omega)X_2 = \frac{F_2}{k_1 m_r}$$

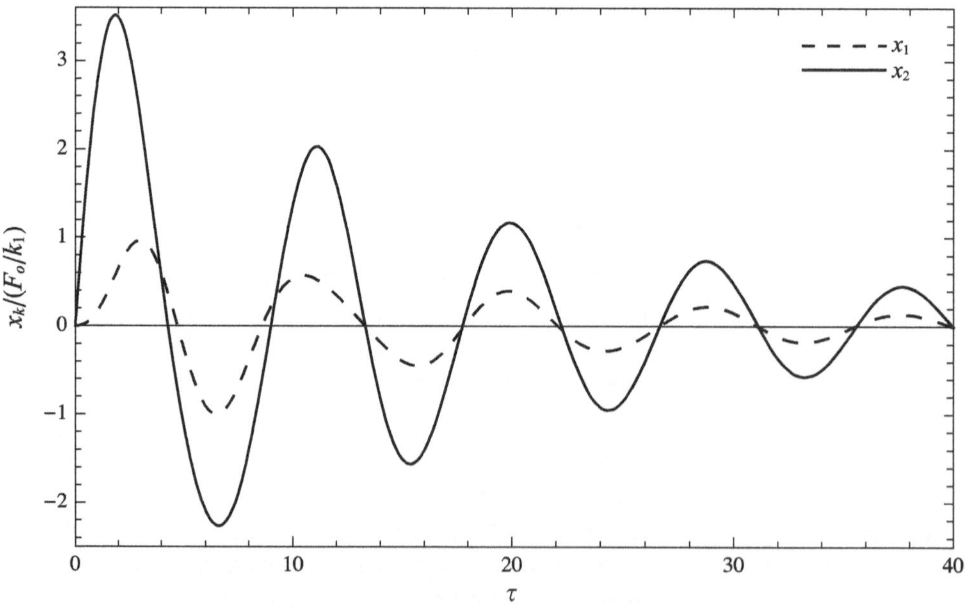

FIGURE 8.4 Displacement responses of the masses of a two degree-of-freedom system to an impulse force on mass m_2 when $m_r = 0.3$ $\omega_r = 0.85$, and $\zeta_1 = \zeta_2 = 0.1$.

where

$$A_1(j\Omega) = -\Omega^2 + 2j(\zeta_1 + \zeta_2 m_r \omega_r)\Omega + 1 + m_r\omega_r^2$$

$$B_1(j\Omega) = 2j\zeta_2 m_r \omega_r \Omega + m_r\omega_r^2$$

$$C_1(j\Omega) = 2j\zeta_2 \omega_r \Omega + \omega_r^2$$

$$E_1(j\Omega) = -\Omega^2 + 2j\zeta_2 \omega_r \Omega + \omega_r^2$$

The solution to this system of equations is

$$X_1 = \frac{F_1 E_1(j\Omega)}{k_1 D_2(j\Omega)} + \frac{F_2 B_1(j\Omega)}{k_1 m_r D_2(j\Omega)}$$
$$X_2 = \frac{F_1 C_1(j\Omega)}{k_1 D_2(j\Omega)} + \frac{F_2 A_1(j\Omega)}{k_1 m_r D_2(j\Omega)} \quad \text{(d)}$$

where

$$D_2(j\Omega) = \Omega^4 - j\left[2\zeta_1 + 2\zeta_2\omega_r m_r + 2\zeta_2\omega_r\right]\Omega^3$$
$$-\left[1 + m_r\omega_r^2 + \omega_r^2 + 4\zeta_1\zeta_2\omega_r\right]\Omega^2$$
$$+ j\left[2\zeta_2\omega_r + 2\zeta_1\omega_r^2\right]\Omega + \omega_r^2$$

It is seen that $A_1(j\Omega)$ can be obtained from $A(s)$ by replacing s (which is non-dimensional in this case) with $j\Omega$; the same is true for obtaining $B_1(j\Omega)$ from $B(s)$, $C_1(j\Omega)$ from $C(s)$, $E_1(j\Omega)$ from $E(s)$, and $D_2(j\Omega)$ from $D_1(s)$. For this reason, the Laplace transform is sometimes referred to as the one-sided Fourier transform. In other words, in contextually appropriate situations one can go from the Laplace transform domain to the frequency domain by replacing s with $j\Omega$. In the dimensional case, one would replace s (now a dimensional quantity) with $j\omega$, where ω is the frequency in rad/s.

Example 8.14: One-dimensional heat conduction in rectangular coordinates with nonhomogeneous boundary conditions and an initial condition

Consider the heat equation, which in non-dimensional form is given as

$$\frac{d\Theta}{d\tau} = \frac{\partial^2\Theta}{\partial\eta^2} \quad \text{(a)}$$

It is assumed that the initial condition is

$$\Theta(\eta, 0) = 1 \quad \text{(b)}$$

and the boundary conditions are

$$\frac{\partial \Theta(0,\tau)}{\partial \eta} = a$$

$$\Theta(\infty,\tau) = \text{finite} \tag{c}$$

To solve this equation, we take the Laplace transform of Eq. (a) with respect to τ and use the initial condition given by Eq. (b) to obtain

$$\frac{d^2\bar{\Theta}}{d\eta^2} - s\bar{\Theta} = -1 \tag{d}$$

where $\bar{\Theta} = \bar{\Theta}(\eta, s)$ is the Laplace transform of $\Theta(\eta,\tau)$. The boundary conditions given by Eq. (c) become

$$\frac{\partial \bar{\Theta}(0,s)}{\partial \eta} = \frac{a}{s}$$

$$\bar{\Theta}(\infty,s) = \text{finite} \tag{e}$$

The homogeneous solution to Eq. (d) is

$$\bar{\Theta}_h = Ae^{-\eta\sqrt{s}} + Be^{\eta\sqrt{s}} \tag{f}$$

The particular solution to Eq. (d) is obtained by inspection as

$$\bar{\Theta}_p = \frac{1}{s}$$

Then the solution to Eq. (d) is

$$\bar{\Theta} = \bar{\Theta}_h + \bar{\Theta}_p = Ae^{-\eta\sqrt{s}} + Be^{\eta\sqrt{s}} + \frac{1}{s} \tag{g}$$

To satisfy the second boundary condition of Eq. (c), we set $B = 0$. Then, Eq. (g) reduces to

$$\bar{\Theta} = Ae^{-\eta\sqrt{s}} + \frac{1}{s} \tag{h}$$

Substituting Eq. (h) into the first boundary condition in Eq. (e) yields

$$A = -\frac{a}{s^{3/2}}$$

Then, Eq. (h) becomes

$$\bar{\Theta}(\eta,s) = \frac{1}{s} - \frac{ae^{-\eta\sqrt{s}}}{s^{3/2}} \tag{i}$$

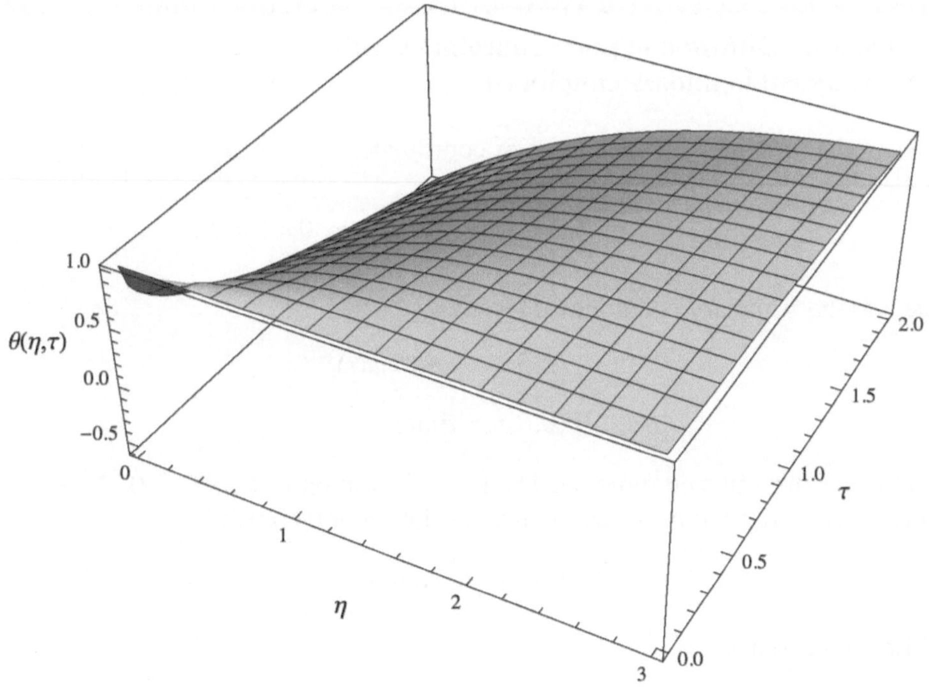

FIGURE 8.5 Numerical evaluation of Eq. (j) of Example 8.14 for a = 1, which depicts the temperature distribution when the outer surface is subject to both a temperature and an initial condition.

The inverse Laplace transform of Eq. (i) is obtained with Mathematica procedure M8.17, which gives

$$\theta(\eta,\tau) = 1 + a\eta \operatorname{erfc}\left(\frac{\eta}{2\sqrt{\tau}}\right) - 2a\sqrt{\frac{\tau}{\pi}}\exp\left(-\frac{\eta^2}{4\tau}\right) \quad \eta > 0 \quad \tau > 0 \tag{j}$$

where

$$\operatorname{erf}(x) = \frac{2}{\sqrt{\pi}}\int_0^x e^{-u^2}\,du$$

$$\operatorname{erfc}(x) = \frac{2}{\sqrt{\pi}}\int_x^\infty e^{-u^2}\,du = 1 - \operatorname{erf}(x)$$

A plot of Eq. (j) for $a = 1$ is given in Figure 8.5, which is obtained with Mathematica procedure M8.18.

Example 8.15: Diffusion in polar coordinates with time-dependent boundary conditions

Consider the diffusion equation in polar coordinates

$$\frac{\partial^2 \Psi}{\partial r^2}+\frac{\partial \Psi}{r\partial r}+\frac{\partial^2 \Psi}{r^2\partial \theta^2}=\frac{1}{\alpha^2}\frac{\partial \Psi}{\partial t} \quad 0<r<a \quad 0<\theta<2\pi \quad t>0 \tag{a}$$

subject to the boundary conditions

$$\begin{aligned}\Psi(a,\theta,t)&=\cos(\theta)u(t)\\ \Psi(0,\theta,t)&=\text{finite}\end{aligned} \tag{b}$$

where $u(t)$ is the unit step function. The initial condition is $\Psi(r,\theta,0) = 0$. Based on the boundary conditions at $r = a$, we assume a solution of the form

$$\Psi(r,\theta,t)=w(r,t)\cos(\theta) \tag{c}$$

and Eq. (a) becomes

$$\frac{\partial^2 w}{\partial r^2}+\frac{\partial w}{r\partial r}-\frac{1}{r^2}w=\frac{1}{\alpha^2}\frac{\partial w}{\partial t} \tag{d}$$

and the boundary conditions given by Eq. (b) become

$$\begin{aligned}w(a,t)&=u(t)\\ w(0,t)&=\text{finite}\end{aligned} \tag{e}$$

and the initial condition becomes $w(r,0) = 0$.

We take the Laplace transform of Eq. (d) with respect to time t and using the fact that the initial condition is zero we obtain

$$r^2\frac{\partial^2 \overline{w}}{\partial r^2}+r\frac{\partial \overline{w}}{\partial r}-\left(\frac{r^2 s}{\alpha^2}+1\right)\overline{w}=0 \tag{f}$$

where $\overline{w}(r,s)$ is the Laplace transform of $w(r,t)$. The Laplace transform of the boundary conditions given by Eq. (e) yields

$$\begin{aligned}\overline{w}(a,s)&=\frac{1}{s}\\ \overline{w}(0,s)&=\text{finite}\end{aligned} \tag{g}$$

Equation (f) is the modified Bessel equation of order 1 whose solution is given by Eq. (5.129) as

$$\overline{w}(r,s)=AI_1\left(r\sqrt{s}/\alpha\right)+BK_1\left(r\sqrt{s}/\alpha\right) \tag{h}$$

However, as indicated in Eq. (5.130), in order for this solution to remain finite at $r = 0$, we must set $B = 0$, and Eq. (h) simplifies to

$$\bar{w}(r,s) = AI_1\left(r\sqrt{s}/\alpha\right) \tag{i}$$

Substituting Eq. (i) into the first boundary condition of Eq. (g), we find that

$$A = \frac{1}{sI_1\left(a\sqrt{s}/\alpha\right)} \tag{j}$$

Substituting Eq. (j) into Eq. (i), the solution in the Laplace transform domain is

$$\bar{w}(r,s) = \frac{I_1\left(r\sqrt{s}/\alpha\right)}{sI_1\left(a\sqrt{s}/\alpha\right)} \tag{k}$$

From Eq. (5.124), the series expansion of $I_1(z)$ is

$$I_1(z) = z\left(\frac{1}{2} + \frac{z^2}{16} + \frac{z^4}{384} + \cdots\right) \tag{l}$$

Hence

$$\frac{I_1\left(r\sqrt{s}/\alpha\right)}{sI_1\left(a\sqrt{s}/\alpha\right)} = \frac{1}{s}\frac{r\sqrt{s}/\alpha\left(\frac{1}{2} + \frac{sr^2/\alpha^2}{16} + \cdots\right)}{a\sqrt{s}/\alpha\left(\frac{1}{2} + \frac{sa^2/\alpha^2}{16} + \cdots\right)} \approx \frac{1}{s}\left(\frac{r}{a}\right) \quad s << 1 \tag{m}$$

and, therefore, $s = 0$ is a simple pole. The remaining poles are at those values of s for which

$$I_1\left(a\sqrt{s}/\alpha\right) = 0 \tag{n}$$

We let $a\sqrt{s}/\alpha = -j\beta$. Since, from Eq. (5.123),

$$I_1(-j\beta) = -jJ_1(\beta)$$

Eq. (n) can be written as

$$J_1(\beta_n) = 0 \tag{o}$$

where β_n are the values that satisfy Eq. (o).

To find the inverse transform of Eq. (k), we use Eq. (8.39) to arrive at

$$w(r,t) = \frac{1}{2\pi j}\int_{\gamma - j\infty}^{\gamma + j\infty} \frac{I_1\left(r\sqrt{z}/\alpha\right)}{zI_1\left(a\sqrt{z}/\alpha\right)} e^{zt}\, dz \tag{p}$$

The residues of Eq. (p) are

$$w(r,t) = \lim_{z\to 0}\left[\frac{zI_1\left(r\sqrt{z}/\alpha\right)}{zI_1\left(a\sqrt{z}/\alpha\right)}e^{zt}\right] + \sum_{n=1}^{\infty}\lim_{z\to z_n}\left[I_1\left(r\sqrt{z}/\alpha\right)e^{zt}\left(z\frac{d}{dz}I_1\left(a\sqrt{z}/\alpha\right)\right)^{-1}\right] \tag{q}$$

However, from Eq. (m), we see that

$$\lim_{z\to 0}\left[\frac{zI_1\left(r\sqrt{z}/\alpha\right)}{zI_1\left(a\sqrt{z}/\alpha\right)}e^{zt}\right] = \lim_{z\to 0}\left[\frac{z}{z}\frac{r}{a}\right] = \frac{r}{a} \tag{r}$$

In addition,

$$\begin{aligned}\sum_{n=1}^{\infty}\lim_{z\to z_n}\left[\frac{I_1\left(r\sqrt{z}/\alpha\right)e^{zt}}{zdI_1\left(a\sqrt{z}/\alpha\right)/dz}\right] &= \sum_{n=1}^{\infty}\lim_{z\to z_n}\left[\frac{4\alpha\sqrt{z}I_1\left(r\sqrt{z}/\alpha\right)e^{zt}}{az\left(I_0\left(a\sqrt{z}/\alpha\right)+I_2\left(a\sqrt{z}/\alpha\right)\right)}\right] \\ &= \sum_{n=1}^{\infty}\frac{4jI_1(-j\beta_n r/a)e^{-\alpha^2\beta_n^2 t/a^2}}{\beta_n\left[I_0(-j\beta_n)+I_2(-j\beta_n)\right]}\end{aligned} \tag{s}$$

where we have used Eq. (5.132), that is,

$$\frac{d}{dz}I_1\left(a\sqrt{z}/\alpha\right) = \frac{a}{4\alpha\sqrt{z}}\left[I_0\left(a\sqrt{z}/\alpha\right)+I_2\left(a\sqrt{z}/\alpha\right)\right]$$

From Eq. (5.126), it is found that

$$I_1(-j\beta_n r/a) = -jJ_1(\beta_n r/a)$$

$$I_0(-j\beta_n r/a) = J_0(\beta_n r/a)$$

$$I_2(-j\beta_n r/a) = -J_2(\beta_n r/a)$$

Then, Eq. (s) becomes

$$4j\sum_{n=1}^{\infty}\frac{I_1(-j\beta_n r/a)e^{-\alpha^2\beta_n^2 t/a^2}}{\beta_n\left[I_0(-j\beta_n)+I_2(-j\beta_n)\right]} = 4\sum_{n=1}^{\infty}\frac{J_1(\beta_n r/a)e^{-\alpha^2\beta_n^2 t/a^2}}{\beta_n\left[J_0(\beta_n)-J_2(\beta_n)\right]} \tag{t}$$

and Eq. (q) reduces to

$$w(r,t) = \frac{r}{a} + 4\sum_{n=1}^{\infty}\frac{J_1(\beta_n r/a)e^{-\alpha^2\beta_n^2 t/a^2}}{\beta_n\left[J_0(\beta_n)-J_2(\beta_n)\right]} \tag{u}$$

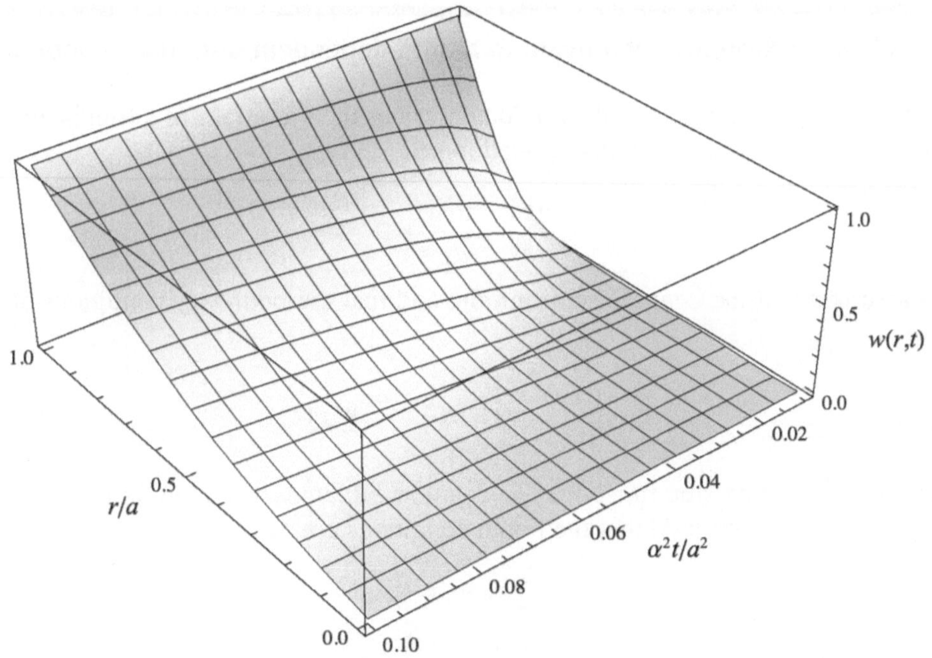

FIGURE 8.6 Numerical evaluation of Eq. (v) of Example 8.15 for $\theta = 0$, which depicts the radial temperature as a function of time when the outer surface is subjected to a step change in temperature.

However, from Eq. (5.103),

$$J_0(\beta_n) + J_2(\beta_n) = \frac{2}{\beta_n} J_1(\beta_n) = 0$$

where we have used Eq. (o). Therefore,

$$J_0(\beta_n) = -J_2(\beta_n)$$

and Eq. (u) becomes

$$w(r,t) = \frac{r}{a} + 2\sum_{n=1}^{\infty} \frac{J_1(\beta_n r / a)}{\beta_n J_0(\beta_n)} e^{-\alpha^2 \beta_n^2 t / a^2}$$

The final result is

$$\Psi(r,\theta,t) = w(r,t)\cos(\theta) = \left(\frac{r}{a} + 2\sum_{n=1}^{\infty} \frac{J_1(\beta_n r / a)}{\beta_n J_0(\beta_n)} e^{-\alpha^2 \beta_n^2 t / a^2} \right) \cos(\theta) \qquad \text{(v)}$$

Equation (v) is plotted in Figure 8.6 for $\theta = 0$ and was obtained with Mathematica procedure M8.19.

Example 8.16: Response of a beam to a time-dependent boundary condition

The following non-dimensional equation describes the vibrations of a thin beam that is not subjected to any externally applied forces

$$\frac{d^4w}{d\eta^4}+\frac{d^2w}{d\tau^2}=0 \tag{a}$$

It is assumed that the beam is semi-infinite and that the boundary conditions at η = 0 are

$$\frac{d^2w(0,\tau)}{d\eta^2}=0 \quad w(0,\tau)=W_ou(\tau) \tag{b}$$

where $u(t)$ is the unit step function. The solution $w(\eta,\tau)$ must remain finite at $\eta \rightarrow \infty$. Furthermore, it is assumed that the initial conditions are zero, that is

$$\begin{aligned} w(\eta,0)&=0 \\ \dot{w}(\eta,0)&=0 \end{aligned} \tag{c}$$

where the over dot indicates the derivative with respect to τ.

To solve Eq. (a), we take its Laplace transform with respect to τ and use Eq. (c) to arrive at

$$\frac{d^4\bar{w}(\eta,s)}{d\eta^4}+4q^4\bar{w}(\eta,s)=0 \tag{d}$$

where $q^4 = s^2/4$. The Laplace transform of the boundary conditions given by Eq. (b) yields

$$\frac{d^2\bar{w}(0,s)}{d\eta^2}=0 \quad \bar{w}(0,s)=\frac{W_o}{s} \tag{e}$$

From Eq. (e) of Example 4.9 or Table 4.2, the solution to Eq. (d) is

$$\bar{w}(\eta,s)=e^{q\eta}\left(b_1\cos q\eta+b_2\sin q\eta\right)+e^{-q\eta}\left(b_3\cos q\eta+b_4\sin q\eta\right) \tag{f}$$

For a bounded solution as $\eta \rightarrow \infty$, we set $b_1 = b_2 = 0$. Then Eq. (f) simplifies to

$$\bar{w}(\eta,s)=e^{-q\eta}\left(b_3\cos q\eta+b_4\sin q\eta\right) \tag{g}$$

It is noted that

$$\frac{d^2\bar{w}(\eta,s)}{d\eta^2}=2e^{-q\eta}q^2\left(b_3\sin q\eta-b_4\cos q\eta\right)$$

This differentiation was performed with Mathematica procedure M8.20.

Then, upon substituting Eq. (g) into the boundary conditions given by Eq. (e), we find that $b_3 = W_o/s$ and $b_4 = 0$. Then, Eq. (g) becomes

$$\bar{w}(\eta,s) = \frac{W_o e^{-q\eta}}{s}\cos q\eta = \frac{W_o e^{-\sqrt{s}\xi}}{s}\cos\sqrt{s}\xi \qquad \text{(h)}$$

where $\xi = \eta/\sqrt{2}$.

The inverse Laplace transform of Eq. (h) is not available. However, it is found that if we rewrite Eq. (h) as

$$\bar{w}(\eta,s) = \left(\frac{W_o}{\sqrt{s}}\right)\left(\frac{e^{-\sqrt{s}\xi}}{\sqrt{s}}\cos\sqrt{s}\xi\right)$$

then the following inverses are available*

$$f(\tau) = L^{-1}\left(\frac{W_o}{\sqrt{s}}\right) = \frac{W_o}{\sqrt{\pi\tau}}$$

$$h(\tau) = L^{-1}\left(\frac{e^{-\sqrt{s}\xi}}{\sqrt{s}}\cos\sqrt{s}\xi\right) = \frac{1}{\sqrt{\pi\tau}}\cos\left(\xi^2/(2\tau)\right)$$

$$= \frac{1}{\sqrt{\pi\tau}}\cos\left(\eta^2/(4\tau)\right)$$

Using the convolution theorem, Eq. (8.26), we obtain

$$w(\eta,\tau) = L^{-1}[F(s)H(s)] = \int_0^\tau f(\mu)h(t-\mu)d\mu$$

$$= \frac{W_o}{\pi}\int_0^\tau \frac{\cos\left(\eta^2/(4\mu)\right)}{\sqrt{\mu}\sqrt{(\tau-\mu)}}d\mu \qquad \text{(i)}$$

$$= W_o\left\{1 - C\left(\eta/\sqrt{2\pi\tau}\right) - S\left(\eta/\sqrt{2\pi\tau}\right)\right\}$$

where

$$C(x) = \int_0^x \cos\left(\pi t^2/2\right)dt, \quad S(x) = \int_0^x \sin\left(\pi t^2/2\right)dt$$

are the Fresnel integrals. The integration was performed with Mathematica procedure M8.21.

A plot of $w(\eta,\tau)/W_o$ as given by Eq. (i) is shown in Figure 8.7, which was obtained with Mathematica procedure M8.22.

---

* Roberts GE, Kaufman H (1966) *Table of Laplace Transforms.* W. B. Saunders, Philadelphia, p. 272 #5.3.2.1 and p. 206 #2.1.

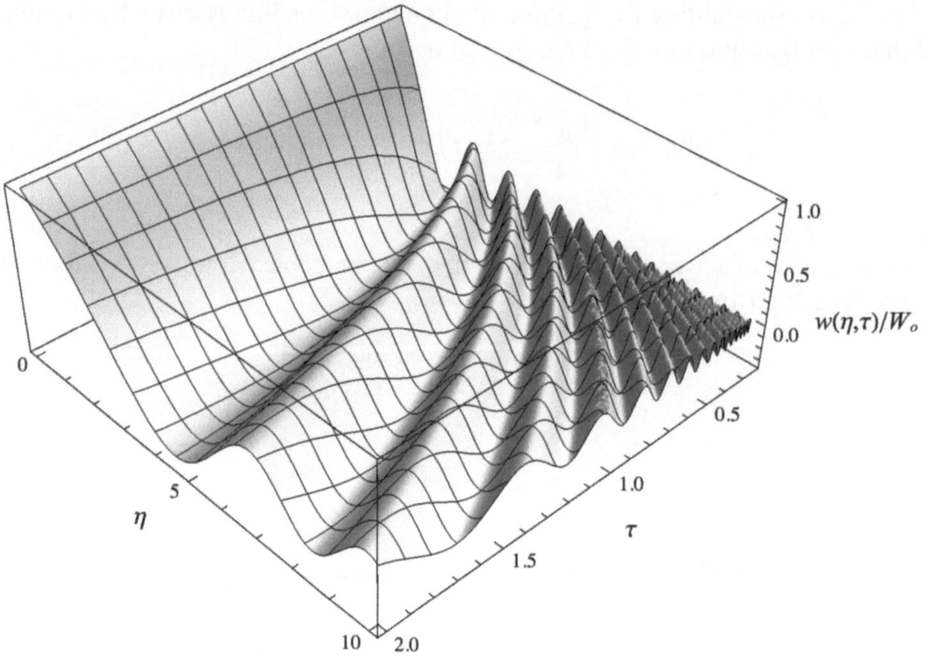

FIGURE 8.7 Numerical evaluation Eq. (i) of Example 8.16, which depicts the displacement response of a semi-infinite beam whose end is subjected to a step displacement.

Example 8.17: Traveling point load on a semi-infinite string

A very long string that is under a tension T (N) and has a density ρ (kg/m) is fixed at each end. A concentrated load of magnitude P (N) is traveling across it at a velocity V (m/s). The governing equation is

$$\frac{\partial^2 y}{\partial t^2} = c^2 \frac{\partial^2 y}{\partial x^2} + \frac{P}{\rho}\delta(Vt - x) \text{ m/s}^2 \tag{a}$$

The quantity $c = \sqrt{T/\rho}$ (m/s) is the speed at which disturbances in the string travel. The boundary conditions are

$$\begin{aligned} y(0,t) &= 0 \\ y(\infty,t) &= \text{finite} \end{aligned} \tag{b}$$

and it is assumed that the initial conditions are zero.

We convert Eqs. (a) and (b) to non-dimensional form by letting $\eta = x/(Vt_o)$ and $\tau = t/t_o$, where t_o is an arbitrary reference time. Then, Eq. (a) becomes

$$\frac{\partial^2 y}{\partial \tau^2} = \alpha^2 \frac{\partial^2 y}{\partial \eta^2} + \hat{P}\delta(\tau - \eta) \text{ m} \tag{c}$$

where $\alpha = c/V$ and

$$\hat{P} = \frac{Pt_o}{V\rho} \rightarrow \frac{\text{Ns}}{(\text{m/s})(\text{kg/m})} = \text{m}$$

The boundary conditions given by Eq. (b) become

$$\begin{aligned} y(0,\tau) &= 0 \\ y(\infty,\tau) &= \text{finite} \end{aligned} \tag{d}$$

To obtain a solution to Eqs. (c) and (d), we take their Laplace transform with respect to τ and obtain

$$\alpha^2 \frac{d^2\bar{y}(\eta,s)}{d\eta^2} - s^2\bar{y}(\eta,s) = -\hat{P}e^{-\eta s} \tag{e}$$

and

$$\begin{aligned} \bar{y}(0,s) &= 0 \\ \bar{y}(\infty,s) &= \text{finite} \end{aligned} \tag{f}$$

where we have used Laplace transform pairs 5 and 13 of Table 8.2 in Appendix 8.1 and the fact that the initial conditions are zero.

The solution to Eq. (e) is

$$\bar{y}(\eta,s) = C_1 e^{s\eta/\alpha} + C_2 e^{-s\eta/\alpha} - \frac{\hat{P}}{s^2\left(\alpha^2 - 1\right)} e^{-\eta s} \tag{g}$$

To satisfy the second boundary condition of Eq. (f), we set $C_1 = 0$. Then, Eq. (g) reduces to

$$\bar{y}(\eta,s) = C_2 e^{-s\eta/\alpha} - \frac{\hat{P}}{s^2\left(\alpha^2 - 1\right)} e^{-\eta s} \tag{h}$$

Upon substituting Eq. (h) into the first boundary condition of Eq. (f), we find that

$$C_2 = \frac{\hat{P}}{s^2\left(\alpha^2 - 1\right)} \tag{i}$$

Then, Eq. (h) becomes

$$\bar{y}(\eta,s) = \frac{\hat{P}}{\alpha^2 - 1}\left(\frac{e^{-s\eta/\alpha} - e^{-\eta s}}{s^2}\right) \tag{j}$$

Using Laplace transform pairs 6 and 17 from Table 8.2 of Appendix 8.1, the inverse Laplace transform of Eq. (j) is

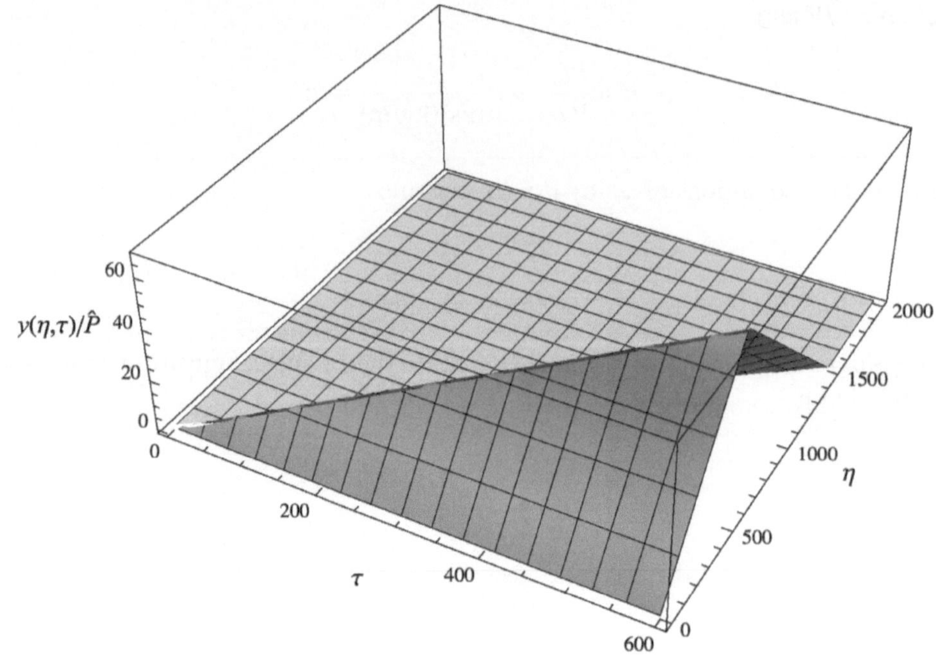

FIGURE 8.8 Numerical evaluation of Eq. (k) of Example 8.17 for $\alpha = 2.5$, which depicts the displacement of a semi-infinite string to a traveling point load.

$$y(\eta,\tau) = \frac{\hat{P}}{\alpha^2 - 1}\left((\tau - \eta/\alpha)u(\tau - \eta/\alpha) - (\tau - \eta)u(\tau - \eta)\right) \text{ m} \tag{k}$$

Equation (k) is verified with Mathematica procedure M8.23 and plotted in Figure 8.8 for $\alpha = 2.5$, which was obtained with Mathematica procedure M8.24.

Example 8.18: Longitudinal vibrations of a bar with a mass attached at a free end

Consider the following non-dimensional equation describing the longitudinal vibrations of a bar (recall Example 7.23)

$$\frac{\partial^2 y}{\partial \eta^2} = \frac{\partial^2 y}{\partial \tau^2} \tag{a}$$

We assume that the boundary conditions are

$$y(0,\tau) = 0$$

$$m\frac{\partial^2 y(1,\tau)}{\partial \tau^2} = -\frac{\partial y(1,\tau)}{\partial \eta} \tag{b}$$

where m is a non-dimensional constant representing the mass attached at the free end of the bar and the initial conditions are

$$
\begin{aligned}
y(\eta,0) &= 0 \\
\frac{\partial y(\eta,0)}{\partial \tau} &= v
\end{aligned}
\tag{c}
$$

We shall use the Laplace transform method to obtain a solution to Eq. (a) subject to the boundary conditions given by Eqs. (b) and the initial conditions given by Eq. (c). Taking the Laplace transform of Eq. (a) with respect to time τ yields

$$\frac{\partial^2 \overline{y}(\eta,s)}{\partial \eta^2} - s^2 \overline{y} = -v \tag{d}$$

where we have used Eq. (c). The boundary conditions given by Eq. (b) become

$$
\begin{aligned}
\overline{y}(0,s) &= 0 \\
ms^2 \overline{y}(1,s) - mv &= -\frac{\partial \overline{y}(1,s)}{\partial \eta}
\end{aligned}
\tag{e}
$$

where we have again used Eq. (c).

To obtain the solution Eq. (d), we use Table 4.2 to obtain the homogeneous solution and inspection to obtain the particular solution and arrive at

$$\overline{y}(\eta,s) = B\cosh s\eta + C\sinh s\eta + \frac{v}{s^2} \tag{f}$$

Substituting Eq. (f) into the first boundary condition given by Eq. (e) results in

$$B = -\frac{v}{s^2}$$

Using this result in Eq. (f), Eq. (f) becomes

$$\overline{y}(\eta,s) = \frac{v}{s^2}\left(1 - \cosh s\eta\right) + C\sinh s\eta \tag{g}$$

Substituting Eq. (g) into the second boundary condition given by Eq. (e), we find that

$$C = \frac{v}{s^2}\frac{\sinh s + ms\cosh s}{ms\sinh s + \cosh s} \tag{h}$$

Therefore, substituting Eq. (h) into Eq. (g), we arrive at

$$\overline{y}(\eta,s) = \frac{v}{s^2}\left\{1 - \frac{ms\sinh\left(s(1-\eta)\right) + \cosh\left(s(1-\eta)\right)}{ms\sinh s + \cosh s}\right\} \tag{i}$$

The inverse of Eq. (i) is obtained by using Eq. (8.39); therefore,

$$y(\eta,\tau) = \frac{v}{2j\pi}\int_{\gamma-j\infty}^{\gamma+j\infty}\left\{\frac{1}{z^2} - \frac{mz\sinh\left(z(1-\eta)\right) + \cosh\left(z(1-\eta)\right)}{z^2\left(mz\sinh z + \cosh z\right)}\right\}e^{z\tau}\,dz \tag{j}$$

It is found by using Mathematica procedure M8.25 that

$$\lim_{z\to 0}\left\{\frac{1}{z^2}-\frac{mz\sinh\big(z(1-\eta)\big)+\cosh\big(z(1-\eta)\big)}{z^2\left(mz\sinh z+\cosh z\right)}\right\}e^{z\tau}=\eta(1+m)-\frac{\eta^2}{2} \tag{k}$$

Therefore, $z = 0$ is an ordinary point. Re-writing Eq. (j) as

$$y(\eta,\tau)=\frac{v}{2j\pi}\int_{\gamma-j\infty}^{\gamma+j\infty}\left\{\frac{mz\sinh z+\cosh z-mz\sinh\big(z(1-\eta)\big)-\cosh\big(z(1-\eta)\big)}{z^2\left(mz\sinh z+\cosh z\right)}\right\}e^{z\tau}\,dz \tag{l}$$

we see that we have a series of simple poles at

$$mz\sinh z+\cosh z=0 \tag{m}$$

If we let $z=\pm j\beta$, then Eq. (m) can be written as

$$\tan\beta_n=\frac{1}{m\beta_n} \tag{n}$$

where, for a given m, β_n are the values that satisfy Eq. (n). Therefore, we compute the residues of Eq. (l) at $z_k=\pm j\beta_k$. Then Eq. (l) becomes

$$\begin{aligned}
y(\eta,\tau)&=v\sum_{k=1}^{\infty}\lim_{z\to z_k}g(z)e^{z\tau}=v\sum_{k=1}^{\infty}g(j\beta_k)e^{j\beta_k\tau}+v\sum_{k=1}^{\infty}g(-j\beta_k)e^{-j\beta_k\tau}\\
&=v\sum_{k=1}^{\infty}g(j\beta_k)\left(\frac{-e^{j\beta_k\tau}+e^{-j\beta_k\tau}}{j}\right)\\
&=2v\sum_{k=1}^{\infty}f_k(\eta)\sin\left(\beta_k\tau\right)
\end{aligned} \tag{o}$$

where

$$g(z)=\frac{mz\sinh z+\cosh z-mz\sinh\big(z(1-\eta)\big)-\cosh\big(z(1-\eta)\big)}{z^2\left(mz\cosh z+(1+m)\sinh z\right)}$$

and

$$f_k(\eta)=\frac{-m\beta_k\sin\big(\beta_k(1-\eta)\big)+\cos\big(\beta_k(1-\eta)\big)}{\beta_k^2\Big(\left(m\beta_k\right)^2+(1+m)\Big)\sin\beta_k} \tag{p}$$

In obtaining Eq. (p), we have used Eq. (n), and the denominator of $g(z)$ was obtained by differentiating the denominator of the integrand of Eq. (l).

A plot of Eq. (p) is shown in Figure 8.9 for $\eta = 1$ and for several values of m. The figure was obtained with Mathematica procedure M8.26.

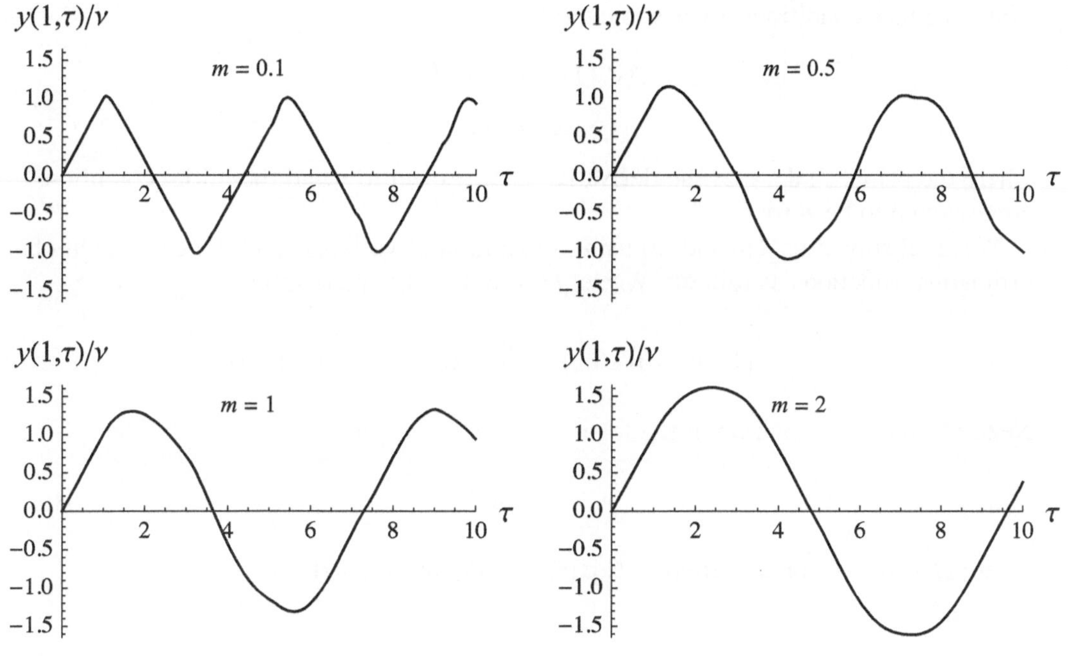

FIGURE 8.9 Numerical evaluation of Eq. (o) of Example 8.18, which depicts the longitudinal displacement of the mass m attached at $\eta = 1$ for several values of m.

Example 8.19: Response of a thin beam with an in-span mass to a dynamic load

Consider a thin elastic beam that has a length L (m), cross-sectional area moment of inertia I (m^4), Young's modulus E (N/m^2), and density ρ (kg/m^3). The beam is subjected to an externally applied load $F(x,t)$ (N/m) and is carrying a mass M_o (kg) that is located at L_o (m), $0 < L_o < L$. If the transverse displacement of the beam is $w = w(x,t)$ (m) and the distance along the beam is x (m), then the governing equation describing the motion of the beam is

$$EI\frac{\partial^4 w}{\partial x^4} + \left(\rho A + M_o\delta(x - L_o)\right)\frac{\partial^2 w}{\partial t^2} = F(x,t) \quad \text{N/m} \quad 0 < x < L \quad t > 0 \qquad \text{(a)}$$

Using a procedure similar to that given in Example 7.26, Eq. (a) is placed in non-dimensional form as

$$\frac{\partial^4 y}{\partial \eta^4} + \left(1 + m_o\delta(\eta - \eta_o)\right)\frac{\partial^2 y}{\partial \tau^2} = f(\eta,\tau) \qquad \text{(b)}$$

where $y = w/L$ and

$$\eta = x/L, \quad \eta_o = L_o/L, \quad \tau = t/t_o, \quad t_o^2 = \frac{\rho A L^4}{EI} \quad \text{s}^2$$

$$m_o = \frac{M_o}{\rho A L}, \quad f(\eta,\tau) = \frac{L^3}{EI}F(\eta,\tau) \qquad \text{(c)}$$

The boundary conditions are assumed to be

$$\begin{aligned} y(0,\tau) &= y'(0,\tau) = 0 \\ y(1,\tau) &= y'(1,\tau) = 0 \end{aligned} \tag{d}$$

where the prime (′) denotes the derivative with respect to η and the initial conditions are assumed to be zero.

We shall solve Eqs. (b) and (d) using separation of variables and the generation of orthogonal functions as follows. We set $f = 0$ in Eq. (b) and obtain

$$\frac{\partial^4 y_h}{\partial \eta^4} + \left(1 + m_o\delta(\eta - \eta_o)\right)\frac{\partial^2 y_h}{\partial \tau^2} = 0 \quad 0 < \eta < 1 \quad \tau > 0 \tag{e}$$

Next, we assume a solution of the form

$$y_h = Y(\eta)e^{j\Omega^2\tau} \tag{f}$$

where $\Omega^2 = \omega t_o$. Upon substituting Eq. (f) into Eq. (e), we arrive at

$$\frac{d^4 Y}{d\eta^4} - \left(1 + m_o\delta(\eta - \eta_o)\right)\Omega^4 Y = 0 \quad 0 < \eta < 1 \tag{g}$$

and the boundary conditions given by Eq. (d) become

$$\begin{aligned} Y(0) &= Y'(0) = 0 \\ Y(1) &= Y'(1) = 0 \end{aligned} \tag{h}$$

We see that Eqs. (g) and (h) form a fourth-order Sturm–Liouville system, the solution to which will have orthogonal eigenfunctions corresponding to the eigenvalues Ω_n. Recall Section 6.1.5.

We shall determine the solution to Eq. (g) by taking the Laplace transform with respect to η. Upon performing this operation and re-arranging the results, we arrive at

$$\begin{aligned} \bar{Y}(s) = {} & \frac{s^3}{s^4 - \Omega^4}Y(0) + \frac{s^2}{s^4 - \Omega^4}Y'(0) + \frac{s}{s^4 - \Omega^4}Y''(0) + \frac{1}{s^4 - \Omega^4}Y'''(0) \\ & + m_o\Omega^4 Y(\eta_o)\frac{e^{-s\eta_o}}{s^4 - \Omega^4} \end{aligned} \tag{i}$$

The inverse Laplace transform of Eq. (i) is obtained from Laplace transform pairs 6 and 27 to 30 of Table 8.2 of Appendix 8.1, that is,

$$\frac{s^3}{s^4-\Omega^4} \Leftrightarrow Q(\Omega\eta)$$

$$\frac{s^2}{s^4-\Omega^4} \Leftrightarrow \frac{R(\Omega\eta)}{\Omega}$$

$$\frac{s}{s^4-\Omega^4} \Leftrightarrow \frac{S(\Omega\eta)}{\Omega^2} \tag{j}$$

$$\frac{1}{s^4-\Omega^4} \Leftrightarrow \frac{T(\Omega\eta)}{\Omega^3}$$

$$\frac{e^{-s\eta_o}}{s^4-\Omega^4} \Leftrightarrow \frac{T(\Omega[\eta-\eta_o])}{\Omega^3}u(\eta-\eta_o)$$

where $u(\eta - \eta_o)$ is the unit step function and

$$Q(\Omega\eta) = \frac{1}{2}\left[\cos(\Omega\eta) + \cosh(\Omega\eta)\right]$$

$$R(\Omega\eta) = \frac{1}{2}\left[\sin(\Omega\eta) + \sinh(\Omega\eta)\right]$$

$$S(\Omega\eta) = \frac{1}{2}\left[-\cos(\Omega\eta) + \cosh(\Omega\eta)\right] \tag{k}$$

$$T(\Omega\eta) = \frac{1}{2}\left[-\sin(\Omega\eta) + \sinh(\Omega\eta)\right]$$

Then the inverse Laplace transform of Eq. (i) is

$$\begin{aligned} Y(\eta) = Q(\Omega\eta)Y(0) + \frac{R(\Omega\eta)}{\Omega}Y'(0) + \frac{S(\Omega\eta)}{\Omega^2}Y''(0) + \frac{T(\Omega\eta)}{\Omega^3}Y'''(0) \\ + m_o\Omega Y(\eta_o)T(\Omega[\eta-\eta_o])u(\eta-\eta_o) \end{aligned} \tag{l}$$

Using the first set of boundary conditions of Eq. (h) reduces Eq. (l) to

$$Y(\eta) = \frac{S(\Omega\eta)}{\Omega^2}Y''(0) + \frac{T(\Omega\eta)}{\Omega^3}Y'''(0) + m_o\Omega Y(\eta_o)T(\Omega[\eta-\eta_o])u(\eta-\eta_o) \tag{m}$$

Before proceeding, we note the following relations

$$Q'(\Omega\eta) = \Omega T(\Omega\eta)$$

$$R'(\Omega\eta) = \Omega Q(\Omega\eta)$$

$$S'(\Omega\eta) = \Omega R(\Omega\eta) \tag{n}$$

$$T'(\Omega\eta) = \Omega S(\Omega\eta)$$

where the prime (′) denotes the derivative with respect to η. Then, using Eq. (n), we find that

$$Y'(\eta)=\frac{R(\Omega\eta)}{\Omega}Y''(0)+\frac{S(\Omega\eta)}{\Omega^2}Y'''(0)+m_o\Omega^2Y(\eta_o)S(\Omega[\eta-\eta_o])u(\eta-\eta_o) \quad \text{(o)}$$

Upon using Eqs. (m) and (o) in the second set of boundary conditions given by Eq. (h), we obtain

$$\begin{pmatrix}\Omega S(\Omega) & T(\Omega)\\ \Omega R(\Omega) & S(\Omega)\end{pmatrix}\begin{Bmatrix}Y''(0)\\ Y'''(0)\end{Bmatrix}=\begin{Bmatrix}-m_o\Omega^4Y(\eta_o)T(\Omega[1-\eta_o])\\ -m_o\Omega^4Y(\eta_o)S(\Omega[1-\eta_o])\end{Bmatrix} \quad \text{(p)}$$

Solving for $Y''(0)$ and $Y'''(0)$ yields

$$\begin{aligned}Y''(0)&=\frac{m_o\Omega^3Y(\eta_o)}{D_o(\Omega)}\hat{C}_1(\Omega,\eta_o)\\ Y'''(0)&=\frac{m_o\Omega^4Y(\eta_o)}{D_o(\Omega)}\hat{C}_2(\Omega,\eta_o)\end{aligned} \quad \text{(q)}$$

where

$$D_o(\Omega)=S^2(\Omega)-T(\Omega)R(\Omega)$$

$$\hat{C}_1(\Omega,\eta_o)=T(\Omega)S(\Omega[1-\eta_o])-S(\Omega)T(\Omega[1-\eta_o])$$

$$\hat{C}_2(\Omega,\eta_o)=R(\Omega)T(\Omega[1-\eta_o])-S(\Omega)S(\Omega[1-\eta_o])$$

Substituting Eq. (q) into Eq. (m), we arrive at

$$\begin{aligned}Y(\eta)=\frac{m_o\Omega Y(\eta_o)}{D_o(\Omega)}\Big(&\hat{C}_1(\Omega,\eta_o)S(\Omega\eta)+\hat{C}_2(\Omega,\eta_o)T(\Omega\eta)\\ &+D_o(\Omega)T(\Omega[\eta-\eta_o])u(\eta-\eta_o)\Big)\end{aligned} \quad \text{(r)}$$

Since Eq. (r) is valid at $\eta=\eta_o$, we set $\eta=\eta_o$ in Eq. (r) and find that the characteristic equation is

$$D_o(\Omega_n)-m_o\Omega\left(\hat{C}_1(\Omega_n,\eta_o)S(\Omega_n\eta_o)+\hat{C}_2(\Omega_n,\eta_o)T(\Omega_n\eta_o)\right)=0 \quad \text{(s)}$$

where Ω_n are the eigenvalues that satisfy Eq. (s). Corresponding to these eigenvalues are the orthogonal eigenfunctions

$$Y_n(\eta)=\frac{\hat{C}_1(\Omega_n,\eta_o)}{D_o(\Omega_n)}S(\Omega_n\eta)+\frac{\hat{C}_2(\Omega_n,\eta_o)}{D_o(\Omega_n)}T(\Omega_n\eta)+T(\Omega_n[\eta-\eta_o])u(\eta-\eta_o)) \quad \text{(t)}$$

where we have ignored the scale factor $m_o\Omega Y(\eta_o)$.

Thus, we have a set of orthogonal eigenfunctions that satisfy the homogeneous ordinary differential equation

$$\frac{d^4Y_n}{d\eta^4}-\left(1+m_o\delta(\eta-\eta_o)\right)\Omega_n^4Y_n=0 \quad \text{(u)}$$

and the homogeneous boundary conditions

$$\begin{aligned} Y_n(0) &= Y_n'(0) = 0 \\ Y_n(1) &= Y_n'(1) = 0 \end{aligned} \tag{v}$$

We now use these results to solve Eq. (b) by assuming a solution to Eq. (b) of the form

$$y(\eta,\tau) = \sum_{k=1}^{\infty} a_k(\tau) Y_k(\eta) \tag{w}$$

Upon substituting Eq. (w) into Eq. (b), we arrive at

$$\sum_{n=1}^{\infty}\left[a_n(\tau)\frac{\partial^4 Y_n(\eta)}{\partial\eta^4} + \left(1 + m_o\delta(\eta-\eta_o)\right)Y_n(\eta)\frac{\partial^2 a_n(\tau)}{\partial\tau^2}\right] = f(\eta,\tau)$$

Using Eq. (u), this expression becomes

$$\sum_{n=1}^{\infty}\left(1 + m_o\delta(\eta-\eta_o)\right)Y_n(\eta)\left[\frac{\partial^2 a_n(\tau)}{\partial\tau^2} + \Omega_n^4 a_n(\tau)\right] = f(\eta,\tau) \tag{x}$$

Since $Y_n(\eta)$ is an orthogonal function, we multiply Eq. (x) by $Y_m(\eta)$ and integrate over the interval to obtain

$$\begin{aligned} \sum_{n=1}^{\infty}\int_0^1 \left(1 + m_o\delta(\eta-\eta_o)\right)Y_n(\eta)Y_m(\eta)d\eta\left[\frac{\partial^2 a_n(\tau)}{\partial\tau^2} + \hat{\Omega}_n^4 a_n(\tau)\right] &= \int_0^1 f(\eta,\tau)Y_m(\eta)d\eta \\ \frac{\partial^2 a_n(\tau)}{\partial\tau^2} + \hat{\Omega}_n^4 a_n(\tau) &= g_n(\tau) \end{aligned} \tag{y}$$

where

$$\begin{aligned} g_n(\tau) &= \frac{1}{N_n}\int_0^1 f(\eta,\tau)Y_n(\eta)d\eta \\ N_n &= \int_0^1 \left(1 + m_o\delta(\eta-\eta_o)\right)Y_n^2(\eta)d\eta = \int_0^1 Y_n^2(\eta)d\eta + m_o Y_n^2(\eta_o) \end{aligned} \tag{z}$$

We note from Eq. (6.102) that $R(x) \rightarrow R(\eta) = 1 + m_o\delta(\eta-\eta_o)$. The solution to Eq. (y) is given by Eq. (c) of Example 4.20 (or Table 4.3), which can be used once $f(\eta,\tau)$ has been specified.

To illustrate these results, we assume that an impulse force is applied at $\eta = \eta_f$. Then,

$$f(\eta,\tau) = \delta(\tau)\delta(\eta-\eta_f)$$

From Eq. (z), we find that

$$g_n(\tau) = \delta(\tau) G_n$$

where

$$G_n = \frac{1}{N_n}\int_0^1 \delta(\eta - \eta_f) Y_n(\eta) d\eta = \frac{Y_n(\eta_f)}{N_n}$$

Then, Eq. (y) becomes

$$\frac{\partial^2 a_n(\tau)}{\partial \tau^2} + \hat{\Omega}_n^4 a_n(\tau) = \delta(\tau) G_n \tag{aa}$$

From Eq. (c) of Example 4.20, it is straightforward to show that when the initial conditions are zero the solution to Eq. (aa) is

$$a_n(\tau) = \frac{G_n}{\hat{\Omega}_n^2}\sin\left(\hat{\Omega}_n^2 \tau\right)$$

Using this result in Eq. (w), we find that the solution is

$$y(\eta,\tau) = \sum_{n=1}^{\infty} \frac{Y_n(\eta_f)}{N_n \hat{\Omega}_n^2} Y_n(\eta) \sin\left(\hat{\Omega}_n^2 \tau\right) \tag{bb}$$

Equation (bb) is evaluated for the case where $\eta_f = 0.45$, $\eta_o = 0.72$, and $m_o = 0.5$. The results are shown in Figure 8.10 for a four-term solution and were obtained with Mathematica procedure M8.27.

Example 8.20: Response of a Timoshenko beam to a dynamic load

Referring to Eq. (h) of Example 7.27 with $S = m_a = 0$, we consider the following set of coupled non-dimensional partial differential equations describing the forced transverse motion of a Timoshenko beam of constant cross section

$$\begin{gathered}\frac{\partial}{\partial \eta}\left(\frac{\partial y}{\partial \eta} - \psi\right) - \gamma_{bs} R_o^2 \frac{\partial^2 y}{\partial \tau^2} = -\gamma_{bs} R_o^2 \hat{f}_a(\eta,\tau) \\ \frac{\partial y}{\partial \eta} - \psi + \gamma_{bs} R_o^2 \frac{\partial^2 \psi}{\partial \eta^2} - \gamma_{bs} R_o^4 \frac{\partial^2 \psi}{\partial \tau^2} = 0\end{gathered} \tag{a}$$

We shall assume that both ends of the beam are clamped. For this case, the boundary conditions are

$$\begin{gathered}y(0,\tau) = \psi(0,\tau) = 0 \\ y(1,\tau) = \psi(1,\tau) = 0\end{gathered} \tag{b}$$

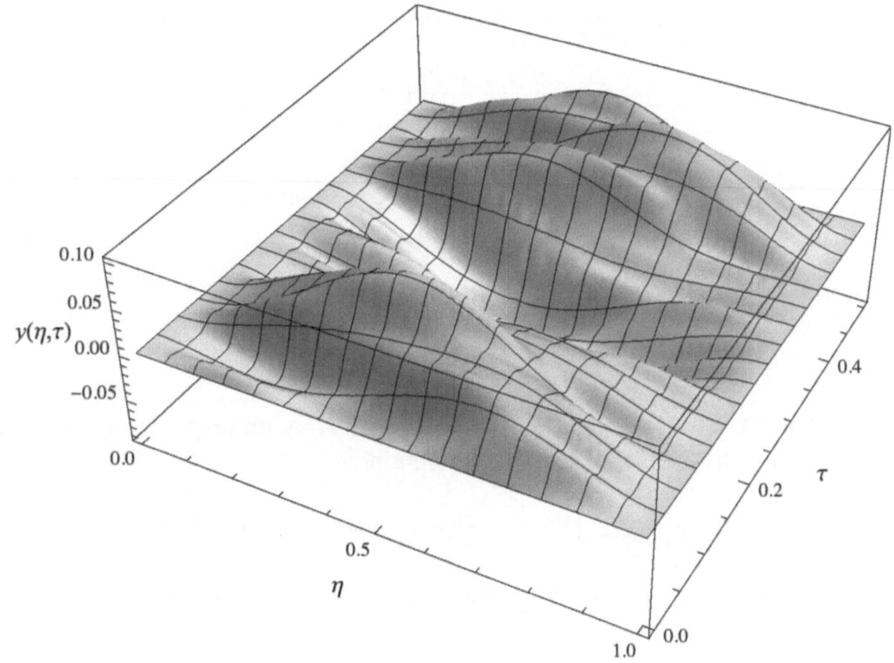

FIGURE 8.10 Displacement response of a beam clamped at each end when an in-span mass $m_o = 0.5$ is attached at $\eta_o = 0.72$ and an impulse force is applied at $\eta_f = 0.45$.

In addition, we shall assume that all initial conditions are zero. It is seen upon comparing Eq. (a) with Eq. (6.154) that $u_1 = y$, $u_2 = \psi$, and

$$L_1[y,\psi] = \frac{\partial}{\partial\eta}\left(\frac{\partial y}{\partial\eta} - \psi\right), \quad r_1(\eta) = \gamma_{bs}R_o^2$$
$$L_2[y,\psi] = \frac{\partial y}{\partial\eta} - \psi + \gamma_{bs}R_o^2\frac{\partial^2\psi}{\partial\eta^2}, \quad r_2(\eta) = \gamma_{bs}R_o^4 \tag{c}$$

It is mentioned that these equations and boundary conditions can be derived from the system's total energy, which is a symmetric quadratic. In addition, the boundary conditions given by Eq. (b) are homogeneous. Consequently, when the eigenfunctions are determined they will be orthogonal eigenfunctions. Therefore, to start the solution process, we first obtain the eigenvalues and eigenfunctions. To do this, we set $\hat{f}_a = 0$ and assume that

$$y(\eta,\tau) = Y(\eta)e^{j\Omega^2\tau}$$
$$\psi(\eta,\tau) = \Psi(\eta)e^{j\Omega^2\tau} \tag{d}$$

where $\Omega^2 = \omega t_o$, ω (rad/s) is the excitation frequency, and from Example 7.27

$$t_o = \sqrt{\frac{\rho A L^4}{EI}} \text{ s}$$

Then, substitution of Eq. (d) into Eqs. (a) and (b), respectively, yields

$$\frac{d^2Y}{d\eta^2} - \frac{d\Psi}{d\eta} + \gamma_{bs} R_o^2 \Omega^4 Y = 0$$
$$\frac{dY}{d\eta} - \Psi + \gamma_{bs} R_o^2 \frac{d^2\Psi}{d\eta^2} + \gamma_{bs} R_o^4 \Omega^4 \Psi = 0 \tag{e}$$

and

$$Y(0) = \Psi(0) = 0$$
$$Y(1) = \Psi(1) = 0 \tag{f}$$

We shall solve Eq. (e) by using the Laplace transform with respect to η. Thus, upon taking the Laplace transform of Eq. (e), we obtain

$$\left(s^2 + \gamma_{bs} R_o^2 \Omega^4\right)\bar{Y}(s) - s\bar{\Psi}(s) = sY(0) + Y'(0) - \Psi(0)$$
$$s\bar{Y}(s) + \left(\gamma_{bs} R_o^2 s^2 + \gamma_{bs} R_o^4 \Omega^4 - 1\right)\bar{\Psi}(s) = Y(0) + \gamma_{bs} R_o^2 \left(s\Psi(0) + \Psi'(0)\right) \tag{g}$$

Upon using the first set of equations of Eq. (f), Eq. (g) reduces to

$$\left(s^2 + \gamma_{bs} R_o^2 \Omega^4\right)\bar{Y}(s) - s\bar{\Psi}(s) = Y'(0)$$
$$s\bar{Y}(s) + \left(\gamma_{bs} R_o^2 s^2 + \gamma_{bs} R_o^4 \Omega^4 - 1\right)\bar{\Psi}(s) = \gamma_{bs} R_o^2 \Psi'(0) \tag{h}$$

Solving Eq. (h) for $\bar{Y}(s)$ and $\bar{\Psi}(s)$ gives

$$\bar{Y}(s) = \frac{1}{\gamma_{bs} R_o^2 D_o(s)}\left(\gamma_{bs} R_o^2 s\Psi'(0) + \left(\gamma_{bs} R_o^2 s^2 + \gamma_{bs} R_o^4 \Omega^4 - 1\right)Y'(0)\right)$$
$$\bar{\Psi}(s) = \frac{1}{\gamma_{bs} R_o^2 D_o(s)}\left(\gamma_{bs} R_o^2 \left(\gamma_{bs} R_o^2 \Omega^4 + s^2\right)\Psi'(0) - sY'(0)\right) \tag{i}$$

where

$$D_o(s) = \left(s^2 - \alpha^2\right)\left(s^2 + \beta^2\right)$$

and

$$\beta^2 = \frac{\Omega^2}{2}\left(\Omega^2 R_o^2 \left(1 + \gamma_{bs}\right) + \sqrt{4 + \Omega^4 R_o^4 \left(1 - \gamma_{bs}\right)^2}\right)$$

$$\alpha^2 = \frac{\Omega^2}{2}\left(-\Omega^2 R_o^2 \left(1 + \gamma_{bs}\right) + \sqrt{4 + \Omega^4 R_o^4 \left(1 - \gamma_{bs}\right)^2}\right)$$

In addition, we restrict Ω such that

$$\Omega^4 < \frac{1}{\gamma_{bs} R_o^4}$$

This will ensure that $\alpha^2 > 0$.

The inverse transforms of Eq. (i) are obtained from pairs 31 to 34 in Table 8.2 of Appendix 8.1 as

$$\begin{aligned}
\frac{s^3}{(s^2-\alpha^2)(s^2+\beta^2)} &\Leftrightarrow Q(\eta,\alpha,\beta)=\frac{1}{\alpha^2+\beta^2}\left[\beta^2\cos(\beta t)+\alpha^2\cosh(\alpha t)\right]\\
\frac{s^2}{(s^2-\alpha^2)(s^2+\beta^2)} &\Leftrightarrow R(\eta,\alpha,\beta)=\frac{1}{\alpha^2+\beta^2}\left[\beta\sin(\beta t)+\alpha\sinh(\alpha t)\right]\\
\frac{s}{(s^2-\alpha^2)(s^2+\beta^2)} &\Leftrightarrow S(\eta,\alpha,\beta)=\frac{1}{\alpha^2+\beta^2}\left[-\cos(\beta t)+\cosh(\alpha t)\right]\\
\frac{1}{(s^2-\alpha^2)(s^2+\beta^2)} &\Leftrightarrow T(\eta,\alpha,\beta)=\frac{1}{\alpha^2+\beta^2}\left[-\frac{1}{\beta}\sin(\beta\eta)+\frac{1}{\alpha}\sinh(\alpha\eta)\right]
\end{aligned} \tag{j}$$

Then, taking the inverse Laplace transform of Eq. (i) yields

$$\begin{aligned}
Y(\eta)&=g_1(\eta,\Omega)\Psi'(0)+g_2(\eta,\Omega)Y'(0)\\
\Psi(\eta)&=g_3(\eta,\Omega)\Psi'(0)+g_4(\eta,\Omega)Y'(0)
\end{aligned} \tag{k}$$

where

$$\begin{aligned}
g_1(\eta,\Omega)&=S(\eta,\alpha,\beta)\\
g_2(\eta,\Omega)&=R(\eta,\alpha,\beta)+\frac{\left(\gamma_{bs}R_o^4\Omega^4-1\right)}{\gamma_{bs}R_o^2}T(\eta,\alpha,\beta)\\
g_3(\eta,\Omega)&=\gamma_{bs}R_o^2\Omega^4T(\eta,\alpha,\beta)+R(\eta,\alpha,\beta)\\
g_4(\eta,\Omega)&=\frac{g_1(\eta)}{\gamma_{bs}R_o^2}
\end{aligned}$$

We now substitute Eq. (k) into the second set of boundary conditions given by Eq. (f) to obtain

$$\begin{pmatrix} g_1(1,\Omega) & g_2(1,\Omega)\\ g_3(1,\Omega) & g_4(1,\Omega)\end{pmatrix}\begin{Bmatrix}\Psi'(0)\\ Y'(0)\end{Bmatrix}=0 \tag{l}$$

The eigenvalues Ω_n are those values of Ω for which the determinant of the coefficients of Eq. (l) equals zero, that is,

$$g_1(1,\Omega_n)g_4(1,\Omega_n)-g_2(1,\Omega_n)g_3(1,\Omega_n)=0 \tag{m}$$

Upon using Eq. (6.36), the corresponding eigenfunctions are obtained from the first rows of Eq. (l) and Eq. (k) as

$$\begin{aligned}
Y_n(\eta)&=C_ng_1(\eta,\Omega_n)+g_2(\eta,\Omega_n)\\
\Psi_n(\eta)&=C_ng_3(\eta,\Omega_n)+g_4(\eta,\Omega_n)
\end{aligned} \tag{n}$$

where

$$C_n = \frac{\Psi'(0)}{Y'(0)} = -\frac{g_2(\eta,\Omega_n)}{g_1(\eta,\Omega_n)}$$

and for convenience we have set $Y'(0) = 1$.

Thus, $Y_n(\eta)$ and $\Psi_n(\eta)$ are solutions to

$$\begin{aligned} &\frac{\partial}{\partial\eta}\left(\frac{\partial Y_n}{\partial\eta} - \Psi_n\right) + \gamma_{bs} R_o^2 \Omega_n^4 Y_n = 0 \\ &\frac{\partial Y_n}{\partial\eta} - \Psi_n + \gamma_{bs} R_o^2 \frac{\partial^2 \Psi_n}{\partial\eta^2} + \gamma_{bs} R_o^4 \Omega_n^4 \Psi_n = 0 \end{aligned} \tag{o}$$

and the boundary conditions

$$\begin{aligned} Y_n(0) &= \Psi_n(0) = 0 \\ Y_n(1) &= \Psi_n(1) = 0 \end{aligned} \tag{p}$$

We now proceed with obtaining the solution to Eq. (a) by assuming separable solutions of the form

$$\begin{aligned} y(\eta,\tau) &= \sum_{n=1}^{\infty} Y_n(\eta)\varphi_n(\tau) \\ \psi(\eta,\tau) &= \sum_{n=1}^{\infty} \Psi_n(\eta)\varphi_n(\tau) \end{aligned} \tag{q}$$

where $\varphi_n(\tau)$ is to be determined. Upon substituting Eq. (q) into Eq. (a), we obtain

$$\begin{aligned} &\sum_{n=1}^{\infty}\left\{\frac{\partial}{\partial\eta}\left(\frac{\partial Y_n}{\partial\eta} - \Psi_n\right)\varphi_n(\tau) - \gamma_{bs} R_o^2 \frac{\partial^2 \varphi_n(\tau)}{\partial\tau^2} Y_n\right\} = -\gamma_{bs} R_o^2 \hat{f}_a(\eta,\tau) \\ &\sum_{n=1}^{\infty}\left\{\left(\frac{\partial Y_n}{\partial\eta} - \Psi_n + \gamma_{bs} R_o^2 \frac{\partial^2 \Psi_n}{\partial\eta^2}\right)\varphi_n(\tau) - \gamma_{bs} R_o^4 \frac{\partial^2 \varphi_n(\tau)}{\partial\tau^2} \Psi_n\right\} = 0 \end{aligned} \tag{r}$$

Next, we use Eq. (o) in Eq. (r) to arrive at

$$\begin{aligned} &\sum_{n=1}^{\infty} Y_n \left\{\Omega_n^4 \varphi_n(\tau) + \frac{\partial^2 \varphi_n(\tau)}{\partial\tau^2}\right\} = \hat{f}_a(\eta,\tau) \\ &\sum_{n=1}^{\infty} \Psi_n \left\{\Omega_n^4 \varphi_n(\tau) + \frac{\partial^2 \varphi_n(\tau)}{\partial\tau^2}\right\} = 0 \end{aligned} \tag{s}$$

To take advantage of the fact that $Y_n(\eta)$ and $\Psi_n(\eta)$ are orthogonal functions, we note that the orthogonality condition for this system is given by Eq. (6.162), that is,

$$\gamma_{bs}R_o^2\int_0^1\left[Y_n(\eta)Y_m(\eta)+R_o^2\Psi_n(\eta)\Psi_m(\eta)\right]d\eta=\delta_{nm}N_n \tag{t}$$

where

$$N_n=\gamma_{bs}R_o^2\int_0^1\left[Y_n^2(\eta)+R_o^2\Psi_n^2(\eta)\right]d\eta \tag{u}$$

Therefore, we multiply the first equation of Eq. (s) by $\gamma_{bs}R_o^2Y_m(\eta)$ and the second equation by $\gamma_{bs}R_o^4\Psi_m(\eta)$, integrate each of them over the length of the beam, and add them to obtain

$$\sum_{n=1}^{\infty}\left\{\frac{\partial^2\varphi_n(\tau)}{\partial\tau^2}+\Omega_n^4\varphi_n(\tau)\right\}\gamma_{bs}R_o^2\int_0^1\left[Y_n(\eta)Y_m(\eta)+R_o^2\Psi_n(\eta)\Psi_m(\eta)\right]d\eta=h_m(\tau) \tag{v}$$

where

$$h_m(\tau)=\gamma_{bs}R_o^2\int_0^1 Y_m(\eta)\hat{f}_a(\eta,\tau)d\eta \tag{w}$$

However, from Eq. (t), we find that Eq. (v) becomes

$$\frac{\partial^2\varphi_n(\tau)}{\partial\tau^2}+\Omega_n^4\varphi_n(\tau)=\frac{h_n(\tau)}{N_n} \tag{x}$$

where N_n is given by Eq. (u). The solution to Eq. (x) is obtained from Table 4.3 for zero initial conditions and $\zeta = 0$ as

$$\varphi_n(\tau)=\frac{1}{N_n\Omega_n^2}\int_0^\tau h_n(\xi)\sin\left(\Omega_n^2(\tau-\xi)\right)d\xi \tag{y}$$

Substitution of Eq. (w) into Eq. (y) and the result into Eq. (q) gives the following solution in terms of an arbitrary force $\hat{f}_a(\eta,\tau)$

$$\begin{aligned}y(\eta,\tau)&=\gamma_{bs}R_o^2\sum_{n=1}^{\infty}\frac{Y_n(\eta)}{N_n\Omega_n^2}\int_0^\tau\left(\int_0^1 Y_n(\eta)\hat{f}_a(\eta,\xi)d\eta\right)\sin\left(\Omega_n^2(\tau-\xi)\right)d\xi\\ \psi(\eta,\tau)&=\gamma_{bs}R_o^2\sum_{n=1}^{\infty}\frac{\Psi_n(\eta)}{N_n\Omega_n^2}\int_0^\tau\left(\int_0^1 Y_n(\eta)\hat{f}_a(\eta,\xi)d\eta\right)\sin\left(\Omega_n^2(\tau-\xi)\right)d\xi\end{aligned} \tag{z}$$

MATHEMATICA PROCEDURES

Note: In order to make the Mathematica procedures that follow a little more readable, the `Clear` statement and many of the graphics-enhancing statements have been omitted.

```
(*M8.1*)
Simplify[LaplaceTransform[UnitStep[t-a],t,s],Assumptions->a>0]

(*M8.2*)
LaplaceTransform[t^k,t,s]

(*M8.3*)
LaplaceTransform[Sin[a t],t,s]
LaplaceTransform[Cos[a t],t,s]
LaplaceTransform[Sinh[a t],t,s]
LaplaceTransform[Cosh[a t],t,s]

(*M8.4*)
LaplaceTransform[Exp[a t],t,s]

(*M8.5*)
LaplaceTransform[Exp[b t] Sin[a t],t,s]
LaplaceTransform[Exp[b t] Cos[a t],t,s]
LaplaceTransform[Exp[b t] t^m,t,s]

(*M8.6*)
Expand[InverseLaplaceTransform[s/(s^2+1)/(s+2)^2,s,t]]

(*M8.7*)
g=(s^3+(eps^2-delta^2) s)/((s^2-delta^2) (s^2+eps^2));
FullSimplify[InverseLaplaceTransform[g,s,t]]

(*M8.8*)
InverseLaplaceTransform[1/s^2/(s-a),s,t]

(*M8.9*)
InverseLaplaceTransform[1/(s^2+k^2)^2,s,t]

(*M8.10*)
sol=LaplaceTransform[r i[t]+Lo i'[t]+1/c*
   Integrate[i[x],{x,0,t}]-v[t],t,s]/.
   {LaplaceTransform[i[t],t,s]->is,
   LaplaceTransform[v[t],t,s]->vs};
is/.Solve[sol==0,is][[1]]

(*M8.11*)
trin[t_,n_]:=(an=2 n; bn=2 n+2;
   (-1)^n (t-an)UnitStep[t-an]-(-1)^n (t-bn)UnitStep[t-bn])
Plot[Total[Table[trin[t,n],{n,0,4}]],{t,0,8}]
```

```
(*M8.12*)
Residue[Exp[z t]/(z Sinh[a z]),{z,0}]+
   Total[Table[Simplify[ComplexExpand[Residue[
  Exp[z t]/(z Sinh[a z]),{z,-I n Pi/a}]+
  Residue[Exp[z t]/(z Sinh[a z]),{z,I n Pi/a}]]],{n,1,4}]]

(*M8.13*)
so=Residue[Exp[z t]Sinh[a z]/z^2/Cosh[b z],{z,0}]+
   Total[Table[Simplify[ComplexExpand[Residue[Exp[z t]*
   Sinh[a z]/z^2/Cosh[b z],{z,I Pi/b (2 n-1)/2}]+
   Residue[Exp[z t]Sinh[a z]/z^2/Cosh[b z],
   {z,-I Pi/b (2 n-1)/2}]]],{n,1,4}]]
so/.{a->1,b->1}
fgn[t_]:=1-8/Pi^2 Total[Table[Cos[(2 n-1) t Pi/2]/((2 n-1))^2,
   {n,1,25}]]
Plot[fgn[t],{t,0,8}]

(*M8.14*)
der=Simplify[D[Sqrt[z] Exp[z t] Cosh[eta Sqrt[z]]/
   Sinh[b Sqrt[z]],z]]
Limit[der,z->0]

(*M8.15*)
es=0.35;  as=es (1-es)^2;
bs=(1+2 es) (1-es)^2;
yfcn[eta_,Ks_]:=(ys=es^2 (1-es)^2/24/(1-Ks(-as/2 es^2+bs/6 es^3));
   Ks ys (-as/2 eta^2+bs/6 eta^3-(eta-es)^3/6 UnitStep[eta-es])+
   eta^2 (eta-1)^2/24)
Plot[{-yfcn[eta,0],-yfcn[eta,100],-yfcn[eta,900],
   -yfcn[eta,100000]},{eta,0,1}]

(*M8.16*)
mr=0.3; wr=0.85; z1=0.1; z2=0.1;
D1=s^4+(2 z1+2 z2 wr mr+2 z2 wr) s^3+
   (1+mr wr^2+wr^2+4 z1 z2 wr) s^2+(2 z2 wr+2 z1 wr^2) s+wr^2;
Bs=2 z2 mr wr s+mr wr^2;
As=s^2+2 (z1+z2 mr wr) s+1+mr wr^2;
X1=InverseLaplaceTransform[Bs/D1/mr,s,t];
X2=InverseLaplaceTransform[As/D1/mr,s,t];
Plot[{Chop[X1],Chop[X2]},{t,0,40},PlotRange->All]

(*M8.17*)
InverseLaplaceTransform[1/s-a Exp[-eta Sqrt[s]]/s^(3/2),s,t]

(*M8.18*)
th[eta_,t_,a_]:=1+a eta Erfc[eta/2/Sqrt[t]]-
   2 a Sqrt[t/Pi] Exp[-eta^2/4/t]
Plot3D[th[eta,t,1],{eta,0.01,3},{t,0,2}]
```

```
(*M8.19*)
nroot=5;
rt1=x/.FindRoot[BesselJ[1,x],{x,4}];
rte=Table[x/.FindRoot[BesselJ[1,x],{x,rt1+(n-1) Pi}],{n,2,nroot}];
b=Flatten[{rt1,rte}];
w[r_,t_]:=r+2 Total[Table[BesselJ[1,b[[n]] r] Exp[-t b[[n]]^2]/
   (b[[n]] BesselJ[0,b[[n]]]),{n,1,nroot}]]
Plot3D[w[x,t],{x,0,1},{t,0.01,0.1}]

(*M8.20*)
Simplify[D[Exp[-q eta] (b3 Cos[q eta]+ b4 Sin[q eta]),eta,eta]]

(*M8.21*)
Integrate[Cos[eta^2/(4 x)]/Sqrt[x]/Sqrt[(t-x)],{x,0,t}]/Pi

(*M8.22*)
Plot3D[1-FresnelC[eta/Sqrt[2 Pi t]]-FresnelS[eta/Sqrt[2 Pi t]],
   {t,0,2},{eta,0,10}]

(*M8.23*)
InverseLaplaceTransform[(Exp[-s eta/alpha]-Exp[-eta s])/s^2,s,t]

(*M8.24*)
alpha=2.5;
Plot3D[((t-eta/alpha) UnitStep[t-eta/alpha]-(t-eta)*
   UnitStep[t-eta])/(alpha^2-1),{t,0,600},{eta,0,2000}]

(*M8.25*)
Limit[Exp[s z] (1-(m s Sinh[s (1-n)]+
   Cosh[s (1-n)])/(m s Sinh[s]+Cosh[s]))/s^2,s->0]

(*M8.26*)
tant[x_,mm_]:=mm x Tan[x]-1
fcn2[b_,t_,m_,n_]:=2 Total[Table[(-m b[[nn]] Sin[b[[nn]] (1-n)]+
   Cos[b[[nn]] (1-n)]) Sin[b[[nn]] t]/(b[[nn]]^2 ((m b[[nn]])^2+
   (1+m)) Sin[b[[nn]]]),{nn,1,nroot}]]
plt[mm_,eta_]:=(nroot=10;
rt1=x/.FindRoot[tant[x,mm],{x,0.5}];
rte=Table[x/.FindRoot[tant[x,mm],{x,rt1+(n-1) Pi}],{n,2,nroot}];
rt=Flatten[{rt1,rte}];
Plot[fcn2[rt,t,mm,eta],{t,0,10}])
GraphicsGrid[{{plt[0.1,1],plt[0.5,1]},{plt[1,1],plt[2,1]}}]

(*M8.27*)
Q[Om_,eta_]:=0.5 (Cos[Om eta]+Cosh[Om eta])
R[Om_,eta_]:=0.5 (Sin[Om eta]+Sinh[Om eta])
S[Om_,eta_]:=0.5 (-Cos[Om eta]+Cosh[Om eta])
T[Om_,eta_]:=0.5 (-Sin[Om eta]+Sinh[Om eta])
C1bar[Om_,eta_,etao_]:= T[Om,eta] S[Om,1-etao]-
   S[Om,eta] T[Om,1-etao]
C2bar[Om_,eta_,etao_]:= R[Om,eta] T[Om,1-etao]-
   S[Om,eta] S[Om,1-etao]
```

```
doo[Om_,eta_]:=S[Om,eta]^2-T[Om,eta] R[Om,eta]
char[Om_,eta_,mo_,etao_]:=doo[Om,eta]-
   mo Om (C1bar[Om,eta,etao] S[Om,etao]+
   C2bar[Om,eta,etao] T[Om,etao])
Yn[eta_,Om_,etao_]:=C1bar[Om,1,etao]/doo[Om,1] S[Om,eta]+
   C2bar[Om,1,etao]/doo[Om,1] T[Om,eta]+
   T[Om,eta-etao] UnitStep[eta-etao]
etao=0.72; mo=0.5; etaf=0.45;
rte=x/.FindRoot[char[x,1,mo,etao],{x,{4,8,12,14}}]//Quiet;
mx=Table[Max[Table[Abs[Yn[x,rte[[n]],etao]],{x,0,1,0.01}]],
   {n,1,Length[rte]}];
Nn=Table[NIntegrate[(Yn[x,rte[[n]],etao]/mx[[n]])^2,{x,0,1}]+
   mo (Yn[etao,rte[[n]],etao]/mx[[n]])^2,{n,1,Length[rte]}];
yet[eta_,t_,etao_,etaf_]:=Table[(Yn[etaf,rte[[n]],etao]/mx[[n]])^2
   /(Nn[[n]]rte[[n]]^2) Sin[rte[[n]]^2 t] Yn[eta,rte[[n]],etao]/
   mx[[n]],{n,1,Length[rte]}]
Plot3D[Total[yet[eta,t,etao,etaf]],{eta,0,1},{t,0,0.5},
   Exclusions->None]
```

EXERCISES

Section 8.1.2

8.1 Obtain the Laplace transform of

$$\cosh t \sin t$$

using Eq. (8.16).

8.2 Obtain the Laplace transform of

$$e^{-t} t \cos t$$

8.3 Use Eqs. (8.6) and (8.7) to obtain the solution to the following equation

$$x\frac{d^2y}{dx^2} - 2(x+1)\frac{dy}{dx} + (x+2)y = 0$$

Section 8.1.3

8.4 Use partial fractions to determine the inverse Laplace transform of

$$L^{-1}\left[\frac{a^3}{s(s+a)^3}\right]$$

8.5 Use partial fractions to determine the inverse Laplace transform of

$$F(s) = \frac{6+(s+1)(s^2-5s+11)}{s(s-1)(s-2)(s+1)}$$

8.6 Determine the inverse Laplace transform of

$$L^{-1}\left[\frac{e^{-\pi s}}{1+s^2}\right]$$

8.7 Determine the Laplace transform of

$$e^{-t}\int_0^t e^{-\tau}\tau\cos[\omega(t-\tau)]d\tau$$

8.8 Determine the inverse Laplace transform of

$$\frac{s}{\left(s^2+k^2\right)^2}$$

Section 8.1.6

8.9 Obtain the Laplace transform of the following periodic waveform (sawtooth wave) of period T

$$f(t)=t/T \quad 0\le t\le T$$

8.10 Obtain the Laplace transform of the following periodic waveform (triangular wave) of period T

$$\begin{aligned} f(t) &= 2t/T \quad 0\le t\le T/2 \\ &= 2(1-t/T) \quad T/2\le t\le T \end{aligned}$$

Section 8.1.7

8.11 Use the Bromwich inversion integral to determine the inverse Laplace transform of

$$F(s)=\frac{1}{(s+b)\cosh(as)}$$

8.12 Use the Bromwich inversion integral to determine the inverse Laplace transform of

$$F(s)=\frac{\cosh(x\sqrt{s/a})}{s\cosh(k\sqrt{s/a})}$$

8.13 Use the Bromwich inversion integral to determine the inverse Laplace transform of

$$F(s)=\frac{\sinh(a\sqrt{s})}{s\sinh(\sqrt{s})}$$

8.14 Use the Bromwich inversion integral to determine the inverse Laplace transform of

$$F(s) = \frac{\sinh\left(a\sqrt{s}\right)}{s\left(m_o\sqrt{s}\cosh\left(a\sqrt{s}\right) + \sinh\left(a\sqrt{s}\right)\right)} \tag{a}$$

Section 8.2

8.15 For arbitrary initial conditions, use the Laplace transform to determine the general solution to

$$\frac{d^2y}{dt^2} - 2k\frac{dy}{dt} + k^2y = f(t)$$

8.16 The non-dimensional form of the one-dimensional heat flow equation in a solid is given by

$$\frac{d\theta}{d\tau} = \frac{\partial^2\theta}{\partial\eta^2} \tag{a}$$

It is assumed that the boundary conditions are

$$\frac{\partial\theta(0,\tau)}{\partial\eta} = 0, \quad \theta(1,\tau) = \theta_1 \tag{b}$$

and the initial condition are

$$\theta(\eta,0) = 1 \tag{c}$$

Use the Laplace transform with respect to τ to obtain a solution to Eq. (a) subject to the conditions given by Eqs. (b) and (c).

8.17 Consider a hanging chain of length L that is free at $x = 0$ and fixed at $x = L$. The governing equation of motion for this system is

$$\frac{\partial}{\partial x}\left(x\frac{\partial u}{\partial x}\right) = \frac{1}{g}\frac{\partial^2 u}{\partial t^2} \tag{a}$$

where g is the gravity constant. The boundary conditions are

$$\begin{aligned} u(L,t) &= A\sin\omega t \\ u(0,t) &= \text{finite} \end{aligned} \tag{b}$$

that is, one end is oscillating harmonically at a frequency ω. The initial conditions are assumed to be zero. Use the Laplace transform with respect to τ to obtain a solution to Eq. (a) subject to the conditions given by Eq. (b).

8.18 Consider the non-dimensional equation governing the static displacement of a beam having an in-span elastic support at η_s, $0 < \eta_s < 1$, and subjected to an externally applied load

$$q(\eta) = q_o\delta(\eta - \eta_o), \quad 0 < h_o < 1$$

$$\frac{d^4w}{d\eta^4} + K_s\delta(\eta - \eta_s)y = q_o\delta(\eta - \eta_o) \quad 0 \le \eta \le 1$$

The boundary condition at $\eta = 0$ are

$$w'''(0) = -K_L w(0), \quad w''(0) = K_{tL} w'(0)$$

and those at $\eta = 1$ are

$$w'''(1) = -K_R w(1), \quad w''(1) = K_{tR} w'(1)$$

where the prime indicates the derivative with respect to η and the K_s are constants that can vary between 0 and ∞. Using the Laplace transform method, obtain the general solution for $w(\eta)$ in terms of $w(\eta_s)$, K_α, and q_o. Judiciously define groups of terms at appropriate solution stages in the solution process; this will greatly simplify the algebra.

8.19 Take the Laplace transform with respect to τ to obtain a solution to the following non-dimensional equation

$$\frac{\partial u}{\partial \tau} = \frac{\partial^2 u}{\partial \eta^2} \quad 0 \le \eta \le 1 \quad \tau > 0$$

with zero initial conditions and with the boundary conditions

$$\frac{\partial u(0,\tau)}{\partial \eta} = -1 \quad \frac{\partial u(1,\tau)}{\partial \eta} = 0$$

8.20 Take the Laplace transform with respect to τ to obtain a solution to the following partial differential equation

$$\frac{\partial u}{\partial \tau} = \frac{\partial^2 u}{\partial \eta^2} + \frac{\partial u}{\eta \partial \eta} \quad 0 \le \eta \le 1 \quad \tau > 0$$

with the boundary conditions

$$\lim_{\eta \to 0} |u(\eta,\tau)| < \infty$$

$$u(1,\tau) = \sin \Omega\tau$$

and with the initial condition $u(\eta,0) = 0$.

8.21 Take the Laplace transform with respect to t to obtain a solution to the following partial differential equation

$$\frac{\partial u}{\partial t} = a^2 \frac{\partial^2 u}{\partial x^2} \quad 0 \le x \le L \quad t > 0$$

with the boundary conditions

$$\frac{\partial u(0,t)}{\partial x} = 0$$

$$\kappa \frac{\partial u(L,t)}{\partial x} = \alpha(1 - \beta t)$$

and with zero initial condition.

8.22 Take the Laplace transform with respect to t to obtain a solution to the following partial differential equation

$$\frac{\partial u}{\partial t} = a^2 \frac{\partial^2 u}{\partial x^2} \quad 0 \le x < \infty \quad t > 0$$

with the boundary conditions that u is finite as $x \to \infty$, $u(0,t) = f(t)$, and the initial condition is zero.

8.23 Take the Laplace transform with respect to t to obtain a solution to the following partial differential equation

$$\frac{\partial u}{\partial t} = \frac{\partial^2 u}{\partial x^2} - a^2 u \quad -1 \le x < 1 \quad t > 0$$

with the boundary conditions

$$u(-1,t) = 1$$

$$u(1,t) = 0$$

and zero initial condition, that is, $u(x,0) = 0$.

APPENDIX 8.1

Laplace Transform Pairs

TABLE 8.2
Laplace Transform Pairs

	$g(t)\ (t \geq 0)$	$G(s)$	Comment
1	$ag(at)$	$G(s/a)$	Scaling of variable
2	$a_1g_1(t)+a_2g_2(t)$	$a_1G_1(s)+a_2G_2(s)$	Linearity theorem
3	$g(t)e^{at}$	$G(s-a)$	s-domain shifting
4	$\frac{1}{a}f(t/a)e^{bt/a}$	$F(as-b)$	Scaling and shifting
5	$g^n(t)=\frac{d^ng}{dt^n}$	$s^nG(s)-\sum_{k=1}^{n}s^{n-k}g^{k-1}(0)$	nth-order derivative, $n = 1, 2, \ldots$
6	$g(t-t_o)\,u(t-t_o)$	$e^{-t_os}G(s)$	Time shifting, $u(t-t_o)$ is unit step function, $t_o \geq 0$
7	$\int_0^t g(\eta)h(t-\eta)d\eta=\int_0^t g(t-\eta)h(\eta)d\eta$	$G(s)H(s)$	Convolution
8	$\lim_{t\to0^+} g(t)$	$\lim_{s\to\infty} sG(s)$	
9	$\lim_{t\to\infty} g(t)$	$\lim_{s\to0} sG(s)$	
10	e^{at}	$\frac{1}{s-a}$	Exponential
11	$\int_0^t f(\xi)d\xi$	$\frac{F(s)}{s}$	Integral
12	$u(t-t_o)$	$\frac{e^{-st_o}}{s}$	Unit step function
13	$\delta(t-t_o)$	e^{-st_o}	Delta function
14	$\delta(t-t_o)f(t)$	$e^{-st_o}f(t_o)$	From definition of delta function
15	$\frac{d}{dt}\delta(t-t_o)$	se^{-st_o}	Derivative of delta function
16	$\frac{t^a}{\Gamma(a+1)}$	$\frac{1}{s^{a+1}}$	$\alpha > -1$; $\Gamma(x)$ = gamma function
17	$\frac{t^n}{n!}$	$\frac{1}{s^{n+1}}$	$n = 0, 1, 2, \ldots$
18	$t^n f(t)$	$(-1)^n\frac{d^nF(s)}{ds^n}$	$n = 0, 1, 2, \ldots$
19	$\frac{f(t)}{t}$	$\int_s^\infty F(s)ds$	

(Continued)

TABLE 8.2 (CONTINUED)
Laplace Transform Pairs

	g(t) (t ≥ 0)	G(s)	Comment
20	$e^{bt}\, t^m$	$\frac{m!}{(s-b)^{m+1}}$	$m = 0,1,2, \ldots$
21	$\frac{1}{(b-a)}[e^{-at} - e^{-bt}]\quad a \neq b$ $te^{-bt}\quad a = b$	$\frac{1}{(s+a)(s+b)}$	
22	$\frac{1}{(b-a)}[be^{-bt} - ae^{-at}]\quad a \neq b$ $(1-bt)e^{-bt}\quad a = b$	$\frac{s}{(s+a)(s+b)}$	
23	$e^{bt}\cos(\omega t)$	$\frac{s-b}{(s-b)^2+\omega^2}$	
24	$e^{bt}\sin(\omega t)$	$\frac{\omega}{(s-b)^2+\omega^2}$	
25	$e^{bt}\cosh(\omega t)$	$\frac{s-b}{(s-b)^2-\omega^2}$	
26	$e^{bt}\sinh(\omega t)$	$\frac{\omega}{(s-b)^2-\omega^2}$	
27	$\frac{1}{2a^3}(\sinh(at) - \sin(at))$	$\frac{1}{s^4-a^4}$	Special case of 31
28	$\frac{1}{2a^2}(\cosh(at) - \cos(at))$	$\frac{s}{s^4-a^4}$	Special case of 32
29	$\frac{1}{2a}(\sinh(at) + \sin(at))$	$\frac{s^2}{s^4-a^4}$	Special case of 33
30	$\frac{1}{2}(\cosh(at) + \cos(at))$	$\frac{s^3}{s^4-a^4}$	Special case of 34
31	$\frac{1}{\alpha^2+\beta^2}\left[-\frac{1}{\beta}\sin(\beta t) + \frac{1}{\alpha}\sinh(\alpha t)\right]$	$\frac{1}{(s^2-\alpha^2)(s^2+\beta^2)}$	
32	$\frac{1}{\alpha^2+\beta^2}\left[-\cos(\beta t) + \cosh(\alpha t)\right]$	$\frac{s}{(s^2-\alpha^2)(s^2+\beta^2)}$	
33	$\frac{1}{\alpha^2+\beta^2}\left[\beta\sin(\beta t) + \alpha\sinh(\alpha t)\right]$	$\frac{s^2}{(s^2-\alpha^2)(s^2+\beta^2)}$	
34	$\frac{1}{\alpha^2+\beta^2}\left[\beta^2\cos(\beta t) + \alpha^2\cosh(\alpha t)\right]$	$\frac{s^3}{(s^2-\alpha^2)(s^2+\beta^2)}$	
35	$\frac{1}{\omega_d}e^{-\zeta\omega_n t}\sin(\omega_d t)$	$\frac{1}{s^2+2\zeta\omega_n s+\omega_n^2}$	$\omega_d = \omega_n\sqrt{1-\zeta^2}\quad 0 \leq \zeta < 1$

(Continued)

TABLE 8.2 (CONTINUED)
Laplace Transform Pairs

	$g(t)$ $(t \ge 0)$	$G(s)$	Comment
36	$1-\frac{\omega_n}{\omega_d}e^{-\zeta\omega_n t}\sin(\omega_d t+\phi)$	$\frac{\omega_n^2}{s\left(s^2+2\zeta\omega_n s+\omega_n^2\right)}$	$\phi=\cos^{-1}\zeta \quad 0\le\zeta<1$ $\omega_d=\omega_n\sqrt{1-\zeta^2}$
37	$-\frac{\omega_n}{\omega_d}e^{-\zeta\omega_n t}\sin(\omega_d t-\phi)$	$\frac{s}{s^2+2\zeta\omega_n s+\omega_n^2}$	$\phi=\cos^{-1}\zeta \quad 0\le\zeta<1$ $\omega_d=\omega_n\sqrt{1-\zeta^2}$
38	$\frac{\omega_n}{\omega_d}e^{-\zeta\omega_n t}\sin(\omega_d t+\phi)$	$\frac{s+2\zeta\omega_n}{s^2+2\zeta\omega_n s+\omega_n^2}$	$\phi=\cos^{-1}\zeta \quad 0\le\zeta<1$ $\omega_d=\omega_n\sqrt{1-\zeta^2}$
39	$u(t)-u(t-t_o)$	$\frac{1-e^{-st_o}}{s}$	Rectangular pulse of duration t_o 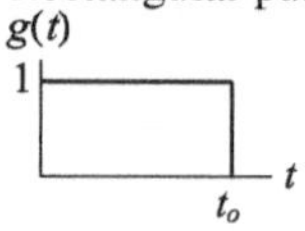
40	$u(t)-2u(t-t_o)+u(t-2t_o)$	$\frac{\left(1-e^{-t_o s}\right)^2}{s}$	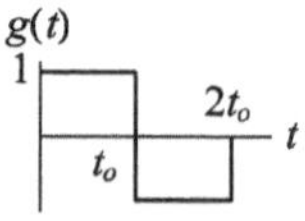
41	$\sin(\omega_o t)[u(t)-u(t-\pi/\omega_o)]$	$\frac{\omega_o(1+e^{-\pi s/\omega_o})}{s^2+\omega_o^2}$	Half sine wave of frequency ω_o 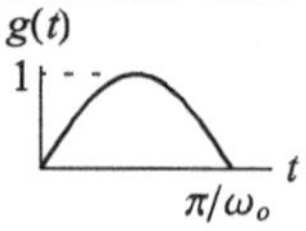
42	$(t/t_o)-(t-t_o)/t_o u(t-t_o)-(t$ $-t_1)/t_o u(t-t_1)+(t-t_1-t_o)/t_o)u(t$ $-t_1-t_o)]$	$\frac{1}{t_o s^2}\left(1-e^{-st_o}\right)\left(1-e^{-st_1}\right)$	Double ramp 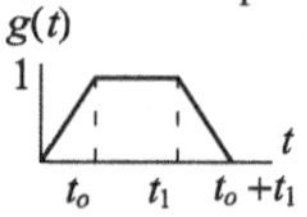
43	$t/t_o)[u(t)-u(t-t_o)]+(-t/t_o+2)$ $[u(t-t_o)-u(t-2t_o)]$	$\frac{1}{t_o s^2}\left(1-e^{-st_o}\right)^2$	Triangle 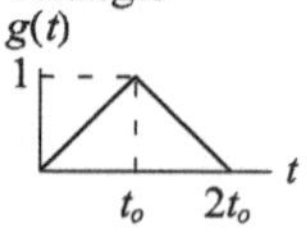
44	$(1-t/t_o)[u(t)-u(t-t_o)]$	$\frac{e^{-t_o s}+t_o s-1}{t_o s^2}$	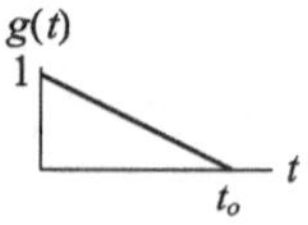

9 Putting It All Together—Examples from the Literature

9.1 INTRODUCTION

In this chapter, we shall apply many of the results of the preceding chapters to solve a variety of problems from different engineering disciplines. In particular, we have selected a set of multi-physics applications taken from fluid mechanics, solid mechanics, electromagnetics, and heat transfer. Both linear and nonlinear examples have been selected. The examples have been chosen to emphasize the usefulness and applicability of techniques and methods presented in the previous chapters.

The applications chosen and their corresponding section numbers are as follows:

- 9.2 Squeeze film air damping and the effects on the response of a single-degree-of-freedom system
- 9.3 Viscous fluid damping and the effects on the response of a single-degree-of-freedom system
- 9.4 Natural frequencies of a cantilever beam with an in-span spring–mass system
- 9.5 Piezoelectric energy harvester using a single-degree-of-freedom system
- 9.6 Determination of the onset of flutter
- 9.7 Thermal runaway in microwave heating of ceramics

A summary of equations and boundary conditions and/or external disturbances for each of these examples is presented in Table 9.1.

9.2 SQUEEZE FILM AIR DAMPING

9.2.1 Introduction

At the micromechanical scale, it has been found that the reaction forces generated by partially restricted viscous airflow can significantly affect a system's response. This phenomenon is called *squeeze film damping.*

Squeeze film air damping is caused by the entrapment of air between two parallel surfaces that are moving relative to each other in the normal direction. The reactive pressure brought about by the relative movement of the surfaces consists of two components. One component is due to the viscous flow of air that is squeezed out of the volume between the two surfaces, and the second component is due to compression of the air between the two surfaces. It will be shown that the former component results in a frequency-dependent damping and the latter component a frequency-dependent stiffness. We shall consider the case of a rigid rectangular surface. The governing equation that is used to determine the pressure between the surfaces is the linearized Reynolds lubrication equation. The force in the air gap is determined by integrating the pressure over the area of the surface. When one

TABLE 9.1
Summary of Equations and Boundary Conditions of Systems Examined in Chapter 9

Section	Description	Equations	Boundary Conditions and/or External Disturbance
9.2	Squeeze film damping	$\frac{\partial^2 \hat{p}}{\partial \eta^2}+\frac{\partial^2 \hat{p}}{\partial \xi^2}-\sigma\frac{\partial \hat{p}}{\partial \tau}=-\sigma\delta\sin\tau$	$\hat{p}(\pm\beta/2,\xi,\tau)=0$ $\hat{p}(\eta,\pm 1/2,\tau)=0$
9.3	Viscous fluid damping	$\nabla^2_{r\theta}\left(\nabla^2_{r\theta}\psi-\frac{\rho_f}{\mu_f}\frac{\partial\psi}{\partial t}\right)=0$ $\nabla^2_{r\theta}=\frac{\partial^2}{\partial r^2}+\frac{1}{r}\frac{\partial}{\partial r}+\frac{1}{r^2}\frac{\partial^2}{\partial\theta^2}$	$u_r(b/2,\theta,t)=u_o\cos\theta e^{j\omega t}$ $u_\theta(b/2,\theta,t)=-u_o\sin\theta e^{j\omega t}$
9.4	Beam with single-degree-of-freedom system	$\frac{\partial^4 y}{\partial \eta^4}+K_s\left(y-\hat{z}(\tau)\right)\delta(\eta-\eta_s)+\frac{\partial^2 y}{\partial \tau^2}=\hat{f}(\eta,\tau)$ $\frac{1}{\Omega_s^4}\frac{d^2\hat{z}}{d\tau^2}+\hat{z}=y(\eta_s,\tau)$	$y(0,\tau)=y'(0,\tau)=0$ $y''(1,\tau)=y'''(1,\tau)=0$
9.5	Energy harvester	$m\frac{d^2w}{dt^2}+c\frac{dw}{dt}+\left(k+K_p\right)w+e_{33}\frac{A}{h}V=-m\frac{d^2y}{dt^2}$ $C_pR_L\frac{dV}{dt}+V-e_{33}\frac{AR_L}{h}\frac{dw}{dt}=0$	$y=Y_oe^{j\omega t}$
9.6	Flutter frequencies	$EI\frac{\partial^4 w}{\partial z^4}+m\frac{\partial^2 w}{\partial t^2}-x_m m\frac{\partial^2\theta}{\partial t^2}=L(w,\theta,t)$ $-GJ\frac{\partial^2\theta}{\partial z^2}+I_o\frac{\partial^2\theta}{\partial t^2}-mx_m\frac{\partial^2 w}{\partial t^2}=M(w,\theta,t)$	$w(0,t)=w'(0,t)=0$ $\theta(0,t)=0$ $w''(L_a,t)=w'''(L_a,t)=0$ $\theta'(L_a,t)=0$
9.7	Microwave heating and thermal runaway	$\frac{d^2u}{d\xi^2}+K_s^2\left(1-jc_se^{\gamma\theta}\right)u=0$ $\frac{d^2\theta}{d\xi^2}+pc_se^{\gamma\theta}\lvert u\rvert^2=0$	$-j2K_m=u'(0)-jK_mu(0)$ $0=u'(1)+jK_mu(1)$ $\theta'(0)=Bi\left\{\theta(0)+R\left[\left(\theta(0)+1\right)^4-1\right]\right\}$ $\theta'(1)=-Bi\left\{\theta(1)+R\left[\left(\theta(1)+1\right)^4-1\right]\right\}$

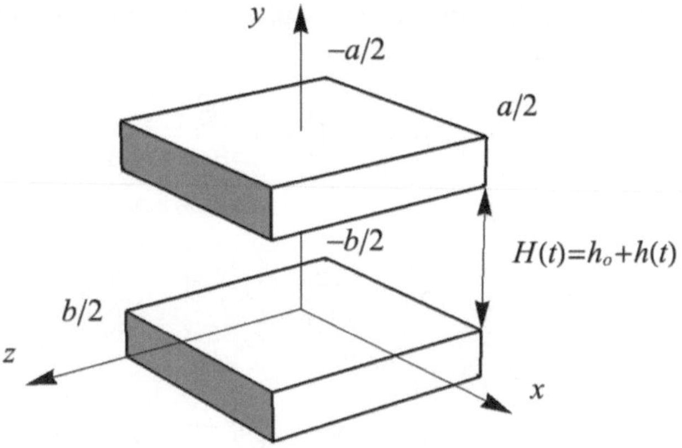

FIGURE 9.1 Squeeze film damping geometry for rigid parallel rectangular surfaces (not to scale).

of the surfaces is the surface of the mass of a single-degree-of-freedom system, the single-degree-of-freedom system is said to be subjected to squeeze film damping.

9.2.2 Squeeze Film Damping for Parallel Rectangular Surfaces Subject to Harmonic Excitation*

We shall be considering the forced harmonic oscillations of the moveable surface shown in Figure 9.1. The gas under consideration resides in the volume between the two surfaces. The governing equation used to determine the pressure $p = p(x,z,t)$ (N/m^2) between two parallel surfaces is the nonlinear Reynolds equation given by

$$\frac{\partial}{\partial x}\left(p\frac{\partial p}{\partial x}\right)+\frac{\partial}{\partial z}\left(p\frac{\partial p}{\partial z}\right)-\frac{12\mu}{H^3}\frac{\partial(pH)}{\partial t}=0 \quad \text{N}^2/\text{m}^6 \tag{9.1}$$

where μ (Ns/m^2) is the dynamic viscosity and H (m) is small compared to the dimensions of the surface. Equation (9.1) is linearized by assuming that: (i) $H = h_o + h(t)$ (m), where h_o is the constant thickness of the gas in the y-direction and the magnitude of the time-dependent change in height about h_o is such that $h(t) << h_o$; and (ii) the pressure p is independent of y, that is, $p = p(x,z,t)$ (N/m^2) and $p(x,z,t) = p_a + P(x,z,t)$, where p_a is the atmospheric pressure and is a constant value and the magnitude of $P << p_a$. The notation that $h = h(t)$ indicates that the opposing surfaces are parallel. With these assumptions, Eq. (9.1) becomes

$$\frac{\partial^2 P}{\partial x^2}+\frac{\partial^2 P}{\partial z^2}-\frac{12\mu}{h_o^3}\frac{\partial h}{\partial t}-\frac{12\mu}{p_a h_o^2}\frac{\partial P}{\partial t}=0 \quad \text{N/m}^4 \tag{9.2}$$

We introduce into Eq. (9.2) the following non-dimensional quantities

$$\hat{p}=\frac{P}{p_a},\quad \eta=\frac{x}{b},\quad \xi=\frac{z}{b},\quad \hat{h}=\frac{h}{h_o} \tag{9.3}$$

* Blech JJ (1983) On Isothermal Squeeze Films. *Journal of Lubrication Technology* 105: 615–620. See also: Crandall IB (1918) The Air-Damped Vibrating System: Theoretical Calibration of the Condenser Transmitter. *Physical Review* 11(6): 449–460; Perez MA, Shkel AM (2008); The Effect of Squeeze Film Constriction on Bandwidth Improvement in Interferometric Accelerometers. *Journal of Micromechanics and Microengineering* 18: 055031; and Veijola T, Kuisma H, Lahdenpera J, Ryhänen T (1995) Equivalent-Circuit Model of the Squeezed Gas Film in a Silicon Accelerometer. *Sensors and Actuators A* 48: 239–248.

where b (m) and a (m) are the dimensions of the plates as shown in Figure 9.1. Substitution of Eq. (9.3) into Eq. (9.2) yields

$$\nabla^2_{\eta\xi}\hat{p} - \sigma_o \frac{\partial \hat{p}}{\partial \tau} = \sigma_o \frac{\partial \hat{h}}{\partial \tau} \tag{9.4}$$

where

$$\sigma_o = \frac{12\mu b^2}{h_o^2 p_a} \quad \text{s} \tag{9.5}$$

and

$$\nabla^2_{\eta\xi} = \frac{\partial^2}{\partial \eta^2} + \frac{\partial^2}{\partial \xi^2} \tag{9.6}$$

The boundary conditions along the perimeter of the surface are that the film pressure is at atmospheric pressure, that is, $p = p_a$ and, therefore, $P = 0$ $(p = 0)$. Thus, if we place the coordinate system at the center of the surface, the boundary conditions are expressed as

$$\begin{aligned} \hat{p}(\pm\beta/2, \xi, t) &= 0 \\ \hat{p}(\eta, \pm 1/2, t) &= 0 \end{aligned} \tag{9.7}$$

where $\beta = a/b$.

We shall consider a system where one surface is fixed, and the other surface is undergoing harmonic oscillations at a frequency ω rad/s of the form

$$h = \varepsilon \cos \omega t \tag{9.8}$$

where ε (m) is the magnitude of the displacement. In terms of non-dimensional quantities, Eq. (9.8) is written as

$$\hat{h} = \delta \cos \omega t \tag{9.9}$$

where $\delta = \varepsilon/h_o$. Thus, Eq. (9.4) becomes

$$\nabla^2_{\eta\xi}\hat{p} - \sigma_o \frac{\partial \hat{p}}{\partial t} = -\sigma_o \delta \sin \omega t \tag{9.10}$$

In view of Eq. (9.10), we assume a solution of the form (recall Example 4.17)

$$\hat{p}(\eta, \xi, \tau) = P_1(\eta, \xi)\cos \omega t + P_2(\eta, \xi)\sin \omega t \tag{9.11}$$

Substituting Eq. (9.11) into Eq. (9.10), we obtain

$$\left(\nabla^2_{\eta\xi} P_1 - \sigma P_2\right)\cos \omega t + \left(\nabla^2_{\eta\xi} P_2 + \sigma P_1 + \sigma\delta\right)\sin \omega t = 0 \tag{9.12}$$

where $\sigma = \sigma_o\omega$ is called the *squeeze number*. Therefore, for arbitrary t, one must satisfy the following coupled partial differential equations

$$\nabla^2_{\eta\xi} P_1 = \sigma P_2 \tag{9.13}$$

and

$$\nabla^2_{\eta\xi} P_2 = -\sigma\left(P_1 + \delta\right) \tag{9.14}$$

Upon using Eq. (9.11) in Eq. (9.7), the boundary conditions become

$$\begin{aligned} &P_i(\pm\beta/2,\xi)=0 \\ &P_i(\eta,\pm 1/2)=0 \quad i=1,2 \end{aligned} \tag{9.15}$$

The quantity of interest is the force of the upper surface that is acting on the film. This is determined from the solutions to Eqs. (9.13) and (9.14) that satisfy the boundary conditions given by Eq. (9.15). The force acting on the gas is obtained from the integration of the pressure over the area of the surface, that is,

$$\begin{aligned} F &= \int_{-b/2}^{b/2}\int_{-a/2}^{a/2} P(x,y,t)dxdy = p_a b^2 \int_{-1/2}^{1/2}\int_{-\beta/2}^{\beta/2} p(\eta,\xi,\tau)d\eta d\xi \\ &= F_1\cos\omega t + F_2 \sin\omega t \quad \text{N} \end{aligned} \tag{9.16}$$

where we have used Eq. (9.11) and

$$F_i = p_a b^2 \int_{-1/2}^{1/2}\int_{-\beta/2}^{\beta/2} P_i(\eta,\xi)d\eta d\xi \quad \text{N} \quad i=1,2 \tag{9.17}$$

To solve Eqs. (9.13) and (9.14), we select the following solution composed of a set of orthogonal functions that satisfy the boundary conditions given by Eq. (9.15)

$$P_i(\eta,\xi)=\sum_{n=1,3,5}^{\infty}\sum_{m=1,3,5}^{\infty} A_{inm}\cos\left(n\pi\eta/\beta\right)\cos\left(m\pi\xi\right) \quad i=1,2 \tag{9.18}$$

The A_{inm} will be determined from the satisfaction of Eqs. (9.13) and (9.14). Substituting Eq. (9.18) into Eq. (9.17), we obtain

$$\begin{aligned} F_i &= p_a b^2 \sum_{n=1,3,5}^{\infty}\sum_{m=1,3,5}^{\infty} A_{inm} \int_{-\beta/2}^{\beta/2}\cos\left(n\pi\eta/\beta\right)d\eta \int_{-1/2}^{1/2}\cos\left(m\pi\xi\right)d\xi \\ &= \frac{4p_a b^2\beta}{\pi^2}\sum_{n=1,3,5}^{\infty}\sum_{m=1,3,5}^{\infty}\frac{A_{inm}}{nm}(-1)^{(n-1)/2}(-1)^{(m-1)/2} \quad i=1,2 \end{aligned} \tag{9.19}$$

where we have used the fact that

$$\begin{aligned} &\int_{-\beta/2}^{\beta/2}\cos\left(n\pi\eta/\beta\right)d\eta = \frac{2\beta}{n\pi}(-1)^{(n-1)/2} \\ &\int_{-1/2}^{1/2}\cos\left(m\pi\xi\right)d\xi = \frac{2}{m\pi}(-1)^{(m-1)/2} \end{aligned} \tag{9.20}$$

Since Eq. (9.18) is composed of orthogonal functions over the region indicated, we can expand the magnitude δ of the displacement of the oscillating surface appearing in Eq. (9.14) in terms of these functions. Thus, we let

$$\delta = \sum_{n=1,3,5}^{\infty} \sum_{m=1,3,5}^{\infty} d_{nm} \cos(n\pi\eta/\beta)\cos(m\pi\xi) \tag{9.21}$$

To obtain d_{nm}, we multiply Eq. (9.21) by $\cos(kn\pi\eta/\beta)\cos(q\pi\xi)$, where q and k are odd integers, and integrate over the plate area to arrive at

$$\begin{aligned} &\delta \int_{-\beta/2}^{\beta/2} \cos(k\pi\eta/\beta)d\eta \int_{-1/2}^{1/2} \cos(q\pi\xi)d\xi \\ &= \sum_{n=1,3,5}^{\infty} \sum_{m=1,3,5}^{\infty} d_{nm} \int_{-\beta/2}^{\beta/2} \cos(k\pi\eta/\beta)\cos(n\pi\eta/\beta)d\eta \int_{-1/2}^{1/2} \cos(m\pi\xi)\cos(q\pi\xi)d\xi \end{aligned} \tag{9.22}$$

However, from orthogonality,

$$\begin{aligned} &\int_{-\beta/2}^{\beta/2} \cos(k\pi\eta/\beta)\cos(n\pi\eta/\beta)d\eta = \frac{\beta}{2}\delta_{kn} \\ &\int_{-1/2}^{1/2} \cos(m\pi\xi)\cos(q\pi\xi)d\xi = \frac{1}{2}\delta_{mq} \end{aligned} \tag{9.23}$$

where δ_{ij} is the Kronecker delta. Then, using Eqs. (9.20) and (9.23) in Eq. (9.22), we obtain

$$\begin{aligned} &\sum_{n=1,3,5}^{\infty} \sum_{m=1,3,5}^{\infty} d_{nm} \left(\frac{\beta}{2}\delta_{kn}\right)\left(\frac{1}{2}\delta_{mq}\right) = \delta\left(\frac{2\beta}{k\pi}(-1)^{(k-1)/2}\right)\left(\frac{2}{q\pi}(-1)^{(q-1)/2}\right) \\ &d_{nm} = \frac{16\delta}{nm\pi^2}(-1)^{(n-1)/2}(-1)^{(m-1)/2} \qquad n,m = 1,3,5,\ldots \end{aligned} \tag{9.24}$$

Upon substituting Eq. (9.24) into Eq. (9.21) and the result, in turn, into Eqs. (9.13) and (9.14), we obtain

$$\begin{aligned} &\sum_{n=1,3,5}^{\infty} \sum_{m=1,3,5}^{\infty} \left[\alpha_{nm}A_{1nm} + \sigma A_{2nm}\right]\cos(n\pi\eta/\beta)\cos(m\pi\xi) = 0 \\ &\sum_{n=1,3,5}^{\infty} \sum_{m=1,3,5}^{\infty} \left[\alpha_{nm}A_{2nm} - \sigma\left(A_{1nm} + d_{nm}\right)\right]\cos(n\pi\eta/\beta)\cos(m\pi\xi) = 0 \end{aligned} \tag{9.25}$$

where

$$\alpha_{nm} = \left(\frac{n\pi}{\beta}\right)^2 + (m\pi)^2 \tag{9.26}$$

For Eq. (9.25) to be satisfied for arbitrary η and ξ,

$$\begin{aligned} A_{1nm}\alpha_{nm} + \sigma A_{2nm} &= 0 \\ -\sigma A_{1nm} + A_{2nm}\alpha_{nm} &= \sigma d_{nm} \end{aligned} \tag{9.27}$$

Solving Eq. (9.27) for A_{inm}, $i = 1, 2$, we obtain

$$A_{1nm} = \frac{\sigma^2 d_{nm}}{\alpha_{nm}^2 + \sigma^2} = \frac{-16\delta\sigma^2}{nm\pi^2\left(\alpha_{nm}^2 + \sigma^2\right)}(-1)^{(n-1)/2}(-1)^{(m-1)/2}$$

$$A_{2nm} = \frac{\sigma\alpha_{nm} d_{nm}}{\alpha_{nm}^2 + \sigma^2} = \frac{16\delta\sigma\alpha_{nm}}{nm\pi^2\left(\alpha_{nm}^2 + \sigma^2\right)}(-1)^{(n-1)/2}(-1)^{(m-1)/2} \tag{9.28}$$

where we have used Eq. (9.24).

Substituting Eq. (9.28) into Eq. (9.19), we arrive at

$$F_1 = -ab\delta\, p_a S_1(\beta,\sigma) \ \text{ N}$$

$$F_2 = ab\delta\, p_a S_2(\beta,\sigma) \ \text{ N} \tag{9.29}$$

where

$$S_1(\beta,\sigma) = \frac{64\sigma^2}{\pi^8} \sum_{n=1,3,5}^{\infty} \sum_{m=1,3,5}^{\infty} \frac{1}{D_{mn}(\beta,\sigma)}$$

$$S_2(\beta,\sigma) = \frac{64\sigma}{\pi^6} \sum_{n=1,3,5}^{\infty} \sum_{m=1,3,5}^{\infty} \frac{m^2 + (n/\beta)^2}{D_{mn}(\beta,\sigma)} \tag{9.30}$$

$$D_{mn}(\beta,\sigma) = m^2 n^2 \left\{\left(m^2 + (n/\beta)^2\right)^2 + \sigma^2/\pi^4\right\}$$

and we have used the fact that

$$(-1)^{(n-1)/2}(-1)^{(m-1)/2}(-1)^{(n-1)/2}(-1)^{(m-1)/2} = (-1)^{(n-1)}(-1)^{(m-1)} = 1 \quad m,n = 1,3,5\ldots$$

Using Eq. (9.29) in Eq. (9.16), the force acting on the fluid is

$$F = -k_o S_1(\beta,\sigma)\varepsilon\cos\omega t + k_o S_2(\beta,\sigma)\varepsilon\sin\omega t \ \text{ N} \tag{9.31}$$

where

$$k_o = \frac{abp_a}{h_o} \ \text{ N/m} \tag{9.32}$$

Therefore, the reaction force, the force acting on the surface is

$$F_r = -F = k_o S_1(\beta,\sigma)\varepsilon\cos\omega t - k_o S_2(\beta,\sigma)\varepsilon\sin\omega t \ \text{ N} \tag{9.33}$$

It is recalled that the applied motion to the surface is given by Eq. (9.8). Therefore, we can rewrite Eq. (9.33) as

$$F_r = k_{rf} h(t) + c_{rf}\dot{h}(t) \ \text{ N} \tag{9.34}$$

where

$$k_{rf} = k_o S_1(\beta,\sigma) \ \text{ N/m}$$

$$c_{rf} = \frac{k_o}{\omega} S_2(\beta,\sigma) \ \text{ Ns/m} \tag{9.35}$$

and the over dot indicates the derivative with respect to t. Thus, it is seen that the reaction force is composed of two frequency-dependent components, one that is proportional to displacement for which the constant of proportionality is k_{rf} and one that is proportional to velocity for which the constant of proportionality is c_{rf}. Therefore, k_{rf} represents the film's stiffness and c_{rf} the film's viscous damping. The curves of k_{rf}/k_o and $c_{rff}/(k_o/\omega)$ as a function of squeeze number σ are given in Figure 9.2. The value of σ at the intersection of these curves is defined as the critical squeeze number, σ_c.

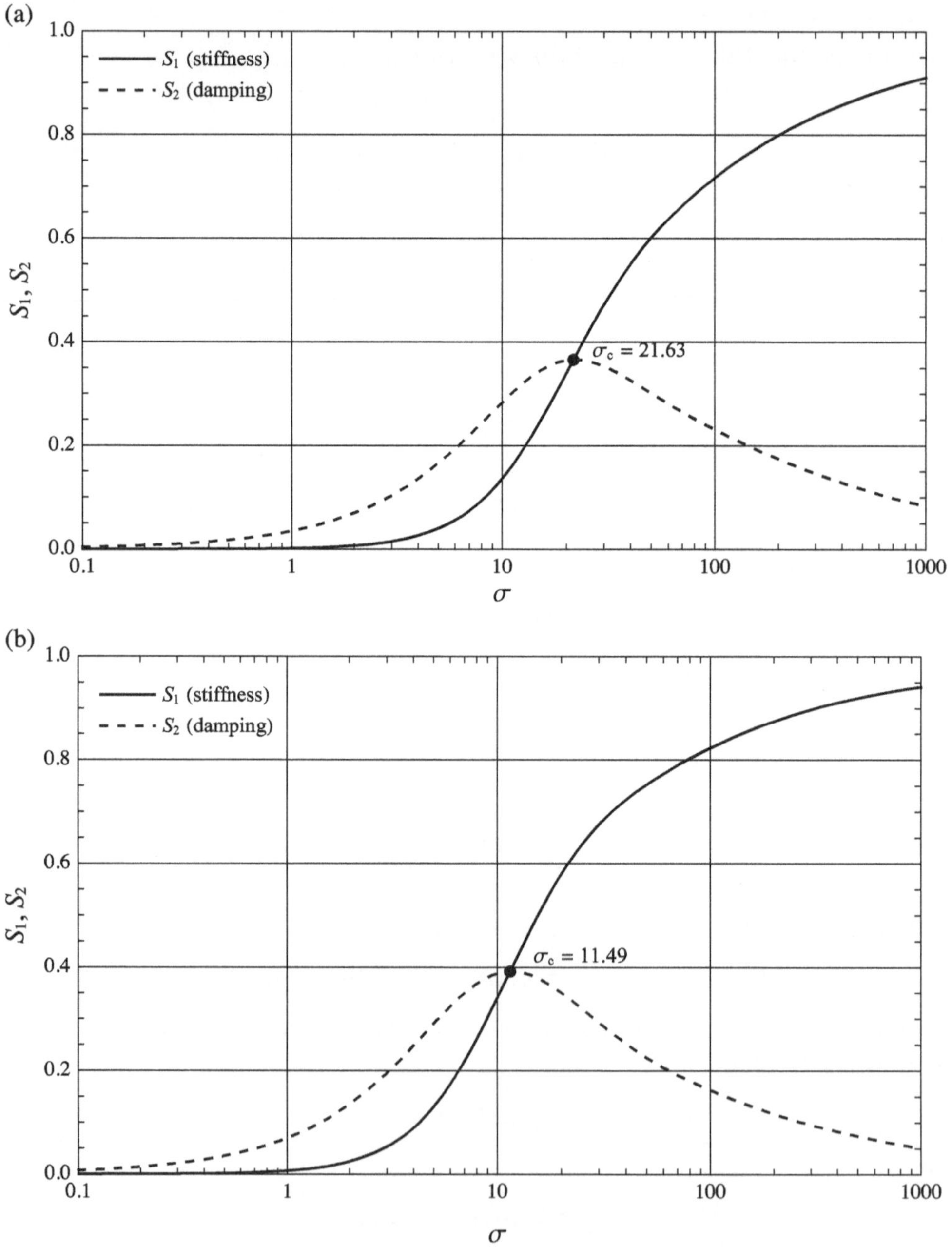

FIGURE 9.2 Squeeze film's stiffness and damping parameters as a function of squeeze number for (a) $\beta = 1$, and (b) $\beta = 4$.

9.2.3 Base-Excited Single-Degree-of-Freedom System with Squeeze Film Air Damping

A base-excited single-degree-of-freedom system with a mass m_s (kg), stiffness k_s (N/m), and viscous damping c_s (Ns/m) is shown in Figure 9.3. The displacement of the mass is $x(t)$ (m). The base is subject to a given displacement $y(t)$ (m), and the mass is subjected to an applied force $f(t)$ and a reaction force F_r. The governing equation for this system is

$$m_s \frac{d^2 z}{dt^2} + c_s \frac{dz}{dt} + k_s z + F_r = f(t) - m_s \frac{d^2 y}{dt^2} \quad \text{N} \tag{9.36}$$

where

$$z(t) = x(t) - y(t) \tag{9.37}$$

The reaction force, F_r, is given by Eq. (9.34). However, in order to use this reaction force in Eq. (9.36), we do so with the understanding that $f(t)$ and $y(t)$ will be harmonic. For our purposes, we assume that $f(t) = 0$. Realizing that $h(t) = x(t)$, we can substitute Eq. (9.34) into Eq. (9.36) and obtain

$$m_s \frac{d^2 z}{dt^2} + \left(c_s + c_{rf}\right)\frac{dz}{dt} + \left(k_s + k_{rf}\right) z = -m_s \frac{d^2 y}{dt^2} \quad \text{N} \tag{9.38}$$

Upon introducing the non-dimensional time $\tau = \omega_n t$, and the definitions

$$\omega_n = \sqrt{\frac{k_s}{m_s}} \quad \text{rad/s} \quad 2\zeta = \frac{c_s}{\sqrt{k_s m_s}} \tag{9.39}$$

we find that Eq. (9.36) can be written as

$$\frac{d^2 z}{d\tau^2} + \left(2\zeta + \frac{c_{rf}\omega_n}{k_s}\right)\frac{dz}{d\tau} + \left(1 + \frac{k_{rf}}{k_s}\right) z = -\frac{d^2 y}{d\tau^2} \quad \text{m} \tag{9.40}$$

The base is excited harmonically as

$$y(t) = Y_o e^{j\Omega\tau} \tag{9.41}$$

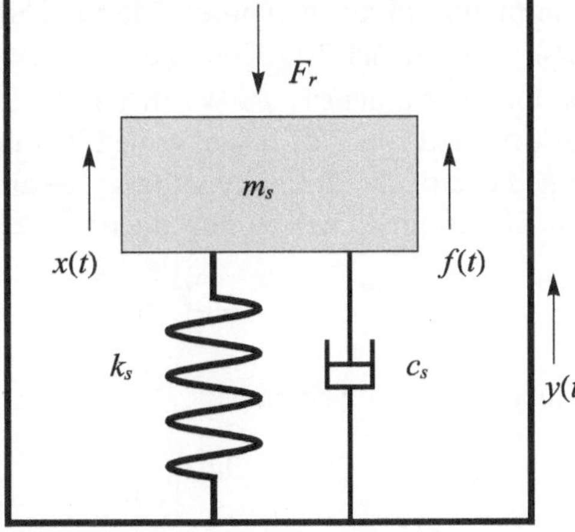

FIGURE 9.3 Base-excited single-degree-of-freedom system.

where $\Omega = \omega/\omega_n$ and ω (rad/s) is the excitation frequency. Based on the form of Eq. (9.41), we assume a solution for $z(t)$ as

$$z(t) = Z_o e^{j\Omega\tau} \tag{9.42}$$

Upon substituting Eqs. (9.41) and (9.42) into Eq. (9.40) and solving for Z_o, we obtain

$$\frac{Z_o}{Y_o} = \frac{\Omega^2}{r_B} e^{j(\Omega\tau - \varphi(\Omega))} \tag{9.43}$$

where

$$\begin{aligned} r_B &= \sqrt{g_1^2 + g_2^2} \\ \varphi(\Omega) &= \tan^{-1}\frac{g_2}{g_1} \end{aligned} \tag{9.44}$$

and

$$g_1 = 1 + \frac{k_o}{k_s} S_1(\beta, \sigma_n, \Omega) - \Omega^2$$

$$g_2 = \Omega\left(2\zeta + \frac{k_o}{k_s \Omega} S_2(\beta, \sigma_n, \Omega)\right)$$

$$\sigma_n = \frac{12\mu\omega_n b^2}{h_o^2 p_a}$$

Equation (9.43) is plotted in Figure 9.4 for $\beta = 1$ (square surfaces), $k_o/k_s = 0.25$, $\zeta = 0$, and for several values of σ_n.

9.3 VISCOUS FLUID DAMPING

9.3.1 Forces on a Submerged Harmonically Oscillating Rigid Cylinder in a Viscous Fluid*

We start by considering the motion of an incompressible fluid in the (r,θ)-plane that at a point has a velocity u_r (m/s) in the radial direction and a velocity u_θ (m/s) perpendicular to the radial direction. The fluid has a density ρ_f (kg/m^3) and a dynamic viscosity μ_f (Ns/m^2) that are independent of time and space. If it is assumed that in the fluid the convective accelerations terms can be neglected, the Stokes hypothesis (assumption) can be employed, and the stream function $\psi = \psi(r,\theta,t)$ (m^2/s) can be introduced such that

$$u_r = \frac{1}{r}\frac{\partial \psi}{\partial \theta}, \quad u_\theta = -\frac{\partial \psi}{\partial r} \text{ m/s} \tag{9.45}$$

Then the Navier–Stokes equation can be written as

$$\nabla_{r\theta}^2\left(\nabla_{r\theta}^2 \psi - \frac{\rho_f}{\mu_f}\frac{\partial \psi}{\partial t}\right) = 0 \tag{9.46}$$

* Stokes SS (1901) *Mathematical and Physical Papers*, Vol. III. Cambridge University Press, pp. 38–47.

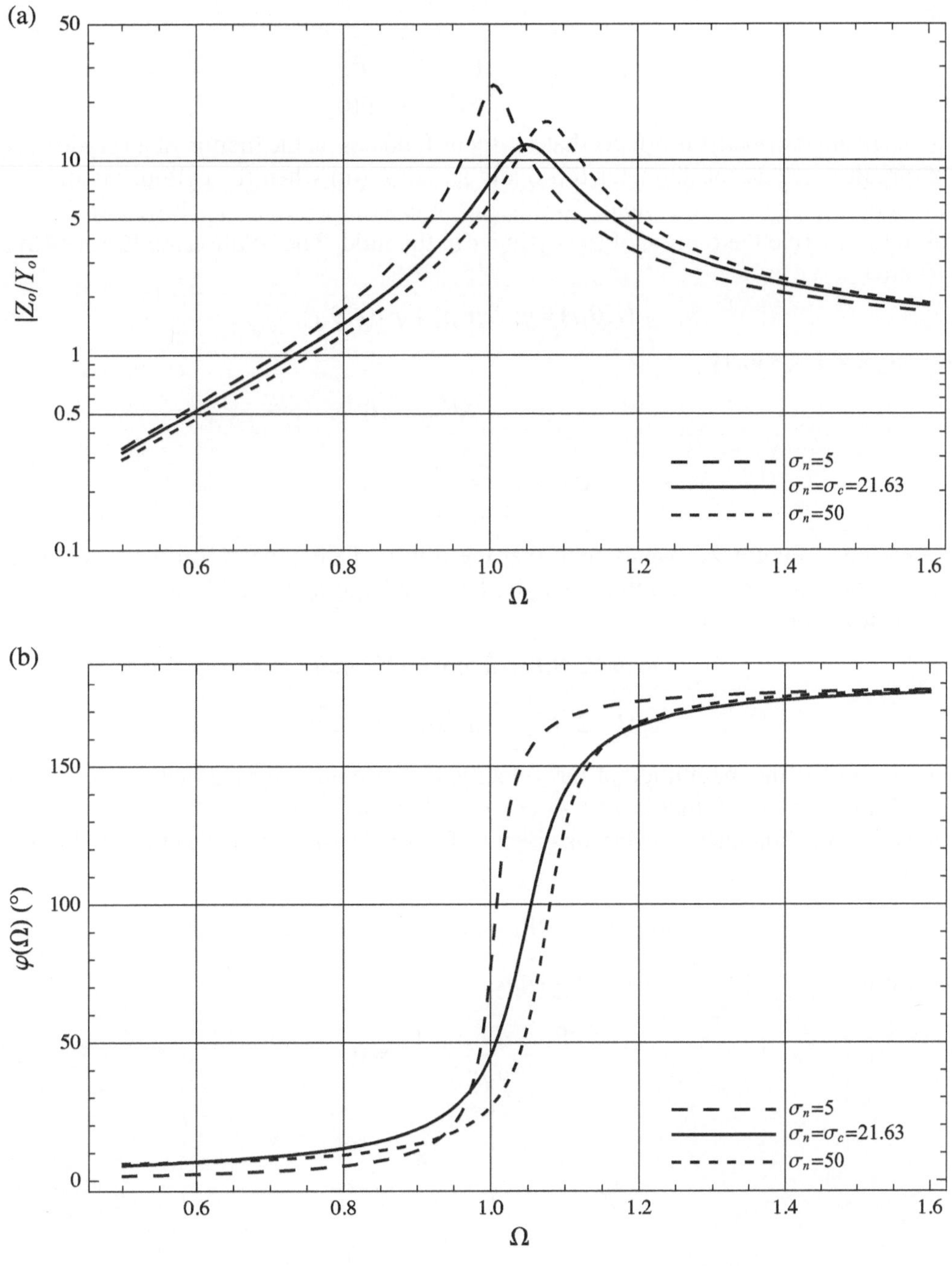

FIGURE 9.4 Effect of squeeze film stiffness and damping on a single-degree-of-freedom system for $\beta = 1$, $k_o/k_s = 0.25$, $\zeta = 0$, and the parameters indicted in the graphs: (a) magnitude, (b) phase.

or

$$\left(\nabla_{r\theta}^2 - \frac{\rho_f}{\mu_f}\frac{\partial}{\partial t}\right)\nabla_{r\theta}^2\psi = 0 \tag{9.47}$$

where

$$\nabla^2_{r\theta} = \frac{\partial^2}{\partial r^2} + \frac{1}{r}\frac{\partial}{\partial r} + \frac{1}{r^2}\frac{\partial^2}{\partial \theta^2} \tag{9.48}$$

We shall approximate the effects that a viscous fluid has on the motion of a very long rigid solid circular cylinder of diameter b (m) and length L (m) when the cylinder is immersed in an incompressible viscous fluid. To this end, it is assumed that the above equations can be used to describe the motion of the cylinder in the fluid. The solutions to Eqs. (9.46) and (9.47) are

$$\psi(r,\theta,t) = \psi_1(r,\theta,t) + \psi_2(r,\theta,t) \tag{9.49}$$

where ψ_k, $k = 1,2$, satisfy

$$\begin{aligned} \nabla^2_{r\theta}\psi_1 &= 0 \\ \nabla^2_{r\theta}\psi_2 - \frac{\rho_f}{\mu_f}\frac{\partial \psi_2}{\partial t} &= 0 \end{aligned} \tag{9.50}$$

Referring to Figure 9.5, we assume that the origin of the polar coordinate system is at the center of the cylinder, and that the cylinder is undergoing forced harmonic oscillations at frequency ω of the form

$$\begin{aligned} u_r(b/2,\theta,t) &= u_o \cos\theta e^{j\omega t} \quad \text{m/s} \\ u_\theta(b/2,\theta,t) &= -u_o \sin\theta e^{j\omega t} \quad \text{m/s} \end{aligned} \tag{9.51}$$

where u_o (m/s) is the magnitude of the velocity. It is further assumed that the surrounding fluid is infinite in extent, that is, as $r \to \infty$, $u_r \to 0$ and $u_\theta \to 0$.

In view of the boundary conditions given by Eq. (9.51), we assume solutions to Eq. (9.50) of the form

$$\psi_k(r,\theta,t) = \Psi_k(r)\sin\theta e^{j\omega t} \quad k = 1,2 \tag{9.52}$$

Substituting Eq. (9.52) into Eq. (9.50) yields

$$\frac{d^2\Psi_1}{dr^2} + \frac{d\Psi_1}{rdr} - \frac{\Psi_1}{r^2} = 0 \tag{9.53}$$

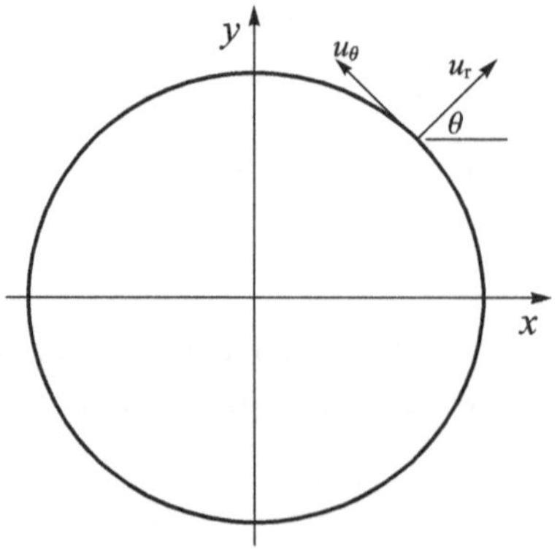

FIGURE 9.5 Velocities acting on the boundary of a rigid cylinder when the cylinder is oscillating in the x-direction.

and

$$r^2 \frac{d^2\Psi_2}{dr^2} + r\frac{d\Psi_2}{dr} - \left(\left(\frac{2\lambda r}{b}\right)^2 + 1\right)\Psi_2 = 0 \tag{9.54}$$

where

$$\lambda = \sqrt{j\,\mathrm{Re}} \quad \mathrm{Re} = \frac{\rho_f \omega b^2}{4\mu_f} \tag{9.55}$$

and Re is the Reynolds number.

Using Eq. (4.58), the solution to Eq. (9.53) is

$$\Psi_1 = \frac{C_1}{r} + C_2 r \tag{9.56}$$

To satisfy the requirements that as $r \to \infty$, $u_r \to 0$ and $u_\theta \to 0$, we must set $C_2 = 0$. Then, Eq. (9.56) becomes

$$\Psi_1 = \frac{C_1}{r} \tag{9.57}$$

From Eqs. (5.121) and (5.129), the solution to Eq. (9.54) is

$$\Psi_2 = C_3 I_1(2\lambda r / b) + C_4 K_1(2\lambda r / b) \tag{9.58}$$

where $I_1(x)$ is the modified Bessel function of the first kind of order 1 and $K_1(x)$ is the modified Bessel function of the second kind of order 1. To satisfy the requirements that as $r \to \infty$, $u_r \to 0$ and $u_\theta \to 0$, we use Eq. (5.131) and find that we must set $C_3 = 0$. Then, Eq. (9.58) becomes

$$\Psi_2 = C_4 K_1(2\lambda r / b) \tag{9.59}$$

Thus, from Eqs. (9.49), (9.57), and (9.59),

$$\psi = \left(\frac{C_1}{r} + C_4 K_1(2\lambda r / b)\right)\sin\theta e^{j\omega t} \tag{9.60}$$

Upon using Eqs. (9.45) and (9.60), we find that the velocities are given by

$$\begin{aligned} u_r &= \frac{1}{r}\left(\frac{C_1}{r} + C_4 K_1(2\lambda r / b)\right)\cos\theta e^{j\omega t} \\ u_\theta &= \left(\frac{C_1}{r^2} + \frac{2\lambda C_4}{b}\left[\frac{b}{2\lambda r}K_1(2\lambda r / b) + K_0(2\lambda r / b)\right]\right)\sin\theta e^{j\omega t} \end{aligned} \tag{9.61}$$

The constants C_1 and C_4 are determined by substituting Eq. (9.61) into the boundary conditions given by Eq. (9.51), which results in

$$\begin{aligned} u_o &= \frac{4C_1}{b^2} + \frac{2}{b}C_4 K_1(\lambda) \\ -u_o &= \frac{4C_1}{b^2} + \frac{2\lambda C_4}{b}\left[\frac{1}{\lambda}K_1(\lambda) + K_0(\lambda)\right] \end{aligned} \tag{9.62}$$

Solving Eq. (9.62) for C_1 and C_4, we obtain

$$C_1 = \frac{u_o b^2}{4}\left(\frac{2}{\lambda}\frac{K_1(\lambda)}{K_0(\lambda)}+1\right)$$
$$C_4 = -\frac{u_o b}{\lambda K_0(\lambda)} \qquad (9.63)$$

Then Eqs. (9.57) and (9.59), respectively, become

$$\Psi_1 = \frac{u_o b^2}{4r}\left(\frac{2}{\lambda}\frac{K_1(\lambda)}{K_0(\lambda)}+1\right) \quad \text{m}^2/\text{s} \qquad (9.64)$$

and

$$\Psi_2 = -u_o b\frac{K_1(2\lambda r/b)}{\lambda K_0(\lambda)} \quad \text{m}^2/\text{s} \qquad (9.65)$$

It has been shown that the component of the force F that the fluid exerts on the cylinder of length L at $r = b/2$ in the x-direction is*

$$F = -L\rho_f \frac{b}{2}\int_0^{2\pi}\frac{\partial}{\partial t}\left(r\frac{\partial \psi_1}{\partial r}+\psi_2\right)_{r=b/2}\sin\theta d\theta \quad \text{N} \qquad (9.66)$$

Upon substituting Eqs. (9.52), (9.64), and (9.65) into Eq. (9.66), we arrive at

$$F = j\omega u_o e^{j\omega t}M_f\Gamma_{\text{cyl}} \quad \text{N} \qquad (9.67)$$

where

$$M_f = \rho_f \pi\frac{b^2}{4}L \quad \text{kg}$$
$$\Gamma_{\text{cyl}} = \left(1+\frac{4}{\lambda}\frac{K_1(\lambda)}{K_0(\lambda)}\right) \qquad (9.68)$$

The quantity M_f is the mass of the fluid displaced by the cylinder and Γ_{cyl} is a complex quantity called a *hydrodynamic function*. For Re >> $R_{\min}$, Eq. (9.68) can be expressed as

$$\Gamma_{\text{cyl}} = 1+\frac{4}{\sqrt{2\,\text{Re}}} - j\frac{4}{\sqrt{2\,\text{Re}}} \quad \text{Re} > R_{\min} \qquad (9.69)$$

where it has been found numerically that for Real[Γ_{cyl}] the error is less than 5% when $R_{\min}$ > 0.6 and for Imag[Γ_{cyl}] the error is less than 5% when $R_{\min}$ > 200.

The force acting on the cylinder is obtained by taking the real part of F, which is

$$F_r = -\omega u_o M_f\left[\text{Real}\left(\Gamma_{\text{cyl}}\right)\sin(\omega t)+\text{Imag}\left(\Gamma_{\text{cyl}}\right)\cos(\omega t)\right] \quad \text{N} \qquad (9.70)$$

If the harmonic displacement of the cylinder in the x-direction is given by $x = X_o\sin(\omega t)$, then Eq. (9.70) can be written as

$$F_r = m_a\ddot{x}+c_v\dot{x} \quad \text{N} \qquad (9.71)$$

* Stokes, ibid., p. 46.

where the over dot indicates the derivative with respect to time,

$$m_a = M_f \text{Real}\left(\Gamma_{\text{cyl}}\right) \quad \text{kg}$$
$$c_v = -\omega M_f \text{Imag}\left(\Gamma_{\text{cyl}}\right) \quad \text{kg/s (or Ns/m)} \tag{9.72}$$

and we have recognized that for harmonic oscillations $u_o = \omega X_o$. As seen from Eq. (9.69), $\text{Imag}(\Gamma_{\text{cyl}}) < 0$ and, therefore, $c_v > 0$.

9.3.2 Mass-Excited Single-Degree-of-Freedom System Subject to Viscous Fluid Damping

We again consider the single-degree-of-freedom system shown in Figure 9.3. In the present case, we fix the base; that is, we set $y(t) = 0$ and therefore, $z(t) = x(t)$. We assume that the mass m_s is submerged in a viscous fluid with a dynamic viscosity μ_f (Ns/m^2) and density ρ_f (kg/m^3). It is further assumed that the mass is a solid circular cylinder of density ρ_s (kg/m^3) with cross-sectional area A and length L so that $m_s = \rho_s AL$ (kg). We assume that the mass is moving harmonically so that $f(t) = F_o e^{j\omega t}$. Then, Eq. (9.36) becomes

$$\left(m_s + m_a\right)\frac{d^2x}{dt^2} + \left(c_s + c_v\right)\frac{dx}{dt} + k_s x = F_o e^{j\omega t} \quad \text{N} \tag{9.73}$$

where we have used Eq. (9.71).

If, in Eq. (9.73), we set $\tau = \omega_n t$ and use Eq. (9.39), then Eq. (9.73) can be written as

$$m_e(\Omega)\frac{d^2x}{d\tau^2} + c_e(\Omega)\frac{dx}{d\tau} + x = \frac{F_o}{k_s}e^{j\Omega\tau} \quad \text{m} \tag{9.74}$$

where $\Omega = \omega/\omega_n$,

$$m_e(\Omega) = 1 + \frac{\rho_f}{\rho_s}\text{Real}\left[\Gamma_{\text{cyl},n}\right]$$
$$c_e(\Omega) = 2\zeta - \frac{\rho_f}{\rho_s}\Omega\text{Imag}\left[\Gamma_{\text{cyl},n}\right] \tag{9.75}$$

and

$$\Gamma_{\text{cyl},n} = \left(1 + \frac{4}{\lambda}\frac{K_1(\lambda_n)}{K_0(\lambda_n)}\right) \quad \lambda_n = \sqrt{j\Omega \text{Re}_n} \quad \text{Re}_n = \frac{\rho_f \omega_n b^2}{4\mu_f} \tag{9.76}$$

Based on the form of the nonhomogeneous term in Eq. (9.74), we assume a solution to Eq. (9.74) as (recall Example 4.18)

$$x(t) = X_o e^{j\Omega\tau} \quad \text{m} \tag{9.77}$$

Substitution of Eq. (9.77) into Eq. (9.74) yields

$$X_o = \frac{F_o}{k_s r_v}e^{j(\Omega\tau - \psi(\Omega))} \quad \text{m} \tag{9.78}$$

where

$$r_v = \sqrt{\left(1 - m_e(\Omega)\Omega^2\right)^2 + \left(c_e(\Omega)\Omega\right)^2}$$

$$\psi(\Omega) = \tan^{-1}\frac{c_e(\Omega)\Omega}{1 - m_e(\Omega)\Omega^2}$$

Equation (9.78) is plotted in Figure 9.6 for $\rho_f/\rho_s = 0.3$, $\zeta = 0$, and for several values of Re_n.

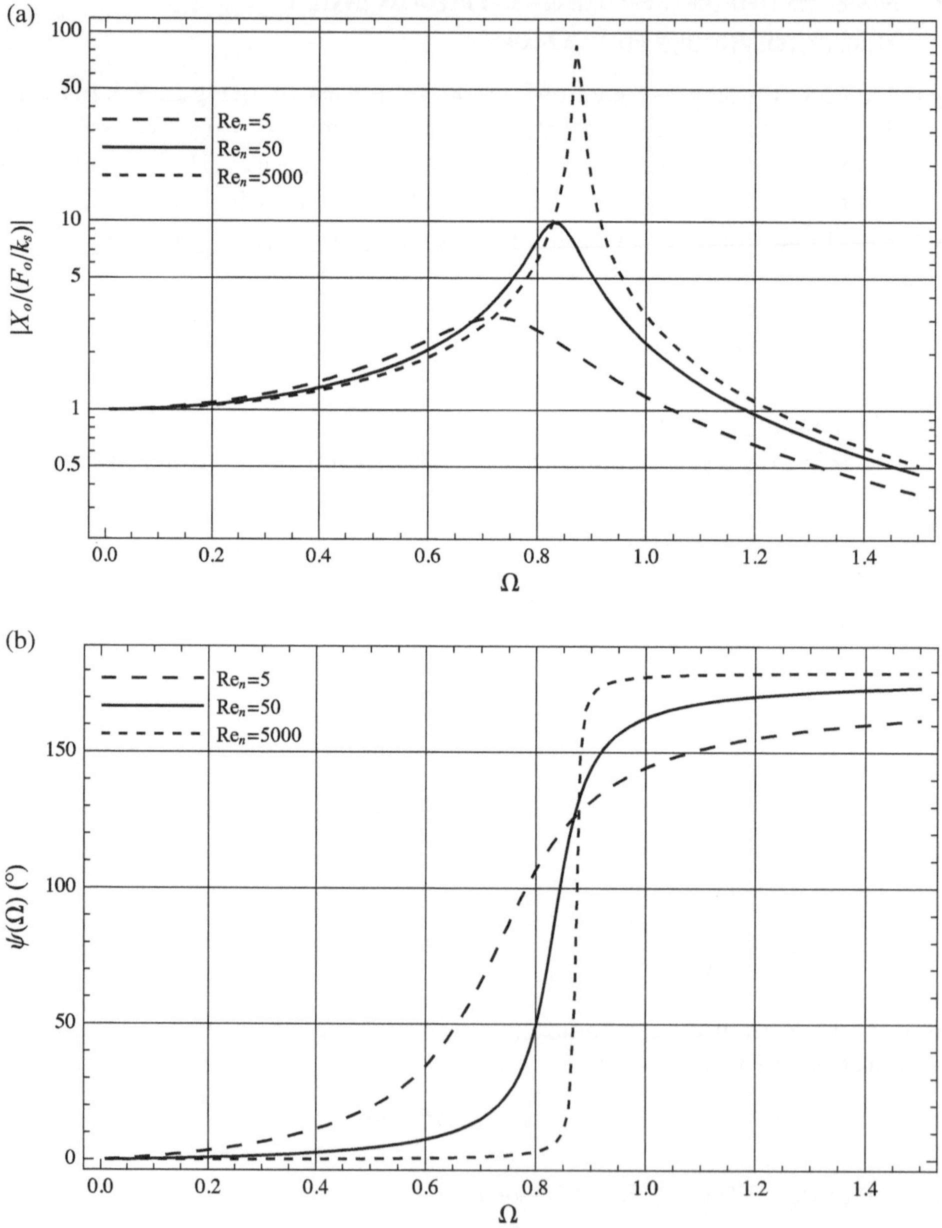

FIGURE 9.6 Effect of viscous fluid damping on a single degree-of-freedom system for $\rho_f/\rho_s = 0.3$, $\zeta = 0$, and for the parameters indicted in the graphs: (a) magnitude, (b) phase.

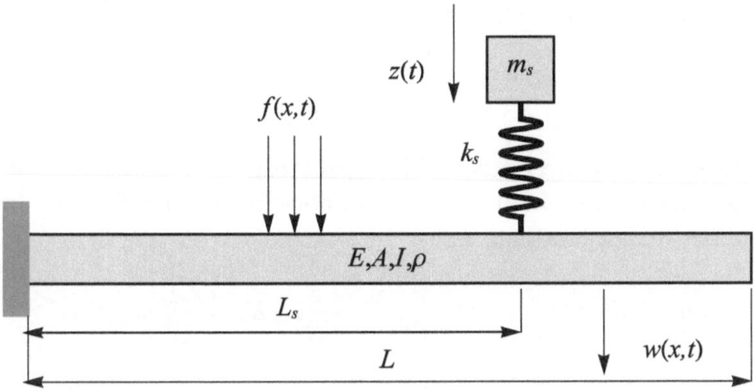

FIGURE 9.7 Nomenclature for a cantilever beam with externally applied force and in-span spring–mass system.

9.4 NATURAL FREQUENCIES OF A CANTILEVER BEAM WITH AN IN-SPAN SPRING–MASS SYSTEM

9.4.1 INTRODUCTION

Consider a cantilever beam of length L (m) that has a constant cross section along the length of the beam with area A (m^2), area moment of inertia I (m^4), density ρ (kg/m^3), and Young's modulus E (N/m^2). As shown in Figure 9.7, the beam has positioned at an in-span location L_s a single-degree-of-freedom system with mass m_s and spring with stiffness k_s. The displacement of the mass of the spring–mass system is $z = z(t)$ and the externally applied force applied to the beam is $f(x,t)$. The transverse displacement of the beam is denoted $w = w(x,t)$ (m). The governing equation for the beam with an in-span spring–mass system is*

$$EI\frac{\partial^4 w}{\partial x^4} + k_s\left(w(x,t) - z(t)\right)\delta(x - L_s) + \rho A\frac{\partial^2 w}{\partial t^2} = f(x,t) \quad \text{N/m} \tag{9.79}$$

and the governing equation for the attached spring–mass system is

$$m_s\frac{d^2 z}{dt^2} + k_s z = k_s w(L_s, t) \quad \text{N} \tag{9.80}$$

A cantilever beam has one end clamped, which means that the displacement and the slope at that end are zero. The other end of a cantilever beam is free, which means that the moment and shear force at this end are zero. Then, the boundary conditions at the clamped end of the beam, which is at $x = 0$, are

$$\begin{aligned} w(0,t) &= 0 \\ w'(0,t) &= 0 \end{aligned} \tag{9.81}$$

* The derivation of this equation and a more general analysis of beam systems can be found in Magrab EB (2012) *Vibrations of Elastic Systems: With Application to MEMS and NEMS*, Solid Mechanics and Its Applications 184. Springer Science + Business Media B.V., New York, Section 3.5.1.

and those at the free end, which are at $x = L$, are

$$M = EIw''(L,t) = 0$$
$$V = EIw'''(L,t) = 0 \tag{9.82}$$

where the prime denotes the derivative with respect to x.

We shall now convert Eqs. (9.79) to (9.82) to a non-dimensional form by introducing the following quantities (recall Example 7.24)

$$\eta = \frac{x}{L}, \quad \tau = \frac{t}{t_o}, \quad y = \frac{w}{L}, \quad \hat{z} = \frac{z}{L}, \quad K_s = \frac{k_s L^3}{EI},$$
$$\eta_s = \frac{L_s}{L}, \quad M_s = \frac{m_s}{m_b}, \quad \Omega_s^4 = \omega_s^2 t_o^2 = \frac{K_s}{M_s}, \tag{9.83}$$
$$t_o = L^2\sqrt{\frac{A\rho}{EI}} \ \text{s}, \quad m_b = \rho AL \ \text{kg}, \quad \omega_s = \sqrt{\frac{k_s}{m_s}} \ \text{rad/s}$$

where t_o is a characteristic time of the beam. Then, using Eq. (9.83) in Eq. (9.79), we arrive at

$$\frac{\partial^4 y}{\partial \eta^4} + K_s\left(y - \hat{z}(\tau)\right)\delta(\eta - \eta_s) + \frac{\partial^2 y}{\partial \tau^2} = \hat{f}(\eta,\tau) \tag{9.84}$$

where

$$\hat{f}(\eta,\tau) = \frac{f(\eta,\tau)L^3}{EI_o} \tag{9.85}$$

and from Eq. (B.4) of Appendix B, it is shown that $\delta(x) = \delta(L\eta) = \delta(\eta)/|L|$.

The equation for the single-degree-of-freedom system given by Eq. (9.80) becomes

$$\frac{1}{\Omega_s^4}\frac{d^2\hat{z}}{d\tau^2} + \hat{z} = y(\eta_s,\tau) \tag{9.86}$$

The boundary conditions given by Eq. (9.81) at $\eta = 0$ become

$$y(0,\tau) = 0$$
$$y'(0,\tau) = 0 \tag{9.87}$$

and those given by Eq. (9.82) at $\eta = 1$ become

$$y''(1,\tau) = 0$$
$$y'''(1,\tau) = 0 \tag{9.88}$$

where the prime now denotes the derivative with respect to η.

9.4.2 Determination of Natural Frequencies and Mode Shapes

To determine the natural frequencies, we set $f(x,t) = 0$ and assume solutions of the form

$$y(\eta,\tau) = Y(\eta)e^{j\Omega^2\tau}$$
$$z(\tau) = Z_s(\eta)e^{j\Omega^2\tau} \tag{9.89}$$

where

$$\Omega^2 = \omega t_o \tag{9.90}$$

and ω (rad/s) is the frequency. Then Eqs. (9.84) and (9.86), respectively, become

$$\frac{d^4Y(\eta)}{d\eta^4} + K_s\left(Y(\eta) - Z_s(\eta)\right)\delta(\eta - \eta_s) - \Omega^4 Y(\eta) = 0 \tag{9.91}$$

and

$$Z_s(\eta_s) = Y(\eta_s)\left(1 - \frac{\Omega^4}{\Omega_s^4}\right)^{-1} \tag{9.92}$$

Combining Eqs. (9.91) and (9.92), we arrive at

$$\frac{d^4Y(\eta)}{d\eta^4} - R(\eta,\Omega)Y(\eta) = 0 \tag{9.93}$$

where

$$R(\eta,\Omega) = \left(G(\Omega)\delta(\eta - \eta_s) + \Omega^4\right) \tag{9.94}$$

and

$$G(\Omega) = \frac{M_s\Omega^4}{\left(1 - \Omega^4/\Omega_s^4\right)} = \frac{M_s\Omega^4}{\left(1 - M_s\Omega^4/K_s\right)} = \frac{K_s\Omega^4}{\left(K_s/M_s - \Omega^4\right)} \tag{9.95}$$

From Eqs. (9.94) and (9.95) it is seen that $R(\eta,\Omega)$ is not proportional to Ω^4; that is, Ω^4 cannot be factored. Therefore, based on Eq. (6.91), we see that Eq. (9.93) is not a fourth-order Sturm–Liouville equation. Consequently, the eigenfunctions will *not* form an orthogonal set.

Equation (9.95) has two limiting cases that are of interest. The first is when the spring becomes infinitely stiff, that is, when $K_s \to \infty$. In this case, $G(\Omega) \to M_s\Omega^4$, which is the case of a beam with a concentrated mass attached at $\eta = \eta_s$. The second case is when $M_s \to \infty$; that is, the end of the spring attached to the mass is fixed. In this case, we find that $G(\Omega) \to -K_s$, which is the case of a beam with a spring attached at $\eta = \eta_s$ when its other end is fixed.

The boundary conditions given by Eq. (9.87) at $\eta = 0$ become

$$\begin{aligned} Y(0) &= 0 \\ Y'(0) &= 0 \end{aligned} \tag{9.96}$$

and the boundary conditions given by Eq. (9.88) at $\eta = 1$ become

$$\begin{aligned} Y''(1) &= 0 \\ Y'''(1) &= 0 \end{aligned} \tag{9.97}$$

There are several methods that can be used to solve Eq. (9.93) subject to the boundary conditions given by Eqs. (9.96) and (9.97). For this particular system, we shall use the Laplace transform with respect to the spatial variable η. The Laplace transform can handle easily the delta function in the governing equation and, as will become evident shortly, we will be able to reduce immediately the solution with four unknown solution constants to a solution with two unknown constants, which greatly simplifies the subsequent algebra.

Taking the Laplace transform of Eq. (9.93) with respect to η, we obtain (recall Examples 6.2, 8.12, and 8.19)

$$\bar{Y}(s) = \frac{1}{s^4 - \Omega^4}\left[s^3 Y(0) + s^2 Y'(0) + sY''(0) + Y'''(0) + G(\Omega)Y(\eta_s)e^{-s\eta_s}\right] \tag{9.98}$$

where s is the Laplace transform parameter, $\bar{Y}(s)$ is the Laplace transform of $Y(\eta)$, the prime indicates the derivative with respect to η, and we have used Laplace transform pairs 5 and 13 from Table 8.2 of Appendix 8.1.

It is noted that in Eq. (9.98) $Y(0)$ is the displacement at $\eta = 0$, $Y'(0)$ is the slope at $\eta = 0$, $Y''(0)$ is proportional to the moment at $\eta = 0$, and $Y'''(0)$ is proportional to the shear force at $\eta = 0$. These four quantities and the value of Ω represent the five unknown quantities that have to be determined.

To obtain the inverse Laplace transform, we use Laplace transform pairs 27 to 30 from Table 8.2 of Appendix 8.1 and find that

$$\begin{aligned} Y(\eta) &= Y(0)Q(\Omega\eta) + Y'(0)R(\Omega\eta)/\Omega + Y''(0)S(\Omega\eta)/\Omega^2 \\ &+ Y'''(0)T(\Omega\eta)/\Omega^3 \\ &+ G(\Omega)Y(\eta_s)T(\Omega[\eta - \eta_s])u(\eta - \eta_s)/\Omega^3 \end{aligned} \tag{9.99}$$

where $u(\eta)$ is the unit step function, $Y(\eta_s)$ is the displacement at $\eta = \eta_s$, and

$$\begin{aligned} Q(\Omega\eta) &= \frac{1}{2}\left[\cos(\Omega\eta) + \cosh(\Omega\eta)\right] \\ R(\Omega\eta) &= \frac{1}{2}\left[\sin(\Omega\eta) + \sinh(\Omega\eta)\right] \\ S(\Omega\eta) &= \frac{1}{2}\left[-\cos(\Omega\eta) + \cosh(\Omega\eta)\right] \\ T(\Omega\eta) &= \frac{1}{2}\left[-\sin(\Omega\eta) + \sinh(\Omega\eta)\right] \end{aligned} \tag{9.100}$$

From the boundary conditions at $\eta = 0$, which are given by Eq. (9.96), it is seen that $Y(0) = 0$ and $Y'(0) = 0$. Thus, Eq. (9.99) simplifies to

$$\begin{aligned} Y(\eta) &= Y''(0)S(\Omega\eta)/\Omega^2 + Y'''(0)T(\Omega\eta)/\Omega^3 \\ &+ G(\Omega)Y(\eta_s)T(\Omega[\eta - \eta_s])u(\eta - \eta_s)/\Omega^3 \end{aligned} \tag{9.101}$$

The remaining two constants, $Y''(0)$ and $Y'''(0)$, are determined by substituting Eq. (9.101) into the boundary conditions at $\eta = 1$, which are given by Eq. (9.97). Performing the required differentiations for this substitution (see Eq. (e) of Example 6.2), we arrive at the following system of equations

$$\begin{bmatrix} Q(\Omega) & R(\Omega) \\ \Omega T(\Omega) & \Omega Q(\Omega) \end{bmatrix} \begin{Bmatrix} Y''(0) \\ Y'''(0) \end{Bmatrix} = \begin{Bmatrix} -G(\Omega)Y(\eta_s)R(\Omega[1-\eta_s])/\Omega \\ -G(\Omega)Y(\eta_s)Q(\Omega[1-\eta_s]) \end{Bmatrix} \tag{9.102}$$

Solving Eq. (9.102) for the unknown quantities $Y''(0)$ and $Y'''(0)$, it is found that

$$Y''(0)=\frac{G(\Omega)Y(\eta_s)}{\Omega D(\Omega)}\big(R(\Omega)Q(\Omega[1-\eta_s])-Q(\Omega)R(\Omega[1-\eta_s])\big)$$
$$Y'''(0)=\frac{G(\Omega)Y(\eta_s)}{D(\Omega)}\big(T(\Omega)R(\Omega[1-\eta_s])-Q(\Omega)Q(\Omega[1-\eta_s])\big) \tag{9.103}$$

where

$$D(\Omega)=Q^2(\Omega)-R(\Omega)T(\Omega) \tag{9.104}$$

Upon substituting Eq. (9.103) into Eq. (9.101), we obtain

$$Y(\eta)=\frac{G(\Omega)Y(\eta_s)}{\Omega^3 D(\Omega)}\big(H(\Omega,\eta,\eta_s)+D(\Omega)T(\Omega[\eta-\eta_s])u(\eta-\eta_s)\big) \tag{9.105}$$

where

$$H(\Omega,\eta,\eta_s)=S(\Omega\eta)\big(R(\Omega)Q(\Omega[1-\eta_s])-Q(\Omega)R(\Omega[1-\eta_s])\big)$$
$$+T(\Omega\eta)\big(T(\Omega)R(\Omega[1-\eta_s])-Q(\Omega)Q(\Omega[1-\eta_s])\big) \tag{9.106}$$

To obtain an expression from which the natural frequency coefficients $\Omega = \Omega_n$ can be determined, we note that Eq. (9.105) must be satisfied at $\eta = \eta_s$ (recall Example 8.12). Therefore, setting $\eta = \eta_s$ in Eq. (9.106) and using Eq. (9.95) yields

$$D(\Omega_n)-\frac{M_s\Omega_n}{\left(1-\Omega_n^4/\Omega_s^4\right)}H(\Omega_n,\eta_s,\eta_s)=0 \tag{9.107}$$

since, from Eq. (9.100), it is seen that $T(0) = 0$.

The eigenfunctions (called mode shapes in vibration analysis) are obtained from Eq. (9.105) as

$$Y_n(\eta)=\frac{H(\Omega_n,\eta,\eta_s)}{D(\Omega_n)}+T(\Omega_n[\eta-\eta_s])u(\eta-\eta_s) \tag{9.108}$$

where, for convenience, the collection of common terms multiplying this result has been set to unity. As mentioned previously, these eigenfunctions are *not* orthogonal functions. However, for the three special cases of Eq. (9.107) discussed below, the eigenfunctions will be orthogonal.

No in-span attachment, M_s = 0: In this case, Eq. (9.107) becomes

$$D(\Omega_n)=0 \tag{9.109}$$

and the corresponding mode shape is an orthogonal function.

Concentrated mass at $\eta = \eta_s$, $K_s \to \infty$: Using Eq. (9.95) in Eq. (9.107), Eq. (9.107) becomes

$$D(\Omega_n)-M_s\Omega_n H(\Omega_n,\eta_s,\eta_s)=0 \tag{9.110}$$

and the corresponding mode shape is an orthogonal function.

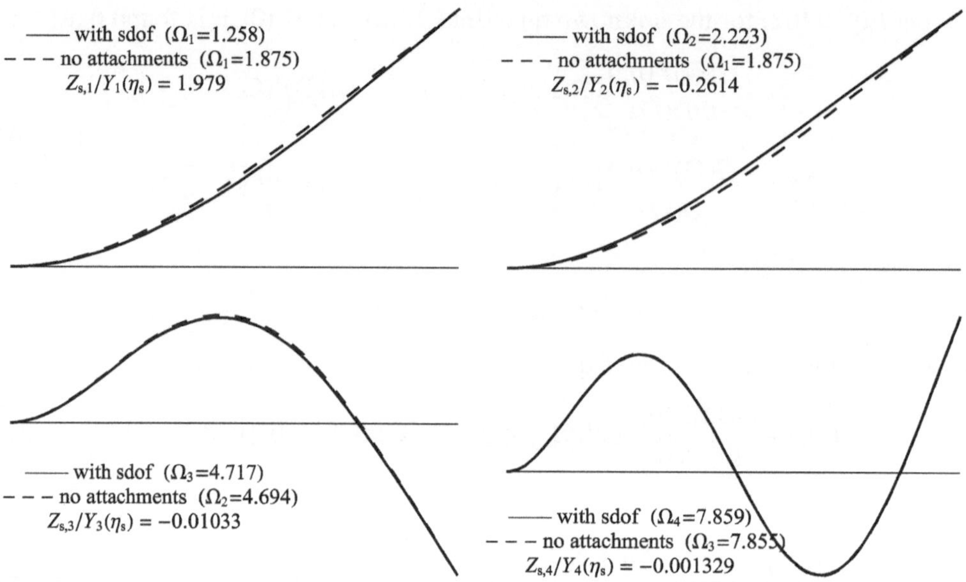

FIGURE 9.8 The lowest four natural frequency coefficients and associated mode shapes for a cantilever beam with an in-span single-degree-of-freedom system attached at $\eta_s = 0.99$ for $M_s = 0.5$ and $\Omega_s = 1.5$.

Spring placed at $\eta = \eta_s$, $M_s \to \infty$: Using Eq. (9.95) in Eq. (9.107), Eq. (9.107) becomes

$$D(\Omega_n) + \frac{K_s}{\Omega_n^3} H(\Omega_n, \eta_s, \eta_s) = 0 \tag{9.111}$$

and the corresponding mode shape is an orthogonal function.

From Eq. (9.92), the amplitude of the mass of the single-degree-of-freedom system relative to the amplitude of the beam at the attachment point at $\Omega = \Omega_n$ is

$$\frac{Z_{s,n}}{Y_n(\eta_s)} = \left(1 - \frac{\Omega_n^4}{\Omega_s^4}\right)^{-1} \tag{9.112}$$

It is seen from Eq. (9.112) that when Ω_n is in the vicinity of Ω_s the relative amplitude of the mass of the attached spring–mass system is large.

The lowest four natural frequency coefficients and corresponding eigenfunctions for the cantilever beam are plotted in Figure 9.8 for $\eta_s = 0.99$, $M_s = 0.5$, and $\Omega_s = 1.5$. Included in the figure are the ratios $Z_{s,n}/Y_n(\eta_s)$ and the eigenvalue and eigenfunction for the beam without the single-degree-of-freedom system.

9.5 PIEZOELECTRIC ENERGY HARVESTER: SINGLE-DEGREE-OF-FREEDOM SYSTEM

9.5.1 Piezoelectric Generator

The purpose of the energy harvester is to convert mechanical kinetic energy into electrical energy. There are many candidate solutions that have been explored to do this. We shall obtain the electrical energy by using piezoelectric materials, which have the ability to convert directly applied strain energy into usable electric energy.

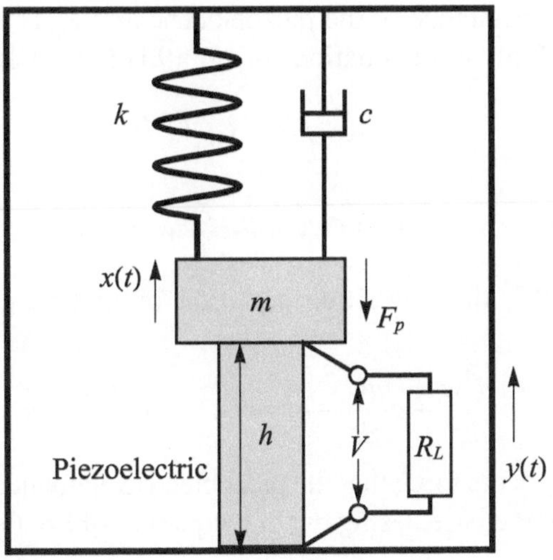

FIGURE 9.9 Piezoelectric element and a single-degree-of-freedom energy-harvesting device when the output of the piezoelectric element is connected to a resister R_L.

Consider the base-excited single-degree-of-freedom system shown in Figure 9.9. The mass of the single-degree-of-freedom system is attached to one end of a piezoelectric material such that a compression or extension of the piezoelectric material creates a voltage. The piezoelectric element obeys the following constitutive relations for one-dimensional motion*

$$\begin{aligned} T_3 &= c_{33}S_3 - e_{33}E_3 \quad \text{N/m}^2 \\ D_3 &= e_{33}S_3 + \varepsilon_{33}E_3 \quad \text{C/m}^2 \end{aligned} \tag{9.113}$$

where† D_3 is the total electric charge per area (C/m^2), E_3 is the electric field strength (N/C = V/m), T_3 is the stress (N/m^2), S_3 is the total strain, e_{33} is an electromechanical coupling constant that relates the strain to the applied field (C/m^2), c_{33} is the elastic stiffness of the material measured at constant electric field (N/m^2), and ε_{33} is the dielectric constant (permittivity) measured at constant strain (F/m).

Regarding Eq. (9.113), if A (m^2) is the area of the piezoelectric element that is in contact with the mass and h its height, then by definition $V = -E_3h$ (V) is the voltage created by its deformation, $w = S_3h$ (m) is the displacement resulting from the application of the stress on the area A, $F_p = AT_3$ (N) is the force due to the deformation, and $q = AD_3$ (C) is the charge. In addition, we introduce the definitions

$$\begin{aligned} C_p &= A\varepsilon_{33}/h \quad \text{F} \\ K_p &= Ac_{33}/h \quad \text{N/m} \end{aligned} \tag{9.114}$$

---

* Tiersten HF (1969) *Linear Piezoelectric Plate Vibrations.* Plenum Press, NY, pp. 34–36.

† The following units may prove useful in the subsequent material. If C = coulomb, J = joule, V = volt, A = ampere, W = watt, H = henry, and F = farad, then J = Ws = Nm, V = J/C, A = C/s, F = C/V, ohm = Js/C^2, and H = Js^2/C^2. Also note the distinction in the use of the symbol C: C is coulomb when not italicized and *C* italicized denotes capacitance.

where C_p is the loaded capacitance of the piezoelectric and K_p is the stiffness of the piezoelectric element. We multiply both equations of Eq. (9.113) by A and obtain

$$\begin{aligned} F_p &= e_{33}\frac{A}{h}V + K_p w \quad \text{N} \\ q &= -C_p V + e_{33}\frac{A}{h}w \quad \text{C} \end{aligned} \tag{9.115}$$

Referring to Figure 9.9, the governing equation of motion of a base-excited single-degree-of-freedom is

$$m\frac{d^2w}{dt^2} + c\frac{dw}{dt} + kw + F_p = -m\frac{d^2y}{dt^2} \quad \text{N} \tag{9.116}$$

where $w(t) = x(t) - y(t)$, F_p is the force that the piezoelectric element exerts on the mass m, and $y(t)$ is the displacement of the base. Using the first equation of Eq. (9.115) in Eq. (9.116) yields

$$m\frac{d^2w}{dt^2} + c\frac{dw}{dt} + \left(k + K_p\right)w + e_{33}\frac{A}{h}V = -m\frac{d^2y}{dt^2} \quad \text{N} \tag{9.117}$$

When the piezoelectric element connects to a resistive load R_L, then

$$\frac{dq}{dt} = \frac{V}{R_L} \quad \text{A} \tag{9.118}$$

We differentiate the second equation of Eq. (9.115) with respect to time and use Eq. (9.118) to obtain*

$$C_p R_L\frac{dV}{dt} + V - e_{33}\frac{AR_L}{h}\frac{dw}{dt} = 0 \quad \text{V} \tag{9.119}$$

Hence, the coupled equations governing the piezoelectric harvester/generator and the single-degree-of-freedom system are given by Eqs. (9.117) and (9.119).

We consider the case of harmonic oscillations of the form

$$y(t) = Y_o e^{j\omega t} \quad \text{m} \tag{9.120}$$

where Y_o is known and ω is the excitation frequency. Therefore, we assume solutions to Eqs. (9.117) and (9.119) of the form

$$\begin{aligned} w(t) &= W_o e^{j\omega t} \quad \text{m} \\ V(t) &= V_o e^{j\omega t} \quad \text{V} \end{aligned} \tag{9.121}$$

In addition, to simplify things a little, we assume that $k << K_p$ and, therefore, can be ignored. Substituting Eqs. (9.120) and (9.121) into Eqs. (9.117) and (9.119), respectively, we arrive at

$$\begin{aligned} \left(1 - \Omega^2 + j2\zeta\Omega^2\right)W_o + \frac{e_{33}}{c_{33}}V_o &= \Omega^2 Y_o \quad \text{m} \\ \left(1 + j\Omega r_L\right)V_o - j\Omega\frac{e_{33}}{\varepsilon_{33}}r_L W_o &= 0 \quad \text{V} \end{aligned} \tag{9.122}$$

* Renno JM, Daqaq MF, Inman, DJ (2009) On the Optimal Energy Harvesting from a Vibration Source. *Journal of Sound and Vibration* 320: 386–405.

where

$$\Omega = \frac{\omega}{\omega_n} \quad 2\zeta = \frac{c}{m\omega_n} \quad \omega_n = \sqrt{\frac{K_p}{m}} \text{ rad/s} \tag{9.123}$$

and

$$r_L = \omega_n C_p R_L \tag{9.124}$$

Solving Eq. (9.122) for W_o and V_o, it is found that

$$\begin{aligned} W_o &= \frac{Y_o\Omega^2\left(1 + j\Omega r_L\right)}{a + jb} \text{ m} \\ V_o &= \frac{j\Omega^3 r_L V_{\text{ref}}}{a + jb} \text{ V} \end{aligned} \tag{9.125}$$

where

$$\begin{aligned} a &= 1 - \Omega^2 - 2\zeta\Omega^2 r_L \\ b &= \Omega\left[2\zeta + r_L\left(1 - \Omega^2 + k_e^2\right)\right] \\ V_{\text{ref}} &= \frac{Y_o e_{33}}{\varepsilon_{33}^S} \text{ V} \end{aligned} \tag{9.126}$$

and

$$k_e^2 = \frac{e_{33}}{c_{33}\varepsilon_{33}} \tag{9.127}$$

which is a non-dimensional quantity.

9.5.2 Maximum Average Power of a Piezoelectric Generator

To determine the conditions under which one can transfer the maximum average power from the piezoelectric element to the load resistor R_L, we use Eq. (9.125) to obtain

$$P_{\text{avg}} = \frac{V_{\text{rms}}^2}{R_L} = \frac{|V_o|^2}{2R_L} = \frac{\Omega^6 r_L k_e^2 P_{\text{ref}}}{2\left(a^2 + b^2\right)} \text{ W} \tag{9.128}$$

where for a harmonically varying voltage we used the relation $V_{\text{rms}} = V_o / \sqrt{2}$, where V_{rms} is the root mean square (rms) voltage, and

$$P_{\text{ref}} = \omega_n Y_o^2 K_p \text{ W}$$

To find the value of r_L that maximizes P_{avg} given by Eq. (9.128), we determine the value of $r_L = r_{L,\text{opt}}$ that satisfies

$$\frac{dP_{\text{avg}}}{dr_L} = 0 \tag{9.129}$$

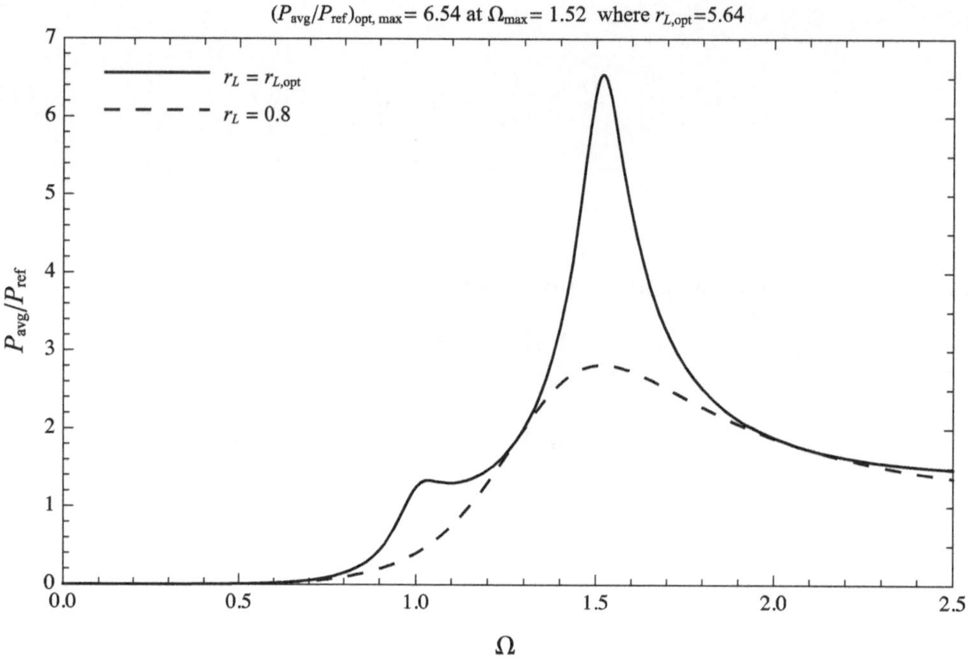

FIGURE 9.10 Maximum average power response for $r_{L,\text{opt}}(\Omega)$ and for $r_L=0.8$ when $\zeta=0.05$, $k_K=0$, and $k_e=1.134$.

Performing the operation indicated in Eq. (9.129) yields

$$r_{L,\text{opt}}(\Omega) = \frac{1}{\Omega}\sqrt{\frac{\Omega^4 + \left(4\zeta^2 - 2\right)\Omega^2 + 1}{\Omega^4 + \left(4\zeta^2 - 2\left(1 + k_e^2\right)\right)\Omega^2 + \left(1 + k_e^2\right)^2}}. \tag{9.130}$$

The maximum average power given by Eq. (9.128) using Eq. (9.130) is shown in Figure 9.9. The solid line is the value of P_{avg} at Ω for $r_{L,\text{opt}}(\Omega)$. For any other value of r_L at this frequency, the individual power response will never exceed the optimal power response that is obtained by evaluating P_{avg} with the optimum value of $r_{L,\text{opt}}$ at that frequency. This observation is illustrated in Figure 9.10 for $\zeta=0.03$ and $k_e=1.134$, where we have plotted $P_{\text{avg}}\left(r_{L,\text{opt}}(\Omega),\Omega\right)$ and $P_{\text{avg}}\left(0.8,\Omega\right)$.

9.6 DETERMINATION OF THE ONSET OF FLUTTER

9.6.1 Governing Equations

The effects of flow-induced forces on aircraft wings are of great practical interest. A typical model of the aircraft wing is a cantilever beam subject to an aerodynamically induced lift force and an aerodynamically induced moment. The beam cross section is such that the neutral axis does not coincide with the center of gravity. Therefore, a simultaneous combination of a force and a moment couples the torsion and bending response of the beam. Under certain conditions, this combination of loading has the effect of producing flutter, which is a dynamic instability that is a function of the aircraft speed and the bending and torsional properties of the beam.

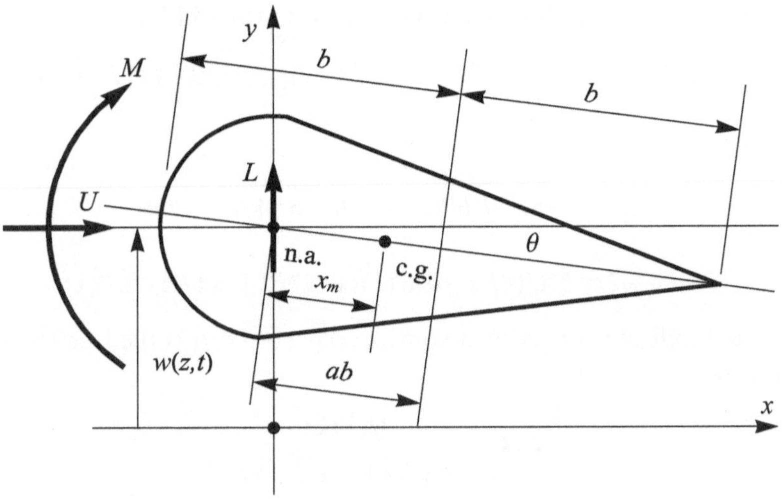

FIGURE 9.11 Nomenclature for the bending and twisting of the cross section of an airfoil of length L_a subjected to a lift force L and a twisting moment M; n.a. = neutral axis, c.g. = center of gravity.

Consider the wing model as shown in Figure 9.11. Air is flowing over the beam (wing) at a velocity U (m/s). The center of gravity of the wing cross section is a distance x_m (m) away from the neutral axis of the beam. The displacement of the beam at any point along its length L_a (m) is denoted $w = w(z,t)$ (m) and the rotation (twist) about this axis is denoted $\theta = \theta(z,t)$ (rad). The lift force per unit length L (N/m) acts at the plane containing the neutral axis, and the moment per unit length M (Nm/m) acts about this axis. The mass of the beam per unit length is m (kg/m). The beam has the following properties: Tensile modulus E (N/m^2), shear modulus G (N/m^2), area moment of inertia about the neutral axis I (m^4), and torsion constant for the cross section J (m^4). The mass moment of inertia of the cross section about its center of gravity is denoted J_{cg} (kg m^2).

The governing equations for this geometry are*

$$EI\frac{\partial^4 w}{\partial z^4} + m\frac{\partial^2 w}{\partial t^2} - x_m m\frac{\partial^2 \theta}{\partial t^2} = L \quad \text{N/m} \tag{9.131}$$

and

$$-GJ\frac{\partial^2 \theta}{\partial z^2} + I_o\frac{\partial^2 \theta}{\partial t^2} - m x_m\frac{\partial^2 w}{\partial t^2} = M \quad \text{N} \tag{9.132}$$

where

$$I_o = J_{cg} + m x_m^2 \tag{9.133}$$

* The original theoretical work for obtaining the expressions for L and M was done by T. Theodorsen in a 1935 NCAA Report 496 titled "General Theory of Aerodynamic Instability and the Mechanism of Flutter", pp. 291–311. This work was then incorporated with a beam in Bisplinghoff RL, Ashley H, Halfman RL (1955) *Aeroelasticity.* Addison-Wesley Publishing Company, Cambridge, MA, pp. 188–272. See also: Prasad SN, Nemat-Nasser S, Herrmann G (1967) Bending-Torsional Flutter of a Swept Wing in a High-Density, Low-Speed Flow. *AIAA Journal* **5**(2): 316–321; and Feldt WT, Herrmann G (1974) Bending-Torsional Flutter of a Cantilevered Wing Containing a Tip Mass and Subjected to a Transverse Follower Force. *Journal of the Franklin Institute* pp. 468–478.

is the mass moment of inertia about the neutral axis. In Eq. (9.131)

$$L = -\pi\rho b^2\left(\ddot{w} + U\dot{\theta} - ba\ddot{\theta}\right) - 2\pi\rho UbC(k)\left(\dot{w} + U\theta + b(1/2 - a)\dot{\theta}\right) \tag{9.134}$$

and in Eq. (9.132)

$$\begin{aligned} M &= \pi\rho b^2\left(ba\ddot{w} - Ub(1/2 - a)\dot{\theta} - b^2(1/8 + a^2)\ddot{\theta}\right) \\ &+ 2\pi\rho b^2 U(1/2 + a)C(k)\left(\dot{w} + U\theta + b(1/2 - a)\dot{\theta}\right) \end{aligned} \tag{9.135}$$

where the over dot indicates the derivative with respect to t, ρ is the free stream density of the air (kg/m^3), and

$$C(k) = \frac{H_1^{(2)}(k)}{H_1^{(2)}(k) + jH_0^{(2)}(k)} \tag{9.136}$$

In Eq. (9.136), $H_n^{(2)}(k)$ is the Hankel function of the second kind of order n given by Eq. (5.95), and $k = \omega b/U$ is called the reduced frequency where ω is the excitation frequency. It is noted that a is a non-dimensional quantity.

To reduce the number of parameters in Eqs. (9.131) to (9.136), we introduce the following definitions

$$\begin{aligned} &\eta = \frac{z}{L_a}, \quad \bar{w} = \frac{w}{b}, \quad \bar{x}_m = \frac{x_m}{b}, \quad \mu = \frac{\pi\rho b^2}{m}, \quad \tau = \frac{t}{t_o} \\ &t_o^2 = \frac{mL_a^4}{EI} \ \text{s}^2, \quad \bar{U} = \frac{Ut_o}{b}, \quad \gamma_o = \frac{I_o}{mb^2}, \quad \varepsilon = \frac{EIb^2}{GJL_a^2} \end{aligned} \tag{9.137}$$

Then combining Eqs. (9.131) and (9.134) and using Eq. (9.137), we arrive at

$$\begin{aligned} &\frac{\partial^4 \bar{w}}{\partial \eta^4} + (1+\mu)\frac{\partial^2 \bar{w}}{\partial \tau^2} - (\bar{x}_m + \mu a)\frac{\partial^2 \theta}{\partial \tau^2} \\ &+ \mu\bar{U}\left(1 + C(k)(1-2a)\right)\frac{\partial \theta}{\partial \tau} + 2\mu\bar{U}C(k)\frac{\partial \bar{w}}{\partial \tau} + 2\mu\bar{U}^2 C(k)\theta = 0 \end{aligned} \tag{9.138}$$

Similarly, combining Eqs. (9.132) and (9.135) and using Eq. (9.137), we obtain

$$\begin{aligned} &-\varepsilon\frac{\partial^2 \theta}{\partial \eta^2} + \left(\gamma_o + \mu(1/8 + a^2)\right)\frac{\partial^2 \theta}{\partial \tau^2} - (\bar{x}_m + a\mu)\frac{\partial^2 \bar{w}}{\partial \tau^2} \\ &+ \mu\bar{U}(1/2 - a)\left(1 - (1+2a)C(k)\right)\frac{\partial \theta}{\partial \tau} \\ &- \mu\bar{U}(1+2a)C(k)\frac{\partial \bar{w}}{\partial \tau} - \mu\bar{U}^2(1+2a)C(k)\theta = 0 \end{aligned} \tag{9.139}$$

9.6.2 Determination of Flutter Frequencies

The boundary conditions for a cantilever beam that is clamped at $\eta = 0$ are

$$\bar{w}(0,\tau) = \bar{w}'(0,\tau) = \theta(0,\tau) = 0 \tag{9.140}$$

and those at the free end $\eta = 1$ are

$$\bar{w}''(1,\tau) = \bar{w}'''(1,\tau) = \theta'(1,\tau) = 0 \tag{9.141}$$

where the prime denotes the derivative with respect to η.

We shall now determine the frequencies at which flutter can occur by assuming that

$$\begin{aligned} \bar{w}(\eta,\tau) &= W(\eta)e^{j\Omega\tau} \\ \theta(\eta,\tau) &= \Theta(\eta)e^{j\Omega\tau} \end{aligned} \tag{9.142}$$

where $\Omega = \omega t_o$. With this definition of Ω, we see that

$$k = \frac{\omega b}{U} = \frac{\omega t_o b}{U t_o} = \frac{\Omega}{\bar{U}} \tag{9.143}$$

and, therefore, $C(k) \to C(\Omega / \bar{U})$.

Upon substituting Eq. (9.142) into Eqs. (9.138) and (9.139), respectively, we arrive at

$$\frac{d^4 W}{d\eta^4} + a_1 W + b_1 \Theta = 0 \tag{9.144}$$

and

$$-\varepsilon \frac{d^2\Theta}{d\eta^2} + b_2\Theta + a_2 W = 0 \tag{9.145}$$

where

$$\begin{aligned} a_1 &= j2\mu\bar{U}\Omega C(\Omega/\bar{U}) - (1+\mu)\Omega^2 \\ a_2 &= (\bar{x}_m + a\mu)\Omega^2 - \mu\bar{U}(1+2a)C(\Omega/\bar{U}) \\ b_1 &= (\bar{x}_m + \mu a)\Omega^2 + j\mu\bar{U}\Omega\big(1 + (1-2a)C(\Omega/\bar{U})\big) + 2\mu\bar{U}^2 C(\Omega/\bar{U}) \\ b_2 &= -\Omega^2\big(\gamma_o + \mu(1/8 + a^2)\big) + j\Omega\mu\bar{U}(1/2 - a)\big(1 - (1+2a)C(\Omega/\bar{U})\big) \\ &\quad - \mu\bar{U}^2(1+2a)C(\Omega/\bar{U}) \end{aligned} \tag{9.146}$$

The boundary conditions given by Eqs. (9.140) and (9.141), respectively, become

$$W(0) = W'(0) = \Theta(0) = 0 \tag{9.147}$$

and

$$W''(1) = W'''(1) = \Theta'(1) = 0 \tag{9.148}$$

The solutions to Eqs. (9.144) and (9.145) are of the form

$$\begin{aligned} W(\eta) &= Ae^{\lambda\eta} \\ \Theta(\eta) &= Be^{\lambda\eta} \end{aligned} \tag{9.149}$$

where A and B are to be determined. Substituting Eq. (9.149) into Eqs. (9.144) and (9.145), we obtain

$$\begin{aligned}(\lambda^4 + a_1)A + b_1 B &= 0\\ a_2 A + (b_2 - \varepsilon\lambda^2)B &= 0\end{aligned} \tag{9.150}$$

The values of λ for which the determinant of the coefficients of Eq. (9.150) is zero will provide the solution, that is, the values of λ that satisfy

$$-\varepsilon\lambda^6 + b_2\lambda^4 - \varepsilon a_1\lambda^2 + a_1 b_2 - b_1 a_2 = 0 \tag{9.151}$$

Hence, there are six roots to Eq. (9.151) denoted λ_i and, therefore, the solutions given by Eq. (9.149) become

$$\begin{aligned}W(\eta) &= \sum_{i=1}^{6} A_i e^{\lambda_i \eta}\\ \Theta(\eta) &= \sum_{i=1}^{6} B_i e^{\lambda_i \eta}\end{aligned} \tag{9.152}$$

From the second equation of Eq. (9.150), we see that at $\lambda = \lambda_i$,

$$B_i = K_i A_i \tag{9.153}$$

where

$$K_i = \frac{a_2}{\varepsilon\lambda_i^2 - b_2} \tag{9.154}$$

Then Eq. (9.152) can be written as

$$\begin{aligned}W(\eta) &= \sum_{i-1}^{6} A_i e^{\lambda_i \eta}\\ \Theta(\eta) &= \sum_{i-1}^{6} K_i A_i e^{\lambda_i \eta}\end{aligned} \tag{9.155}$$

We now substitute Eq. (9.155) into the boundary conditions given by Eqs. (9.147) and (9.148), respectively, and obtain

$$\begin{aligned}W(0) &= \sum_{i-1}^{6} A_i = 0\\ W'(0) &= \sum_{i-1}^{6} \lambda_i A_i = 0\\ \Theta(0) &= \sum_{i-1}^{6} K_i A_i = 0\end{aligned} \tag{9.156}$$

and

$$W''(1) = \sum_{i-1}^{6} \lambda_i^2 A_i e^{\lambda_i} = 0$$

$$W'''(1) = \sum_{i-1}^{6} \lambda_i^3 A_i e^{\lambda_i} = 0 \qquad (9.157)$$

$$\Theta'(1) = \sum_{i-1}^{6} \lambda_i K_i A_i e^{\lambda_i} = 0$$

Placing Eqs. (9.156) and (9.157) in matrix form, we arrive at

$$\begin{pmatrix} 1 & 1 & 1 & 1 & 1 & 1 \\ \lambda_1 & \lambda_2 & \lambda_3 & \lambda_4 & \lambda_5 & \lambda_6 \\ K_1 & K_2 & K_3 & K_4 & K_5 & K_6 \\ \lambda_1^2 e^{\lambda_1} & \lambda_2^2 e^{\lambda_2} & \lambda_3^2 e^{\lambda_3} & \lambda_4^2 e^{\lambda_4} & \lambda_5^2 e^{\lambda_5} & \lambda_6^2 e^{\lambda_6} \\ \lambda_1^3 e^{\lambda_1} & \lambda_2^3 e^{\lambda_2} & \lambda_3^3 e^{\lambda_3} & \lambda_4^3 e^{\lambda_4} & \lambda_5^3 e^{\lambda_5} & \lambda_6^3 e^{\lambda_6} \\ \lambda_1 K_1 e^{\lambda_1} & \lambda_2 K_2 e^{\lambda_2} & \lambda_3 K_3 e^{\lambda_3} & \lambda_4 K_4 e^{\lambda_4} & \lambda_5 K_5 e^{\lambda_5} & \lambda_6 K_6 e^{\lambda_6} \end{pmatrix} \begin{Bmatrix} A_1 \\ A_2 \\ A_3 \\ A_4 \\ A_5 \\ A_6 \end{Bmatrix} = 0 \qquad (9.158)$$

Note: The form of this solution assumes that there are no repeated roots. [Recall Eq. (4.43).]

A nontrivial solution to Eq. (9.158) requires that the determinant of the coefficients, denoted Δ, equals zero. Thus, for a given set of parameters a, $\bar{x}_m$, μ, $\bar{U}$, γ_o, and ε, the complex values of $\Omega = \Omega_n$ are determined. If $\Omega_n = \Omega_{1n} + j\Omega_{2n}$, then from Eq. (9.142) it is seen that by examining the time-varying part of the solution, we find

$$e^{j\Omega\tau} = e^{j(\Omega_{1n} + j\Omega_{2n})\tau} = e^{-\Omega_{2n}\tau + j\Omega_{1n}\tau} \qquad (9.159)$$

It is seen from Eq. (9.159) that a negative value of Ω_{2n} will make the exponential function increase without bounds as τ increases. In addition, the frequency given by Ω_{1n} must be such that $\Omega_{1n} > 0$. Thus, if it is assumed that for a set of geometric parameters we can vary $\bar{U}$, then the smallest value of $\bar{U}$ that makes Ω_{2n} negative, keeps $\Omega_{1n} > 0$, and for which $\Delta = 0$ is defined as the lowest flutter frequency coefficient.

An example of the determination of the flutter frequency, it is assumed that $\gamma_o = 0.16$, $\bar{x}_m = 0.07$, $a = 0.1$, $\varepsilon = 0.17$, and $\mu = 0.39$. For these parameters, it is found that flutter occurs in the region $4.995 < \bar{U} < 5.0$ with the corresponding value of $\Omega_{1n} \approx 0.916$.

9.7 THERMAL RUNAWAY IN MICROWAVE HEATING OF CERAMICS

9.7.1 Introduction

Microwave heating has been applied in a variety of applications including sintering and joining of ceramics and chemical vapor infiltration of fiber-reinforced ceramic composites. One of the major hurdles faced in the application of microwave processing of ceramics is the occurrence of thermal runaway. This phenomenon corresponds to a situation in which a small change in the thickness of the sample or microwave power causes a very large increase in the temperature of the material being heated. Thermal runaway is due to the coupling of the thermal and electric fields and a positive feedback mechanism such that a small increase in temperature increases the dissipation rate and leads to a further increase in temperature. Under these conditions, the steady-state temperature of the material is not a single-valued function of the applied power.

The focus of describing thermal runaway will be on how it depends on sample thickness, frequency of the microwaves, the loss modulus of the material, and the temperature sensitivity of the loss modulus.

9.7.2 Heat Equation and Boundary Conditions

The governing equation in Cartesian coordinates for steady-state one-dimensional heat conduction in the z-direction in a material with constant properties and a thickness b is

$$\frac{d^2T}{dz^2}+\frac{Q''}{k_{tc}}=0 \quad \text{K/m}^2 \tag{9.160}$$

where Q'' (W/m³) represents the heat produced in the material per unit volume and k_{tc} (W/(Km)) is the thermal conductivity of the material. Prior to obtaining a solution to Eq. (9.160), we must obtain an expression for Q'', which is due to the microwave heating. This quantity will be obtained in the next section.

The boundary conditions are assumed to include convection and radiation. Hence, if $0\le\bar{\varepsilon}\le 1$ is the emissivity of the surfaces, s = 5.6703 × 10⁻⁸ (W/(m²K⁴)) is the Stefan–Boltzmann constant, and h (W/(m²K)) is the coefficient of convective heat transfer at the walls, then with a few simplifying assumptions, the boundary condition at $z = 0$ is

$$k_{tc}T'(0)=h\left(T(0)-T_o\right)+\bar{\varepsilon}s\left(T^4(0)-T_o^4\right) \quad \text{W/m}^2 \tag{9.161}$$

and that at $z = b$ is

$$-k_{tc}T'(b)=h\left(T(b)-T_o\right)+\bar{\varepsilon}s\left(T^4(b)-T_o^4\right) \quad \text{W/m}^2 \tag{9.162}$$

where the prime (′) denotes the derivative with respect to z.

In anticipation of what is to follow, we introduce the following definitions (recall Example 7.25)

$$\xi=\frac{z}{b} \quad \theta=\frac{T-T_o}{T_o} \tag{9.163}$$

Then, using Eq. (9.163), Eq. (9.160) becomes

$$\frac{d^2\theta}{d\xi^2}+\frac{Q''}{p_s}=0 \tag{9.164}$$

where

$$p_s=\frac{k_{tc}T_o}{b^2} \quad \text{W/m}^3 \tag{9.165}$$

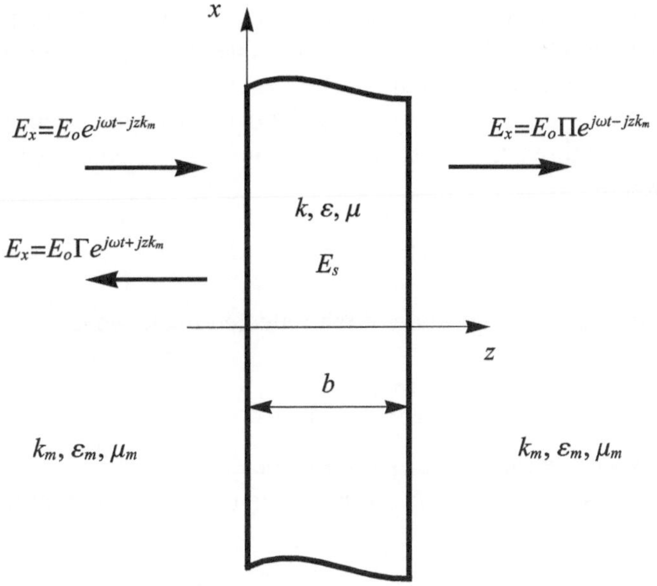

FIGURE 9.12 Microwave plane wave impinging upon a slab of thickness b. The plane wave E_x is parallel to the (x,y)-plane.

The quantity p_s is the heat in the slab per unit volume. Using Eq. (9.163) in Eqs. (9.161) and (9.162), the boundary conditions can be written as

$$\begin{aligned} \theta'(0) &= Bi\theta(0) + R\left[\left(\theta(0)+1\right)^4 - 1\right] \\ -\theta'(0) &= Bi\theta(1) + R\left[\left(\theta(1)+1\right)^4 - 1\right] \end{aligned} \tag{9.166}$$

where the prime (′) now denotes the derivative with respect to ξ and

$$Bi = \frac{hb}{k_{tc}} \quad R = \frac{\bar{\varepsilon}\, sbT_o^3}{k_{tc}} \tag{9.167}$$

The quantity Bi is the Biot number.

9.7.3 Steady-State Microwave Heating of a Slab*

We shall determine the temperature increase in a slab of material that has a thickness b (m) in the z-direction as shown in Figure 9.12 when the slab is subjected to a harmonically varying microwave plane wave. The slab is infinite in extent in the xy-plane. We shall consider the case of a wave whose electric field intensity $\boldsymbol{E}$ (V/m) is harmonic and of the following vector form

$$\boldsymbol{E} = E_o e^{-jkz}\mathbf{i} \tag{9.168}$$

* Kriegsmann GA (1992) Thermal Runaway in Microwave Heated Ceramics: A One-Dimensional Model. *Journal of Applied Physics* 71(4): 1960–1966 and Gupta N, Midha V, Balakotaiah V, Economou DJ (1999) Bifurcation Analysis of Thermal Runaway in Microwave Heating of Ceramics. *Journal of the Electrochemical Society* 146(12): 4659–4665.

where the negative sign in the exponent indicates a wave traveling in the positive z-direction and E_o (V/m) is the magnitude of the field intensity. The quantity $k = \omega/c$ is the wave number where c is the wave speed in the medium at frequency ω.

From the assumption implied in Eq. (9.168), the non-zero component of the electric field is denoted E_x. There are three regions to consider. The first region is $z \le 0$ in which the wave propagation is given by

$$E_x = E_o\left(e^{-jk_m z} + \Gamma e^{jk_m z}\right) \quad z \le 0 \tag{9.169}$$

where ΓE_o $(0 < \Gamma < 1)$ is the magnitude of the reflected plane wave,

$$k_m = \frac{\omega}{c_m} \tag{9.170}$$

and c_m is the wave speed in the medium at frequency ω.

The second region is for $0 \le z \le b$. In this region, we assume that the wave motion is given by

$$E_s = U(z)e^{j\omega t} \quad 0 \le z \le b \tag{9.171}$$

where $U(z)$ is, after a suitable reduction of the Maxwell equations for harmonic oscillations, the solution to

$$\frac{d^2U}{dz^2} + \varepsilon\mu\omega^2 U = 0 \tag{9.172}$$

where ε (F/m = As/(Vm)) is the permittivity and μ (N/A^2) the permeability of the slab's material. If we assume that

$$\varepsilon = \varepsilon' - j\varepsilon'' \quad \text{F/m} \tag{9.173}$$

where the imaginary part signifies the dissipation of energy in the medium in the presence of a time-varying electric field, then Eq. (9.172) can be written as*

$$\frac{d^2U}{dz^2} + k_s^2\left(1 - j\frac{\sigma_s}{\omega\varepsilon'}\right)U = 0 \tag{9.174}$$

where

$$k_s^2 = \frac{\varepsilon'}{\varepsilon_0}\frac{\omega^2}{c^2} \quad \text{m}^{-2}$$

$$\sigma_s = \omega\varepsilon'' \quad \text{A/(Vm)} \tag{9.175}$$

In Eq. (9.175), $\varepsilon_0 = 8.854 \times 10^{-12}$ (F/m=As/(Vm)) is the permittivity in a vacuum, c is the speed of light in a vacuum (3×10^8 m/s), and we have used the relation $\mu = 1/(\varepsilon_0 c^2)$.

The third region is for $z \ge b$, where the portion of the wave that leaves the slab at $z = b$ is traveling in the positive z-direction and is given by

$$E_x = \Pi E_o e^{-jk_m z} \quad z \ge b \tag{9.176}$$

* This form of the equation assumes that the static conductivity is zero.

where Π $(0 < \Pi < 1)$ is the fraction of the original magnitude of the wave that entered the slab.

Lastly, for this one-dimensional model, the power dissipated in a nonmagnetic material is determined from

$$P_{\text{diss}} = \frac{1}{2}\sigma_s |E_s|^2 \text{ W/m}^3 \tag{9.177}$$

According to our definition of σ_s, its value varies linearly with ω; therefore, any results for σ_s will be valid only at a specific frequency. We see that Q'' of Eq. (9.164) is given by P_{diss}. Hence, our objective is to determine P_{diss}. It is seen that P_{diss} is a function of E_s which, in turn, is a function of $U(z)$.

To determine $U(z)$, we start by establishing its continuity conditions, which are that the electric field and its derivative of the waves at the boundary of the slab and the medium external to the slab are equal at $z = 0$ and at $z = b$. Hence, at $z = 0$, we require

$$\begin{aligned} E_x(0) &= E_o + \Gamma E_o = U(0) \\ E_x'(0) &= jk_m\left(-E_o + \Gamma E_o\right) = U'(0) \end{aligned} \tag{9.178}$$

where the prime (′) denotes the derivative with respect to z and we have used Eq. (9.169). Using the first equation of Eq. (9.178), we find that

$$\Gamma E_o = U(0) - E_o \tag{9.179}$$

and, therefore, the second boundary condition at $z = 0$ becomes

$$U'(0) - jk_m U(0) = -j2k_m E_o \tag{9.180}$$

In a similar manner, at $z = b$ we find from Eq. (9.176) that

$$\begin{aligned} E_x(b) &= \Pi E_o e^{-jk_m b} = U(b) \\ E_x'(b) &= -jk_m \Pi E_o e^{-jk_m b} = U'(b) \end{aligned} \tag{9.181}$$

and, therefore, the boundary condition at $z = b$ becomes

$$U'(b) + jk_m U(b) = 0 \tag{9.182}$$

For materials in which thermal runaway has been observed, it has been found that ε'' increases exponentially with temperature. Hence, we shall assume that

$$\sigma_s(T) = ae^{qT} \tag{9.183}$$

In Eq. (9.183), the constants a (A/(Vm)) and q (K^{-1}) are determined from experiment. Then using Eq. (9.183), we can form the quantity

$$\frac{\sigma_s(T)}{\omega\varepsilon'} \rightarrow \frac{\sigma_s(\theta)}{\omega\varepsilon'} = c_s e^{\gamma\theta} \tag{9.184}$$

where θ is given by Eq. (9.163),

$$c_s = \frac{ae^{\gamma}}{\omega\varepsilon'} \quad \gamma = qT_o \tag{9.185}$$

The quantities c_s and γ are non-dimensional.

Then, using (9.184) and introducing the following definitions

$$\xi = \frac{z}{b}, \quad u = \frac{U}{E_o}, \quad K_m = k_m b, \quad K_s = k_s b \tag{9.186}$$

Eq. (9.174) can be written as

$$\frac{d^2u}{d\xi^2} + K_s^2\left(1 - jc_s e^{\gamma\theta}\right)u = 0 \tag{9.187}$$

and the boundary conditions given by Eqs. (9.180) and (9.182) can be written as

$$\begin{aligned} u'(0) - jK_m u(0) &= -j2K_m \\ u'(1) + jK_m u(1) &= 0 \end{aligned} \tag{9.188}$$

Equations (9.187) and (9.188) are in terms of non-dimensional quantities. In addition, from Eq. (9.177), we find that

$$Q'' = P_{\text{diss}} = p_i c_s e^{\gamma\theta} |u|^2 \tag{9.189}$$

where

$$p_i = \frac{1}{2}\omega\varepsilon' E_o^2 \quad \text{W/m}^3 \tag{9.190}$$

The temperature in the slab can now be determined from Eq. (9.164). In this case, using Eq. (9.189), Eq. (9.164) becomes

$$\frac{d^2\theta}{d\xi^2} + pc_s e^{\gamma\theta} |u|^2 = 0 \tag{9.191}$$

where $p = p_i/p_s$. The quantity p_i is the heat input to the slab per unit volume, and p_s is the heat in the slab per unit volume. Notice that Eq. (9.191) is in terms of real quantities only. The boundary conditions are those given by Eq. (9.166).

Thus, the coupled system requires the simultaneous solution to Eq. (9.187) with its boundary conditions given by Eq. (9.188) and Eq. (9.191) with its boundary conditions given by Eq. (9.166). All these equations and boundary conditions are in terms of non-dimensional quantities.

9.7.4 Outline to Obtain Numerical Results

To solve these equations, we convert Eq. (9.187) and the corresponding boundary conditions given by Eq. (9.188) into equations containing only real variables. Equation (9.191) is always real. To this end, we let $u = u_r + ju_i$. Then Eq. (9.187) becomes

$$\frac{d^2}{d\xi^2}(u_r + ju_i) + K_s^2\left(1 - jc_s e^{\gamma\theta}\right)(u_r + ju_i) = 0$$

Upon separating this equation into its real and imaginary parts, we arrive at

$$\begin{aligned} \frac{d^2u_r}{d\xi^2} + K_s^2 u_r + c_s K_s^2 e^{\gamma\theta} u_i &= 0 \\ \frac{d^2u_i}{d\xi^2} + K_s^2 u_i - c_s K_s^2 e^{\gamma\theta} u_r &= 0 \end{aligned} \tag{9.192}$$

The boundary conditions given by Eq. (9.188) become

$$u_r'(0) + K_m u_i(0) = 0, \qquad u_i'(0) - K_m u_r(0) + 2K_m = 0$$
$$u_r'(1) - K_m u_i(1) = 0, \qquad u_i'(1) + K_m u_r(1) = 0$$

The third equation is given by Eq. (9.191) and its boundary conditions are given by Eq. (9.166). The solution technique to solve these equations and some results are given in the cited references.

Appendix A Series Expansions

Polynomials

$$
\begin{aligned}
(1 \pm x)^{-1} &= 1 \mp x + x^2 \mp x^3 + \cdots \quad x^2 < 1 \\
(1 \pm x)^{-1/2} &= 1 \mp \frac{x}{2} + \frac{1 \times 3}{2 \times 4} x^2 \mp \frac{1 \times 3 \times 5}{2 \times 4 \times 6} x^3 + \cdots \quad x^2 < 1 \\
(1 \pm x)^{1/2} &= 1 \pm \frac{x}{2} - \frac{1}{2 \times 4} x^2 \pm \frac{1 \times 3}{2 \times 4 \times 6} x^3 + \cdots \quad x^2 < 1 \\
(1 \pm x)^{n} &= 1 \pm nx + \frac{n(n-1)}{2!} x^2 \pm \frac{n(n-1)(n-2)}{3!} x^3 + \cdots \quad x^2 < 1
\end{aligned} \tag{A.1}
$$

Trigonometric

$$
\begin{aligned}
\sin x &= \sum_{n=0}^{\infty} \frac{(-1)^n x^{2n+1}}{(2n+1)!} = x - \frac{x^3}{3!} + \frac{x^5}{5!} - \cdots \\
\cos x &= \sum_{n=0}^{\infty} \frac{(-1)^n x^{2n}}{(2n)!} = 1 - \frac{x^2}{2!} + \frac{x^4}{4!} - \cdots \\
\sin(a + x) &= \sin a + x \cos a - \frac{1}{2} x^2 \sin a + \cdots \\
\cos(a + x) &= \cos a - x \sin a - \frac{1}{2} x^2 \cos a + \cdots
\end{aligned} \tag{A.2}
$$

Exponential

$$
\begin{aligned}
e^x &= \sum_{n=0}^{\infty} \frac{x^n}{n!} = 1 + x + \frac{1}{2} x^2 + \frac{1}{6} x^3 \cdots \\
a^x &= e^{x \ln a} = \sum_{n=0}^{\infty} \frac{(x \ln a)^n}{n!} = 1 + x \ln a + \frac{1}{2} (x \ln a)^2 + \frac{1}{6} (x \ln a)^3 \cdots
\end{aligned} \tag{A.3}
$$

Euler's formula

$$
\begin{aligned}
e^{j\theta} &= \sum_{n=0}^{\infty} \frac{(j\theta)^n}{n!} = 1 + j\theta - \frac{\theta^2}{2!} - j\frac{\theta^3}{3!} + \frac{\theta^4}{4!} + j\frac{\theta^5}{5!} \\
&= \left(1 - \frac{\theta^2}{2!} + \frac{\theta^4}{4!} + \cdots\right) + j\left(\theta - \frac{\theta^3}{3!} + \frac{\theta^5}{5!} - \cdots\right) \\
&= \sum_{n=0}^{\infty} \frac{(-1)^n \theta^{2n}}{(2n)!} + j \sum_{n=0}^{\infty} \frac{(-1)^n \theta^{2n+1}}{(2n+1)!} \\
&= \cos\theta + j \sin\theta
\end{aligned} \tag{A.4}
$$

Hyperbolic

$$\sinh x = \sum_{n=0}^{\infty} \frac{x^{2n+1}}{(2n+1)!} = x + \frac{x^3}{3!} + \frac{x^5}{5!} + \cdots$$

$$\cosh x = \sum_{n=0}^{\infty} \frac{x^{2n}}{(2n)!} = 1 + \frac{x^2}{2!} + \frac{x^4}{4!} + \cdots \tag{A.5}$$

Logarithmic

$$\ln(1+x) = \sum_{n=0}^{\infty} \frac{(-1)^n x^{n+1}}{n+1} = x - \frac{1}{2}x^2 + \frac{1}{3}x^3 \cdots$$

$$\begin{aligned} \ln(a+x) &= \ln\left(a\left(1+x/a\right)\right) = \ln(a) + \ln\left(1+x/a\right) \\ &= \ln(a) + \sum_{n=0}^{\infty} \frac{(-1)^n (x/a)^{n+1}}{n+1} \\ &= \ln(a) + \frac{x}{a} - \frac{1}{2}\left(\frac{x}{a}\right)^2 + \frac{1}{3}\left(\frac{x}{a}\right)^3 \cdots \end{aligned} \tag{A.6}$$

Appendix B Delta Function

B.1 DELTA FUNCTION

The delta function $\delta(x)$ is called a generalized function; it is zero everywhere except where its argument is zero such that

$$\int_{-\infty}^{\infty} \delta(x-c)\,dx = 1 \tag{B.1}$$

Note that the units of the delta function are the reciprocal of those of its argument. If a function $f(x)$ is continuous at $x = c$, then $\delta(x)$ has the property that

$$\int_{-\infty}^{\infty} f(x)\delta(x-c)\,dx = f(c) \tag{B.2}$$

Consider the following modification to Eq. (B.2)

$$\int_{-\infty}^{\infty} f(\eta)\delta(\alpha[\eta-\eta_1])\,d\eta \tag{B.3}$$

If, in Eq. (B.3), we let $u = \alpha\eta - \alpha\eta_1$, then

$$\int_{-\infty}^{\infty} f(\eta)\delta(\alpha[\eta-\eta_1])\,d\eta = \int_{-\infty}^{\infty} \frac{f(u/\alpha+\eta_1)}{\alpha}\delta(u)\,du = \frac{f(\eta_1)}{\alpha}$$

When $u = -\alpha\eta + \alpha\eta_1$, then

$$\int_{-\infty}^{\infty} f(\eta)\delta(-\alpha[\eta-\eta_1])\,d\eta = \int_{\infty}^{-\infty} \frac{f(\eta_1-u/\alpha)}{-\alpha}\delta(u)\,du = \frac{f(\eta_1)}{\alpha}$$

Therefore,

$$\int_{-\infty}^{\infty} f(\eta)\delta(\alpha[\eta-\eta_1])\,d\eta = \frac{f(\eta_1)}{|\alpha|} \tag{B.4}$$

When the delta function appears in differential equations, it means that integration is pending; that is, during the solution process integration with respect to the independent variable will be performed. The significance of Eq. (B.4) is that it provides a means to make the delta function non-dimensional. Consider the case of $\delta(x - L)$. Based on Eq. (B.4), the argument can be written as $\delta(L[\eta - 1])$ which, in turn, can be written as $\delta(\eta - 1)/|L|$, where $\eta = x/L$ is a non-dimensional quantity.

The derivative of the delta function is defined as

$$\int_{-\infty}^{\infty} f(x)\delta'(x-c)\,dx = -f'(c) \tag{B.5}$$

where it is assumed that $f(x)$ is differentiable and the prime denotes the derivative with respect to x. For higher-order derivatives, the definition for the nth derivative is extended as

$$\int_{-\infty}^{\infty} f(x)\delta^{(n)}(x-c)\,dx = (-1)^n f^{(n)}(c) \tag{B.6}$$

B.2 DELTA FUNCTION AND THE HEAVISIDE FUNCTION

Consider the indefinite integral

$$\int_{-\infty}^{x} \delta(x-c)\,dx = H(x-c) \tag{B.7}$$

where $H(x)$ is the Heaviside function and the indefinite integral

$$\int_{-\infty}^{x} f(x)\delta(x-c)\,dx = f(c)H(x-c) \tag{B.8}$$

Equation (B.7) is considered the definition of the Heaviside function where $H(x-c) = 1$ when $x > c$ and $H(x-c) = 0$ when $x < c$.

We also note that

$$\int_{-\infty}^{x} H(x)\,dx = \begin{cases} 0 & x < 0 \\ x & x > 0 \end{cases} \tag{B.9}$$

Therefore,

$$\int_{-\infty}^{x} H(x)dx = xH(x) \tag{B.10}$$

and, in general,

$$\int_{-\infty}^{x} x^n H(x)dx = \frac{x^{n+1}}{(n+1)!}H(x) \tag{B.11}$$

B.3 DELTA FUNCTION IN TWO AND THREE DIMENSIONS AND IN DIFFERENT COORDINATE SYSTEMS

Cartesian Coordinates

For a point (x_0,y_0,z_0) in Cartesian coordinates, Eq. (B.1) is extended as

$$\int_{-\infty}^{\infty}\int_{-\infty}^{\infty}\int_{-\infty}^{\infty}\delta(x-x_0)\delta(y-y_0)\delta(z-z_0)dxdydz=1 \tag{B.12}$$

We shall now extend this result to orthogonal curvilinear coordinates.

Orthogonal Curvilinear Coordinates

We can use a short-hand notation by letting the position in a coordinate system be replaced by a position vector $\boldsymbol{r}$. Then,

$$\begin{aligned}\int_V \delta(\boldsymbol{r}-\boldsymbol{r}_0)JdV &=1 \quad \text{if } \boldsymbol{r}_0 \text{ is in } V\\ &=0 \quad \text{otherwise}\end{aligned} \tag{B.13}$$

where J is the Jacobian matrix defined as

$$J=\det\begin{pmatrix}\frac{\partial x}{\partial \xi_1} & \frac{\partial x}{\partial \xi_2} & \frac{\partial x}{\partial \xi_2}\\ \frac{\partial y}{\partial \xi_1} & \frac{\partial y}{\partial \xi_2} & \frac{\partial y}{\partial \xi_3}\\ \frac{\partial z}{\partial \xi_1} & \frac{\partial z}{\partial \xi_2} & \frac{\partial z}{\partial \xi_3}\end{pmatrix}=\det\begin{pmatrix}\frac{\partial u}{\partial \xi_1} & \frac{\partial u}{\partial \xi_2} & \frac{\partial u}{\partial \xi_2}\\ \frac{\partial v}{\partial \xi_1} & \frac{\partial v}{\partial \xi_2} & \frac{\partial v}{\partial \xi_3}\\ \frac{\partial w}{\partial \xi_1} & \frac{\partial w}{\partial \xi_2} & \frac{\partial w}{\partial \xi_3}\end{pmatrix}$$

since, in general, $x=u(\xi_1, \xi_2, \xi_3)$, $y=v(\xi_1, \xi_2, \xi_3)$, and $z=w(\xi_1, \xi_2, \xi_3)$ and, therefore, $J=J(\xi_1, \xi_2, \xi_3)$. For Cartesian coordinates, $\xi_1=x$, $\xi_2=y$, and $\xi_3=z$ and

$$J=\det\begin{pmatrix}1 & 0 & 0\\ 0 & 1 & 0\\ 0 & 0 & 1\end{pmatrix}=1$$

In extending the Cartesian integral given by Eq. (B.13), we note that in order for the following relationship to hold for J different from 1, that is, in order for

$$\begin{aligned}\int_V \delta(\boldsymbol{r}-\boldsymbol{r}_0)JdV &=1 \quad \text{if } \boldsymbol{r}_0 \text{ is in } V\\ &=0 \quad \text{otherwise}\end{aligned}$$

we must replace

$$\delta(\boldsymbol{r}-\boldsymbol{r}_0) \rightarrow \frac{\delta(\boldsymbol{r}-\boldsymbol{r}_0)}{J}$$

so that

$$\int_V \delta(\boldsymbol{r}-\boldsymbol{r}_0)dV = 1 \quad \text{if } \boldsymbol{r}_0 \text{ is in } V$$
$$= 0 \quad \text{otherwise}$$

We now examine these relations in cylindrical and in spherical coordinate systems.

Cylindrical Coordinates

For cylindrical coordinates, $\xi_1 = r$, $\xi_2 = \theta$, and $\xi_3 = z$ and $dV = dr d\theta dz$. In addition,

$$u = r\cos\theta$$
$$v = r\sin\theta$$
$$w = z$$

Then the Jacobian is

$$J = \det\begin{pmatrix} \frac{\partial u}{\partial \xi_1} & \frac{\partial u}{\partial \xi_2} & \frac{\partial u}{\partial \xi_2} \\ \frac{\partial v}{\partial \xi_1} & \frac{\partial v}{\partial \xi_2} & \frac{\partial v}{\partial \xi_3} \\ \frac{\partial w}{\partial \xi_1} & \frac{\partial w}{\partial \xi_2} & \frac{\partial w}{\partial \xi_3} \end{pmatrix} = \det\begin{pmatrix} \frac{\partial u}{\partial r} & \frac{\partial u}{\partial \theta} & \frac{\partial u}{\partial z} \\ \frac{\partial v}{\partial r} & \frac{\partial v}{\partial \theta} & \frac{\partial v}{\partial z} \\ \frac{\partial w}{\partial r} & \frac{\partial w}{\partial \theta} & \frac{\partial w}{\partial z} \end{pmatrix}$$

$$= \det\begin{pmatrix} \cos\theta & -r\sin\theta & 0 \\ \sin\theta & r\cos\theta & 0 \\ 0 & 0 & 1 \end{pmatrix} \tag{B.14}$$

$$= r\cos^2\theta + r\sin^2\theta = r$$

Then in cylindrical coordinates,

$$\delta(\boldsymbol{r}-\boldsymbol{r}_0) = \frac{\delta(r-r_0)}{r}\delta(\theta-\theta_0)\delta(z-z_0) \tag{B.15}$$

which is used to describe a point in the cylindrical coordinate system and, therefore,

$$\int_V \frac{\delta(r-r_0)}{r}\delta(\theta-\theta_0)\delta(z-z_0) r dr d\theta dz = \int_0^\infty \delta(r-r_0)dr \int_{-\infty}^\infty \delta(z-z_0)dz \int_0^{2\pi} \delta(\theta-\theta_0)d\theta \tag{B.16}$$
$$= 1$$

When the coordinate transformation is independent of θ, we must, because of the last integral in Eq. (B.16), modify the delta function as follows

$$\delta(\boldsymbol{r}-\boldsymbol{r}_0)=\frac{\delta(r-r_0)}{2\pi r}\delta(z-z_0) \tag{B.17}$$

This form of the delta function is used to describe a ring in the cylindrical coordinate system. Then

$$\int_V \frac{\delta(r-r_0)}{2\pi r}\delta(z-z_0)rdrd\theta dz=\frac{1}{2\pi}\int_0^\infty \delta(r-r_0)dr\int_{-\infty}^{\infty}\delta(z-z_0)dz\int_0^{2\pi}d\theta \tag{B.18}$$
$$=1$$

When the coordinate transformation is independent of θ and z we must again, because of the last integral in Eq. (B.18), modify the delta function as follows

$$\delta(\boldsymbol{r}-\boldsymbol{r}_0)=\frac{\delta(r-r_0)}{2\pi r} \tag{B.19}$$

Then,

$$\int_A \frac{\delta(r-r_0)}{2\pi r}rdrd\theta=\frac{1}{2\pi}\int_0^\infty \delta(r-r_0)dr\int_0^{2\pi}d\theta$$
$$=1$$

Spherical Coordinates

For spherical coordinates, $\xi_1 = r$, $\xi_2 = \theta$, and $\xi_3 = \varphi$ and $dV = drd\theta d\varphi$. In addition,

$$u = r\sin\theta\cos\varphi$$
$$v = r\sin\theta\sin\varphi$$
$$w = r\cos\theta$$

Hence the Jacobian is

$$J=\det\begin{pmatrix}\frac{\partial u}{\partial \xi_1} & \frac{\partial u}{\partial \xi_2} & \frac{\partial u}{\partial \xi_2}\\ \frac{\partial v}{\partial \xi_1} & \frac{\partial v}{\partial \xi_2} & \frac{\partial v}{\partial \xi_3}\\ \frac{\partial w}{\partial \xi_1} & \frac{\partial w}{\partial \xi_2} & \frac{\partial w}{\partial \xi_3}\end{pmatrix}=\det\begin{pmatrix}\frac{\partial u}{\partial r} & \frac{\partial u}{\partial \theta} & \frac{\partial u}{\partial \varphi}\\ \frac{\partial v}{\partial r} & \frac{\partial v}{\partial \theta} & \frac{\partial v}{\partial \varphi}\\ \frac{\partial w}{\partial r} & \frac{\partial w}{\partial \theta} & \frac{\partial w}{\partial \varphi}\end{pmatrix}$$

$$=\det\begin{pmatrix}\sin\theta\cos\varphi & r\cos\theta\cos\varphi & -r\sin\theta\sin\varphi\\ \sin\theta\sin\varphi & r\cos\theta\sin\varphi & r\sin\theta\cos\varphi\\ \cos\theta & -r\sin\theta & 0\end{pmatrix}$$

$$=r^2\sin\theta$$

Then in spherical coordinates,

$$\delta(\boldsymbol{r}-\boldsymbol{r}_0)=\frac{\delta(r-r_0)}{r^2\sin\theta}\delta(\theta-\theta_0)\delta(\varphi-\varphi_0) \tag{B.20}$$

which is used to describe a point in the spherical coordinate system. Then

$$\begin{aligned}&\int_V \frac{\delta(r-r_0)}{r^2\sin\theta}\delta(\theta-\theta_0)\delta(\varphi-\varphi_0)r^2\sin\theta drd\theta d\varphi\\&=\int_0^\infty \delta(r-r_0)dr\int_0^{2\pi}\delta(\varphi-\varphi_0)d\varphi\int_0^\pi \delta(\theta-\theta_0)d\theta=1\end{aligned} \tag{B.21}$$

When the coordinate transformation is independent of φ, we must, because of the middle integral in Eq. (B.21), modify the delta function as follows

$$\delta(\boldsymbol{r}-\boldsymbol{r}_0)=\frac{\delta(r-r_0)}{2\pi r^2\sin\theta}\delta(\theta-\theta_0) \tag{B.22}$$

This form of the delta function is used to describe a ring in the spherical coordinate system. Then

$$\begin{aligned}\int_V \frac{\delta(r-r_0)}{r^2\sin\theta}\delta(\theta-\theta_0)r^2\sin\theta drd\theta d\varphi&=\frac{1}{2\pi}\int_0^\infty \delta(r-r_0)dr\int_0^{2\pi}d\varphi\int_0^\pi \delta(\theta-\theta_0)d\theta\\&=1\end{aligned}$$

When the coordinate transformation is independent of θ, we must, because of the last integral in Eq. (B.21), modify the delta function as follows

$$\delta(\boldsymbol{r}-\boldsymbol{r}_0)=\frac{\delta(r-r_0)}{2r^2}\delta(\varphi-\varphi_0) \tag{B.23}$$

and

$$\begin{aligned}\int_V \frac{\delta(r-r_0)}{2r^2}\delta(\varphi-\varphi_0)r^2\sin\theta drd\theta d\varphi&=\frac{1}{2}\int_0^\infty \delta(r-r_0)dr\int_0^{2\pi}\delta(\varphi-\varphi_0)d\varphi\int_0^\pi \sin\theta d\theta\\&=1\end{aligned}$$

When the coordinate transformation is independent of φ and θ, we must, because of the last two integrals in Eq. (B.21), modify the delta function as follows

$$\delta(\boldsymbol{r}-\boldsymbol{r}_0)=\frac{\delta(r-r_0)}{4\pi r^2} \tag{B.24}$$

and

$$\int_V \frac{\delta(r-r_0)}{4\pi r^2} r^2 \sin\theta dr d\theta d\varphi = \frac{1}{4\pi}\int_0^{\infty} \delta(r-r_0)dr \int_0^{2\pi} d\varphi \int_0^{\pi} \sin\theta d\theta$$

$$=1$$

Appendix C Gamma Function

The gamma function is defined as

$$\Gamma(x) = \int_0^\infty t^{x-1} e^{-t} dt \quad \text{Re}(x) > 0 \tag{C.1}$$

Thus,

$$\Gamma(1) = \int_0^\infty e^{-t} dt = -e^{-t}\Big|_0^\infty = 1 \tag{C.2}$$

In addition,

$$\begin{aligned} \Gamma(x+1) &= \int_0^\infty t^x e^{-t} dt = t^x \left(-e^{-t}\right)\Big|_0^\infty - \int_0^\infty x t^{x-1} \left(-e^{-t}\right) dt \\ &= x \int_0^\infty t^{x-1} e^{-t} dt = x\Gamma(x) \end{aligned} \tag{C.3}$$

The gamma function is shown in Figure C.1.

We now use this result as follows

$$\begin{aligned} &\Gamma(x+1) = x\Gamma(x) \\ &\Gamma(x+2) = \Gamma(x+1+1) = (x+1)\Gamma(x+1) = x(x+1)\Gamma(x) \\ &\Gamma(x+3) = \Gamma(x+2+1) = (x+2)\Gamma(x+2) = x(x+1)(x+2)\Gamma(x) \\ &\vdots \\ &\Gamma(x+N) = x(x+1)(x+2)\cdots(x+N-1)\Gamma(x) \end{aligned} \tag{C.4}$$

From this result, we obtain the following two relations. When $x = 1$,

$$\Gamma(1+N) = 1(2)(3)\cdots(N)\Gamma(1) = N! \tag{C.5}$$

and when $x = v + 1$

$$\frac{\Gamma(x+N)}{\Gamma(x)} \to \frac{\Gamma(v+1+N)}{\Gamma(v+1)} = (v+1)(v+2)(v+3)\cdots(v+N) \tag{C.6}$$

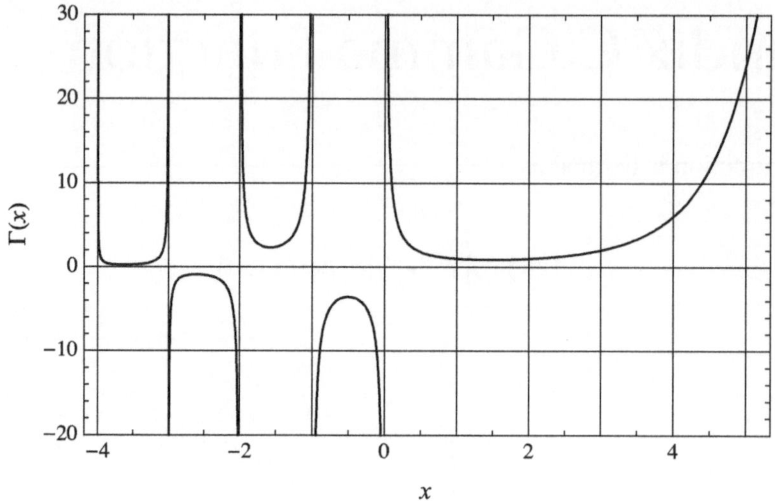

FIGURE C.1 The gamma function.

The recursion relationship can also be used to determine the gamma function of negative quantities. If $x = -z$ and z is not an integer, then we can use Eq. (C.4) by finding the smallest integer value of N for which $N - z > 0$. In this case, upon rearrangement we arrive at

$$\Gamma(-z) = \frac{\Gamma(-z+N)}{-z(-z+1)(-z+2)\cdots(-z+N-1)} \tag{C.7}$$

Since $N - z > 0$, the gamma function on the righthand side exists and the denominator is simply algebraic; therefore, the gamma function of a negative quantity can be determined.

A useful relation is Euler's reflection formula, which is given by

$$\Gamma(1-z)\Gamma(z) = \frac{\pi}{\sin \pi z} \tag{C.8}$$

When $z = 1/2$, we find from Eq. (C.8) that

$$\begin{aligned} \left[\Gamma(1/2)\right]^2 &= \frac{\pi}{\sin \pi/2} = \pi \\ \Gamma(1/2) &= \sqrt{\pi} \end{aligned} \tag{C.9}$$

In addition, when $z = -z$, we see from Eq. (C.8) that

$$\begin{aligned} \Gamma(1+z)\Gamma(-z) &= -\frac{\pi}{\sin \pi z} \\ \Gamma(-z) &= -\frac{\pi}{\Gamma(z+1)\sin \pi z} = -\frac{\pi}{z\Gamma(z)\sin \pi z} \end{aligned} \tag{C.10}$$

which is a simpler formula to use for negative arguments, especially when z is a large negative number.

Another useful relation is called the duplication formula, which states that

$$\Gamma(z)\Gamma\left(z+\frac{1}{2}\right)=\frac{2\sqrt{\pi}}{2^{2z}}\Gamma(2z) \tag{C.11}$$

When $z = n$, an integer, Eq. (C.11) gives

$$\begin{aligned}\Gamma(n)\Gamma\left(n+\frac{1}{2}\right)&=\frac{2\sqrt{\pi}}{2^{2n}}\Gamma(2n)\\ \Gamma\left(n+\frac{1}{2}\right)&=\frac{2\sqrt{\pi}}{2^{2n}}\frac{\Gamma(2n)}{\Gamma(n)}=\frac{2\sqrt{\pi}}{2^{2n}}\left[\frac{\Gamma[2n+1]}{2n}\right]\left[\frac{n}{\Gamma[n+1]}\right]\\ &=\frac{\sqrt{\pi}}{2^{2n}}\frac{(2n)!}{n!}\end{aligned} \tag{C.12}$$

and

$$\begin{aligned}\Gamma(n+1)\Gamma\left(n+\frac{3}{2}\right)&=\frac{2\sqrt{\pi}}{2^{2n+2}}\Gamma(2n+2)\\ \Gamma\left(n+\frac{3}{2}\right)&=\frac{\sqrt{\pi}}{2^{2n+1}}\frac{\Gamma([2n+1]+1)}{\Gamma(n+1)}=\frac{\sqrt{\pi}}{2^{2n+1}}\frac{(2n+1)!}{n!}\end{aligned} \tag{C.13}$$

Equation (C.12) can be extended to negative n as follows.

$$\begin{aligned}\Gamma(-n)\Gamma\left(-n+\frac{1}{2}\right)&=\frac{2\sqrt{\pi}}{2^{-2n}}\Gamma(-2n)\\ \Gamma\left(\frac{1}{2}-n\right)&=2^{1+2n}\sqrt{\pi}\frac{\Gamma(-2n)}{\Gamma(-n)}\\ &=2^{1+2n}\sqrt{\pi}\left[-\frac{\pi}{2n\Gamma(2n)\sin 2\pi n}\right]\left[\frac{n\Gamma(n)\sin\pi n}{-\pi}\right]\\ &=2^{2n}\sqrt{\pi}\left[\frac{\Gamma(n)}{\Gamma(2n)}\right]\left[\frac{\sin\pi n}{\sin 2\pi n}\right]=2^{2n}\sqrt{\pi}\left[\frac{\Gamma(n)}{\Gamma(2n)}\right]\frac{\sec n\pi}{2}\\ &=(-1)^n 2^{2n-1}\sqrt{\pi}\left[\frac{2n}{(2n)!}\right]\left[\frac{n!}{n}\right]=(-1)^n 2^{2n}\sqrt{\pi}\left[\frac{n!}{(2n)!}\right]\end{aligned} \tag{C.14}$$

The derivative of the gamma function is

$$\frac{d}{dx}\Gamma(x)=\frac{d}{dx}\int_0^\infty t^{x-1}e^{-t}dt=\int_0^\infty t^{x-1}\ln te^{-t}dt=\Gamma(x)\psi(x) \tag{C.15}$$

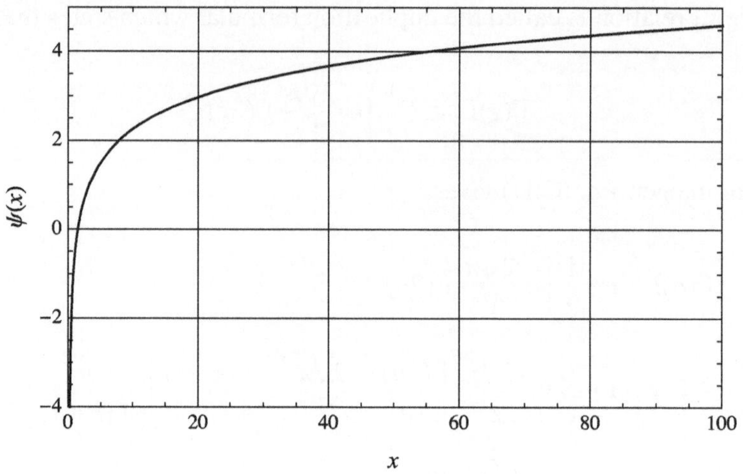

FIGURE C.2 The digamma function.

where

$$\psi(x) = -\gamma + \sum_{n=0}^{\infty}\left(\frac{1}{n+1} - \frac{1}{n+x}\right) \quad x \neq 0, -1, -2, -3, \ldots \tag{C.16}$$

is the digamma function and $\gamma = 0.5772156649$ is Euler's constant determined from

$$\gamma = \lim_{n\to\infty}\left[\sum_{k=1}^{n}\frac{1}{k} - \ln n\right] \tag{C.17}$$

The digamma function is shown in Figure C.2.

When $x = m =$ integer,

$$\begin{aligned}\psi(1) &= -\gamma \\ \psi(m) &= -\gamma + \sum_{n=1}^{m-1}\frac{1}{n}\end{aligned} \tag{C.18}$$

The second relation in Eq. (C.18) is obtained by noting that

$$\begin{aligned}\sum_{n=0}^{\infty}\left(\frac{1}{n+1} - \frac{1}{n+m}\right) &= \left(1 + \frac{1}{2} + \frac{1}{3} + \cdots + \frac{1}{k} + \frac{1}{k+1} + \frac{1}{k+2} + \cdots\right) \\ &\quad - \left(\frac{1}{m} + \frac{1}{m+1} + \frac{1}{m+2} + \cdots\right) \\ &= \left(1 + \frac{1}{2} + \frac{1}{3} + \cdots + \frac{1}{m-1}\right) \text{ when } k \text{ becomes equal to } m \\ &= \sum_{n=1}^{m-1}\frac{1}{n}\end{aligned}$$

Furthermore,

$$\frac{d}{dx}\left(\frac{1}{\Gamma(x)}\right) = -\frac{\psi(x)}{\Gamma(x)} \tag{C.19}$$

Bibliography

ENGINEERING MATHEMATICS: GENERAL

Greenberg MD (1988) *Advanced Engineering Mathematics.* Prentice Hall.
Jeffery A (2002) *Advanced Engineering Mathematics.* Harcourt/Academic Press.
Kreyszig E (2011) *Advanced Engineering Mathematics*, 10th ed. John Wiley & Sons.
O'Neil PV (2003) *Advanced Engineering Mathematics*, 5th ed. Thomson Brooks/Cole.
Riley KF, Hobson MP, Bence SJ (1997) *Mathematical Methods for Physics and Engineering: A Comprehensive Guide.* Cambridge University Press.
Turyn L (2013) *Advanced Engineering Mathematics.* CRC Press.
Wiley CR, Barrett LC (1995) *Advanced Engineering Mathematics*, 6th ed. McGraw-Hill.

ENGINEERING MATHEMATICS: ADVANCED

Duffy DG (2003) *Advanced Engineering Mathematics with MATLAB*, 2nd ed. Chapman & Hall/CRC Press.
Harmon TL, Dabney J, Richert N (2000) *Advanced Engineering Mathematics with MATLAB*, 2nd ed. Thomson Brooks/Cole.
Hayek SI (2001) *Advanced Mathematical Methods in Science and Engineering.* Marcel Dekker.
Zill DG, Cullen MR (2000) *Advanced Engineering Mathematics.* Jones and Bartlett Publishers.

BOUNDARY VALUE PROBLEMS

Asmar NH (2005) *Partial Differential Equations: With Fourier Series and Boundary Value Problems*, 2nd ed. Pearson/Prentice Hall.
Boyce WE, DiPrima RC (1977) *Elementary Differential Equations and Boundary Value Problems*, 3rd ed. John Wiley & Sons.
Brown JW, Churchill RV (1993) *Fourier Series and Boundary Value Problems*, 5th ed. McGraw-Hill.
Edwards CH, Penny DE (2000) *Differential Equations and Boundary Value Problems: Computers and Modeling*, 2nd ed. Prentice Hall.
Hanna JR, Rowland JH (1990) *Fourier Series, Transforms, and Boundary Value Problems*, 2nd ed. John Wiley & Sons.
Powers DL (1999) *Boundary Value Problems*, 4th ed. Harcourt/Academic Press.
Trench WF (2013) *Elementary Differential Equations with Boundary Value Problems*, Free ed. Available at http://ramanujan.math.trinity.edu/wtrench/texts/index.shtml.

LAPLACE AND FOURIER TRANSFORMS

Churchill RV (1958) *Operational Mathematics*, 2nd ed. McGraw-Hill.
Debnath L (1995) *Integral Transforms and Their Applications.* CRC Press.
Duffy DG (2004) *Transform Methods for Solving Partial Differential Equations*, 2nd ed. Chapman & Hall/CRC Press.

HANDBOOKS

NIST Digital Library of Mathematical Functions: https://dlmf.nist.gov. Companion to: Olver WJ, Lozier DW, Boisvert RF, Clark CW (2010) *NIST Handbook of Mathematical Functions.* Cambridge University Press.

Polyanin AD (2002) *Handbook of Linear Partial Differential Equations for Engineers and Scientists.* Chapman & Hall/CRC Press.

Roberts GE, Kaufman H (1966) *Table of Laplace Transforms.* W. B. Saunders Co.

Index

H

I

K

L

M

N

O

P